Reviews in Computational Chemistry Volume 28

Reviews in Computational Chemistry 28

Edited by

Abby L. Parrill
Kenny B. Lipkowitz

Abby L. Parrill,
Department of Chemistry
The University of Memphis
Memphis, TN 38152, U.S.A.
aparrill@memphis.edu

Kenny B. Lipkowitz,
Office of Naval Research
875 North Randolph Street
Arlington, VA 22203-1995, U.S.A.
kenny.lipkowitz@navy.mil

Published by John Wiley & Sons, Inc., Hoboken, New Jersey
Published simultaneously in Canada

For general information on our other products and services or for technical support, please contact our Customer Care Department within the United States at (800) 762-2974, outside the United States at (317) 572-3993 or fax (317) 572-4002.

Wiley also publishes its books in a variety of electronic formats. Some content that appears in print may not be available in electronic formats. For more information about Wiley products, visit our web site at www.wiley.com.

Library of Congress Cataloging-in-Publication Data:

ISBN: 978-1-1118-40777-6
ISSN: 1069-3599

Typeset in 10/12pt SabonLTStd by Laserwords Private Limited, Chennai, India.

Printed in the United States of America

10 9 8 7 6 5 4 3 2 1

1   2015

Contents

Preface

As *Reviews in Computational Chemistry* begins its 24th year with this 28th volume, it is interesting to reflect on how it has integrated with and influenced my own career over this time period, now that I have joined Kenny as a new series editor. The pedagogically driven reviews focused on computational chemistry were a tremendous resource to me during my graduate studies in the mid-1990s. The series was such an asset in so many ways that none of the volumes could ever be found in the university library, but were always on loan to one of the members of my research group and could be found right in our lab on someone's desk. I have continued to use the series as a first resource when moving into new areas of computational chemistry, as well as a first place to refer my own graduate students when they begin learning how to function as computational chemists. I hope you have enjoyed and utilized this series in these and other ways throughout your own computational chemistry career.

This volume of *Reviews in Computational Chemistry* begins with a tutorial on the theory and practice of metadynamics for free-energy calculations. Metadynamics is one of the growing number of methodologies to improve sampling of rare events in order to study chemical processes that occur at timescales outside the scope of unbiased molecular dynamics simulations. As with many biased sampling methods, the choices of the user have tremendous influence on the quality of results. Giovanni Bussi and Davide Branduardi discuss results of different types of metadynamics simulations on the alanine dipeptide with different choices of collective variables to illustrate practical usage of the theories they describe. Users interested in learning how to perform metadynamics calculations will appreciate the sample input files provided in Appendix A.

Chapter 2 addresses a different modeling challenge, namely improving the accuracy of force fields with regard to their treatment of the environmental dependence of charge polarization in molecular systems. Traditional molecular mechanics force fields have relied upon fixed charges that can only represent averaged polarization. This leads to inaccuracies that prevent fixed-charge force fields from correctly modeling the phase transitions of molecules as small as water or accurately computing the barrier height to ion permeation across lipid membranes. Yue Shi, Pengyu Ren, Michael Schnieders, and Jean-Philip Piquemal provide a historical reflection on the development of polarizable force fields and a thorough review of recent developments. Specific applications of polarizable force fields from water simulations to crystal structure prediction demonstrate the current capabilities of polarizable force fields.

In Chapter 3, Clare-Louise Towse and Valerie Daggett demonstrate how the simple principle of microscopic reversibility can be utilized to simplify investigations of protein folding. In particular, a variety of methods to accelerate the unfolding process are described in order to address two of the major challenges faced when simulating protein folding by molecular dynamics. First, the protein unfolding pathway can be initiated from a well-defined structure for any protein that has been characterized by X-ray crystallography. This is not the case for the protein folding pathway, for which starting structures are not known. Second, the timescale of protein folding can be as long as minutes, a timescale that is not tractable with unbiased and unaccelerated simulations. Applications of protein unfolding simulations and the insights they provide regarding protein folding demonstrate the value of this approach.

Assessment is an integral part of demonstrating the value of any new computational method or application area. Chapter 4 illustrates one of the ongoing 'big science' assessment initiatives, CAPRI (Critical Assessment of Predicted Interactions). CAPRI competitions over a 12-year period have been possible through the cooperation of structural biologists, who delayed publication of crystallographic structures of the protein assemblies that served as the basis for blind predictions of the assembled complex using crystallographic structures of individual members of the assembly as input. Joël Janin, Shoshana J. Wodak, Marc F. Lensink, and Sameer Velankar describe the assessment metrics, results and impacts of the CAPRI experiment on the field of protein–protein interaction prediction.

Chapter 5 draws attention to kinetic Monte Carlo simulations of electrochemical systems, systems with substantial commercial significance due to the relevance of such systems to our future energy needs. Kinetic Monte Carlo simulations are suitable for timescales and length scales between those typically modeled using molecular dynamics and those typically modeled using mesoscale modeling. Given a set of states and transition rates between those states, kinetic Monte Carlo simulations can predict both the system time evolution behavior and thermodynamic averages of the system at equilibrium. Applications of kinetic Monte Carlo to understand problems ranging from passive layer formation at the solid electrolyte interphase to electrochemical dealloying demonstrate the value of simulations in the design of next-generation materials and operational advances to improve device performance.

Chapter 6 focuses on liquid interfaces, at which reactivity and dynamics play a role in applications ranging from solar cell operation to ion channel function. Throughout this chapter, Ilan Benjamin balances the theoretical underpinnings and practical aspects of simulating the reactivity and dynamics of liquid interfaces with a review of the experimental data used to assess the accuracy of the simulated behaviors.

Chapter 7 is tightly focused on clathrate hydrates. As in the previous two chapters, commercial significance is linked with energy due to the role of clathrate hydrates in natural gas pipeline obstructions and safety concerns as well as the untapped potential of naturally occurring methane hydrates to serve as a valuable energy source capable of fueling current energy consumption rates for several hundred years. Advances are described that allow computation of clathrate hydrate features from thermodynamic stability to hydrate nucleation and growth.

The volume concludes with a chapter by John Herbert defining the challenges inherent in modeling systems containing loosely-bound electrons (such as solvated electrons) and methods to meet those challenges. Studies on systems involving loosely-bound electrons have the potential to expand our understanding of low-energy electron-induced reactions, that exhibit highly selective cleavage of specific bonds, which are often not the thermodynamically weakest bonds in the molecule. Quantum calculations must accommodate the highly diffuse nature of the weakly-bound electron through the use of an "ultra-diffuse" one-particle basis set. More specialized techniques are required in the case of metastable and formally unbound anions.

Reviews in Computational Chemistry continues to be a valuable resource to the scientific community entirely due to the contributions of the authors whom we have contacted to provide the pedagogically driven reviews that have made this ongoing book series so popular. We are grateful for their contributions.

The most recent volumes of *Reviews in Computational Chemistry* are available in an online form through Wiley InterScience. Please consult the Web (http://www.interscience.wiley.com/onlinebooks) or contact reference@wiley.com for the latest information.

We thank the authors of this and previous volumes for their excellent chapters.

Abby L. Parrill
Memphis

Kenny B. Lipkowitz
Washington
January 2015

List of Contributors

Ilan Benjamin, Department of Chemistry and Biochemistry, University of California, Santa Cruz, CA 95064, USA

Davide Branduardi, Theoretical Molecular Biophysics Group, Max Planck Institute of Biophysics, Frankfurt am Main, D-60438, Germany

Giovanni Bussi, Statistical and Molecular Biophysics Group, International School for Advanced Studies (SISSA), Trieste, IT-34136, Italy

Valerie Daggett, Department of Bioengineering, University of Washington, Seattle, WA 98195-5013, USA

Brett I. Dunlap, Chemistry Division, US Naval Research Laboratory, Washington, DC, 20375–5342, USA

Lev D. Gelb, Department of Materials Science and Engineering, University of Texas at Dallas, Richardson, TX, 75080, USA

John M. Herbert, Department of Chemistry and Biochemistry, The Ohio State University, Columbus, OH 43210, USA

Joël Janin, IBBMC, Université Paris-Sud, Orsay 91405, France

Marc F. Lensink, UGSF, CNRS UMR8576, University Lille North of France, Villeneuve d'Ascq, 59658, France

Jean-Philip Piquemal, Laboratoire de Chimie Théorique (UMR 7616), UPMC, Sorbonne Universités, Paris, Cedex 05 75252, France

Pengyu Ren, Department of Biomedical Engineering, The University of Texas at Austin, Austin, TX 78712, USA

Michael Schnieders, Department of Biomedical Engineering, College of Engineering, Department of Biochemistry, Carver College of Medicine, The University of Iowa, Iowa City, IA 52242, USA

Yue Shi, Department of Biomedical Engineering, The University of Texas at Austin, Austin, TX 78712, USA

Clare-Louise Towse, Department of Bioengineering, University of Washington, Seattle, WA 98195-5013, United States

John S. Tse, Department of Physics and Engineering Physics, University of Saskatchewan, Saskatoon, Saskatchewan S7N 5B2, Canada

C. Heath Turner, Department of Chemical and Biological Engineering, The University of Alabama, Tuscaloosa, AL 35487–0203, USA

Sameer Velankar, European Bioinformatics Institute, Hinxton, Cambridgeshire, CB10 1SD, UK

Shoshana J. Wodak, VIB Structural Biology Research Center, VUB Building E Pleinlaan 2 1050, Brussel, Belgium

Zhongtao Zhang, Department of Chemical and Biological Engineering, The University of Alabama, Tuscaloosa, AL 35487–0203, USA

Contributors to Previous Volumes

Volume 1 (1990)

David Feller and Ernest R. Davidson, Basis Sets for Ab Initio Molecular Orbital Calculations and Intermolecular Interactions.

James J. P. Stewart, Semiempirical Molecular Orbital Methods.

Clifford E. Dykstra, Joseph D. Augspurger, Bernard Kirtman, and David J. Malik, Properties of Molecules by Direct Calculation.

Ernest L. Plummer, The Application of Quantitative Design Strategies in Pesticide Design.

Peter C. Jurs, Chemometrics and Multivariate Analysis in Analytical Chemistry.

Yvonne C. Martin, Mark G. Bures, and Peter Willett, Searching Databases of Three-Dimensional Structures.

Paul G. Mezey, Molecular Surfaces.

Terry P. Lybrand, Computer Simulation of Biomolecular Systems Using Molecular Dynamics and Free Energy Perturbation Methods.

Donald B. Boyd, Aspects of Molecular Modeling.

Donald B. Boyd, Successes of Computer-Assisted Molecular Design.

Ernest R. Davidson, Perspectives on Ab Initio Calculations.

Volume 2 (1991)

Andrew R. Leach, A Survey of Methods for Searching the Conformational Space of Small and Medium-Sized Molecules.

John M. Troyer and **Fred E. Cohen,** Simplified Models for Understanding and Predicting Protein Structure.

J. Phillip Bowen and **Norman L. Allinger,** Molecular Mechanics: The Art and Science of Parameterization.

Uri Dinur and **Arnold T. Hagler,** New Approaches to Empirical Force Fields.

Steve Scheiner, Calculating the Properties of Hydrogen Bonds by Ab Initio Methods.

Donald E. Williams, Net Atomic Charge and Multipole Models for the Ab Initio Molecular Electric Potential.

Peter Politzer and **Jane S. Murray,** Molecular Electrostatic Potentials and Chemical Reactivity.

Michael C. Zerner, Semiempirical Molecular Orbital Methods.

Lowell H. Hall and **Lemont B. Kier,** The Molecular Connectivity Chi Indexes and Kappa Shape Indexes in Structure-Property Modeling.

I. B. Bersuker and **A. S. Dimoglo,** The Electron-Topological Approach to the QSAR Problem.

Donald B. Boyd, The Computational Chemistry Literature.

Volume 3 (1992)

Tamar Schlick, Optimization Methods in Computational Chemistry.

Harold A. Scheraga, Predicting Three-Dimensional Structures of Oligopeptides.

Andrew E. Torda and **Wilfred F. van Gunsteren,** Molecular Modeling Using NMR Data.

David F. V. Lewis, Computer-Assisted Methods in the Evaluation of Chemical Toxicity.

Volume 4 (1993)

Jerzy Cioslowski, Ab Initio Calculations on Large Molecules: Methodology and Applications.

Michael L. McKee and Michael Page, Computing Reaction Pathways on Molecular Potential Energy Surfaces.

Robert M. Whitnell and Kent R. Wilson, Computational Molecular Dynamics of Chemical Reactions in Solution.

Roger L. DeKock, Jeffry D. Madura, Frank Rioux, and Joseph Casanova, Computational Chemistry in the Undergraduate Curriculum.

Volume 5 (1994)

John D. Bolcer and Robert B. Hermann, The Development of Computational Chemistry in the United States.

Rodney J. Bartlett and John F. Stanton, Applications of Post-Hartree–Fock Methods: A Tutorial.

Steven M. Bachrach, Population Analysis and Electron Densities from Quantum Mechanics.

Jeffry D. Madura, Malcolm E. Davis, Michael K. Gilson, Rebecca C. Wade, Brock A. Luty, and J. Andrew McCammon, Biological Applications of Electrostatic Calculations and Brownian Dynamics Simulations.

K. V. Damodaran and Kenneth M. Merz Jr., Computer Simulation of Lipid Systems.

Jeffrey M. Blaney and J. Scott Dixon, Distance Geometry in Molecular Modeling.

Lisa M. Balbes, S. Wayne Mascarella, and Donald B. Boyd, A Perspective of Modern Methods in Computer-Aided Drug Design.

Volume 6 (1995)

Christopher J. Cramer and Donald G. Truhlar, Continuum Solvation Models: Classical and Quantum Mechanical Implementations.

Gernot Frenking, Iris Antes, Marlis Böhme, Stefan Dapprich, Andreas W. Ehlers, Volker Jonas, Arndt Neuhaus, Michael Otto, Ralf Stegmann, Achim Veldkamp, and Sergei F. Vyboishchikov, Pseudopotential Calculations of Transition Metal Compounds: Scope and Limitations.

Thomas R. Cundari, Michael T. Benson, M. Leigh Lutz, and Shaun O. Sommerer, Effective Core Potential Approaches to the Chemistry of the Heavier Elements.

Jan Almlöf and Odd Gropen, Relativistic Effects in Chemistry.

Donald B. Chesnut, The Ab Initio Computation of Nuclear Magnetic Resonance Chemical Shielding.

Volume 9 (1996)

James R. Damewood, Jr., Peptide Mimetic Design with the Aid of Computational Chemistry.

T. P. Straatsma, Free Energy by Molecular Simulation.

Robert J. Woods, The Application of Molecular Modeling Techniques to the Determination of Oligosaccharide Solution Conformations.

Ingrid Pettersson and Tommy Liljefors, Molecular Mechanics Calculated Conformational Energies of Organic Molecules: A Comparison of Force Fields.

Gustavo A. Arteca, Molecular Shape Descriptors.

Volume 10 (1997)

Richard Judson, Genetic Algorithms and Their Use in Chemistry.

Eric C. Martin, David C. Spellmeyer, Roger E. Critchlow Jr., and Jeffrey M. Blaney, Does Combinatorial Chemistry Obviate Computer-Aided Drug Design?

Robert Q. Topper, Visualizing Molecular Phase Space: Nonstatistical Effects in Reaction Dynamics.

Raima Larter and Kenneth Showalter, Computational Studies in Nonlinear Dynamics.

Chung F. Wong, Tom Thacher, and Herschel Rabitz, Sensitivity Analysis in Biomolecular Simulation.

Paul Verwer and Frank J. J. Leusen, Computer Simulation to Predict Possible Crystal Polymorphs.

Jean-Louis Rivail and Bernard Maigret, Computational Chemistry in France: A Historical Survey.

Volume 13 (1999)

Thomas Bally and Weston Thatcher Borden, Calculations on Open-Shell Molecules: A Beginner's Guide.

Neil R. Kestner and Jaime E. Combariza, Basis Set Superposition Errors: Theory and Practice.

James B. Anderson, Quantum Monte Carlo: Atoms, Molecules, Clusters, Liquids, and Solids.

Anders Wallqvist and Raymond D. Mountain, Molecular Models of Water: Derivation and Description.

James M. Briggs and Jan Antosiewicz, Simulation of pH-dependent Properties of Proteins Using Mesoscopic Models.

Harold E. Helson, Structure Diagram Generation.

Volume 14 (2000)

Michelle Miller Francl and Lisa Emily Chirlian, The Pluses and Minuses of Mapping Atomic Charges to Electrostatic Potentials.

T. Daniel Crawford and Henry F. Schaefer III, An Introduction to Coupled Cluster Theory for Computational Chemists.

Bastiaan van de Graaf, Swie Lan Njo, and Konstantin S. Smirnov, Introduction to Zeolite Modeling.

Sarah L. Price, Toward More Accurate Model Intermolecular Potentials For Organic Molecules.

Christopher J. Mundy, Sundaram Balasubramanian, Ken Bagchi, Mark E. Tuckerman, Glenn J. Martyna, and Michael L. Klein, Nonequilibrium Molecular Dynamics.

Donald B. Boyd and Kenny B. Lipkowitz, History of the Gordon Research Conferences on Computational Chemistry.

Mehran Jalaie and Kenny B. Lipkowitz, Appendix: Published Force Field Parameters for Molecular Mechanics, Molecular Dynamics, and Monte Carlo Simulations.

Volume 15 (2000)

F. Matthias Bickelhaupt and Evert Jan Baerends, Kohn-Sham Density Functional Theory: Predicting and Understanding Chemistry.

Michael A. Robb, Marco Garavelli, Massimo Olivucci, and Fernando Bernardi, A Computational Strategy for Organic Photochemistry.

Larry A. Curtiss, Paul C. Redfern, and David J. Frurip, Theoretical Methods for Computing Enthalpies of Formation of Gaseous Compounds.

Russell J. Boyd, The Development of Computational Chemistry in Canada.

Volume 16 (2000)

Richard A. Lewis, Stephen D. Pickett, and David E. Clark, Computer-Aided Molecular Diversity Analysis and Combinatorial Library Design.

Keith L. Peterson, Artificial Neural Networks and Their Use in Chemistry.

Jörg-Rüdiger Hill, Clive M. Freeman, and Lalitha Subramanian, Use of Force Fields in Materials Modeling.

M. Rami Reddy, Mark D. Erion, and Atul Agarwal, Free Energy Calculations: Use and Limitations in Predicting Ligand Binding Affinities.

Volume 17 (2001)

Ingo Muegge and Matthias Rarey, Small Molecule Docking and Scoring.

Lutz P. Ehrlich and Rebecca C. Wade, Protein-Protein Docking.

Christel M. Marian, Spin-Orbit Coupling in Molecules.

Lemont B. Kier, Chao-Kun Cheng, and Paul G. Seybold, Cellular Automata Models of Aqueous Solution Systems.

Kenny B. Lipkowitz and Donald B. Boyd, Appendix: Books Published on the Topics of Computational Chemistry.

Volume 18 (2002)

Geoff M. Downs and John M. Barnard, Clustering Methods and Their Uses in Computational Chemistry.

Hans-Joachim Böhm and Martin Stahl, The Use of Scoring Functions in Drug Discovery Applications.

Steven W. Rick and Steven J. Stuart, Potentials and Algorithms for Incorporating Polarizability in Computer Simulations.

Dmitry V. Matyushov and Gregory A. Voth, New Developments in the Theoretical Description of Charge-Transfer Reactions in Condensed Phases.

George R. Famini and Leland Y. Wilson, Linear Free Energy Relationships Using Quantum Mechanical Descriptors.

Sigrid D. Peyerimhoff, The Development of Computational Chemistry in Germany.

Donald B. Boyd and Kenny B. Lipkowitz, Appendix: Examination of the Employment Environment for Computational Chemistry.

Volume 19 (2003)

Robert Q. Topper, David, L. Freeman, Denise Bergin and Keirnan R. LaMarche, Computational Techniques and Strategies for Monte Carlo Thermodynamic Calculations, with Applications to Nanoclusters.

David E. Smith and Anthony D. J. Haymet, Computing Hydrophobicity.

Lipeng Sun and William L. Hase, Born-Oppenheimer Direct Dynamics Classical Trajectory Simulations.

Gene Lamm, The Poisson-Boltzmann Equation.

Volume 20 (2004)

Sason Shaik and **Philippe C. Hiberty,** Valence Bond Theory: Its History, Fundamentals and Applications. A Primer.

Nikita Matsunaga and **Shiro Koseki,** Modeling of Spin Forbidden Reactions.

Stefan Grimme, Calculation of the Electronic Spectra of Large Molecules.

Raymond Kapral, Simulating Chemical Waves and Patterns.

Costel Sârbu and **Horia Pop,** Fuzzy Soft-Computing Methods and Their Applications in Chemistry.

Sean Ekins and **Peter Swaan,** Development of Computational Models for Enzymes, Transporters, Channels and Receptors Relevant to ADME/Tox.

Volume 21 (2005)

Roberto Dovesi, Bartolomeo Civalleri, Roberto Orlando, Carla Roetti and **Victor R. Saunders,** Ab Initio Quantum Simulation in Solid State Chemistry.

Patrick Bultinck, Xavier Gironés and **Ramon Carbó-Dorca,** Molecular Quantum Similarity: Theory and Applications.

Jean-Loup Faulon, Donald P. Visco, Jr. and **Diana Roe,** Enumerating Molecules.

David J. Livingstone and **David W. Salt,** Variable Selection- Spoilt for Choice.

Nathan A. Baker, Biomolecular Applications of Poisson-Boltzmann Methods.

Baltazar Aguda, Georghe Craciun and **Rengul Cetin-Atalay,** Data Sources and Computational Approaches for Generating Models of Gene Regulatory Networks.

Volume 22 (2006)

Patrice Koehl, Protein Structure Classification.

Emilio Esposito, Dror Tobi and **Jeffry Madura,** Comparative Protein Modeling.

Jeetain Mittal, William P. Krekelberg, Jeffrey R. Errington, and Thomas M. Truskett, Computing Free Volume, Structured Order, and Entropy of Liquids and Glasses.

Laurence E. Fried, The Reactivity of Energetic Materials at Extreme Conditions.

Julio A. Alonso, Magnetic Properties of Atomic Clusters of the Transition Elements.

Laura Gagliardi, Transition Metal- and Actinide-Containing Systems Studied with Multiconfigurational Quantum Chemical Methods.

Hua Guo, Recursive Solutions to Large Eigenproblems in Molecular Spectroscopy and Reaction Dynamics.

Hugh Cartwright, Development and Uses of Artificial Intelligence in Chemistry.

Volume 26 (2009)

C. David Sherrill, Computations of Noncovalent π Interactions.

Gregory S. Tschumper, Reliable Electronic Structure Computations for Weak Noncovalent Interactions in Clusters.

Peter Elliott, Filip Furche and Kieron Burke, Excited States from Time-Dependent Density Functional Theory.

Thomas Vojta, Computing Quantum Phase Transitions.

Thomas L. Beck, Real-Space Multigrid Methods in Computational Chemistry.

Francesca Tavazza, Lyle E. Levine and Anne M. Chaka, Hybrid Methods for Atomic-Level Simulations Spanning Multi-Length Scales in the Solid State.

Alfredo E. Cárdenas and Eric Bath, Extending the Time Scale in Atomically Detailed Simulations.

Edward J. Maginn, Atomistic Simulation of Ionic Liquids.

Volume 27 (2011)

Stefano Giordano, Allessandro Mattoni, Luciano Colombo, Brittle Fracture: From Elasticity Theory to Atomistic Simulations.

Igor V. Pivkin, Bruce Caswell, George Em Karniadakis, Dissipative Particle Dynamics.

Peter G. Bolhuis and Christoph Dellago, Trajectory-Based Rare Event Simulation.

Douglas L. Irving, Understanding Metal/Metal Electrical Contact Conductance from the Atomic to Continuum Scales.

Max L. Berkowitz and James Kindt, Molecular Detailed Simulations of Lipid Bilayers.

Sophya Garaschuk, Vitaly Rassolov, Oleg Prezhdo, Semiclassical Bohmian Dynamics.

Donald B. Boyd, Employment Opportunities in Computational Chemistry.

Kenny B. Lipkowitz, Appendix: List of Computational Molecular Scientists.

## CHAPTER 1

Free-Energy Calculations with Metadynamics: Theory and Practice

Giovanni Bussi,[a] **and Davide Branduardi**[b]

[a]*Statistical and Molecular Biophysics Group, International School for Advanced Studies (SISSA), Trieste, IT 34136, Italy*
[b]*Theoretical Molecular Biophysics Group, Max Planck Institute of Biophysics, Frankfurt am Main D-60438, Germany*

INTRODUCTION

Molecular dynamics (MD) is a powerful tool in modern chemistry that allows one to describe the time evolution of a computational model for a complex molecular system.[1-3] Typical models range from being highly accurate where energy and forces are computed with advanced and expensive quantum chemistry methods to faster but less accurate empirically parameterized force fields at atomistic or coarser resolution. The power of these techniques lies in their ability to reproduce experimental observable quantities accurately while, at the same time, giving access to the mechanistic details of chemical reactions or conformational changes at very high spatial resolution – typically at atomistic scale. For this reason, MD is often used to complement experimental investigations and to help in interpreting experiments and in designing new ones. Moreover, thanks to new parallelization algorithms and to the continuous improvements in computer hardware driven by Moore's law, the range of application of these techniques has grown exponentially in the past decades and can be expected to continue growing.

Reviews in Computational Chemistry, Volume 28, First Edition.
Edited by Abby L. Parrill and Kenny B. Lipkowitz.
© 2015 John Wiley & Sons, Inc. Published 2015 by John Wiley & Sons, Inc.

In spite of its success, however, MD is still limited to the study of events on a very short timescale. Indeed, depending on the required accuracy and on the available computational resources, MD can provide trajectories for events happening on the timescale of picoseconds (quantum chemistry) to microseconds (empirical force fields). Thus, many interesting phenomena, namely, chemical reactions, protein folding and aggregation, and macromolecular rearrangement are still out of reach of direct investigation using straightforward MD trajectories. Besides the optimization of computer software (e.g., Ref. 4) and/or hardware (e.g. Refs. 5, 6), it is a possible complementary strategy to alleviate this issue by using algorithms where the time evolution is modified to sample more frequently the event under investigation. Then, appropriate postprocessing techniques are necessary to recover unbiased properties from the accelerated trajectories.

Many algorithms to accelerate MD simulations have been designed in the past decades, and a discussion of all of them is out of the scope of this chapter. Some of these algorithms are based on increasing the temperature of the simulated system (e.g., parallel tempering[7] and solute tempering[8]), while others are based on exploiting an *a priori* knowledge of the investigated transition to design a proper order parameter to both describe and accelerate it. This last class includes umbrella sampling,[9] adaptive biasing force,[10] metadynamics,[11] self-healing umbrella sampling,[12] and other methods that keep the selected order parameters at an artificially high temperature.[13–15] This chapter focuses on metadynamics, which was first introduced in 2002[11] and then improved with several variants in the past decade. Metadynamics has been employed successfully in several fields, ranging from chemical reactions[16] to protein folding[17] and aggregation,[18] molecular docking,[19] crystal structure prediction,[20] and nucleation.[21] A further push in the diffusion of metadynamics application has been its availability in a few widespread molecular dynamics codes[22–24] and in open-source plugins.[25–27]

The main goal of this chapter is to provide an entry-level tutorial for metadynamics. In Section "Molecular Dynamics and Free-Energy Estimation" we provide an introduction to the basic concepts of molecular dynamics and of free-energy calculations. In Section "A Toy Model: Alanine Dipeptide" we introduce a toy model that will then be used for subsequent examples. Section "Biased Sampling" is devoted to the introduction of biased sampling. In Sections "Adaptive Biasing with Metadynamics" and "Well-Tempered Metadynamics" metadynamics is introduced, and Section "Metadynamics How-To" provides a practical how-to for performing a free-energy calculation with metadynamics. For all the simulations described in that section a sample input file for the open-source package PLUMED 2[26] is given in the Appendix. In the remaining sections, a quick overview of some of the latest improvements in the field is given, followed by a concluding section.

MOLECULAR DYNAMICS AND FREE-ENERGY ESTIMATION

Molecular Dynamics

In classical MD,[1-3] the Hamilton equations of motion are solved numerically to follow in real time the propagation of a collection of atoms. For a system of N_{at} atoms with coordinates q_i, momenta p_i, and masses m_i, a potential energy $U(q)$ should be defined. Notice that we use here q, without subscript, meaning the full $3N_{at}$-dimensional vector containing all the atomic positions. The Hamilton equations of motion will then read

$$\dot{q}_i = \frac{p_i}{m_i} \qquad [1a]$$

$$\dot{p}_i = -\frac{\partial U(q)}{\partial q_i} \qquad [1b]$$

Here with $\dot{x}$ we mean the derivative with respect to the time of the variable x. The potential energy function $U(q)$ describes the interatomic interactions. These interactions are sometimes defined in terms of empirically parameterized force fields, which provide a cheap and reasonably accurate approximation for $U(q)$, and sometimes the interactions are obtained by solving the Schrödinger equation for the electrons (*ab initio* calculations), to allow studying phenomena such as electron transfer and chemical reactions. In our examples we will only use empirical potentials. However, the specific definition of U is totally irrelevant for what concerns the discussed methodologies, which often rely on *ab initio* calculations.

For a system evolving according to the Hamilton equations [1a] and [1b], the total energy $H(p,q) = \sum_i \frac{p_i^2}{2m_i} + U(q)$ is conserved, so that only configurations that have a total energy exactly equal to the initial one are explored. This will provide the correct properties for an isolated system. On the contrary, whenever a system is coupled to an external bath, the transfer of energy between the system and the bath implies that different values of the total energy are accessible. More precisely, the phase space point (p,q) will be explored with a probability $P(p,q)$, which, in the case of a thermal bath, corresponds to the canonical ensemble:

$$P(p,q)dpdq \propto \exp\left(-\frac{\sum_i \frac{p_i^2}{2m_i} + U(q)}{k_B T}\right)dqdp \qquad [2]$$

where k_B is the Boltzmann constant and T is the temperature of the thermal bath. Within MD, this is typically done by adding a so-called "thermostat" to the Hamilton equations.[2] Strictly speaking, a thermostat alters the dynamical properties, so that the latter could lose their physical meaning. This in turn depends a lot on the details of the adopted thermostat. Nonetheless, irrespective of the thermostat, MD can always be used to generate configurations according to the canonical distribution.

A crucial aspect of using MD for sampling the canonical distribution is the so-called *ergodic hypothesis*: if a system is simulated long enough all the states pertaining to the canonical ensemble will be explored, each with its own correct statistical weight. Unfortunately, this hypothesis cannot be proven for most of the systems and, even when verified, the length of a simulation necessary for this hypothesis to be exploited in calculating ensemble averages is often far out of reach for numerical simulations. This has profound consequences and led in the past decades to the development of several enhanced sampling algorithms that were designed to alleviate this difficulty.

Free-Energy Landscapes

The canonical distribution of a system sampled via MD carries full information about its thermodynamic properties. However, this probability distribution is of very little use. Indeed, the space on which it is defined (i.e., the set of all possible positions and velocities of all the atoms in a system) is huge – it is a $6N_{at}$ dimensional space – so that this function is completely unintelligible. For this reason, molecular systems are often analyzed in terms of collective variables (CVs) rather than atomic coordinates. A CV is a function of the atomic coordinates that is capable of describing the physics behind the process under investigation. As an example, for an isomerization process a reasonable CV could be a torsional angle, whereas for a proton transfer two reasonable CVs could be the distances between the hydrogen and each of the two involved heavy atoms. Because the CVs are functions of the atomic coordinates, we shall indicate them as $s_\alpha(q)$, with $\alpha = 1, \ldots, N_{CV}$ and let N_{CV} be equal to the number of CVs used. In short, the CVs represent a sort of coarse description of the system, which can be used to analyze a given process in low dimensionality. A basic requirement for a CV is being able to distinguish all the peculiar states of interest without lumping together states that are very different physicochemically.

When analyzing a molecular system using CVs $s(q)$, a role equivalent to that of the potential energy is played by the Helmholtz free energy $F(s)$. $F(s)$ is defined in such a manner that the probability of observing a given value of s is

$$P(s)ds \propto \exp\left(-\frac{F(s)}{k_B T}\right) ds \qquad [3]$$

It can be shown that the relationship between $U(q)$ and $F(s)$ can be written explicitly as

$$F(s_0) = -k_B T \log \int dq \delta(s(q) - s_0) \exp\left(-\frac{U(q)}{k_B T}\right) + C \qquad [4]$$

where the Dirac δ selects all the microscopic configurations corresponding to a specific value s_0 of the CV and C is an arbitrary constant.

A typical free-energy landscape for an activated event is shown in Figure 1. Here one can appreciate a stable state denoted by "A" (low free energy, thus high probability), a metastable one denoted by "B" (slightly higher free energy, thus lower probability), and an intermediate region (very high free energy, thus very small probability). A proper definition of metastability is out of the scope of this chapter. The height of the free-energy barrier compared with the value of the thermal energy $k_B T$ affects the probability of observing the system in the intermediate region and thus the typical time required to go from "A" to "B" or vice versa. When CVs are appropriately chosen, the transition rate between minima "A" and "B" can be estimated according to transition state theory by an Arrhenius-like formula:

$$\nu_{A \to B} = \nu_0 \exp^{-\frac{\Delta F^{\ddagger}}{k_B T}} \qquad [5]$$

where $\Delta F^{\ddagger}$ is the free-energy difference between the starting minimum and the transition state and ν_0 is a prefactor. Notably, the ratio between forward ($\nu_{A \to B}$) and backward ($\nu_{B \to A}$) rates is equal to $e^{\frac{F_A - F_B}{k_B T}}$, according to detailed balance. Thus, when appropriate CVs are used, the free-energy landscape provides a quantitative picture of the transition in terms of reactants and products stability and transition rates.

Evaluating the free-energy landscape defined by Eq. [4] is usually a daunting task, as it would require the calculation of a multidimensional integral in $3N_{at}$ dimensions. For this reason, the typical approach employed to compute

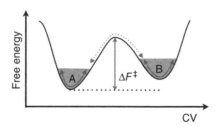

Figure 1 A model double-well potential. The stable state is denoted by "A" and the metastable one is denoted by "B." The gray shaded regions are those where the system fluctuates because of its thermal energy. The region in between is very unlikely to be explored, therefore making the transition from "A" to "B" less probable to occur.

free-energy landscapes is based on conformational sampling by means of MD or Monte Carlo (MC) methods. Indeed, if one is capable of producing a sequence of conformations (i.e., a trajectory) that are distributed according to the canonical ensemble, the free energy can be computed from the histogram of the visited conformations as

$$F(s) = -k_B T \log N(s) \qquad [6]$$

where $N(s)$ counts how many times the value s of the CVs has been explored. Because typical CVs are noninteger numbers, the histogram $N(s)$ is typically accumulated using a binning procedure. As discussed earlier, MD simulations allow one to produce such distribution via the ergodic hypothesis although, as we will show in the following section, this might be problematic even for very simple cases.

A TOY MODEL: ALANINE DIPEPTIDE

In this section we introduce a simple toy model that presents all the features of a typical real-life molecular system but still is small enough to allow the reader to readily test the concepts by means of inexpensive molecular dynamics simulations. This system is alanine dipeptide (ACE-ALA-NME), which is a small peptide in nonzwitterionic form (see Figure 2).

We describe interatomic interactions using the functional form $U(q)$ defined by the CHARMM 27 force field,[29] a standard force field available in many biomolecular-oriented molecular dynamics codes. For the sake of simplicity, all the simulations are carried out in vacuum as the free-energy landscape in such conditions presents several interesting features and has been studied previously using many different free-energy methods (see, e.g., Refs. 30–35) thus being a perfect test bed for the discussion.

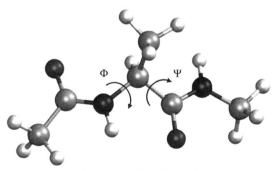

Figure 2   Molecular representation of alanine dipeptide. The two Ramachandran dihedral angles are denoted with Φ and Ψ. All the molecular representations are produced with VMD.[28]

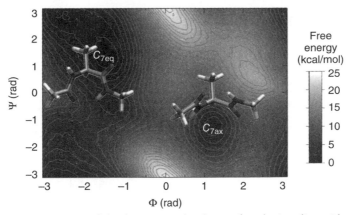

Figure 3 Representation of the free-energy landscape for alanine dipeptide as a function of the two Ramachandran dihedral angles Φ and Ψ. Each isoline accounts for 1 kcal/mol difference in free energy. The two main minima, namely, C_{7eq} and C_{7ax}, are labeled.

Alanine dipeptide in vacuum presents a peculiar free-energy landscape that can be rationalized in terms of Ramachandran dihedral angles. It displays two main minima: C_{7eq} and C_{7ax} (see Figure 3), placed around $\Phi \simeq -1.41$ rad, $\Psi \simeq 1.25$ rad and $\Phi \simeq 1.26$ rad, $\Psi \simeq -1.27$ rad, respectively. These two basins are separated by barriers around 8 kcal/mol that are remarkably higher than thermal energy at 300 K ($k_B T = 0.597$ kcal/mol), therefore presenting a typical case of metastability. In such a small system the differences in potential energy landscape are comparable with free-energy differences. Therefore, the Arrhenius equation can be used as a model to roughly estimate the rate of transition from one state to the other

$$\nu = \nu_0 \exp^{-\frac{\Delta U^{\ddagger}}{k_B T}} \qquad [7]$$

where $\Delta U^{\ddagger}$ is the potential energy difference between one metastable state, say C_{7eq}, and one of the transition states toward C_{7ax}. By assuming that the prefactor ν_0, as an upper bound, corresponds to the carbon–carbon vibration frequency in the force field we get $\nu_0 = 5 \times 10^9$/s. By using a barrier of 8 kcal/mol and $k_B T = 0.597$ kcal/mol we obtain a rate of 7×10^3 events per second, thus each barrier crossing could take about a millisecond. Unfortunately, such timescales are inaccessible to MD simulations that nowadays can reach several microseconds at most. Therefore, it is easy to understand why the problem of acquiring statistics of events that allow us to estimate the relative probability of metastable states is one of the grand challenges in computer simulations. For this reason some different techniques, also called *enhanced sampling* methods, have been devised through the years. Some of them will be the subject of the next sections.

BIASED SAMPLING

We have seen in Section "A Toy Model: Alanine Dipeptide" that even for a very small and simple molecule some transitions could be hindered by free-energy barriers. If the system is going to sample only minimum "A" or minimum "B" in Figure 1 and no transitions are observed, then Eq. [6] is completely useless to evaluate the free-energy difference between "A" and "B." Because the low transition rate is due to the fact that a nonlikely region (the barrier) is in between the two minima, it is intuitive that increasing the probability of sampling the barrier could alleviate this problem. Furthermore, because the barrier height affects the transition rate in an exponential manner (Eq. [5]), it is clear that changing the barrier could easily lead to a dramatic change in the observed rates.

As first step, it is important to analyze what happens if an additional potential $V(s)$, which acts only on the CVs s, is added to the physical one $U(q)$. The resulting potential will be $U(q) + V(s(q))$, so that the explored conformations will be distributed according to a biased canonical distribution

$$P'(p,q) \propto \exp\left(-\frac{\sum_i \frac{p_i^2}{2m_i} + U(q) + V(s(q))}{k_B T}\right) \tag{8}$$

If one tries to evaluate the free-energy landscape from such a biased distribution of conformations, one will end up in a different free energy F', which is related to the original one by

$$F'(s_0) = -k_B T \log \int dq \delta(s(q) - s_0) \exp\left(-\frac{U(q) + V(s(q))}{k_B T}\right) + C'$$

$$= V(s_0) + F(s_0) + C' - C \tag{9}$$

where the part relative to the momenta has been integrated out and C' is another arbitrary constant. The correct free-energy landscape can then be recovered from the biased one up to an arbitrary constant by simply subtracting the bias potential $V(s)$.

Now, by supposing that the free-energy landscape is *a priori* known, at least approximately, it is possible to imagine performing a biased simulation, using $V(s) = -\widetilde{F}(s)$ as a bias where $\widetilde{F}(s)$ is our estimate for the free-energy landscape. The resulting simulation will explore the distribution,

$$P'(s) \propto \exp\left(-\frac{F(s) + V(s)}{k_B T}\right) = \exp\left(-\frac{F(s) - \widetilde{F}(s)}{k_B T}\right) \tag{10}$$

which is an almost flat distribution. More precisely, this distribution is more and more flat if the estimate of the free-energy landscape is more and more accurate. In the ideal case, the free-energy barrier disappears because all the values of *s* are equally likely. This method is known as the *umbrella sampling* method.[9]

The efficiency of umbrella sampling depends on the accuracy of the initial guess of the free-energy landscape. An error of even a few $k_B T$ could provide incorrect statistics because it would be unlikely to overcome the residual free-energy barriers. As we will see in the next section, this problem can be solved using an iterative procedure to build the bias potential.

ADAPTIVE BIASING WITH METADYNAMICS

Several algorithms have been proposed in the literature to progressively build a bias potential suitable for an umbrella sampling calculation.[10,36,37] We focus here on metadynamics.[11,38]

Metadynamics was originally introduced as a coarse dynamics in collective variable space, much in the spirit of Ref. 39. This dynamics was then accelerated by adding a penalty to the already visited states, similarly to the taboo search method[40] and the local elevation method.[41] Later, a version based on an extended Lagrangian formalism was introduced.[16] Presently, the most adopted variant is the continuous one.[38] In this chapter, we will only focus on this latter version.

In metadynamics (MetaD), a history-dependent bias potential is built during the simulation as a sum of repulsive Gaussians in the CV space. This is illustrated in Figure 4. These Gaussians are centered on the explored points in the CV space, have a preassigned width (σ) and height (w_G), and are deposited every τ_G time units as the simulation proceeds. The bias potential at time *t* thus reads

$$V(s,t) = \sum_{i=1}^{t/\tau_G} w_G \, \exp\left[-\sum_{\alpha=1}^{N_{CV}} \frac{\left(s_\alpha - s_\alpha\left(i\tau_G\right)\right)^2}{2\sigma_\alpha^2} \right] \qquad [11]$$

Notice that the width needs to be fixed for each of the CVs (σ_α). These widths determine the binning on each CV, that is, they can be used to state which distance in CV space should be considered as negligible. Intuitively, Gaussians are repeatedly added to the potential according to the explored states, such that they discourage the system from visiting again already visited configurations in the CV space. Notice that this procedure is done in the (low dimensional) CV space. Trying to do so in the full configurational space could not work in practice, because in that representation the system will practically never explore twice the same point. Contrarily, in the CV space, the system has a reasonable chance to explore several times the same value of *s*, albeit at different microscopic configurations *q*.

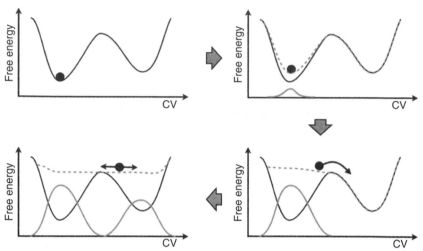

Figure 4   A sketch of the process of metadynamics. First the system evolves according to a normal dynamics, then a Gaussian potential is deposited (solid gray line). This lifts the system and modifies the free-energy landscape (dashed gray line) in which the dynamics evolves. After a while the sum of Gaussian potentials fills up the first metastable state and the system moves into the second metastable basin. After this the second metastable basin is filled, at this point, the system evolves in a flat landscape. The summation of the deposited bias (solid gray profile) provides a first rough negative estimate of the free-energy profile.

Several parameters should be chosen in a MetaD simulation and we will discuss how this is done in Section "Metadynamics How-To". Here we just observe that the w_G and τ_G parameter in Eq. [11] are not independent, and that, if they are chosen within meaningful ranges, what really matters is their rate $\omega = w_G/\tau_G$, also known as the *deposition rate*. In the limit of small τ_G, the expression for the history-dependent bias becomes Eq. [12].

$$V(s,t) = \omega \int_0^t dt' \exp\left[-\sum_{\alpha=1}^{N_{CV}} \frac{\left(s_\alpha - s_\alpha\left(t'\right)\right)^2}{2\sigma_\alpha^2}\right] \qquad [12]$$

Equivalently, it is possible to state that the bias potential $V(s)$ evolves with time according to the following equation of motion.

$$\dot{V}(s,t) = \omega \exp\left[-\sum_{\alpha=1}^{N_{CV}} \frac{\left(s_\alpha - s_\alpha\left(t\right)\right)^2}{2\sigma_\alpha^2}\right] \qquad [13]$$

The combination of this equation of motion with the Hamilton equations [1] describes completely the time evolution of a MetaD simulation. Another important observation is that the Hamiltonian dynamics is perturbed only through the force term produced by $V(s, t)$. Thus, thinking that very wide Gaussians could compensate for the barrier in a faster way is incorrect as these may produce smaller forces.

It can be understood intuitively from Figure 4 that such a procedure not only discourages the exploration of the already visited states in the CV space, but also provides an immediate estimate for the underlying free-energy surface. As more and more Gaussians are added to the bias, the system will explore a larger and larger fraction of the CV space. Because Gaussians are most likely added at points where the effective total free energy $F(s) + V(s)$ is lower, their effect will tend to flatten the $F(s) + V(s)$ function. After a suitable "filling" time, the bias will start growing parallel to itself, and one can expect to directly estimate $F(s)$ as $-V(s)$, but for an additional arbitrary constant. This hand-waving conclusion has been tested accurately on realistic systems.[38] It can be supported by a rigorous analytical derivation that is based on assuming the CVs are evolving by means of a stochastic Langevin dynamics:[42]

$$\dot{s} = -\frac{D}{k_B T}\frac{\partial F}{\partial s} + \sqrt{2D}\xi \qquad [14]$$

where D is the diffusion coefficient of the CV and ξ a random noise. Within this framework, the error made in using $-V(s)$ as an estimator for $F(s)$ can be quantified from the function

$$\varepsilon(s) = F(s) + V(s) - \frac{1}{\Omega}\int ds[F(s) + V(s)] \qquad [15]$$

where the integral is taken over the CV space and Ω is a normalization factor equal to the total volume of the integration region. Recalling that $F(s)$ is defined up to an arbitrary constant, the integral in Eq. [15] allows this arbitrary constant to be removed. For Langevin dynamics, it can be shown that $\varepsilon(s)$ is not exactly zero, but its difference from zero fluctuates in time. The fluctuations have squared amplitude $\langle\varepsilon^2\rangle \propto \omega/D$. This leads to two important observations:

- The statistical error grows with the deposition rate. Thus, a compromise should be found: ω should be large enough to quickly fill the basins in the free-energy landscape, but should also be small enough to limit the error in the free-energy estimation.
- The statistical error is larger for slower CVs (i.e., smaller D) because for such variables Gaussians are deposited on top of one another thus acting like an effective higher ω. This influences the compromise discussed above and implies that for slower descriptors one must choose a smaller deposition rate.

Because the free-energy estimate can be shown to be free from systematic errors, at least for model systems, it is always possible to increase its accuracy by taking a time average of the result, as first proposed by Micheletti et al.[43] It is thus in principle sufficient to prolong a simulation enough to decrease the error above any desired threshold. This, however, can lead to another problem of MetaD: because MetaD is a flat histogram method it tries to sample the whole CV space. Consequently, the simulated system can be pushed toward states with nonphysically high free energy and might drift the simulation toward thermodynamically nonrelevant configurations. Additional restraining potentials on the CVs may alleviate this effect as discussed, for example, in Ref. 44. Nevertheless, because restraining potentials can add small artifacts in the free-energy estimation at the border, a better strategy was introduced in Ref. 45 and is based on the idea of neglecting forces due to the Gaussian potential when the CV is outside the desired range.

Reweighting

One of the drawbacks of metadynamics, which is shared with all the methods based on biasing CVs, is that those CVs should be chosen before performing the simulation, and their choice typically affects the accuracy of the final result. However, it is sometimes very useful to compute free energies as functions of CVs that differs from the biased CV. This can be done by an *a posteriori* analysis. This kind of analysis on MetaD simulations has been introduced in Ref. 46. It is based on the weighted histogram analysis method.[47] Although typically used in the framework of bias exchange metadynamics (see Section "Bias Exchange Metadynamics") where a large number of CVs is used, this technique can be applied straightforwardly to normal MetaD.

WELL-TEMPERED METADYNAMICS

Standard MetaD, introduced in the previous section, has two well-known problems:

- Its estimate for the free-energy landscape does not converge but fluctuates around an estimate that, at least for simplified systems, can be demonstrated to be unbiased.
- Because it is a flat histogram method, it tries to sample the whole CV space. This can push the simulated system toward states with nonphysically high free energy and might drift the simulation toward thermodynamically nonrelevant configurations.

As discussed, both these problems have been recognized and tackled respectively by taking time averages[43] and by using restraining potentials.[44,45] An alternative method that addresses both the problems in an elegant fashion, called well-tempered metadynamics (WTMetaD), has been introduced in Ref. 48.

In WTMetaD, the rule for the bias update (Eq. [13]) is modified slightly to:

$$\dot{V}(s,t) = \omega \exp\left[-\frac{V(s(t),t)}{k_B \Delta T}\right] \exp\left[-\sum_{\alpha=1}^{N_{CV}} \frac{\left(s_\alpha - s_\alpha(t)\right)^2}{2\sigma_\alpha^2}\right] \quad [16]$$

Equivalently, each Gaussian, when deposited, is scaled down by a factor $e^{-\frac{V(s(t),t)}{k_B \Delta T}}$, where the bias potential has been evaluated at the same point where the Gaussian is centered and ΔT is an input parameter measured in temperature units. Equation [16] implies that, after the initial filling, Gaussians of different height are added in different regions of the CV space. In particular, on top of deep wells, where a sizable bias has been already accumulated, the additional Gaussians have small height. In contrast, at the border of the explored region, where the bias is still small, the additional Gaussians have large height. In the long time limit, when the bias potential tends to grow parallel to itself, the simulated system should systematically spend more time on the regions where smaller Gaussians are used, that is, on top of deeper wells. This disrupts the flat histogram properties of the method and in turn implies that the sum $V(s) + F(s)$ is no longer encouraged to become flat.

To estimate this deviation from the flat histogram behavior quantitatively, one should assume that the added Gaussians are narrower than the features of the underlying free-energy landscape and thus can be considered equivalent to δ functions with the proper normalization:

$$\exp\left[-\sum_{\alpha=1}^{N_{CV}} \frac{\left(s_\alpha - s_\alpha(t)\right)^2}{2\sigma_\alpha^2}\right] \approx \delta(s - s(t)) \prod_{\alpha=1}^{N_{CV}} (\sqrt{2\pi}\sigma_\alpha) \quad [17]$$

This allows us to approximate the bias update rule (Eq. [16]) as

$$\dot{V}(s,t) = \omega \exp\left(-\frac{V(s(t),t)}{k_B \Delta T}\right) \prod_{i=\alpha}^{N_{CV}} (\sqrt{2\pi}\sigma_\alpha)\delta(s - s(t)) \quad [18]$$

In the long time limit, the CV s will be distributed according to a biased canonical distribution $P(s) \propto e^{-\frac{F(s)+V(s,t)}{k_B T}}$, and the bias will grow according to

$$\dot{V}(s,t) \propto \exp\left(-\frac{V(s(t),t)}{k_B \Delta T} - \frac{V(s,t) + F(s)}{k_B T}\right) \quad [19]$$

By direct substitution, one finds the condition for the bias to grow uniformly, that is, for $\dot{V}(s,t)$ to be independent of s:

$$V(s) = -\frac{\Delta T}{T + \Delta T}(F(s) - C(t)) \quad [20]$$

This last equation indicates that in WTMetaD the bias does not tend to become the negative of the free energy but is instead a fraction $\frac{\Delta T}{T+\Delta T}$ of it. Thus, it only partially compensates existing free-energy barriers by an *a priori* known scaling factor. This factor can be tuned adjusting the ΔT input parameter. Moreover, in the long time limit, the system will explore the biased canonical distribution:

$$P(s) \propto e^{-\frac{F(s)+V(s,t)}{k_B T}} \propto e^{-\frac{F(s)}{k_B(T+\Delta T)}} \qquad [21]$$

As a consequence of the bias potential, the collective variable is exploring the canonical ensemble at an effective temperature $T + \Delta T$. Notice that the other microscopic variables are still sampled using a thermostatted MD at temperature T. In this sense, WTMetaD is related to other methods where an equivalent effect is obtained by extending the configurational space and by exploiting adiabatic separation of the CVs from the microscopic fluctuations.[13−15]

In short, WTMetaD allows performing a simulation where, in the long time limit, the effective temperature of one or more selected CVs is kept at an arbitrarily high value $T + \Delta T$. For $\Delta T \to \infty$, standard metadynamics is recovered (see Section "Adaptive Biasing with Metadynamics"). For $\Delta T = 0$, unbiased sampling is recovered.

This last feature of WTMetaD is of great advantage as it allows limiting the exploration of the CV space only to regions of reasonable free energy. Indeed, by fixing ΔT according to the height of the typical free-energy barrier for the problem under consideration, one will avoid overcoming barriers that are much higher than that.

Reweighting

As discussed in Section "Reweighting," it is sometimes useful to compute free energy as a function of *a posteriori* chosen CVs different from the biased ones. This is typically done in WTMetaD using a reweighting scheme introduced in Ref. 49. Here the time derivative of the partition function is estimated on the fly and allows consistent combination of data obtained at different stages of the simulation. An alternative has also been recently proposed[35] on the basis of the classical umbrella sampling formula.[9] Both these techniques are specific for WTMetaD.

METADYNAMICS HOW-TO

When adopting a new technique, the most frequent questions novice users ask are about how to choose the right input parameters. In MetaD in its various flavors there are three aspects that should be taken into account, namely:

- The choice of CV(s).

- The width of the deposited Gaussian potential.
- The energy rate at which the Gaussian potential is grown and, for WTMetaD, the parameter ΔT that determines the schedule for decreasing it along the simulation.

All of these decisions are deeply intertwined and we briefly outline here the main criteria for their choice, pushing on the practical implications and error diagnostics that allow one to decide whether the simulation parameters are well chosen or not.

The choice of the various parameters in MetaD generally relies on the following formula[38,42] that determines the error of a MetaD simulation under the assumption that the dynamics can be approximated by the Langevin equation:

$$\bar{\varepsilon} = C(d)\sqrt{\frac{S\delta s}{D}\omega \ k_B T} \qquad [22]$$

Here $C(d)$ is a prefactor that depends on the dimensionality of the problem (i.e., the number of CVs included), S is the dimension of the domain to be explored, δs is the width of the Gaussian potentials, D is the diffusion coefficient of the variable in the chosen space, and ω is the energy deposition rate for the Gaussian potential. This equation has several nuances that need to be explained, and here we will do it by considering alanine dipeptide as a "real-life" example.

The Choice of the CV(s)

Understanding the system one wants to study is pivotal in devising the correct CV(s) one needs. So, at any level, an accurate bibliographic search is very important in understanding the key factors in performing any kind of MetaD calculation. As an example, in computational biology problems, a mutation experiment that inhibits folding might be a signature of an important side-chain interaction that one needs to take into account.

More importantly it is worth understanding a key point of all the enhanced sampling techniques involving biasing one or more collective coordinate: a good descriptor might not be a good biasing coordinate. As a matter of fact, it may happen that different conformations might be distinguishable by using a specific dihedral as CV but if this is the outcome of a more complex chain of events and not the cause, one might end up by being disappointed by the result obtained by MetaD or other biased-sampling methods. Therefore, it is very important to understand that the lack of convergence of a free-energy calculation is often due to a suboptimal choice of CVs rather than the specific choice of the enhanced sampling technique.

In Section "Advanced Collective Variables" we will discuss *ad hoc* solutions for finding CVs that are optimal for specific problems. For the time being, we refer to the example of alanine dipeptide and try to identify the relevant order parameters. Assume that we have no prior knowledge of the importance of the Ramachandran dihedral angles, and we intend to describe the transition between C_{7eq} and C_{7ax} with one single order parameter. Intuitively, the C_{7eq} conformer is more extended than is the C_{7ax} conformation (see Figure 3). Therefore, "gyration radius," defined as

$$r_{gyr} = \sqrt{\frac{1}{N}\sum_{i=1}^{N} q_i^2 - \left(\frac{1}{N}\sum_{i=1}^{N} q_i\right)^2} \qquad [23]$$

where the summation runs on N heavy atoms, could be considered as a reasonable guess.

A simple molecular dynamics run of 200 ps in both basins can show that this descriptor is able to distinguish one state from the other (see Figure 5) because most of the sampling for C_{7eq} is concentrated at high values, whereas C_{7ax} exhibits low values of radius of gyration.

It is worth mentioning here that a better order parameter should be able to map the two minima onto two completely different regions of the CV space. The implication of having a partial overlap of the populations for the two basins, when they are structurally dissimilar as in this case, is that the kinetics cannot be simply retrieved through the free-energy landscape but requires the calculation of the transmission coefficient.[2] Nevertheless, for the sake of argument, we keep the choice as simple as possible to reproduce the possible pathological behaviors that one can encounter when experimenting with metadynamics.

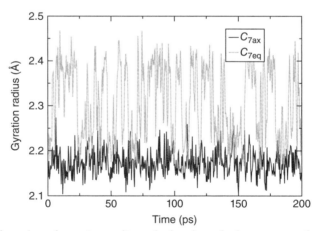

Figure 5   The value of gyration radius calculated on the heavy atoms for a trajectory of 200 ps of MD for alanine dipeptide in the two minima. Note that the average values of the gyration radius in the two cases are rather different, so this can be considered a possible choice for a CV.

The Width of the Deposited Gaussian Potential

The width(s) of the deposited Gaussian potentials is/are usually adjusted to adapt to the underlying free-energy landscape. Therefore, a preliminary unbiased MD run is generally performed, and the shape of the initial free-energy well is estimated. In this way one might figure out how wide the metastable basin is in the space of CVs. In a large system, to avoid performing extra heavy computations, the final part of the equilibration run can be postprocessed for this purpose. More specifically, for each CV, the probability density along the sampled CV range can be computed and, by assuming that the underlying free-energy landscape is a quadratic minimum, a Gaussian can be fitted on it. Under such assumptions, the addition of a Gaussian potential having height $k_B T$ and a width equal to the standard deviation of the fitted Gaussian can perfectly compensate the free-energy landscape so that the resulting landscape in the neighborhood should be flat. It should be taken into account that metadynamics does not aim at compensating perfectly the free-energy landscape at each step, though it aims at doing it on average in the long run (either completely, in normal MetaD of Section "Adaptive Biasing with Metadynamics," or partially, in WTMetaD of Section "Well-Tempered Metadynamics"); therefore, the Gaussian width σ of MetaD is generally set as a fraction of the standard deviation of such fitted Gaussian. One-half or one-third is generally suitable. Choosing a smaller width would capture better the free-energy features but would also give slower filling time and produce a steeper force that could interfere with integrator stability and therefore produce irrecoverable errors. Using a Gaussian fitting is done here with didactic purpose only: very often one just calculates the standard deviation out of the trajectory. This is illustrated in Figure 6.

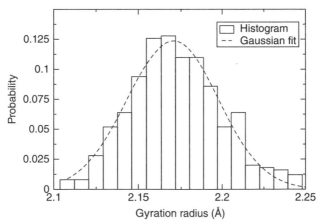

Figure 6 The probability of gyration radius calculated on the heavy atoms for a trajectory of 200 ps of MD for alanine dipeptide for the C_{7eq} minimum. The fitted normal distribution is centered in 2.17 Å and has standard deviation of 0.15 Å.

Moreover, it is worth noting that the chosen width for one region may not be optimal for other regions. This is particularly important with those CVs that present an anisotropic compression of phase space (see below in Section "Adaptive Gaussians" for a discussion of this issue and a possible solution).

The Deposition Rate of the Gaussian Potential

The deposition rate for the Gaussian potential can be expressed as the rate between the Gaussian height and the time interval between subsequent Gaussian depositions, that is, $\omega = w_G/\tau_G$. This rate can thus be tuned by adjusting both these parameters.

The choice of ω is crucial in metadynamics simulations because this parameter affects both the error and the speed at which the free-energy basins will be filled. It is wise to make it small enough to reduce the error but on the other side it must be sufficiently large to allow a reasonable filling rate of the free-energy landscape and to allow resampling the same point in CV many times. This balance will eventually produce accurate statistics that correspond to a good estimate of the free energy.

To avoid abrupt changes in the total potential when adding new Gaussians, one typically chooses the Gaussian height w_G as a fraction of the thermal energy $k_B T$. A typical choice is on the order of 0.1 kcal/mol for systems simulated at room temperature. Once this is fixed, the deposition time τ_G is typically adjusted to tune the deposition rate $\omega = w_G/\tau_G$.

As a rule of thumb, this can be done through a short free dynamics as shown in Figure 7. It is evident that the typical autocorrelation time of the CV is of the order of a picosecond or less. This sets a rough estimate for τ_G, which,

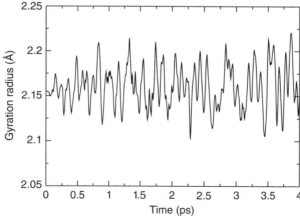

Figure 7   The value of gyration radius calculated on the heavy atoms for a trajectory of 4 ps of MD for alanine dipeptide in the C_{7ax} minimum. The values are acquired at each timestep of 2 fs.

for the following examples, will be chosen as 1.2 ps. By choosing the τ_G in this manner, the system is able to relax between one deposition and the other, to avoid placing the next Gaussian potential on top of the previously deposited one. This latter situation would produce an effective Gaussian potential with twice the height. In other words, this information encodes to some extent an estimate of the diffusion coefficient in the collective variable space that appears in Eq. [22].

A First Test Run Using Gyration Radius

With the parameters obtained from the analysis carried out above we can start a MetaD run. The parameters are $\sigma = 0.07$ Å, $w_G = 0.1$ kcal/mol, $\tau_G = 1.2$ ps. Using these parameters, we performed a 4-ns-long simulation. The time evolution of the CV is shown in Figure 8a. In a typical MetaD evolution the system spends a bit of time in a single basin around 2.2 Å of gyration radius, which is progressively filled, and this allows jumping after a few ps to the next minimum at 2.4 Å. The system is progressively biased and can reach very elongated as well as compressed states. While moving between the two extrema it should be noted that, as a result of the already filled free-energy landscape, the system is not stuck anymore in the starting basin: this is exactly the intended purpose of metadynamics.

It is crucial here to observe the so-called "recrossing" events, which means that the putative transition state is being explored by metadynamics many times. This ensures that the interesting states are sampled exhaustively, which is

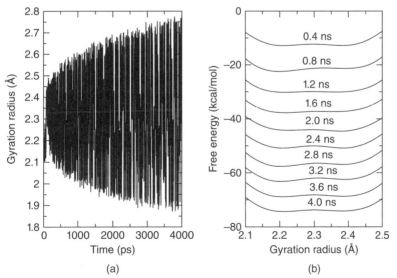

Figure 8 The time evolution of the chosen CV along the metadynamics (a) and the evolution of the bias (b).

important (but not sufficient) to obtain a correct result. In this case, as the two minima identified via straight MD are located at 2.1 Å and 2.4 Å one should look carefully for a number of transition in these regions, which occurs several dozens of times in the presented example. In realistic cases one may expect recrossing events of the order of tens.

In Figure 8b the evolution of the negative values of the bias is shown, which is the best approximation to the free-energy surface at a given time. It is worth noting that one should not worry about the absolute value of the bias to move progressively toward negative values, as the interesting features reside in the free-energy differences within one single profile and not in their absolute change from one time to the other. Interestingly, there is no sizable barrier between the two metastable states. This already casts some doubts about the quality of the CV employed. Indeed, if there is no barrier, one could have expected the transition to happen spontaneously in a few ps of simulation. This is a crucial point because critical judgment of consistency of both biased and unbiased simulations are very important to determine the quality of the results. Very often one can observe cases where an unexpected minimum occurs, which has characteristics not justified by any structural stabilizing feature. In such cases an unbiased MD run starting from a conformation taken from that minimum could allow one to clarify if the metastability is real or if it is an artifact of the CV configuration space coarsening.

An important point is shown in Figure 8a: in standard MetaD the extension of the explored portion of the CV domain increases with time. This generally gives access to higher energies and for many applications this has a negative effect. As an example, in proteins, one can reach energies that allow the unfolding of the secondary structure which, unless the algorithm is able to refold it, completely invalidates the simulation. Very often one has to artificially limit the exploration to prevent such effects by adopting confining potentials as discussed before. In our example the increased range explored implies that the landscape presents small changes from one time to the other in the region of interest, as the Gaussians are now deposited on a much broader range.

In Figure 9 (black lines) we report a set of free energies that are collected every 400 ps. Within 400 ps the system is in fact able to sample the gyration radius domain (2.1 Å < RGyr < 2.5 Å) several times; therefore, each realization of the free energy can be considered as a new estimate of the landscape. Under such conditions we can align all the landscapes by subtracting the average value and it becomes immediately evident that the landscapes may be very different from one another. This nonhomogeneity in the free-energy estimates is a source of error that is referred as "hysteresis" and is generally considered to be the most evident hint of a bad choice of CVs.

Frequently, there exist other significant diagnostics that can highlight problems in the CV choice and may help in devising a more suitable CV. For example, using an independent structural comparison consisting of a different descriptor, such as root mean square deviation (RMSD) after optimal

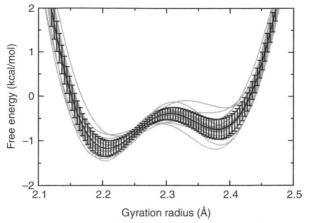

Figure 9 Average free-energy profile (cross-hatched thick profile) and associated error bars for different free-energy estimates aligned on their average value (gray profiles).

fit,[50] can reveal if the used CVs are resampling the expected states or if they are moving the simulation in unwanted phase space regions. Consider as an example alanine dipeptide in Figure 10, where the structural alignment with respect to C_{7ax} and C_{7eq} reference structure is reported. It is evident that C_{7ax} appears only within the first 500 ps (RMSD < 1 Å) of the simulation and never reappears even if the corresponding value of the gyration radius appears frequently in the simulation. In this case MetaD is not accelerating the sampling between the target states although many recrossing events are observed in the CV space.

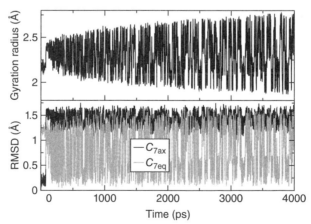

Figure 10 Time evolution of the radius of gyration (upper panel) along with Root Mean Square Deviation (RMSD) after optimal fit onto C_{7ax} and C_{7eq} reference structures.

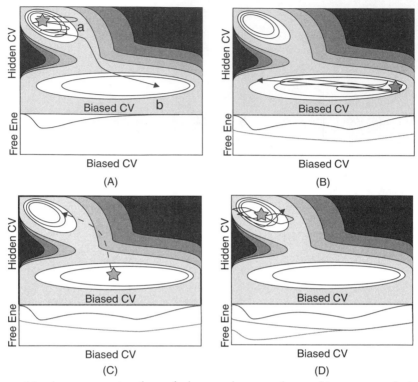

Figure 11   A representative chart of what may happen when making a wrong choice of CVs in MetaD. In the bottom of each panel is shown a sum of the potential to illustrate the effects on the estimated free-energy landscapes.

Figure 11 clarifies what happens in this system. Suppose we are interested in the difference between state a and b in Figure 11, which correspond to the C_{7ax} and C_{7eq} conformers, respectively. When performing MetaD along the variable on the horizontal axis the forces are added only in this direction, and the system is pushed away from minimum a thus reaching minimum b. This jump may have produced a change in a "hidden" CV that could be due to a peculiar free-energy landscape (as shown in Figure 11) or by a fortuitous jump in a parallel valley that is kinetically accessible. Once reaching b, MetaD will push the system in the valley located on the bottom (see Figure 11A) because the MetaD force is acting only in the horizontal direction and cannot control any further jump into the a basin, which is separated by a barrier. As a result, one gets a free-energy profile that resembles the bottom basin almost exclusively, leaving only a minimal trace of the initial basin (see Figure 11B). The system will only jump back to basin a because of an overaccumulation of potential or by a fortuitous thermal fluctuation (see Figure 11C). Whenever this occurs, the free energy changes and becomes similar to that of the projection of the a basin onto the biased CV again (see Figure 11D). In this case the free-energy profile may change considerably at distant times and, consequently, show hysteresis.

This scenario points to the fact that the selected set of CVs is not suitable and a different choice should be made.

A Better Collective Variable: Φ Dihedral Angle

To illustrate the significance of CVs selection in MetaD and in other enhanced sampling schemes, the same calculation is now repeated but using the Ramachandran angle Φ as the CV. The σ was chosen according to the same procedure outlined above to be 0.1 rad while keeping all the other parameters identical.

The outcome using this CV choice is presented in Figure 12 and it is evident, especially in the progress of the deposited bias, that the shape of the deposited potential is conserved and grows approximately parallel to each realization. This results in a more uniform error as shown in Figure 13 where all the free energies are more consistent with one another.

Furthermore, the two metastable states are repeatedly visited as shown in Figure 14 where in both cases there exist structures with an RMSD < 1 Å along the whole trajectory. In the last case, the metadynamics run can be considered to provide a reliable free-energy profile.

Conceptually, the change in CV can be regarded as a sort of warping of the free-energy landscape. This is represented pictorially in Figure 15. In this

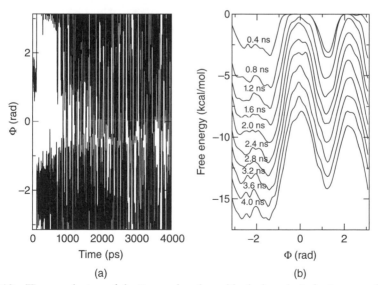

Figure 12 Time evolution of the Ramachandran dihedral angle Φ during metadynamics (a) and evolution of the deposited bias over time (b).

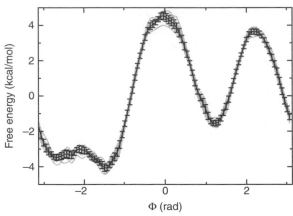

Figure 13   Average free-energy profile when using Φ as the CV in metadynamics. The error is reduced and the free-energy profiles (gray lines) are very coherent with one another.

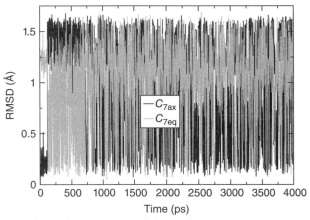

Figure 14   RMSD from the two C_{7ax} and C_{7eq} reference structures during the metadynamics run that uses the Ramachandran dihedral angle Φ as CV in metadynamics.

case no barrier appears on the hidden CV and consequently forces applied by MetaD are now in the direction connecting the two states.

Well-Tempered Metadynamics Using Gyration Radius

We turn here to describe the case of WTMetaD because recently it is becoming widely applied and its error control and associated diagnostics differ slightly from the case of the normal MetaD. In WTMetaD there is an additional parameter to be set, namely, ΔT, which technically regulates the decay factor of the

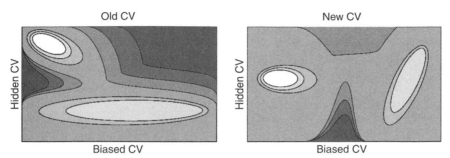

Figure 15 A graphical interpretation of the deformation of the landscape induced by the change of CV. Now no barrier appears on the hidden CV, and the two states do not project on top of each other but instead are well separated.

potential deposition rate. As discussed in Section "Well-Tempered Metadynamics," this can be regarded as a factor that determines the effective temperature of the enhanced CV. Thus, a wise way to choose ΔT is by having a rough estimate of the barrier. In our case, as the barrier is expected to be 8 kcal/mol, which is around 15 $k_B T$ units, we set $(T + \Delta T) = 15T$. This ensures that in the long run the bias will allow passing the barrier or, in other words, that before decreasing the Gaussian height dramatically a potential will be built that is able to overcome the barrier.

Unlike MetaD, WTMetaD provides only an exploration limited to the low free-energy region of the landscape. This is clearly visible from Figure 16 where

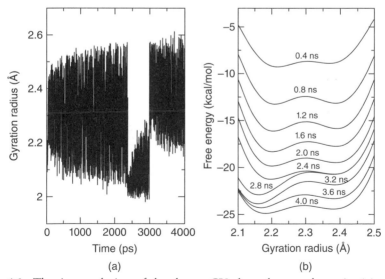

Figure 16 The time evolution of the chosen CV along the metadynamics (a) and the evolution of the bias (b).

the space visited is limited to 2 Å < RGyr < 2.6 Å, which is remarkably less than the non-well-tempered case. This is of course a big advantage over standard MetaD because the exploration can be restricted to only the interesting region where activated processes take place and prevents the sampling of high free-energy regions where irreversible transitions might occur. Another visible difference is that the profiles (Figure 16b) are getting closer to one another as time passes. This is the effect of the Gaussian height diminishing and, therefore, the deposited potential converging. Moreover, because a correct WTMetaD is expected to converge, one typically estimates the free energy from the final bias potential without averaging.

Nevertheless, pathological CVs still show their "hysteresis" issues (see the sudden change in profile at 2.8 ns, Figure 16b), which has a remarkable counterpart in the CV evolution in the Figure 16a. This aberrant behavior generally arises from the system beginning to explore a different free-energy surface after having performed a jump in a so-called "hidden" CV (see Figure 11). At that stage the underlying free-energy landscape changes, and the system stays trapped in a region, therefore adding more and more bias.

From these considerations, it becomes evident that the Gaussian height evolution gives a hint about when this is happening as shown in Figure 17a. Here, while the system seems to decrease the Gaussian height, as it is continuously resampling the same CV space, at 2.4 ns a sudden jump appears. This is usually associated with a jump into a nonsampled region where Gaussian height increases. What happens from the structural point of view is that the state C_{7ax} is occasionally resampled. This is clearly visible from the RMSD, which is reported in Figure 17b. In the schematic of Figure 11 this change corresponds to a jump on the hidden CV from basin b to the target basin of a. Once in C_{7ax}, the system finds a very narrow basin in free energy, which was

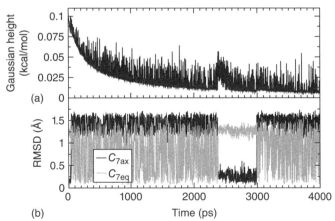

Figure 17 The time evolution of the height of the potential deposited (a) and the RMSD with respect to the two C_{7ax} and C_{7eq} states (b).

not accounted for by the bias potential accumulated by WTMetaD (almost flat in this region), so it starts accumulating potential to escape and eventually it does at 3 ns, where it reaches once again C_{7eq}. At this point the potential starts washing away the changes because of the C_{7ax} basin, and the old landscape is retrieved, leaving the user in doubt about which is the correct landscape.

Once again, a bad choice of CVs makes the transition from C_{7ax} to C_{7eq} rather fortuitous. This indicates that the selected CV is not a good variable for biasing, with either WTMetaD or MetaD.

It may also happen in WTMetaD that the growth of the bias potential is a poor diagnostic tool for detecting "hysteresis," especially when the system shows transitions at very long time and at low Gaussian height. In this respect the probability histogram of the CVs calculated over different trajectory segments can be more indicative of underlying problems. Indeed, in WTMetaD the system should converge to a probability $P(s) \propto e^{-\frac{F(s)}{k_B(T+\Delta T)}}$. Any divergence from this behavior should be looked at with caution as it might be a signature of poor convergence. This is evident in Figure 18 where the trajectory in the CV space is divided in four blocks of 1 ns each, and the probability distribution is calculated for each block. Note that the third and the fourth blocks are remarkably different from the first two and from one another.

Well-Tempered Metadynamics Using Dihedral Angle Φ

We revise here the last example to show how the diagnostics change when the CV is well chosen. As before, we adopt the Φ Ramachandran dihedral angle as a biased CV.

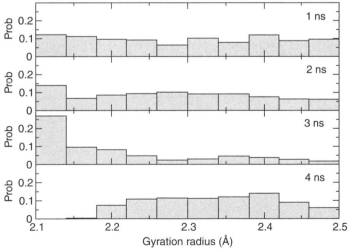

Figure 18 The probability distribution of finding the system in the CV range in four blocks of the simulation. It is evident that this distribution does not converge.

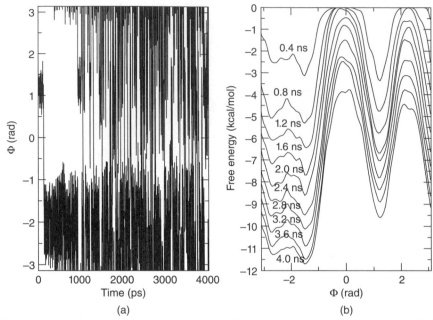

Figure 19   (a) CV evolution with time. (b) The evolution of the best estimate of the free-energy landscape from WTMetaD when using Φ as CV.

While the time evolution of the free-energy landscape (see Figure 19) is similar to the conventional MetaD case (Figure 12) we point out that WTMetaD is depositing less potential than the standard case, because of the diminishing Gaussian height.

The evolution of Gaussian height is also more regular as it can be seen in Figure 20a. Discontinuities in the decay of the Gaussian height are frequent and coincide with the jumps between C_{7ax} and C_{7eq} states, which are repeatedly visited as seen in the RMSD plot (Figure 20b).

Finally, the probability evolution is also more homogeneous with time as one might expect in the case of no hysteresis (see Figure 21).

ADVANCED COLLECTIVE VARIABLES

As discussed in the previous section, one of the most difficult issues in running metadynamics simulations is that of choosing a proper set of CVs. The limited reliability of metadynamics calculations is often mistakenly perceived as a limitation of the method although the real reason is generally due to the descriptors used.

A good CV that is able to distinguish reactants and products may not be sufficient in driving the reaction reversibly. This is due to the fact that a

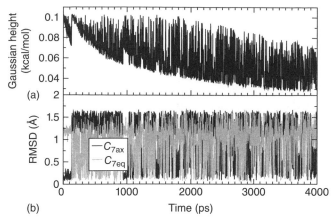

(a)

(b)

Figure 20 (a) The time evolution of the Gaussian height for WTMetaD with Φ CV. (b) RMSD with respect to the two states C_{7ax} and C_{7eq}.

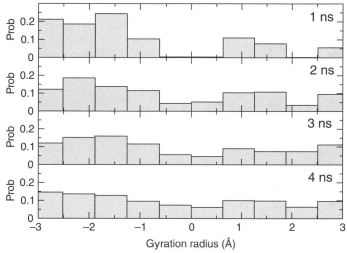

Figure 21 The probability distribution of finding the system in the CV range in four blocks of the simulation. It is evident that this distribution converges smoothly and consistently presents a higher probability for angles corresponding to the two metastable states.

change that is observable during a process might be triggered by a different (and hidden) physical cause. Thus, driving the observable is not guaranteed to facilitate the transition.

We also discussed how important it is to devise a CV that can also produce a force in a direction that permits the system to travel between all relevant configurations.

For these reasons it is typically more useful to explore multiple sets of CVs rather than to tune the other metadynamics parameters. Moreover, moving toward a better sampling by attempting various CV sets always implies a better understanding of the physicochemical process being studied.

Given the inherent difficulty of making such a choice, much effort in the past decade has been focused on developing CVs suitable to describe correctly a variety of conformational changes. We review here a few of the most popular choices used with metadynamics but which can also be used in other schemes like umbrella sampling, thermodynamic integration, and steered molecular dynamics.

Path-Based Collective Variables

The computational cost of metadynamics can scale as $S^{N_{CV}}$ where S is the extent of the CV space and N_{CV} is the number of CVs used. Often, it is inherently difficult to reduce N_{CV} because of its nature of being intrinsically multidimensional. For example, protein folding is due to an interplay of backbone flexibility, hydrogen bonding, hydrophobic collapse, and dehydration intertwined with one another. Another example is chemical reactions in enzymes involving several hydrogen bonds and backbone distortions. Sampling extensively all those degrees of freedom independently is of little help because most of them move in a concerted fashion, along a hypothetical one-dimensional reaction tube that changes character as one progresses from reactant to products. The need to find a free energy along an adapted "reaction tube" was first perceived by Ensing and Klein[51] who performed an explorative metadynamics run, followed by a refinement through an umbrella sampling calculation along the obtained minimum free-energy path. Later on a general functional form was found for this "reaction tube" to be sampled directly through MetaD, which is the purpose of the so-called path collective variables.

By assuming that the user already has M available snapshots $\{\widetilde{q}(i)\}$ (also called "nodes") that are representative of the transition and that can be obtained via high temperature MD or steered MD, a progress variable can then be built:[34]

$$s(q) = \frac{1}{M-1} \frac{\sum_{i=1}^{M}(i-1)\exp(-\lambda\|q-\widetilde{q}(i)\|)}{\sum_{i=1}^{M}\exp(-\lambda\|q-\widetilde{q}(i)\|)} \qquad [24]$$

where λ is a parameter that should be chosen according to the average interframe distance, and $\|q - \widetilde{q}(i)\|$ is the distance between the current molecular dynamics snapshot and the $\widetilde{q}(i)$ node. Intuitively, this variable provides a continuous fractional index ranging from 0 (the reactant) to 1 (the product) that allows one to map the conformations along an ideal "progress along the path" variable.

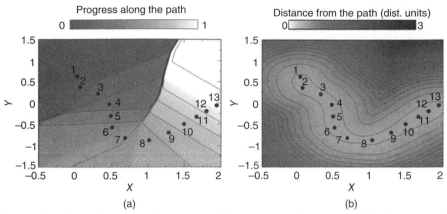

Figure 22 A two-dimensional example of the path variables and the foliation induced by the progress along the path (a) and the distance from the path (b).

Similarly, a complementary variable can be defined that provides the distance from the closest node:[34]

$$z(q) = -\frac{1}{\lambda} \ln \sum_{i=1}^{M} \exp(-\lambda \|q - \tilde{q}(i)\|)$$ [25]

A two-dimensional example is shown for a number of nodes in Figure 22. The foliation induced by Eq. [24] (Figure 22a) is similar to that produced by Voronoi tessellation.[52] The gradient of the variable changes direction along the path to adapt to the reaction character. Nevertheless, it shows some pathological behaviors at specific points where the indexing can change abruptly; this typically happens at large distances from the path. That is why defining the distance from the path by Eq. [25] (Figure 22b) is particularly helpful: the exploration of this variable can be artificially limited to prevent high-distance regions from being sampled. Alternatively, this variable can be used as an extra CV for metadynamics, in an attempt to explore paths that are far from those defined by the nodes.

In this framework the definition of the metrics $\|q - \tilde{q}(i)\|$ plays a key role. An arbitrary reduced representation could be used in the spirit of the string method in CV space.[33] In general, the Cartesian coordinates of a subset of atoms are widely used and compared via mean square deviation after optimal alignment:[50]

$$\|q - \tilde{q}(i)\| = \sum_{k,\alpha} \left[q_{k,\alpha} - cm(q)_\alpha - \sum_\beta \mathbf{R}(q, \tilde{q}(i))_{\alpha,\beta} [\tilde{q}(i)_{k,\beta} - cm(\tilde{q}(i))_\beta] \right]^2$$ [26]

where $\mathbf{R}(q, \tilde{q}(i))_{\alpha,\beta}$ is the optimal alignment matrix that superimposes q to $\tilde{q}(i)$ and $cm(q)_\alpha$ is the α component of the center of mass of q. This metric was used in a variety of contexts including conformational transitions,[34,53] enzymatic catalysis,[54] chemical reactions,[55,56] and folding.[57] Notwithstanding the simplicity in its setup, the choice of the initial nodes must be done carefully. This set of "omnicomprehensive" Cartesian coordinates, although able to include most of the effects, is likely to increase artificially the distance between the molecular dynamics run and the nodes because multiple and irrelevant sources of noise contribute to it. Therefore, other metrics were proved to be useful like the comparison of contact maps,[57] which correctly detected the unfolded versus molten globule transition in folding. Also, the chirality indexes[58] that rely on the internal degrees of freedom associated with backbone conformations were found to be reliable.

An additional benefit from using this set of CVs is that one is not limited by the rough initial input. Instead, an iterative procedure can be set up to refine the frames and end up with an optimal description for the metrics employed.[34] This makes it possible to provide a better description of the saddle point, which can be quantified by computing the committor probability.[59] This procedure can be conveniently simplified by using an iterative steered-MD procedure as shown in Refs. 54, 55. More recently, a new integrated procedure was also introduced[60] on the basis of performing MetaD along the path while simultaneously recording the position orthogonal to the path. At regular intervals along the simulation, the path is evolved. At the end of the procedure it is possible to obtain both the free-energy landscape along the path and the optimized path. This is possible because MetaD allows the free energy along the path to adapt while the definition of the path itself is changed.

Collective Variables Based on Dimensional Reduction Methods

Another way to tackle the problem of defining the CVs is to renounce "general-purpose" CVs (e.g., number of hydrogen bonds, coordination of water, gyration radius in a typical folding problem) and to deduce them directly from the behavior of the system during the simulations.

The objective of these methods is to obtain a representation of the system in terms of a very small number of parameters. Typically, those parameters will be a complex but automatically determined function of the components of an intermediate representation of the system, in which the dimensionality of the problem can involve as many as several hundred degrees of freedom. Examples of such an intermediate representation might be the position of the heavy atoms in alanine dipeptide, the C_α carbons, or all the residue-to-residue distances in a protein.

The most straightforward use of this intermediate coarse-graining representation is through principal component analysis (PCA)[61] of a trajectory.

This produces a set of eigenvectors and eigenvalues that decompose the system motion into different amplitude motions acting on an average conformation. It is then possible to use the projections of the position of the system onto a subset of the eigenvectors as CVs in a MetaD simulation.[62] Generally, projections are performed onto the eigenvectors connected to large amplitude motions to accelerate slow conformational transitions.[63] It is worth noting the strong similarity with conformational flooding[37] in which the potential from the PCA is calculated only once and is not built adaptively as in MetaD.

The system under study can display many basins, each of them with a peculiar average structure and a specific set of eigenvectors, which might be very different from one another. For this reason Tribello and coworkers developed "reconnaissance metadynamics"[64] in which the dynamics is analyzed on the fly and onto which a Gaussian mixture model is fitted. The system is then assigned to one of the basins defined by the Gaussian, and the potential is added in one dimension (the distance from the center of the fitted Gaussian). This creates an "onion-like" potential that builds up to a point where it is detected by its fall into a new basin where the Gaussian mixture fitting is repeated. This method is very effective in sampling the configurational space, but no straightforward procedure exists yet for extracting free-energy profiles from the obtained trajectories.

Other methods exist besides PCA-based approaches that are able to extract a pattern from computed statistics. One of them is the so-called "classical multidimensional scaling method"[65] in which a fictitious, low-dimensional distribution of points is calculated to reproduce the distance relations among the points in the intermediate dimensionality. An example is the map of the distances in RMSD for configurations of dialanine. There, the dihedral angle Φ rotation is very similar to the distances that one would obtain from a suitable distribution of points along a ring. This implies that not all the coordinates of the intermediate representation are useful, because the landscape in this case is effectively two-dimensional.

This idea of using a high-dimensional data to fit a low-dimensional representation was exploited by Spiwok[66] who used this scaled representation to define the following modified path collective variable

$$s_P(q) = \frac{\sum_{i=1}^{M} P_i \exp(-\lambda \|q - \tilde{q}(i)\|)}{\sum_{i=1}^{M} \exp(-\lambda \|q - \tilde{q}(i)\|)} \qquad [27]$$

where the value of P is a general (vectorial) property of the node $\tilde{x}(i)$ that can be chosen arbitrarily. In particular, the authors chose the coordinates P_x, P_y, and P_z from an isomap multidimensional scaling derived from extensive sampling of a small molecule, much in the spirit of the work from Das et al.[67] In this way MetaD on that scaled representation can be performed effectively.

More recently, Ceriotti et al.[68] further developed the multidimensional scaling concepts for molecular representations and proposed a scheme called "sketch-map." Their approach is a filtered multidimensional scaling algorithm that can reproduce the distances of the intermediate representation in a specific window of lengths, while coarsening the distances that are both higher and lower than this window. They proved this makes the problem more tractable and allows for a much broader range of scaling possibilities compared to standard multidimensional scaling. Eventually, they also figured out a novel way to perform MetaD that can be used specifically with sketch-map.[69]

Template-Based Collective Variables

The predictive simulation of protein folding at atomistic resolution is one of the grand challenges in the biophysical community and in statistical mechanics. The problem consists in predicting, with only the knowledge of the residues' sequence, the most stable state, which, in general, is known to be associated with its function. General purposes CVs, as distance or coordination numbers, are ineffective in reversibly folding, even for a small sequence of aminoacids.

Therefore, because proteins generally exist as one of several recurrent structural motifs, it is plausible to use a CV that is a similarity measure with minimal building blocks of one of those motifs. One then evaluates the number of times this building block is present in the actual MD snapshot. This is the central idea behind the template-based collective variables of Pietrucci and Laio.[70]

In detail they first identified an ideal building block for every structural pattern that could be either an α-helix, a parallel β-sheet, or an antiparallel β-sheet. They did that by isolating from the CATH[71] database all segments containing that specific motif using the STRIDE[72] definition and then by clustering them with distance-RMSD (dRMSD) comparison. The choice of dRMSD makes the computational comparison particularly efficient because it avoids calculating the rotation matrices and their derivatives. Once the center of the cluster is identified for each motif, the ideal structural motifs $\mathbf{d}_\alpha$, $\mathbf{d}_{para-\beta}$, $\mathbf{d}_{anti-\beta}$ can be retrieved as a list of pairwise distances. Each template is used to compute three distinct CVs: α-helix content, parallel β-sheet content, and antiparallel β-sheet content. For each MD snapshot the structural motif content of a given consecutive protein segment can be estimated simply by (e.g., for the α-helix case):

$$S_\alpha = \sum_i n[\mathrm{dRMSD}(\mathbf{d}(q_i), \mathbf{d}_\alpha)] \qquad [28]$$

where $\mathbf{d}(q_i)$ includes the atoms of the block of three residues starting with atom q_i. The function n assigns more weight to those structures having a low dRMSD with respect to the template structure and therefore can be seen as a counter. Its functional form is:

$$n[\mathrm{dRMSD}] = \frac{1 - (\mathrm{dRMSD}/r_0)^8}{1 - (\mathrm{dRMSD}/r_0)^{12}} \qquad [29]$$

where r_0 is a radius that discriminates if the segment is similar to the template or not and is generally set to 1.0 Å. It is important to note that in the dRMSD calculation, only C_α, N, C_β, and carbonyl oxygen atoms are included, and that the entire sequence is scrolled by steps of a single residue. The scrolling pattern depends on the type of secondary structure.

In this way the authors successfully produced a varied ensemble of secondary structures. More recently, these variables were used in conjunction with an extensive bias exchange MetaD (see Section "Bias Exchange Metadynamics") on VAL60 to sample, with atomistic resolution, plausible folds containing secondary structures[73] well beyond the variety contained in the PDB database.

Importantly, such strategy could be adapted for use well beyond the actual scope of protein folding and might be useful where other kinds of template are available.

Potential Energy as a Collective Variable

In the last three subsections we saw how carefully designed CV can be used as suitable reaction coordinates. We consider here the possibility of using the potential energy as a CV.

This idea was first proposed by Bartels and Karplus[74] in the context of adaptive umbrella sampling, and later used for plain metadynamics by Micheletti et al.[43] When a flat histogram distribution of potential energy is enforced, the system under investigation is able to freely explore values of the potential energy in a large window. When the potential energy grows, the explored conformations are equivalent to those explored at a higher temperature, so that, roughly speaking, the simulation performs a sort of annealing where the effective temperature is allowed to increase and decrease. At high temperature, the effective barriers are largely decreased, and the system is able to cross free-energy barriers easily. In this sense, achieving a flat histogram distribution of potential energy can be considered as comparable to performing a multicanonical simulation[75] and is also strictly related to Wang-Landau sampling.[76]

In a later work, Bonomi et al.[77] showed that it is possible to perform a WTMetaD simulation on the potential energy by sampling the so-called "well-tempered ensemble." This ensemble is characterized by a parameter γ, which is related to the usual parameter ΔT by $\gamma = \frac{T+\Delta T}{T}$. In the well-tempered ensemble the distribution of the potential energy is not flat but instead is related to the canonical distribution $P(U)$ by $P_{\mathrm{WTE}}(U) \propto P(U)^{1/\gamma}$. The effect of the γ parameter is to increase the fluctuation of the potential energy by a factor γ. This should be compared with a multicanonical ensemble,[75] where the fluctuations of the potential energy are infinite. In Ref. 77 the authors also showed how this amplification of the potential energy fluctuations can be used in practice to enhance the efficiency of parallel tempering simulations and in particular how this method can be used to decrease the number of replicas required for parallel tempering simulations by orders of magnitude. Moreover, thanks to a suitable reweighting scheme (see Section "Reweighting"), it is possible to extract the free-energy landscape as a function of any *a posteriori* chosen CV. The combination of parallel tempering and well-tempered ensemble can also be used in conjunction with plain metadynamics, where standard collective variables such as number of hydrogen bonds or hydrophobic-core size are also biased.[78]

Interestingly, in a recent paper it has been shown that it is possible to dissect the potential energy and use as a CV only a relevant portion of it, namely, the Debye–Hückel interaction between two molecules.[79] Although in that paper this CV was biased using steered molecular dynamics, the extension to umbrella sampling or metadynamics is straightforward.

IMPROVED VARIANTS

Besides optimizing the choice of the CVs, several variants of the basic schemes outlined in Sections "Adaptive Biasing with Metadynamics" and "Well-Tempered Metadynamics" have also been introduced. We will discuss here three of them, which are based on the idea of running multiple metadynamics in parallel with different degrees of coupling. We will also describe a recently introduced scheme aimed at simplifying the choice of Gaussian width and at adapting it on the fly.

Multiple Walkers Metadynamics

As pointed out in Section "Adaptive Biasing with Metadynamics," a critical choice in metadynamics involves the deposition rate. In particular, a higher deposition rate is required for a faster initial filling of the free-energy landscape, but that would affect the final error. The multiple walkers algorithm is designed to exploit concurrent metadynamics simulations to allow a very fast filling albeit using a slow deposition rate.

In multiple walkers metadynamics[80] N_w metadynamics simulations (also referred to as *walkers*) are run concurrently, possibly on different machines. All these simulations contribute to the growth of a unique bias potential, which thus grows at a speed that is N_w times larger than for a single simulation. It has been shown heuristically (see Ref. 80) and analytically for a Langevin system (see Ref. 42) that the resulting error is the same as that expected from a single walker using the same Gaussian height and deposition time. This means that when using N_w walkers a filling time acceleration by a factor N_w can be obtained without increasing the error or, alternatively, the squared error can be decreased by a factor N_w without increasing the filling time.

In multiple walkers metadynamics there is no efficiency gain in computing time, as the same accuracy could be obtained by performing just a single simulation N_w-times longer. However, because the walkers are weakly coupled, this algorithm can be run easily on a parallel machine or even on a weakly interconnected cluster. Only a shared file system is required to allow interwalker communication in typical implementations. Moreover, it is not even necessary to start all the walkers at the same time, and there is no practical problem if one or more of the simulations is interrupted at some point.

Multiple walkers metadynamics can also be used in the well-tempered algorithm.[81] In the long time limit the error prefactor will be decreased when increasing the number of walkers because the walkers will provide independent statistics biased by the same potential.

As a final remark, we note that the theoretically perfect scaling of multiple walkers metadynamics can be reached only if the walkers are started from independent configurations taken from the proper equilibrium distribution. This can be a difficult task, as the free-energy landscape is not known *a priori*.

Replica Exchange Metadynamics

Replica exchange molecular dynamics is a technique where many simulations are performed at the same time using different control parameters. A typical example is parallel tempering,[7] where different temperatures are used. Hamiltonian replica exchange,[82] however, provides a more general formulation where different force field parameterizations are used in different replicas. Replicas are ordered to form a sort of ladder. The first step typically represents experimental conditions whereas the last step represents an artificially modified system (e.g., with a fictitiously high temperature in parallel tempering) where transitions are more likely to be observed. Coordinate swaps are attempted with a chosen time interval. This brings information from highly ergodic modified simulations to the typically frustrated room temperature simulation. To achieve sampling of the canonical ensemble, coordinate exchanges must be accepted according to a Metropolis rule where the acceptance is given by

$$\alpha = \min\left(1, \frac{e^{-\beta_i U_i\left(q_j\right) - \beta_j U_j(q_i)}}{e^{-\beta_i U_i(q_i) - \beta_j U_j(q_j)}}\right) \qquad [30]$$

Here U_i is the potential energy used for replica i, β_i its inverse thermal energy, and q_i represents the coordinates of all the system for the ith replica.

Replica exchange methods are typically expensive as they require many (sometimes hundreds) simulations to be run synchronously, often on a large supercomputer facility. Their advantage is that they allow us to accelerate sampling using a minimal amount of *a priori* knowledge about the specific transition under study. This contrasts with metadynamics, which is very cheap from a computational standpoint but requires physical insight to select proper CVs.

Thanks to their complementarity, metadynamics and replica exchange can nonetheless be optimally combined.[17] To do so one has to perform N_r simultaneous metadynamics simulations and adjust the Metropolis criterion to include the bias potential

$$\alpha = \min\left(1, \frac{e^{-\beta_i\left(U_i\left(q_j\right)+V_i(s(q_j),t)\right)-\beta_j(U_j(q_i)+V_j(s(q_i),t))}}{e^{-\beta_i(U_i(q_i)+V_i(s(q_i),t))-\beta_j(U_j(q_j)+V_j(s(q_j),t))}}\right) \qquad [31]$$

The combined algorithm has the advantage with respect to metadynamics of accelerating all the degrees of freedom of a system without any *a priori* knowledge, albeit being more expensive. Consequently, if the high temperature replica is ergodic enough, hysteresis effects (see Section "A First Test Run Using Gyration Radius") because of a suboptimal choice of the CVs are strongly moderated or even disappear. When compared with replica exchange alone, replica exchange metadynamics improve the sampling of energetically unfavorable points such as metastable minima or transition states, provided they are defined on a CV space that is *a priori* identifiable. In this sense, replica exchange metadynamics takes the best from both worlds.

Notably, metadynamics can also be combined with replica exchange techniques other than parallel tempering, for example, solute tempering.[83]

Bias Exchange Metadynamics

One of the problems in metadynamics is that the maximum number of CVs that can be used in practice is limited. In particular, it is very difficult to converge metadynamics simulations performed using more than three CVs. This limits the applicability of the method to cases where a few appropriate CVs are known *a priori*. However, if one is able to define a super-set of CVs, which is very likely to include the appropriate CVs, multiple metadynamics simulations can be conveniently combined by means of bias exchange metadynamics (BEMD).[84]

In BEMD, several metadynamics simulations acting on different CVs are performed at the same time. Each of the simulations employs one or more CVs, and there might be overlap between the sets of CVs chosen for each simulation. For instance, the first replica could use two-angle CVs ψ and ϕ, the second replica could use only ϕ, and the third only ψ. Then, from time to time, coordinate exchanges are proposed and accepted with the usual Metropolis criterion. Like parallel tempering simulations, if the acceptance rate for the exchange moves is sufficiently high and one replica samples ergodically the configurational space, ergodicity is ensured in all of them. Whereas in parallel tempering the ergodic simulation is typically the one at the highest temperature, in BEMD any of the replicas could be ergodic depending on how virtuous the CV(s) used for that replica is(are) to describe the relevant conformational transitions. In addition, far fewer replicas are needed than for a conventional parallel tempering-metadynamics calculation.

Using BEMD it is thus possible to replace an (unfeasible) metadynamics that would use ten or twenty CVs with ten or twenty weakly coupled one-dimensional metadynamics simulations and retain an accuracy that is determined by the replica containing the optimal CV. Moreover, the statistics of each of the replicas benefits from the exchange. Accordingly, the exploration of phase space by each replica is much higher compared to that of a simple MetaD. We finally point out that in BEMD the statistics produced by all the replicas should be used to calculate the final free-energy landscape via a reweighting procedure.[46]

Adaptive Gaussians

Up to this section, the shape of the repulsive potential adopted by MetaD was assumed to be of the form of Eq. [11] with σ_α fixed at the beginning of the simulation. As presented in Section "Metadynamics How-To," the variance of the CV from a simple MD run is used in devising the values of each σ_α, so, this value may change dramatically depending on the starting configuration. This is what might happen in the case of a protein folding problem, assuming the use of a set of distances as CVs. While the folded state is well defined and the variance measured for many CV might be small, this is not the case for an unfolded state that is slowly diffusing and is moving in a broad and featureless free-energy landscape requiring a larger σ_α.

In addition, the problem of selecting an adequate σ is tightly coupled with the definition of the CVs. For small molecules binding to a protein, coordination numbers are often used. These have a sigmoidal functional form that is tailored to change from one to zero when the small molecule is unbinding, but remains constant at values close to zero when the substrate is unbound. In this last case the free dynamics will provide very small fluctuations in CV toward zero, compared to that calculated for the bound state where the value is changing.

An additional intricacy to be considered is that two variables are often coupled as in the case of the two Ramachandran dihedral angles Φ and Ψ. These angles contain three atoms in common. Therefore, some sort of coupling should be introduced into the scheme to account for their interdependence.

We recently proposed[35] a more general multivariate Gaussian instead of Eq. [11]:

$$V(s,t) = \sum_{i=1}^{t/\tau_G} w_G \, \exp\left[-\frac{1}{2}\sum_{\alpha,\beta}\left(s_\alpha - s_\alpha\left(i\tau_G\right)\right)\sigma_{\alpha,\beta}^{-2}(s_\beta - s_\beta(i\tau_G))\right] \qquad [32]$$

Here α and β runs over the biased CVs and σ is a matrix that, unlike the standard MetaD case, might contain nonzero diagonal elements. Furthermore, in this scheme the σ matrix is not fixed at the beginning of the simulation but adapts on the fly. To find this σ we proposed two approaches: one termed "Dynamically Adapted" (DA) MetaD and the other "Geometry Adapted" (GA) MetaD.

In the DA scheme the σ matrix is determined with an estimate over the space traveled in CV space:

$$\sigma^2_{\alpha,\beta}(t) = \frac{1}{\tau_D} \int_0^t dt' [s_\alpha(t') - \bar{s}_\alpha(t')][s_\beta(t') - \bar{s}_\beta(t')] \exp[-(t-t')/\tau_D] \qquad [33]$$

where the parameter τ_D is a typical time over which the drift is calculated. Intuitively, this is the time span that the Gaussian potential has to keep track of and is the only parameter to be set. $\bar{s}_\alpha(t)$ is the average position over the last portion of the trajectory defined as:

$$\bar{s}_\alpha(t) = \frac{1}{\tau_D} \int_0^t dt' s_\alpha(t') \ \exp[-(t-t')/\tau_D] \qquad [34]$$

The advantage of this scheme is that the system detects the underlying landscape and adapts to it on the fly. A different approach is the GA scheme that takes into account only the compression of the phase space induced by the definition of the CV. In GA the σ matrix depends uniquely on the instantaneous conformation through

$$\sigma^2_{\alpha,\beta}(q) = \sigma^2_G \sum_i \frac{\partial s_\alpha(q)}{\partial q_i} \frac{\partial s_\beta(q)}{\partial q_i} \qquad [35]$$

In this formulation the only parameter to be set is σ_G, which is the effective extension of the Gaussian measured in the Cartesian space of atomic coordinates.

It is worth highlighting that in both formulations only one parameter has to be set and it replaces all the individual σ_α required by standard MetaD, to reduce substantially the number of parameters.

In addition, such schemes change both the width and orientation of the Gaussian potential on the fly thus taking into account the local coupling within variables (for the GA) or the local correlations in the diffusion matrix (for the DA).

In both cases, an estimate of the free energy cannot be done with the usual protocol because changing Gaussian width corresponds to using variable-size bins in the estimation of a probability. The correct free energy can be retrieved through a Torrie–Valleau-like formula[9] applied to the well-tempered case, when, for long simulation times, the bias increases constantly over the whole domain:

$$\tilde{F}(s,t) = -k_B T \log N(s,t) - V(s,t). \qquad [36]$$

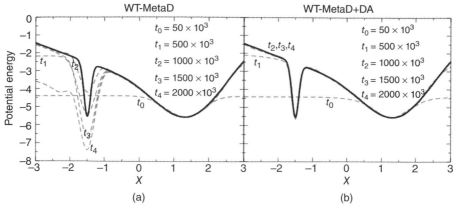

Figure 23 A one-dimensional Langevin model potential (black thick line) on which WTMetaD is applied (a) and on which WTMetaD coupled with the DA scheme is used. In dashed gray line the estimate of the free-energy landscape at different times along the simulation. The DA scheme always delivers a better and faster converging free-energy landscape. (Reprinted (adapted) with permission from Ref. 35. Copyright (2012) American Chemical Society.)

The improvement over standard metadynamics is remarkable as shown in Figure 23 for the case of a one-dimensional Langevin model. While standard WTMetaD may fail whenever σ is chosen too large, the DA scheme automatically tunes to the feature of the energy landscape.

CONCLUSION

Metadynamics is an established method aimed at accelerating molecular dynamics simulations and subsequently recovering free-energy landscapes. Its power and flexibility arise from the fact that it allows us to exploit the chemical and physical insight of the process under investigation to optimally spend the computational effort. However, in spite of its apparent simplicity, using metadynamics is nontrivial due to the input parameters that should be tuned for the specific application and also to the difficulty in choosing the proper collective variables for the simulation. Nevertheless, when done properly it also allows a deep understanding of the process under investigation to be reached, which is the ultimate goal of performing molecular simulations.

This chapter provides an introductory guide that should allow a practitioner to choose the input parameters of a metadynamics simulation and to optimize them for his/her own specific application. A particular focus was placed on avoiding the typical mistakes that can be done when preparing such a simulation and on how they can be solved practically. An assessment of the most widespread variants of the method was also outlined. This chapter is not intended to be complete, as some of the most recent improvements have not been discussed (see, e.g., Refs. 85–90). Nevertheless, this introduction will be of value as a starting point for further explorations in the literature.

ACKNOWLEDGMENTS

The authors are grateful to Alessandro Barducci, Massimiliano Bonomi, Michele Ceriotti, Alessandro Laio, and Fabrizio Marinelli for carefully reading the manuscript and providing several useful suggestions.

APPENDIX A: METADYNAMICS INPUT FILES WITH PLUMED

In this section we want to introduce some typical input files to perform MetaD and WTMetaD by using the PLUMED plugin in its most recent version 2.0.[26] PLUMED is an open-source software that consists of a library enabling users to perform various enhanced sampling methods (among which is MetaD). PLUMED is interfaced and can be combined with a variety of MD codes, including GROMACS,[4] NAMD,[91] LAMMPS,[92] and Quantum-ESPRESSO.[93] These interfaces allow a modeler to use several free-energy methods in a variety of fields thus fostering the applicability, dissemination, and cross-validation of those techniques, at various levels of theory. Moreover, PLUMED provides a single executable that allows PLUMED to be used as a stand-alone tool.

For the examples carried out in Section "Metadynamics How-To" we used the GROMACS 4.5.5[4] classical MD engine. In all the following examples energies are reported in kcal/mol and distances in Å.

PLUMED is not provided with GROMACS code but is separately downloadable from http://www.plumed-code.org. It requires the user to apply a patching procedure and to recompile the MD software according to the procedure reported in its manual. After doing that, GROMACS is simply instructed to use PLUMED on the command line (e.g., `mdrun -plumed input.dat`).

The first example refers to the case where MetaD is performed on gyration radius reported in Section "A First Test Run Using Gyration Radius." Note that in PLUMED one can also monitor other variables that might be helpful in checking the results of the simulation (in this case Φ and Ψ dihedral angles):

Example

```
# choose units - by default PLUMED uses kj/mol, nm, and ps
UNITS ENERGY=kcal/mol LENGTH=A
# these three lines define the atoms involved in the group
# named "all"
# the name of the group is arbitrary. If not defined PLUMED
```

```
# gives a default
all: GROUP ATOMS=1,5,6,7,9,11,15,16,17,19
# this defines the gyration radius using a group
# defined by atoms in "all" and this value will have name "rg"
rg: GYRATION ATOMS=all
# this line sets the metadynamics hills height (HEIGHT) in
# kcal/mol
# and deposition time (PACE) in timestep. Here each
# timestep is 2 fs.
# this produces also a HILLS file containing the
# center and widths of Gaussians
meta: METAD ARG=rg SIGMA=0.07 HEIGHT=0.1 PACE=600
# additional variables can be set to monitor the simulation
# Phi:
t1: TORSION ATOMS=5,7,9,15
# and Psi:
t2: TORSION ATOMS=7,9,15,17
# This produces a COLVAR file that contains the
# values of the variables t1 and t2
# calculated every STRIDE steps, which can be more often than
# HILLS
PRINT ARG=t1,t2 STRIDE=100 FILE=COLVAR
```

At the end of the run one obtains three output files from PLUMED: a log-file, a COLVAR file that is produced by the PRINT command, and a HILLS file that is produced by the METAD. The PLUMED logfile in GROMACS is embedded in the md.log file and the user is strongly encouraged to explore it to understand if PLUMED has interpreted the commands correctly. The COLVAR file reports the time series of the CVs that are included in the input, and the HILLS file reports the centers, the widths, and the heights of the repulsive Gaussian potentials added by MetaD. At the end of the simulation one can calculate the sum of the deposited potential by using the standalone executable of PLUMED with the command plumed sum_hills --hills HILLS.

In the second example we report the input from Section "A Better Collective Variable: Φ Dihedral Angle" where the Φ dihedral angle was used. Please note that adding more arguments in the ARG field and corresponding values in SIGMA allows PLUMED to perform multidimensional metadynamics.

Example

```
UNITS ENERGY=kcal/mol LENGTH=A
all: GROUP ATOMS=1,5,6,7,9,11,15,16,17,19
rg: GYRATION ATOMS=all
t1: TORSION ATOMS=5,7,9,15
# now METAD takes t1 as an argument and the sigma is changed
# according to its fluctuation
```

```
# note that you can give as ARG only a CV which is already
# defined
meta: METAD ARG=t1 SIGMA=0.1 HEIGHT=0.1 PACE=600
# this variable again is only monitored in COLVAR
t2: TORSION ATOMS=7,9,15,17
PRINT ARG=t1,t2 STRIDE=100 FILE=COLVAR
```

Note that the TORSION collective variable is now periodic. The sum_hills tool of PLUMED knows automatically that the domain is periodic.

As a last example we report the use of WTMetaD as discussed in Section "Well-Tempered Metadynamics Using Dihedral Angle Φ."

Example

```
UNITS ENERGY=kcal/mol LENGTH=A
all: GROUP ATOMS=1,5,6,7,9,11,15,16,17,19
rg: GYRATION ATOMS=all
t1: TORSION ATOMS=5,7,9,15
# this is a multiple line command
# BIASFACTOR and TEMP are required to enable
# well-tempered metad
METAD ...
    LABEL=meta
    ARG=t1 SIGMA=0.1 HEIGHT=0.1 PACE=600
    BIASFACTOR=15 TEMP=300
... METAD
t2: TORSION ATOMS=7,9,15,17
PRINT ARG=t1,t2 STRIDE=100 FILE=COLVAR
```

More information regarding PLUMED and the various tools included in it along with some more tutorial examples can be retrieved from http://www.plumed-code.org.

REFERENCES

1. M. P. Allen and D. J. Tildesley, *Computer Simulation of Liquids*, Oxford University Press, Oxford, 1987.

2. D. Frenkel and B. Smit, *Understanding Molecular Simulation*, Academic Press, London, second ed., 2002.

3. M. Tuckerman, *Statistical Mechanics: Theory and Molecular Simulation*, Oxford University Press, Oxford, 2012.

4. B. Hess, C. Kutzner, D. van der Spoel, and E. Lindahl, *J. Chem. Theory Comput.*, 4(3), 435–447 (2008). GROMACS 4: Algorithms for Highly Efficient, Load-Balanced, and Scalable Molecular Simulation.

5. D. Shaw, P. Maragakis, K. Lindorff-Larsen, S. Piana, R. Dror, M. Eastwood, J. Bank, J. Jumper, J. Salmon, Y. Shan, and W. Wriggers, *Science*, **330**, 341–346 (2010). Atomic-Level Characterization of the Structural Dynamics of Proteins.

6. K. Lindorff-Larsen, S. Piana, R. Dror, and D. Shaw, *Science*, **334**, 517–520 (2011). How Fast-Folding Proteins Fold.

7. Y. Sugita and Y. Okamoto, *Chem. Phys. Lett.*, **314**(1–2), 141–151 (1999). Replica-Exchange Molecular Dynamics Method for Protein Folding.

8. P. Liu, B. Kim, R. A. Friesner, and B. J. Berne, *Proc. Natl. Acad. Sci. U. S. A.*, **102**(39), 13749–13754 (2005). Replica Exchange with Solute Tempering: A Method for Sampling Biological Systems in Explicit Water.

9. G. M. Torrie and J. P. Valleau, *J. Comput. Phys.*, **23**(2), 187–199 (1977). Nonphysical Sampling Distributions in Monte Carlo Free-Energy Estimation: Umbrella Sampling.

10. E. Darve and A. Pohorille, *J. Chem. Phys.*, **115**(20), 9169–9183 (2001). Calculating Free Energies Using Average Force.

11. A. Laio and M. Parrinello, *Proc. Natl. Acad. Sci. U. S. A.*, **99**(20), 12562–12566 (2002). Escaping Free-Energy Minima.

12. S. Marsili, A. Barducci, R. Chelli, P. Procacci, and V. Schettino, *J. Phys. Chem. B*, **110**(29), 14011–14013 (2006). Self-Healing Umbrella Sampling: A Non-Equilibrium Approach for Quantitative Free Energy Calculations.

13. J. VandeVondele and U. Rothlisberger, *J. Phys. Chem. B*, **106**(1), 203–208 (2002). Canonical Adiabatic Free Energy Sampling (CAFES): A Novel Method for the Exploration of Free Energy Surfaces.

14. L. Rosso, P. Minary, Z. Zhu, and M. E. Tuckerman, *J. Chem. Phys.*, **116**(11), 4389–4402 (2002). On the Use of the Adiabatic Molecular Dynamics Technique in the Calculation of Free Energy Profiles.

15. L. Maragliano and E. Vanden-Eijnden, *Chem. Phys. Lett.*, **426**, 168–175 (2006). A Temperature Accelerated Method for Sampling Free Energy and Determining Reaction Pathways in Rare Events Simulations.

16. M. Iannuzzi, A. Laio, and M. Parrinello, *Phys. Rev. Lett.*, **90**(23), 238302 (2003). Efficient Exploration of Reactive Potential Energy Surfaces Using Car-Parrinello Molecular Dynamics.

17. G. Bussi, F. L. Gervasio, A. Laio, and M. Parrinello, *J. Am. Chem. Soc.*, **128**(41), 13435–13441 (2006). Free-Energy Landscape for β-Hairpin Folding from Combined Parallel Tempering and Metadynamics.

18. F. Baftizadeh, X. Biarnes, F. Pietrucci, F. Affinito, and A. Laio, *J. Am. Chem. Soc.*, **134**(8), 3886–3894 (2012). Multidimensional View of Amyloid Fibril Nucleation in Atomistic Detail.

19. F. L. Gervasio, A. Laio, and M. Parrinello, *J. Am. Chem. Soc.*, **127**(8), 2600–2607 (2005). Flexible Docking in Solution Using Metadynamics.

20. R. Martoňák, D. Donadio, A. Oganov, and M. Parrinello, *Nat. Mater.*, **5**, 623–626 (2006). Crystal Structure Transformations in SiO_2 from Classical and Ab Initio Metadynamics.

21. F. Trudu, D. Donadio, and M. Parrinello, *Phys. Rev. Lett.*, **97**(10), 105701 (2006). Freezing of a Lennard-Jones Fluid: From Nucleation to Spinodal Regime.

22. CPMD, http://www.cpmd.org/, Copyright IBM Corp 1990–2008, Copyright MPI für Festkörperforschung Stuttgart 1997–2001, 1990–2008.

23. CP2K, http://www.cp2k.org/, 2012.

24. S. Marsili, G. Signorini, R. Chelli, M. Marchi, and P. Procacci, *J. Comput. Chem.*, **31**(5), 1106–1116 (2009). ORAC: A Molecular Dynamics Simulation Program to Explore Free Energy Surfaces in Biomolecular Systems at the Atomistic Level.

25. M. Bonomi, D. Branduardi, G. Bussi, C. Camilloni, D. Provasi, P. Raiteri, D. Donadio, F. Marinelli, F. Pietrucci, R. A. Broglia, and M. Parrinello, *Comput. Phys. Comm*, **180**(10), 1961–1972 (2009). PLUMED: A Portable Plugin for Free-Energy Calculations with Molecular Dynamics.

26. G. A. Tribello, M. Bonomi, D. Branduardi, C. Camilloni, and G. Bussi, *Comput. Phys. Commun.*, **185**(2), 604–613 (2014). PLUMED 2: New Feathers For An Old Bird.

27. G. Fiorin, M. L. Klein, and J. Hénin, *Mol. Phys.*, **111**(22–23), 3345–3362 (2013). Using Collective Variables To Drive Molecular Dynamics Simulations.

28. W. Humphrey, A. Dalke, and K. Schulten, *J. Molec. Graphics*, **14**, 33–38 (1996). VMD – Visual Molecular Dynamics.

29. A. D. MacKerell, Jr., D. Bashford, M. Bellot, R. L. Dunbrack, Jr., J. D. Evanseck, M. J. Field, S. Fischer, J. Gao, H. Guo, S. Ha, D. Joseph-McCarthy, L. Kuchnir, K. Kuczera, F. T. K. Lau, C. Mattos, S. Michnick, T. Ngo, D. T. Nguyen, B. Prodhom, W. E. Reiher III, B. Roux, M. Schlenkrich, J. C. Smith, R. Stote, J. Straub, W. Watanabe, J. Wiorkiewicz-Kunczera, D. Yin, and M. Karplus, *J. Phys. Chem. B*, **102**(18), 3586–3616 (1998). All-Atom Empirical Potential for Molecular Modeling and Dynamics Studies of Proteins.

30. T. Lazaridis, D. J. Tobias, C. Brooks, and M. E. Paulaitis, *J. Chem. Phys.*, **95**(10), 7612–7625 (1991). Reaction Paths and Free Energy Profiles for Conformational Transitions: An Internal Coordinate Approach.

31. D. J. Tobias and C. L. Brooks, *J. Phys. Chem.*, **96**(9), 3864–3870 (1992). Conformational Equilibrium in the Alanine Dipeptide in the Gas Phase and Aqueous Solution: A Comparison of Theoretical Results.

32. C. Bartels and M. Karplus, *J. Comput. Chem.*, **18**(12), 1450–1462 (1997). Multidimensional Adaptive Umbrella Sampling: Application to Main Chain and Side Chain Peptide Conformations.

33. L. Maragliano, A. Fischer, and E. Vanden-Eijnden, *J. Chem. Phys.*, **125**(2), 024106 (2006). String Method in Collective Variables: Minimum Free Energy Paths and Isocommittor Surfaces.

34. D. Branduardi, F. L. Gervasio, and M. Parrinello, *J. Chem. Phys.*, **126**(5), 054103 (2007). From A to B in Free Energy Space.

35. D. Branduardi, G. Bussi, and M. Parrinello, *J. Chem. Theory Comput.*, **8**(7), 2247–2254 (2012). Metadynamics with Adaptive Gaussians.

36. M. Mezei, *J. Comput. Phys.*, **68**(1), 237–248 (1987). Adaptive Umbrella Sampling: Self-Consistent Determination of the Non-Boltzmann Bias.

37. H. Grubmüller, *Phys. Rev. E*, **52**(3), 2893 (1995). Predicting Slow Structural Transitions in Macromolecular Systems: Conformational Flooding.

38. A. Laio, A. Rodriguez-Fortea, F. L. Gervasio, M. Ceccarelli, and M. Parrinello, *J. Phys. Chem. B*, **109**(14), 6714–6721 (2005). Assessing the Accuracy of Metadynamics.

39. C. W. Gear, I. G. Kevrekidis, and C. Theodoropoulos, *Comput. Chem. Eng.*, **26**, 941–963 (2002). "Coarse" Integration/Bifurcation Analysis via Microscopic Simulators: Micro-Galerkin Methods.

40. D. Cvijovic and J. Klinowski, *Science*, **267**, 664–666 (1995). Taboo Search - An Approach To The Multiple Minima Problem.

41. T. Huber, A. E. Torda, and W. F. van Gunsteren, *J. Comput.-Aided Mol. Des.*, **8**, 695–708 (1994). Local Elevation: A Method for Improving the Searching Properties of Molecular Dynamics Simulation.

42. G. Bussi, A. Laio, and M. Parrinello, *Phys. Rev. Lett.*, **96**(9), 090601 (2006). Equilibrium Free Energies from Nonequilibrium Metadynamics.

43. C. Micheletti, A. Laio, and M. Parrinello, *Phys. Rev. Lett.*, **92**(17), 170601 (2004). Reconstructing the Density of States by History-Dependent Metadynamics.

44. Y. Crespo, F. Marinelli, F. Pietrucci, and A. Laio, *Phys. Rev. E*, **81**(5), 055701 (2010). Metadynamics Convergence Law in a Multidimensional System.

45. F. Baftizadeh, P. Cossio, F. Pietrucci, and A. Laio, *Curr. Phys. Chem.*, **2**, 79–91 (2012). Protein Folding and Ligand-Enzyme Binding from Bias-Exchange Metadynamics Simulations.

46. F. Marinelli, F. Pietrucci, A. Laio, and S. Piana, *PLoS Comput. Biol.*, **5**(8), e1000452 (2009). A Kinetic Model of Trp-Cage Folding from Multiple Biased Molecular Dynamics Simulations.

47. S. Kumar, J. Rosenberg, D. Bouzida, R. Swendsen, and P. Kollman, *J. Comput. Chem.*, **13**(8), 1011–1021 (1992). The Weighted Histogram Analysis Method for Free-Energy Calculations on Biomolecules. I. The Method.

48. A. Barducci, G. Bussi, and M. Parrinello, *Phys. Rev. Lett.*, **100**(2), 020603 (2008). Well-Tempered Metadynamics: A Smoothly-Converging and Tunable Free-Energy Method.

49. M. Bonomi, A. Barducci, and M. Parrinello, *J. Comput. Chem.*, **30**(11), 1615–1621 (2009). Reconstructing the Equilibrium Boltzmann Distribution from Well-Tempered Metadynamics.

50. S. K. Kearsley, *Acta Cryst. A*, **45**, 208–210 (1989). On the Orthogonal Transformation Used for Structural Comparison.

51. B. Ensing and M. L. Klein, *Proc. Natl. Acad. Sci. U. S. A.*, **102**(19), 6755–6759 (2005). Perspective on the Reactions Between F^- and CH_3CH_2F: The Free Energy Landscape of the E2 and S_N2 Reaction Channels.

52. E. Vanden-Eijnden and M. Venturoli, *J. Chem. Phys.*, **130**(19), 194103 (2009). Revisiting the Finite Temperature String Method for the Calculation of Reaction Tubes and Free Energies.

53. Y. A. Mantz, D. Branduardi, G. Bussi, and M. Parrinello, *J. Phys. Chem. B*, **113**(37), 12521–12529 (2009). Ensemble of Transition State Structures for the Cis-Trans Isomerization of N-Methylacetamide.

54. A. Lodola, D. Branduardi, M. De Vivo, L. Capoferri, M. Mor, D. Piomelli, and A. Cavalli, *PLoS One*, **7**(2), e32397 (2012). A Catalytic Mechanism for Cysteine N-Terminal Nucleophile Hydrolases, as Revealed by Free Energy Simulations.

55. D. Branduardi, M. De Vivo, N. Rega, V. Barone, and A. Cavalli, *J. Chem. Theory Comput.*, **7**(3), 539–543 (2011). Methyl Phosphate Dianion Hydrolysis in Solution Characterized by Path Collective Variables Coupled with DFT-Based Enhanced Sampling Simulations.

56. G. A. Gallet, F. Pietrucci, and W. Andreoni, *J. Chem. Theory Comput.*, **8**(11), 4029–4039 (2012). Bridging Static and Dynamical Descriptions of Chemical Reactions: An Ab Initio Study of CO2 Interacting with Water Molecules.

57. M. Bonomi, D. Branduardi, F. L. Gervasio, and M. Parrinello, *J. Am. Chem. Soc.*, **130**(42), 13938–13944 (2008). The Unfolded Ensemble and Folding Mechanism of the C-Terminal GB1 Beta-Hairpin.

58. A. Pietropaolo, L. Muccioli, R. Berardi, and C. Zannoni, *Proteins*, **70**(3), 667–677 (2008). A Chirality Index for Investigating Protein Secondary Structures and Their Time Evolution.

59. W. Ren, E. Vanden-Eijnden, and P. Maragakis, *J. Chem. Phys.*, **123**(13), 134109 (2005). Transition Pathways in Complex Systems: Application of the Finite-Temperature String Method to the Alanine Dipeptide.

60. G. Daz Leines and B. Ensing, *Phys. Rev. Lett.*, **109**(2), 020601 (2012). Path Finding on High-Dimensional Free Energy Landscapes.

61. A. Amadei, A. B. M. Linssen, and H. J. C. Berendsen, *Proteins*, **17**(4), 412–425 (1993). Essential Dynamics of Proteins.

62. V. Spiwok, P. Lipovová, and B. Králová, *J. Phys. Chem. B*, **111**(12), 3073–3076 (2007). Metadynamics in Essential Coordinates: Free Energy Simulation of Conformational Changes.

63. L. Sutto, M. D'Abramo, and F. L. Gervasio, *J. Chem. Theory Comput.*, **6**(12), 3640–3646 (2010). Comparing the Efficiency of Biased and Unbiased Molecular Dynamics in Reconstructing the Free Energy Landscape of Met-Enkephalin.

64. G. A. Tribello, M. Ceriotti, and M. Parrinello, *Proc. Natl. Acad. Sci. U. S. A.*, **107**(41), 17509–17414 (2010). A Self-Learning Algorithm for Biased Molecular Dynamics.

65. I. Borg and P. Groenen, *Modern Multidimensional Scaling*, Springer, New York, 2005.

66. V. Spiwok and B. Králová, *J. Chem. Phys.*, **135**(22), 224504 (2011). Metadynamics in the Conformational Space Nonlinearly Dimensionally Reduced by Isomap.

67. P. Das, M. Moll, H. Stamati, L. E. Kavraki, and C. Clementi, *Proc. Natl. Acad. Sci. U. S. A.*, **103**(26), 9885–9890 (2006). Low-Dimensional, Free-Energy Landscapes of Protein-Folding Reactions by Nonlinear Dimensionality Reduction.

68. M. Ceriotti, G. A. Tribello, and M. Parrinello, *Proc. Natl. Acad. Sci. U. S. A.*, **108**(32), 13023–13028 (2011). Simplifying the Representation of Complex Free-Energy Landscapes Using Sketch-Map.

69. G. A. Tribello, M. Ceriotti, and M. Parrinello, *Proc. Natl. Acad. Sci. U. S. A.*, **109**(14), 5196–5201 (2012). Using Sketch-Map Coordinates to Analyze and Bias Molecular Dynamics Simulations.

70. F. Pietrucci and A. Laio, *J. Chem. Theory Comput.*, **5**(9), 2197–2201 (2009). A Collective Variable for the Efficient Exploration of Protein Beta-Sheet Structures: Application to SH3 and GB1.

71. A. Cuff, O. C. Redfern, L. Greene, I. Sillitoe, T. Lewis, M. Dibley, A. Reid, F. Pearl, T. Dallman, A. Todd, R. Garratt, J. Thornton, and C. Orengo, *Structure*, **17**(8), 1051–1062 (2009). The CATH Hierarchy Revisited – Structural Divergence in Domain Superfamilies and the Continuity of Fold Space.

72. D. Frishman and P. Argos, *Proteins: Structure, Function, and Bioinformatics*, **23**(4), 566–579 (1995). Knowledge-Based Protein Secondary Structure Assignment.

73. P. Cossio, A. Trovato, F. Pietrucci, F. Seno, A. Maritan, and A. Laio, *PLoS Comput. Biol.*, **6**(11), e1000957 (2010). Exploring the Universe of Protein Structures Beyond the Protein Data Bank.

74. C. Bartels and M. Karplus, *J. Phys. Chem. B*, **102**(5), 865–880 (1998). Probability Distributions for Complex Systems: Adaptive Umbrella Sampling of the Potential Energy.

75. B. Berg and T. Neuhaus, *Phys. Rev. Lett.*, **68**(1), 9–12 (1992). Multicanonical Ensemble: A New Approach to Simulate First-Order Phase Transitions.

76. F. Wang and D. P. Landau, *Phys. Rev. Lett.*, **86**(10), 2050–2053 (2001). Efficient, Multiple-Range Random Walk Algorithm to Calculate the Density of States.

77. M. Bonomi and M. Parrinello, *Phys. Rev. Lett.*, **104**(19), 190601 (2010). Enhanced Sampling in the Well-Tempered Ensemble.

78. M. Deighan, M. Bonomi, and J. Pfaendtner, *J. Chem. Theory Comput.*, **8**(7), 2189–2192 (2012). Efficient Simulation of Explicitly Solvated Proteins in the Well-Tempered Ensemble.

79. T. N. Do, P. Carloni, G. Varani, and G. Bussi, *J. Chem. Theory Comput.*, **9**(3), 1720–1730 (2013). RNA/Peptide Binding Driven by Electrostatics – Insight From Bidirectional Pulling Simulations.

80. P. Raiteri, A. Laio, F. L. Gervasio, C. Micheletti, and M. Parrinello, *J. Phys. Chem. B*, **110**(8), 3533–3539 (2006). Efficient Reconstruction of Complex Free Energy Landscapes by Multiple Walkers Metadynamics.

81. C. Melis, P. Raiteri, L. Colombo, and A. Mattoni, *ACS Nano*, **5**(12), 9639–9647 (2011). Self-Assembling of Zinc Phthalocyanines on ZnO (1010) Surface through Multiple Time Scales.

82. H. Fukunishi, O. Watanabe, and S. Takada, *J. Chem. Phys.*, **116**(20), 9058–9067 (2002). On the Hamiltonian Replica Exchange Method for Efficient Sampling of Biomolecular Systems: Application to Protein Structure Prediction.

83. C. Camilloni, D. Provasi, G. Tiana, and R. Broglia, *Proteins*, **71**(4), 1647–1654 (2007). Exploring the Protein G Helix Free-Energy Surface by Solute Tempering Metadynamics.

84. S. Piana and A. Laio, *J. Phys. Chem. B*, **111**(17), 4553–4559 (2007). A Bias-Exchange Approach to Protein Folding.

85. L. Zheng, M. Chen, and W. Yang, *Proc. Natl. Acad. Sci. U. S. A.*, **105**(51), 20227–20232 (2008). Random Walk in Orthogonal Space to Achieve Efficient Free-Energy Simulation of Complex Systems.

86. S. Singh, C. Chiu, and J. de Pablo, *J. Stat. Phys.*, **145**(4), 932–945 (2011). Flux Tempered Metadynamics.

87. M. Chen, M. Cuendet, and M. Tuckerman, *J. Chem. Phys.*, **137**(2), 024102 (2012). Heating and Flooding: A Unified Approach for Rapid Generation of Free Energy Surfaces.

88. Q. Zhu, A. Oganov, and A. Lyakhov, *Cryst. Eng. Commun.*, **14**(10), 3596–3601 (2012). Evolutionary Metadynamics: A Novel Method to Predict Crystal Structures.

89. P. Tiwary and M. Parrinello, *Phys. Rev. Lett.*, **111**(23), 230602 (2013). From Metadynamics to Dynamics.

90. M. McGovern and J. de Pablo, *J. Chem. Phys.*, **139**, 084102 (2013). A Boundary Correction Algorithm for Metadynamics in Multiple Dimensions.

91. J. C. Phillips, R. Braun, W. Wang, J. Gumbart, E. Tajkhorshid, E. Villa, C. Chipot, R. D. Skeel, L. Kalé, and K. Schulten, *J. Comput. Chem.*, **26**(16), 1781–802 (2005). Scalable Molecular Dynamics with NAMD.

92. S. Plimpton, *J. Comput. Phys.*, **117**(1), 1–19 (1995). Fast Parallel Algorithms for Short-Range Molecular Dynamics.

93. P. Giannozzi, S. Baroni, N. Bonini, M. Calandra, R. Car, C. Cavazzoni, D. Ceresoli, G. L. Chiarotti, M. Cococcioni, I. Dabo, A. Dal Corso, S. de Gironcoli, S. Fabris, G. Fratesi, R. Gebauer, U. Gerstmann, C. Gougoussis, A. Kokalj, M. Lazzeri, L. Martin-Samos, N. Marzari, F. Mauri, R. Mazzarello, S. Paolini, A. Pasquarello, L. Paulatto, C. Sbraccia, S. Scandolo, G. Sclauzero, A. P. Seitsonen, A. Smogunov, P. Umari, and R. M. Wentzcovitch, *J. Phys.: Cond. Matter*, **21**(39), 395502 (2009). Quantum Espresso: A Modular And Open-Source Software Project For Quantum Simulations Of Materials.

## CHAPTER 2

Polarizable Force Fields for Biomolecular Modeling

Yue Shi,[a] Pengyu Ren,[a] Michael Schnieders,[b] and Jean-Philip Piquemal[c]

[a]*Department of Biomedical Engineering, The University of Texas at Austin, Austin, TX 78712, USA*
[b]*Department of Biomedical Engineering, College of Engineering, Department of Biochemistry, Carver College of Medicine, The University of Iowa, Iowa City, IA 52242, USA*
[c]*Laboratoire de Chimie Théorique (UMR 7616), UPMC, Sorbonne Universités, Paris, Cedex 05 75252, France*

INTRODUCTION

Molecular-mechanics-based modeling has been widely used in the study of chemical and biological systems. The classical potential energy functions and their parameters are referred to as force fields. Empirical force fields for biomolecules emerged in the early 1970s,[1] followed by the first molecular dynamics simulations of the bovine pancreatic trypsin inhibitors (BPTI).[2–4] Over the past 30 years, a great number of empirical molecular mechanics force fields, including AMBER,[5] CHARMM,[6] GROMOS,[7] OPLS,[8] among many others, have been developed. These force fields share similar functional forms, including valence interactions represented by harmonic oscillators, point dispersion–repulsion for van der Waals (vdW) interactions, and an electrostatic contribution based on fixed atomic partial charges. This generation of molecular mechanics force fields has been widely used in the study

Reviews in Computational Chemistry, Volume 28, First Edition.
Edited by Abby L. Parrill and Kenny B. Lipkowitz.
© 2015 John Wiley & Sons, Inc. Published 2015 by John Wiley & Sons, Inc.

of molecular structures, dynamics, interactions, design, and engineering. We refer interested readers to some recent reviews for detailed discussions.[9,10]

Although the fixed charge force fields enjoyed great success in many areas, there remains much room for improvement. In fixed charge based electrostatic models, the atomic partial charges are meant to be "pre-polarized" for condensed phases in an averaged fashion, typically achieved by the fortuitous overestimation of electrostatic charges by low-level *ab initio* quantum mechanics. Such models thus lack the ability to describe the variation in electrostatics because of many-body polarization effects, which have been shown to be a significant component of intermolecular forces.[10-12] With the rapid growth of computational resources, there has been increasing effort to explicitly incorporate many-body induction into molecular mechanics to improve the accuracy of molecular modeling.

Classical electrostatics models that take into account polarization appeared as early as the 1950s. Barker in his 1953 paper "Statistical Mechanics of Interacting Dipoles" discussed the electrostatic energy of molecules in terms of "permanent and induced dipoles."[13] Currently, polarizable models generally fall into three categories: those based on induced point dipoles,[9,14-23] the classical Drude oscillators,[24-26] and fluctuating charges.[27-30] More sophisticated force fields that are "electronic structure-based"[31,32] or use "machine learning methods"[33] also exist, but incur higher computational costs. Discussions of the advantages and disadvantages of each model and their applications are presented in the following sections.

Compared to fixed charge models, the polarizable models are still in a relatively early stage. Only in the past decade or so has there been a systematic effort to develop general polarizable force fields for molecular modeling. A number of reviews have been published to discuss various aspects of polarizable force fields and their development.[9,34-40] Here, we focus on the recent development and applications of different polarizable force fields. We begin with a brief introduction to the basic principles and formulae underlying alternative models. Next, the recent progress of several well-developed polarizable force fields is reviewed. Finally, applications of polarizable models to a range of molecular systems, including water and other small molecules, ion solvation, peptides, proteins, and lipid systems are presented.

MODELING POLARIZATION EFFECTS

Induced Dipole Models

To describe electrostatic interactions involving polarization, we consider a system consisting of a collection of charge distribution sites located at lone pair positions, atomic centers, and/or molecular centers, depending on the resolution of the model. The total charge distribution at site i is the sum of permanent and induced charge

$$M_i = M_i^0 + M_i^{ind} \qquad [1]$$

where M represents the charge distribution. This distribution can be a simple point charge, a point multipole expansion with charge, dipole, quadrupole and/or higher order moments, or a continuous charge distribution. While the principles described below are not limited to any particular representation of charge distribution, we will use point multipoles for convenience.

The electrostatic interaction energy between two charge sites i and j is given by

$$U_{ele} = \frac{1}{2} \sum_i \sum_{j \neq i} M_i^t T_{ij} M_j \qquad [2]$$

where T is the interaction operator and is a function of the distance between i and j. In the case of point charge interactions, T is simply $1/r$. The work (positive energy) needed to polarize a charge distribution also has a quadratic dependence on the induced charge distribution:

$$U_{work} = \frac{1}{2} \sum_i (M_i^{ind})^t \alpha_i^{-1} M_i^{ind} \qquad [3]$$

where α is the polarizability of site i that includes all orders of polarizability, namely, dipole polarizability.[41] Although α is generally treated as an isotropic quantity, as in the Applequist scheme,[41] *ab initio* anisotropic polarizability tensors can be derived from quantum mechanical calculations.[42,43]

The total electrostatic energy is

$$U_{ele} = \frac{1}{2} \sum_i \sum_{j \neq i} M_i^t T_{ij} M_j + \frac{1}{2} \sum_i (M_i^{ind})^t \alpha_i^{-1} M_i^{ind} \qquad [4]$$

The values of the induced moments minimize the total energy, by satisfying

$$\frac{\partial U_{ele}}{\partial M_i^{ind}} = \sum_{j \neq i} T_{ij} M_j + \alpha_i^{-1} M_i^{ind} = 0 \qquad [5]$$

As a result

$$M_i^{ind} = \alpha_i^{-1} \sum_{j \neq i} T_{ij} (M_j^0 + M_j^{ind}) \qquad [6]$$

Equation [6] can be solved iteratively to obtain the induced dipoles. The self-consistent calculation is computationally expensive; however, it can be accelerated with predictors and nonstationary iterative methods.[44]

Substituting $\alpha_i^{-1} M_i^{\text{ind}}$ from Eq. [5] into Eq. [6], the final electrostatic energy becomes

$$U_{\text{ele}} = \frac{1}{2}\sum_i \sum_{j\neq i}(M_i^0)^t T_{ij} M_j^0 + \frac{1}{2}\sum_i \sum_{j\neq i}(M_i^{\text{ind}})^t T_{ij} M_j^0 \tag{7}$$

where the first term is the permanent electrostatic energy and the second term is the polarization energy.

Classic Drude Oscillators

In the Drude oscillator model, the polarization effect is described by a point charge (the Drude oscillator) attached to each nonhydrogen atom via a harmonic spring. The point charge can move relative to the attachment site in response to the electrostatic environment. The electrostatic energy is the sum of the pairwise interactions between atomic charges and the partial charge of the Drude particles

$$E_{\text{ele}} = \sum_{A<B}^{N}\frac{q_C(A)q_C(B)}{r_C(A)-r_C(B)|} + \sum_{A<B}^{N,N_D}\frac{q_D(A)q_C(B)}{|r_D(A)-r_C(B)|} + \sum_{A<B}^{N_D}\frac{q_D(A)q_D(B)}{|r_D(A)-r_D(B)|}$$
$$+\frac{1}{2}\sum_A^{N_D} k_D(r_D(A)-r_C(B))^2 \tag{8}$$

where N_D and N are the number of Drude particles and nonhydrogen atoms, q_D and q_C are the charges on the Drude particle and its parent atom, respectively, r_D and r_C are their respective positions, and k_D is the force constant of the harmonic spring between the Drude oscillator and its parent atom. The last term in Eq. [8] accounts for the cost of polarizing the Drude particles.

The atomic polarizability (α) is a function of both the partial charge on the Drude particle and the force constant of the spring

$$\alpha = \frac{q_D^2(A)}{k_D} \tag{9}$$

Both the induced dipole and Drude oscillator approaches benefit from short-range Thole damping to avoid a polarization catastrophe and to produce an anisotropic molecular polarization response.[45]

Fluctuating Charges

The formalism of the fluctuating charge model is based on the charge equilibration (CHEQ) method,[46] in which the chemical potential is equilibrated via the redistribution of charge density. The charge-dependent energy for a system of M molecules containing N_i atoms per molecule is expressed as

$$E_{\text{CHEQ}}(R,Q) = \sum_{i=1}^{M}\sum_{\alpha=1}^{N}\chi_{i\alpha}Q_{i\alpha} + \frac{1}{2}\sum_{i=1}^{M}\sum_{j=1}^{M}\sum_{\alpha=1}^{N_i}\sum_{\beta=1}^{N_j}J_{i\alpha i\beta}Q_{i\alpha}Q_{j\beta}$$

$$+\frac{1}{2}\sum_{i=1}^{MN'}\sum_{j=1}^{MN'}\frac{Q_i Q_j}{4\pi\varepsilon_0 r_{ij}} + \sum_{j=1}^{M}\lambda_i\left(\sum_{i=1}^{N}Q_{ij} - Q_j^{\text{Total}}\right) \quad [10]$$

where Q_i is the partial charge on atomic site i. The χ describes the atomic electronegativity controlling the directionality of electron flow, and J is the atomic hardness that represents the resistance to electron flow to or from the atom. These parameters are optimized to reproduce molecular dipoles and the molecular polarization response. The charge degrees of freedom are typically propagated via an extended Lagrangian formulation:[47]

$$L = \sum_{i=1}^{M}\sum_{\alpha=1}^{N}\frac{1}{2}m_{i\alpha}\left(\frac{dr_{i\alpha}}{dt}\right)^2 + \sum_{i=1}^{M}\sum_{\alpha=1}^{N}\frac{1}{2}m_{Q,i\alpha}\left(\frac{dQ_{i\alpha}}{dt}\right)^2$$

$$-E(Q,r) - \sum_{i=1}^{M}\lambda_i\sum_{\alpha=1}^{N}Q_{i\alpha} \quad [11]$$

where the first two terms represent the nuclear and charge kinetic energies, the third term is the potential energy, and the fourth term is the molecular charge neutrality constraint enforced on each molecule i via a Lagrange multiplier λ_i. The extended Lagrangian approach can also be applied to the induced dipole and Drude oscillator models described earlier. While the extended Lagrangian seems to be more efficient than the iterative method, fictitious masses and smaller time steps are required to minimize the coupling between the polarization and atomic degrees of freedom, which can never be completely eliminated.[44]

A few general force fields have been developed based on these formulas to explicitly treat the polarization effect. We now discuss development highlights for some of the representative force fields.

RECENT DEVELOPMENTS

AMOEBA

The AMOEBA (atomic multipole optimized energetics for biomolecular applications) force field, developed by Ponder, Ren, and coworkers,[15,18,37] utilizes atomic multipoles to represent permanent electrostatics and induced atomic dipoles for many-body polarization. The valence interactions include bond, angle, torsion, and out-of-plane contributions using typical molecular mechanics functional forms. The vdW interaction is described by a buffered-14-7 function. The atomic multipole moments consist of charge, dipole, and quadrupole moments, which are derived from *ab initio* quantum mechanical calculations using procedures such as Stone's distributed multipole analysis (DMA).[48−50] The higher order moments make possible anisotropic representations of the electrostatic potential outside atoms and molecules. The polarization effect is explicitly taken into account via atomic dipole induction. The combination of permanent atomic multipoles and induced dipoles enables AMOEBA to capture electrostatic interactions in both gas and condensed phase accurately. The vdW parameters of AMOEBA are optimized simultaneously against both *ab initio* gas-phase data and condensed-phase experimental properties.

In the past decade, AMOEBA has been applied to the study of water,[15] monovalent and divalent ions,[51−53] small molecules,[54,55] peptides,[18,56] and proteins.[57−59] AMOEBA demonstrated that a polarizable force field is able to perform well in both gas and solution phases with a single set of parameters. In addition, AMOEBA is the first general-purpose polarizable force field utilized in molecular dynamics simulations of protein–ligand binding and calculation of absolute and relative binding free energies.[58−62] The computed binding free energies between trypsin and benzamidine derivatives suggests significant nonadditive electrostatic interactions as the ligand desolvates from water and enters the protein pocket (see Application, Proteins for further discussion). AMOEBA has recently been extended to biomolecular X-ray crystallography refinement,[63,64] and consistently successful prediction of the structure, thermodynamic stability, and solubility of organic crystals[65] are encouraging.

AMOEBA has been implemented in several widely used software packages including TINKER,[66] OpenMM,[67] AMBER,[68] and Force Field X.[69] The AMOEBA polarizable force field was first implemented within the FORTRAN-based TINKER software package[70] using particle mesh Ewald (PME) for long-range electrostatics. Implementation of the polarizable-multipole Poisson–Boltzmann,[71] which depends on the adaptive Poisson–Boltzmann solver (APBS),[72] and generalized Kirkwood[73] continuum electrostatics models also exist in TINKER, which is now being parallelized using OpenMP. The algorithms in TINKER are also available from within CHARMM using the MSCALE interface.[74,75] Alternative FORTRAN implementations of AMOEBA using PME are available in the Sander and PMEMD molecular dynamics engines of AMBER,[68] with the latter parallelized using MPI. The PME treatment of AMOEBA electrostatics has recently been extended within the Java Runtime Environment (JRE) program *Force Field X* by incorporating explicit support for crystal space group symmetry,[63] parallelizing for heterogeneous computer hardware environments[63] and supporting advanced free energy methods such as the orthogonal space random walk (OSRW) strategy.[65,76] These advancements are critical for applications such as AMOEBA-assisted biomolecular X-ray refinement,[63,77] efficient computation of protein–ligand binding affinity,[57,61] and prediction of the structure, stability, and solubility of organic crystals.[65] Finally, the OpenMM software is working toward a general implementation of AMOEBA using the CUDA GPU programming language.[78]

SIBFA

The SIBFA (sum of interactions between fragments *ab initio* computed) force field for small molecules and flexible proteins, developed by Gresh, Piquemal et al.[79–83] is one of the most sophisticated polarizable force fields because it incorporates polarization, electrostatic penetration,[84] and charge transfer effects.[85]

The polarization is treated with an induced dipole model, in which the distributed anisotropic polarizability tensors[43] are placed on the bond centers and on the heteroatom lone pairs. Quadrupolar polarizabilities are used to treat metal centers. The force field is designed to enable the simultaneous and reliable computation of both intermolecular and conformational energies governing the binding specificities of biologically and pharmacologically relevant molecules. Similar to AMOEBA, permanent multipoles are used for permanent electrostatics in SIBFA. Flexible molecules are modeled by combining the constitutive rigid fragments. SIBFA is formulated on the basis of quantum chemistry and calibrated on energy decomposition analysis, as opposed to AMOEBA, which relies more on condensed-phase experimental data. It aims to produce accurate interaction energy comparable with *ab initio* results. The development of SIBFA emphasizes separability, anisotropy, nonadditivity, and transferability. The analytical gradients for charge transfer energy and solvation contribution are not yet available in SIBFA although molecular dynamics simulations with a simplified potential have been attempted and will be reported in the near future.

SIBFA has been validated on a wide range of molecular systems from water clusters[86] to large complexes like metalloenzymes encompassing Zn(II).[87–92] It has been used to investigate molecular recognition problems including the binding of nucleic acids to metal ions,[93–95] the prediction of oligopeptide conformations,[86,96] and for ligand–protein binding.[97] Most of the SIBFA calculations reproduced closely the quantum chemistry results, including both the interaction energy and the decomposed energy terms. At the same time, electrostatic parameters are demonstrated to be transferable between similar molecules.

A Gaussian-based electrostatic model (GEM) has been explored as an alternative to distributed point multipole electrostatic representation.[98] GEM computes the molecular interaction energies using an approach similar to SIBFA but replacing distributed multipoles by electron densities.[99] GEM better captures the short-range effects on intermolecular interaction energies, and it naturally includes the penetration effect. Calculations on a few simple systems like water clusters[99] have demonstrated GEM's capability to reproduce quantum chemistry results. Furthermore, implementating PME for GEM in a PBC showed reasonable computational efficiency thanks to the use of Hermite Gaussian functions.[100] Therefore, replacing SIBFA's distributed multipoles with the GEM continuous electrostatic model will be a future direction of methodology development.[98]

NEMO

NEMO (nonempirical molecular orbital) is a polarizable potential developed by Karlström and coworkers.[101–103] The NEMO potential energy function is composed of electrostatics, induction, dispersion, and repulsion terms. The induction component is modeled using induced point–dipole moments with recent addition of induced point–quadrupole moments.[22] The electrostatics, previously represented by atomic charges and dipoles, has also been extended to include atomic quadrupole moments leading to notable improvement on formaldehyde. The atomic multipole moments are now obtained from ab *initio* calculation using a LoProp procedure.[104] The LoProp is claimed to provide atomic multipoles and atomic polarizabilities that are less sensitive to basis sets than are other methods such as DMA. Also, NEMO is the only force field that explores the possibility of including interactions between permanent multipoles and higher order induced multipoles involving higher order hyperpolarizabilities.[22]

NEMO has demonstrated its ability to describe accurately both inter- and intramolecular interactions in small systems, including glycine dipeptide conformation profiles,[105] ion–water droplets,[106] and urea transition from nonplanar to planar conformation in water.[107] Its applicability to biomacromolecules is not yet known.

CHARMM-Drude

In addition to the induced dipole model, the classical Drude oscillator model is another popular approach for modeling polarization effects.[39,108] Roux, MacKerell, and their colleagues have been developing a polarizable CHARMM force field based on this approach.[25,26,109,110] Unlike the induced dipole model, which treats the polarization response using point dipoles, the Drude model represents the polarizable centers by a pair of point charges. A point partial charge is tethered via a harmonic spring for each nonhydrogen atom. This point charge (the Drude oscillator) can react to the electrostatic environment and cause the displacement of the local electron density. The atomic polarizability depends on both the Drude particle charge and the harmonic force constant. In MD simulations, the extended Lagrangian is used to evaluate the polarization response, by allowing the Drude particles to move dynamically and experience nonzero forces. Small fictitious masses are assigned to each Drude particle, and independent low-temperature thermostats are applied to the Drude particle degrees of freedom.[111] In case of energy minimization, self-consistent iteration will be required to solve for the polarization.

Determining electrostatic parameters for the Drude oscillator is not as straightforward as for induced dipole models. Masses assigned to the Drude particles are chosen empirically. The values for atomic charges and polarizabilities require a series of calculations of perturbed ESP maps. This force field has been parameterized for water[25,26] and for a series of organic molecules including alkanes,[112] alcohols,[113] aromatics,[114] ethers,[115,116] amides,[109] sulfurs,[117] and ions.[118,119] An attempt has also been made to combine the Drude-based polarizable force field with quantum mechanics in QM/MM methods.[120] It was noted that pair-specific vdW parameters are needed to obtain accurate hydration free energies of small molecules using the polarizable force field. This is likely due to the problematic combining rules used to compute the vdW interactions between unlike atoms.

The Drude model has been implemented in CHARMM[74,121] and in the NAMD package,[122] in which the computational cost is about 1.2–1.8 times greater than that of fixed charge CHARMM.[123]

CHARMM-FQ

The fluctuating charge model (FQ), also known as charge equilibration or electronegativity equalization model, is an empirical approach for calculating charge distributions in molecules. In this formalism, the partial charge on each atom is allowed to change to adapt to different electrostatic environments. The variable partial charges are computed by minimizing the electrostatic energy for a given molecular geometry. Compared with the induced dipole and Drude models, the fluctuating charge models are minimally parameterized and easier to implement because the polarizability is induced without introducing new interactions beyond the point charges. Either extended Lagrangian or self-consistent iteration can be used to compute the fluctuating charges in MD simulations, with similar advantages and disadvantages as discussed above.

The CHARMM-FQ force field,[124,125] developed by Patel, Brooks, and their coworkers, has been parameterized for small molecules,[28] proteins,[28,126] lipids, lipid bilayers,[115,127] and carbohydrates.[124] The force field has been applied to investigate liquid–vapor interfaces in addition to biophysical studies.[128] There are some known limitations for fluctuating charge models; however, such models allow artificial charge transfer between widely separated atoms but that can be controlled with additional constraints. Also, the intramolecular charge flow is limited by the chemical connectivity. It is thus difficult to capture the out-of-plane polarization in molecules such as aromatic benzenes with additional charge sites. The CHARMM-FQ force field has been implemented in the CHARMM software package.[74]

X-Pol

Gao and coworkers proposed the X-Pol framework by combining the fragment-based electronic structure theory with a molecular mechanical force field.[31,32,129] Unlike the traditional force fields, X-Pol does not require bond stretching, angle, and torsion terms because they are represented explicitly by quantum mechanics. The polarization and charge transfer between fragments are also evaluated quantum mechanically.[129] Furthermore, X-Pol can be used to model chemical reactions.

In X-Pol, large molecular systems are divided into small fragments. Electrostatic interactions within the fragments are treated using the electronic structure theory. The electrostatic interactions between fragments are described by the combined quantum mechanical and molecular mechanical (QM/MM) approach. Also, a vdW term is added to the interfragment interaction as a consequence of omitting electron correlation and exchange repulsion. A double self-consistent field (DSCF) procedure is used to converge the total electronic energy of the system as well as the energy within the fragments (this includes the mutual polarization effect).

The X-Pol potential has been applied to MD simulations of liquid water,[130] liquid hydrogen fluoride,[131] and covalently bonded fragments.[132,133] This model was recently used in a molecular dynamics simulation of a solvated protein.[134] As expected the computational efficiency of the X-Pol is in between that of a simple classical force field and a full *ab initio* method. The solvated trypsin required 62.6 h to run a 5 ps simulation on a single 1.5 GHz IBM Power4 processor. A parallel version of X-Pol is being developed.

PFF

Kaminski et al. developed a polarizable protein force field (PFF) based on *ab initio* quantum theory.[135,136] The electrostatic interaction is modeled with induced dipoles and permanent point charges. With the exception of a dispersion parameter, all other parameters, including the electrostatic charges and polarizabilities, are obtained by fitting to quantum chemical binding energy calculations for homodimers. The dispersion parameters are later refined by fitting to the experimental densities of organic liquids.[16] Gas-phase many-body effects, as well as conformational energies, are well reproduced,[136] and MD simulations for real proteins are reasonably accurate at modest computational costs.[16,137]

To reduce the computational cost, a POSSIM (polarizable simulations with second-order interaction model) force field was later proposed, in which the calculation of induced dipoles stops after one iteration.[138,139] The computational efficiency can be improved by almost an order of magnitude by using this formalism. Because the analytical gradients (forces) are unavailable, a Monte-Carlo technique is used in condensed-phase simulations. POSSIM has been validated on selected small model systems, showing good agreement with *ab initio* quantum mechanical and experimental data. Parameters for alanine and protein backbone have been reported.[140]

Polarizable force fields for nonbiological systems also exist. A many-body polarizable force field by Smith and coworkers was developed and applied to the simulations of ion conduction in polyethylene oxide (PEO).[141–143] Cummings and coworkers developed an interesting Gaussian charge polarizable force field for ions and in PEO.[144–146] A polarizable force field for ionic liquids was also reported to provide accurate thermodynamics and transport properties.[147]

APPLICATIONS

Water Simulations

Owing to its important role in life, water is a natural choice for polarizable force field development. After the polarizable (and dissociable) water model of Stillinger and David,[148] more than a dozen polarizable water models have been reported.[149]

Similarly to the polarization models discussed previously, the polarizable water models fall into three major categories. Most belong to the first category, including the Stillinger and David's water model, SPCP,[150] PTIP4P,[151] CKL,[152] NCC,[153] PROL,[154] Dang-Chang,[155] and others. These models all adopted the induced dipole framework to treat polarization, typically using a single polarizable site on water. TTM models[156–159] and the AMOEBA water model[15] utilize an interactive, distributed atomic polarizability with Thole's damping scheme[45] to treat electrostatics and polarization. The Drude oscillator-based water models include SWM4-DP[26] and SWM4-NDP,[25] as well as the charge-on-spring (COS) model[160] and its improved variation.[161] The third group includes the SPC-FQ and TIP4P-FQ[162] water models that utilize the fluctuating charge scheme to model polarization. The partial charges flow from one atom to another, and the total charge of a water molecule need not be zero. Stern et al.[163] proposed a unique water model (POL5) by combining the fluctuating charge with the point induced dipole scheme. Several more sophisticated polarizable water models were developed based on quantum mechanics, including QMPFF,[164] DPP2,[165] and Polarflex.[166] For example, the charge penetration, induction, and charge transfer effects have been incorporated into the DPP2 (distributed point polarizable model) that reproduces well the high-level *ab initio* energetics and structures for large water clusters.

An advantage of a polarizable water model over most nonpolarizable models is the ability to describe the structure and energetics of water in both gas and condensed phases. Water dimer interaction energies, the geometry of water clusters, and the heat of vaporization of neat water can be reproduced well by most polarizable models. Some highly parameterized nonpolarizable force fields such as TIP5P, TIP4P-EW, and TIP4P/2005 actually perform as well or better than some polarizable force fields over a range of liquid properties, including the density–temperature profile, radial distribution function, and diffusion coefficient. However, for water molecules experiencing significant changes in environment, for example, from bulk water to the vicinity of ions or nonpolar molecules, only the polarizable models can capture the change of water dipole, structure, and energetics.[167]

Polarization water models are being extended and applied to other phases as well as to the interface between different phases. Rick et al. recently incorporated charge transfer into their polarizable water model that was then used to study ice/water coexistence properties and properties of the ice Ih phase.[168] The POL3 water model[14,169] was used to study the ice–vapor interface and to calculate the melting point of ice Ih. Bauer and Patel used the TIP4P-QP model to study the liquid–vapor coexistence.[170]

Ion Solvation

Ions are an important component in many chemical and biological systems. Nearly half of all proteins contain metal ions, and they play essential roles in many fundamental biological functions. Some metal ions are critical for both protein structure and function. In enzymes, ions can bind and orient the substrates through electrostatic interactions at the active sites, thus controlling catalytic reaction. Divalent ions are vital in nucleic acid structures. Modeling ion–water and ion–biomolecule interactions accurately is very important.

Owing to high electron density and small sizes of ions, the nonpolarizable models fail to capture the structural details adequately and do not reproduce the atomic dipole of water around the ions.[171–175] Several studies of ion solvation have been reported using different polarizable models[51–53,176,119,177–187] with analyses focused on solvation structures, charge distribution, and binding energies. Note that no straightforward experimental measurement of hydration free energy data exists because the macroscopic system must be neutral. Different assumptions are used to decompose the experimental hydration free energy into single ion contributions. The hydration free energy of some monovalent ions such as Na^+ and K^+ from different sources can vary by as much as 10 kcal/mol. It is more reliable to compare the hydration free energy of the whole salt and the relative energy between cations or anions.

The AMOEBA polarizable force field has been used to model a number of anions and cations, including Na^+, K^+, Mg^{++}, Ca^{++}, Zn^{++}, Cl^-, Br^-, and I^-.[51–53,188] Parameters for these ions, including the vdW parameters and polarization damping coefficients (for divalent ions only), were obtained by fitting to the *ab initio* QM interaction energy profiles of ion–water pairs. Molecular dynamics simulations were then performed to evaluate the ion-cluster solvation enthalpies and solvation free energies.[51–53,188] The excellent agreement between calculated and experimental hydration free energy, often within 1%, demonstrate that polarizable force fields are transferable between phases. *Ab initio* energy decomposition using, for example, the constrained space orbital variations (CSOV) method,[99,189] have also been applied to examine the polarization component of the ion–water interaction energy and to guide the force field parameterization.[53,190] More recently, the AMOEBA force field was used to model the hydration of high-valent Th(IV),[94] and studies on open-shell actinides are underway.

The SIBFA model was used to examine Pb(II),[191] lanthanides (La(III), and Lu(III)) and actinides (Th(IV)) in water.[94] SIBFA-predicted interaction energies generally matched well with the *ab initio* results, including the energy decompositions. Lamoureux and Roux developed the CHARMM polarizable force field for alkali and halide ions based on the Drude oscillator.[177] Hydration free energies, calculated via thermodynamic integration,[192] showed an encouraging agreement with experiment.

Small Molecules

Small molecules are building blocks of biomolecules and serve as substrates and inhibitors. Abundant experimental measurements on various physical and chemical properties exist for common organic molecules, which in turn are used in the parameterization of the force fields. Polarizable and nonpolarizable force fields can usually produce reasonable estimations of physical properties of neat liquids.[193–196] Extensive studies using polarizable force fields, covering major functional group, including alkanes, alcohols, aldehydes, ketones, ethers, acids, aromatic compounds, amines, amides, and some halogen compounds have been reported.[28,36,55,112,114,125,197–199] Calculations of structure, dipole moment, heterodimer binding energy, liquid diffusion constant, density, heat of vaporization, and hydration free energy are usually performed to assess the quality of force field parameters.

The electrostatic multipole parameters in AMOEBA were derived using the DMA procedure. They can be further optimized to the electrostatic potentials of chosen *ab initio* theory and basis sets. The AMOEBA valence parameters were derived from *ab initio* data such as molecular geometries and vibrational frequencies of the gas-phase monomer. The vdW parameters are estimated from gas-phase cluster calculations and subsequently refined in liquid simulations using experimental data (e.g., densities and heats of vaporization). The torsional parameters the last obtained during the parameterization scheme are derived by fitting to *ab initio* QM conformational energy profiles. An automated protocol (PolType) that can generate AMOEBA parameters for small molecules is under development.[200] Because force field parameterization is a tedious process, such an automated tool is convenient and reduces the likelihood of human error.

The CHARMM-Drude force field developers devoted much of their efforts on organic compounds. Their parameterization scheme starts from an initial guess of charge (based on the CHARMM22 force field) and invokes changes at some lone pair sites. Those parameters are then fit to a series of "perturbed" ESP maps. The vdW parameters are then optimized to match neat liquid properties as is done in many other force fields.[117] Overall, a systematic improvement over the CHARMM22 additive force field has been observed for both gas-phase and condensed-phase properties. These studies on small molecules lay the groundwork for developing a Drude-based polarizable force field for proteins and nucleic acids.

Proteins

One of the goals for polarizable force fields is to model accurately protein structures, dynamics, and interactions. Proteins are a ubiquitous class of biopolymers whose functionalities depend on the details of their 3D structures, which, in turn, are largely determined by their amino acid sequences. Fixed charge force fields for proteins, like AMBER, CHARMM, and OPLS-AA, have been developed and for years subjected to various tests and validations. The development of polarizable protein force fields is still in its infancy. Although the importance of including polarization effects was recognized long ago, polarizable protein force fields emerged only in the past decade.[9,21,28,29,37,137,201−205]

The use of polarizable electrostatics in protein simulations dates back to 1976,[1] when Warshel and Levitt simulated lysozyme via single point calculations. Kaminski et al. reported in 2002 an *ab initio* polarizable protein force field (PFF) based on inducible dipoles and point charges.[16,136] Simulations on bovine pancreatic trypsin inhibitor using PDFF showed a satisfactory root-mean-square displacement (RMSD) compared to the experimental crystal structure, and polarization was found to affect the solvation dynamics.[137] The fluctuating charge-based ABEEM/MM force field was used to examine protein systems like trypsin inhibitors[206] and the heme prosthetic group.[207] The SIBFA force field has been used to study the interaction between focal adhesion kinase (FAK) and five pyrrolopyrimidine inhibitors.[208] The energy balances accounting for the solvation/desolvation effects calculated by SIBFA agree with experimental ordering. Water networks in the binding pocket were shown to be critical in terms of binding affinity. Moreover, the polarization contribution was considered as an indispensable component during the molecular recognition. In comparison, the continuum reaction field procedure fails to reproduce these properties. In addition to kinases, the SIBFA protein force field has been used to study a variety of metalloproteins encompassing cations such as Cu^+, Zn^{++}, Ca^{++} or Mg^{++}, as well as enabling inhibition studies.[91,209−211] Future molecular dynamics simulations should extend the applicability of SIBFA to protein–ligand binding.

Ren and coworkers have been systematically developing the AMOEBA protein force field and using it to study several protein systems to understand protein–ligand binding.[57–59,61] More recently, an X-Pol force field for proteins has been developed and demonstrated in a simulation of solvated trypsin.[32]

The first attempt to compute the protein–ligand binding free energy using a polarizable force field was made on the trypsin–benzamidine systems using AMOEBA.[57,61,62] The absolute binding free energy of benzamidine to trypsin and the relative binding free energies for a series of benzamidine analogs were computed using a rigorous alchemical transformation. AMOEBA was successful in evaluating the binding free energies accurately with an average error well within 1.0 kcal/mol. A similar study on trypsin, thrombin, and urokinase was reported using another *ab initio* QM-based polarizable force field.[212] A thermodynamic integration scheme was used to compute the relative binding free energies, which were in excellent agreement with experimental data (root mean square error (RMSE) = 1.0 kcal/mol).

AMOEBA was later used to examine an "entropic paradox" associated with ligand preorganization discovered in a previous study of conformationally constrained phosphorylated-peptide analogs that bind to the SH2 domain of the growth receptor binding protein 2 (Grb2).[59] The paradox refers to the unusual trend in which the binding of unconstrained peptides (assumed to lose more entropy upon binding) is actually more favorable entropically than are the constrained counterparts. AMOEBA correctly reproduced the experimental trend and at the same time repeated a mechanism in which the unconstrained peptide ligands were "locked" by intramolecular nonbonded interactions. The simulations uncovered a crucial caveat that had not been previously acknowledged regarding the general design principle of ligand preorganization, which is presumed by many to have a favorable effect on binding entropy.

More recently, Zhang et al. demonstrated the ability of AMOEBA in dealing with systems with a metal ion.[58] Those authors studied the zinc-containing matrix metalloproteinases (MMPs) in a complex with an inhibitor where the coordination of Zn^{++} was with organic compounds and protein side chains. Polarization was found to play a key role in Zn^{++} coordination geometry in MMP. In addition, the relative binding free energies of selected inhibitors binding with MMP13 were found to be in excellent agreement with experimental results. As with the previous trypsin study, it was found that binding affinities are likely to be overestimated when the polarization between ligands and environments is ignored.

Having a more rigorous physical model for treating polarization, the ability to model protein–ligand interactions has been improved significantly. Systems involving highly charged species, like metal ions, can now be treated with confidence. This, in turn, provides tremendous opportunities for investigating important proteins for drug discovery and for protein engineering.

Lipids

With the rapid development of computational resources, simulations of large systems like lipid bilayers with membrane proteins are feasible.[125,213] Patel and coworkers have been developing a polarizable force field for biomembranes to study the structure and dynamics of ion channel systems.[40,115,127,214] Simulations of solvated DMPC (dimyristoyl phosphatidylcholine) and dipalmitoylphosphatidylcholine (DPPC) bilayers were reported.[115,214] The distribution of the membrane components along the lipid bilayer is similar to that from a fixed charge model. The water dipole moment was found to increase from about 1.9 Debye in the middle of the membrane plane to the average bulk value of 2.5–2.6 Debye. The lipid surface computed with the polarizable force field was not improved from those of nonpolarizable ones, however. In addition, ion permeation in a gramicidin A channel embedded in a DMPC bilayer was investigated.[115] Davis and Patel concluded that including the electronic polarization lowered the ion permeation free energy barrier significantly, from 12 to 6 kcal/mol.

Continuum Solvents for Polarizable Biomolecular Solutes

A continuum solvent replaces explicit atomic details with a bulk, mean-field response. It is possible to demonstrate from statistical mechanics that an implicit solvent potential of mean force (PMF) exists, which preserves exactly the solute thermodynamic properties obtained from explicit solvent.[215] It is possible to formulate a *perfect* implicit solvent in principle, but in practice approximations are necessary to achieve efficiency. This remains an active area of research.[216] An implicit solvent PMF can be formulated via a thermodynamic cycle that discharges the solute in vapor, grows the uncharged (apolar) solute into a solvent $W_{apolar}(X)$, and finally recharges the solute within a continuum dielectric $W_{elec}(X)$

$$W_{PMF}(X) = W_{apolar}(X) + W_{elec}(X) \qquad [12]$$

The continuum electrostatic energy, including mobile electrolytes, can be described by either the nonlinear Poisson–Boltzmann equation (NPBE) or the simplified linearized Poisson–Boltzmann equation (LPBE)

$$\nabla \cdot [\varepsilon(\mathbf{r}) \nabla \phi(\mathbf{r})] - \bar{k}^2(\mathbf{r})\phi(\mathbf{r}) = -4\pi\rho(\mathbf{r}) \qquad [13]$$

where the coefficients are a function of position $\mathbf{r}$, ϕ is the potential, ε is the permittivity, $\bar{k}$ is the modified Debye–Hückel screening factor, and ρ is the solute charge density.[217,218] Implementations of a Poisson–Boltzmann continuum for many-body quantum mechanical potentials have been applied to small molecules for decades. Examples include the polarizable continuum model (PCM),[219,220] COSMO,[221] and the solvent model series (SM*x*).[222] In contrast, applications of biomolecular continuum electrostatics have been limited mainly to fixed partial charge solute descriptions for reasons of computing efficiency and force field availability. However, as a result of increasing computational power and the completion of the polarizable force fields for biomolecules described above, the coupling of classical many-body potentials to continuum electrostatics is now possible.

An important initial demonstration of polarizable biomolecules within a Poisson–Boltzmann continuum used the polarizable force field (PFF) of Maple et al. to model protein–ligand interactions.[223] A second demonstration used the electronic polarization from internal continuum (EPIC), which accounts for intramolecular polarization using a continuum dielectric.[224,225] Finally, the polarizable multipole Poisson–Boltzmann (PMPB) model based on the AMOEBA force field demonstrated that the self-consistent reaction field (SCRF) of proteins within a continuum solvent is consistent with the ensemble average response of explicit solvent.[71] Contrarily, end-state calculations of protein–ligand binding affinity using the PMPB model were shown to not recapitulate explicit solvent alchemical free energies to chemical accuracy.[61] This motivates development of analytic continuum electrostatics (discussed next), which are fast enough to allow binding affinities to be computed using alchemical sampling, rather than merely relying on end states. A key advantage of EPIC is that the biomolecular self-consistent field (SCF) is determined by a single numerical finite-difference (FD) solution of the PBE, unlike the aforementioned atom-centered PFF and PMPB models that require a new solution for each SCF iteration. However, a tradeoff of EPIC's efficiency gain is a reduction in model flexibility because electrostatic masking rules cannot be incorporated into the FD solver (i.e., the permanent field because of 1–2 or 1–3 interactions cannot be neglected). Although masking of short-range bonded interactions is the standard approach used by essentially all biomolecular force fields, this is not possible for an EPIC-style energy model.

The first example of an analytic continuum electrostatic model for polarizable biomolecules is the generalized Kirkwood (GK) model for the AMOEBA force field.[73] The AMOEBA/GK approach has been combined with alchemical sampling to predict trypsin–ligand binding affinity with a correlation coefficient of 0.93. This is a significant improvement over the PMPB end-state approach.[226] A second example, based on the ABEEM$\sigma\pi$ fluctuating charge force field combined with a generalized Born (GB) continuum electrostatic model, showed promising results for the computation of solvation free energies for small organic molecules and peptide fragments.[227]

Macromolecular X-ray Crystallography Refinement

X-ray crystallography is the dominant experimental method for determining the 3-dimensional coordinates of macromolecules. Collected diffraction data is the Fourier transform of the ensemble average electron density of the macro-molecular crystal. While reciprocal space amplitudes of Bragg diffraction peaks are measured, their phases are not. Instead, phase information is derived from the Fourier transform of a model structure that is sufficiently close to the actual experimental ensemble. This is known as molecular replacement (MR). After an initial model has been built into the electron density, further refinement is based on optimizating a target function E_{target} of the form

$$E_{target} = w_A E_{X\text{-ray}} + E_{Force\ Field} \qquad [14]$$

where $E_{X\text{-ray}}$ evaluates the agreement between measured and calculated diffraction amplitudes, $E_{Force\ Field}$ restrains the model using prior knowledge of intra- and intermolecular chemical forces, and w_A weighs the relative strength of the two terms.[77,228] We now focus on the evolution of the prior chemical knowledge used during the X-ray refinement process, and we culminate in the ongoing work using polarizable force fields in combination with PME electrostatics algorithms to obtain the most accurate, informative biomolecular models possible.

The first application of molecular mechanics to macromolecular X-ray crystallography refinement (based on fixed partial charge electrostatics eval-uated using a spherical cutoff) was on influenza virus hemagglutinin by Weis et al. in 1990.[229] This work demonstrated that electrostatics maintained chem-ically reasonable hydrogen bonding, although charged surface residues were sometimes observed to form incorrect salt bridges.[229] The latter observation highlights the importance of accounting for dielectric screening arising from the heterogeneous distribution of solvent within a macromolecular crystal, by using one of the above-described continuum electrostatics models. For example, the generalized Born (GB) model for fixed charge electrostatics has been described, albeit with a spherical cutoff approximation.[230] Comparing refinements with and without GB screening showed that roughly 10% of the amino acid side-chain conformations were altered, with 75% of these side-chain differ-ences because of residues at the macromolecular surface.[230] Although these first applications of fixed charge force field electrostatics were encouraging, the use of spherical cutoffs to approximate crystal lattice sums is now known to be only conditionally convergent and therefore prone to a variety of artifacts.[231]

In 1921, Ewald introduced an absolutely convergent solution to the prob-lem of evaluating electrostatic lattice summations in crystals. He did this by separating the problem into a short-ranged real space sum and a periodic, smoothly varying, long-range sum that can be evaluated efficiently in reciprocal space.[232] This approach, now known as Ewald summation, has been described for both fixed partial charges and atomic multipoles.[233,234] More recently, the efficiency of Ewald summation was improved via the PME algorithm, wherein the reciprocal space summation leverages the fast Fourier transforms (FFT)[235] via b-Spline interpolation[236] for both fixed partial charge and atomic multipole descriptions.[237]

The speed of the PME algorithm has been further improved for crystals by incorporating explicit support for space group symmetry and by parallelization for heterogeneous computer architectures.[63] Combining the polarizable AMOEBA force field with electrostatics evaluated using PME has been shown to improve macromolecular models from X-ray crystallography refinement in a variety of contexts.[64,77,238–240] At high resolution (~1 Å or lower), the information contained within a polarizable atomic multipole force field can be used to formulate the electron density of the scattering model ($E_{X\text{-ray}}$), in addition to contributing chemical restraints ($E_{\text{Force Field}}$).[64,238] The importance of the prior chemical information contained in a polarizable force field is most significant when positioning parts of the model that are not discernable from the experimental electron density, as in the orientation of water hydrogen atoms[239] or secondary structure elements for mid-to-low resolution data sets (~3–4 Å).[63]

Let us consider an example: the AMOEBA-assisted biomolecular X-ray refinement with electrostatics evaluated via PME in the program *Force Field X*. This program was used to re-refine nine mouse and human DNA methyltransferase 1 (Dnmt1) data sets deposited in the protein data bank (PDB). Significant improvements in model quality (presented in Table 1) were achieved as assayed by the MolProbity[241] structure validation tool. The MolProbity score is calibrated to reflect the expected resolution of the X-ray data. After re-refinement, the average MolProbity score was reduced to 2.14, indicating a level of model improvement consistent with collecting data to 0.67 Å higher resolution. For example, the pose of *S*-adenosyl-L-homocysteine (SAH) from mouse (3PT6) and human (3PTA) structures differed by an RMSD of 1.6 Å before re-refinement, but only 0.9 Å afterward (Figure 1).

Table 1 DNA Methyltransferase 1 (Dnmt1) Models Before and After Polarizable X-Ray Refinement with the Program *Force Field X*

| | | Protein Data Bank | | | | Re-Refined with *Force Field X* | | | |
| | | Statistics | | MolProbity | | Statistics | | MolProbity | |
PDB	Res. (Å)	R	R_{free}	Score	(%)	R	R_{free}	Score	(%)
3AV4	2.8	0.232	0.267	2.87	68.0	0.238	0.282	2.25	95.0
3AV5	3.3	0.188	0.264	3.09	79.0	0.216	0.275	2.44	97.0
3AV6	3.1	0.195	0.255	2.99	81.0	0.213	0.265	2.37	97.0
3EPZ	2.3	0.213	0.264	2.27	78.0	0.254	0.292	2.09	87.0
3OS5	1.7	0.211	0.238	2.01	54.0	0.182	0.213	1.77	74.0
3PT6	3.0	0.211	0.266	2.95	78.0	0.207	0.268	1.97	99.0
3PT9	2.5	0.196	0.256	2.72	60.0	0.181	0.248	1.90	97.0
3PTA	3.6	0.257	0.291	3.65	57.0	0.211	0.271	2.41	99.0
3SWR	2.5	0.220	0.272	2.69	62.0	0.204	0.264	2.03	95.0
Mean	2.7	0.214	0.264	2.80	68.6	0.212	0.264	2.14	93.3
Mean improvement								0.67	24.8

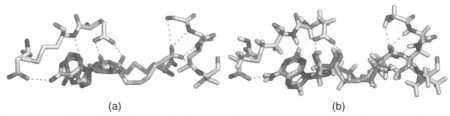

(a)   (b)

Figure 1   Polarizable biomolecular X-ray refinement on two Dnmt1 data sets. (a) The deposited pose of SAH from data sets 3PT6 (mouse, gray) and 3PTA (human, cyan) do not agree (coord. RMSD 1.6 Å). (b) The poses of SAH from mouse and human structures are more consistent (coord. RMSD 0.9 Å) after *Force Field X* refinement. (For a color version of this figure, please see plate 1 in color plate section.)

Prediction of Organic Crystal Structure, Thermodynamics, and Solubility

It was emphasized in 1998 that predicting crystal structures from chemical composition remained a major unsolved challenge.[242] Significant progress has been made since then to address this challenge, as evidenced by successes of the fourth and fifth blind tests of crystal structure prediction (CSP) organized by the Cambridge Crystallographic Data Center (CCDC).[243,244] Prediction of crystal structures is important in the pharmaceutical industry, where extensive experimental screens are necessary to explore the range of stable polymorphs a molecule may form. The unique three-dimensional molecular packing of each polymorph determines its physical properties such as stability and bioavailability. For this reason, both FDA approval and patent protection are awarded to a specific crystal polymorph, rather than to the molecule itself. To illustrate this point, eight companies have filed eleven patents on five possible crystal forms of the molecule cefdinir.[245]

Prediction of thermodynamically stable crystal structures from chemical composition requires a potential energy function capable of distinguishing between large numbers of structures that are closely spaced in thermodynamic stability.[246,247] In this section, we restrict our focus to energy models that explicitly account for electronic polarization classically[65,248,249] and neglect the more expensive electronic structure methods sometimes used to (re)score favorable structures.[250]

The vast majority of CSP has been limited to using intermolecular potentials that lack explicit inclusion of polarization,[249,251] although its importance has become a topic of interest.[35,252–254] Nonpolarizable force fields, based on fixed partial charges or fixed atomic multipoles, must implicitly account for the 20–40% of the lattice energy attributable induction.[249] On the contrary, polarizable models such as the AMOEBA force field for organic molecules [54,255] based on the Thole damping scheme[45] and the Williams–Stone–Misquitta (WSM) method[256,257] for obtaining distributed polarizabilities allow one to include polarization during CSP explicitly.

Beyond polarization, modeling the conformational flexibility and corresponding intermolecular energetics of organic molecules via sampling methods such as molecular dynamics is essential to predicting the thermodynamic properties of crystals.[258] For example, the structure, stability, and solubility of *n*-alkylamide crystals, from acetamide through octanamide, can be predicted by an alchemical sampling method to compute the sublimation/deposition phase transition free energy.[65]

SUMMARY

Significant progress has been made in the past decade in developing general purpose polarizable force fields. Polarizable force fields have exhibited success in disparate research areas including ion solvation, protein–ligand interactions, ion channels and lipids, macromolecular structural refinement, and so on. Yet, there remain plenty of challenges ahead. The importance of polarization still needs to be established systematically for a wide range of biological systems. While polarizable force fields in principle have better transferability than do nonpolarizable force fields, they are also expected to perform better in a broader range of systems, making parameterization a more elaborate process. In addition to polarization, treatment of other physical effects, including high-order permanent charge distributions interactions, short-range electrostatic penetration and charge transfer effects need further improvement to advance the overall quality of classical electrostatic models. Because computational efficiency (including the need for parallelization) has been a major barrier to the adoption of polarizable force fields, better and more efficient algorithms are also required to advance the application of polarizable force fields. A future area for advancement is to combine the polarizable force fields with fixed charge force fields in a multiscale fashion, as is done with QM/MM. Technically, this can be achieved straightforwardly but caution is needed to ensure the interactions across the two resolutions are balanced.

ACKNOWLEDGMENT

The authors are grateful for the support provided by Robert A. Welch Foundation *(F-1691) and the National Institute of Health (GM106137). Support from French CNRS is also acknowledged through the PICS international program.*

REFERENCES

1. A. Warshel and M. Levitt, *J. Mol. Biol.*, **103**(2), 227–249 (1976). Theoretical Studies of Enzymic Reactions - Dielectric, Electrostatic and Steric Stabilization of Carbonium-Ion in Reaction of Lysozyme.

2. J. A. Mccammon, B. R. Gelin, and M. Karplus, *Nature*, **267**(5612), 585–590 (1977). Dynamics of Folded Proteins.

3. M. A. Spackman, *Chem. Phys. Lett.*, **418**(1–3), 158–162 (2006). The Use of the Promolecular Charge Density to Approximate the Penetration Contribution to Intermolecular Electrostatic Energies.

4. D. Nachtigallova, P. Hobza, and V. Spirko, *J. Phys. Chem. A*, **112**(9), 1854–1856 (2008). Assigning the Nh Stretches of the Guanine Tautomers Using Adiabatic Separation: CCSD(T) Benchmark Calculations.

5. W. D. Cornell, P. Cieplak, C. I. Bayly, I. R. Gould, K. M. Merz, D. M. Ferguson, D. C. Spellmeyer, T. Fox, J. W. Caldwell, and P. A. Kollman, *J. Am. Chem. Soc.*, **117**(19), 5179–5197 (1995). A 2nd Generation Force-Field for the Simulation of Proteins, Nucleic-Acids, and Organic-Molecules.

6. A. D. MacKerell, D. Bashford, M. Bellott, R. L. Dunbrack, J. D. Evanseck, M. J. Field, S. Fischer, J. Gao, H. Guo, S. Ha, D. Joseph-McCarthy, L. Kuchnir, K. Kuczera, F. T. K. Lau, C. Mattos, S. Michnick, T. Ngo, D. T. Nguyen, B. Prodhom, W. E. Reiher, B. Roux, M. Schlenkrich, J. C. Smith, R. Stote, J. Straub, M. Watanabe, J. Wiorkiewicz-Kuczera, D. Yin, and M. Karplus, *J. Phys. Chem. B*, **102**(18), 3586–3616 (1998). All-Atom Empirical Potential for Molecular Modeling and Dynamics Studies of Proteins.

7. H. Valdes, K. Pluhackova, M. Pitonak, J. Rezac, and P. Hobza, *Phys. Chem. Chem. Phys.*, **10**(19), 2747–2757 (2008). Benchmark Database on Isolated Small Peptides Containing an Aromatic Side Chain: Comparison between Wave Function and Density Functional Theory Methods and Empirical Force Field.

8. W. L. Jorgensen, D. S. Maxwell, and J. Tirado-Rives, *J. Am. Chem. Soc.*, **118**(45), 11225–11236 (1996). Development and Testing of the OPLS All-Atom Force Field on Conformational Energetics and Properties of Organic Liquids.

9. J. W. Ponder and D. A. Case, *Adv. Protein Chem.*, **66**, 27–85 (2003). Force Fields for Protein Simulations.

10. J. Rezac, P. Jurecka, K. E. Riley, J. Cerny, H. Valdes, K. Pluhackova, K. Berka, T. Rezac, M. Pitonak, J. Vondrasek, and P. Hobza, *Collect. Czech. Chem. Commun.*, **73**(10), 1261–1270 (2008). Quantum Chemical Benchmark Energy and Geometry Database for Molecular Clusters and Complex Molecular Systems (Www.Begdb.Com): A Users Manual and Examples.

11. J. Rezac, K. E. Riley, and P. Hobza, *J. Chem. Theory Comput.*, **7**(8), 2427–2438 (2011). S66: A Well-Balanced Database of Benchmark Interaction Energies Relevant to Biomolecular Structures.

12. K. Berka, R. Laskowski, K. E. Riley, P. Hobza, and J. I. Vondrášek, *J. Chem. Theory Comput.*, **5**(4), 982–992 (2009). Representative Amino Acid Side Chain Interactions in Proteins. A Comparison of Highly Accurate Correlated Ab Initio Quantum Chemical and Empirical Potential Procedures.

13. J. A. Barker, *Proc. Royal Soc. Lond. Ser. A, Math. Phys. Sci.*, **219**(1138), 367–372 (1953). Statistical Mechanics of Interacting Dipoles.

14. J. W. Caldwell and P. A. Kollman, *J. Phys. Chem.*, **99**, 6208–6219 (1995). Structure and Properties of Neat Liquids Using Nonadditive Molecular Dynamics: Water, Methanol, and N-Methylacetamide.

15. P. Y. Ren and J. W. Ponder, *J. Phys. Chem. B*, **107**(24), 5933–5947 (2003). Polarizable Atomic Multipole Water Model for Molecular Mechanics Simulation.

16. G. A. Kaminski, H. A. Stern, B. J. Berne, R. A. Friesner, Y. X. X. Cao, R. B. Murphy, R. H. Zhou, and T. A. Halgren, *J. Comput. Chem.*, **23**(16), 1515–1531 (2002). Development

of a Polarizable Force Field for Proteins Via Ab Initio Quantum Chemistry: First Generation Model and Gas Phase Tests.

17. R. A. Friesner, *Adv. Protein Chem. Struct. Biol.*, **72**, 79–104 (2006). Modeling Polarization in Proteins and Protein-Ligand Complexes: Methods and Preliminary Results.

18. P. Y. Ren and J. W. Ponder, *J. Comput. Chem.*, **23**(16), 1497–1506 (2002). Consistent Treatment of Inter- and Intramolecular Polarization in Molecular Mechanics Calculations.

19. L. F. Molnar, X. He, B. Wang, and K. M. Merz, Jr., *J. Chem. Phys.*, **131**(6), 065102 (2009). Further Analysis and Comparative Study of Intermolecular Interactions Using Dimers from the S22 Database.

20. P. Cieplak, J. Caldwell, and P. Kollman, *J. Comput. Chem.*, **22**(10), 1048–1057 (2001). Molecular Mechanical Models for Organic and Biological Systems Going Beyond the Atom Centered Two Body Additive Approximation: Aqueous Solution Free Energies of Methanol and N-Methyl Acetamide, Nucleic Acid Base, and Amide Hydrogen Bonding and Chloroform/Water Partition Coefficients of the Nucleic Acid Bases.

21. Z. X. Wang, W. Zhang, C. Wu, H. X. Lei, P. Cieplak, and Y. Duan, *J. Comput. Chem.*, **27**(8), 994–994 (2006). Strike a Balance: Optimization of Backbone Torsion Parameters of AMBER Polarizable Force Field for Simulations of Proteins and Peptides.

22. A. Holt and G. Karlström, *J. Comput. Chem.*, **29**(12), 2033–2038 (2008). Inclusion of the Quadrupole Moment When Describing Polarization The Effect of the Dipole-Quadrupole Polarizability.

23. S. Moghaddam, C. Yang, M. Rekharsky, Y. H. Ko, K. Kim, Y. Inoue, and M. K. Gilson, *J. Am. Chem. Soc.*, **133**(10), 3570–81 (2011). New Ultrahigh Affinity Host-Guest Complexes of Cucurbit[7]uril with Bicyclo[2.2.2]octane and Adamantane Guests: Thermodynamic Analysis and Evaluation of M2 Affinity Calculations.

24. D. P. Geerke and W. F. van Gunsteren, *J. Phys. Chem. B*, **111**(23), 6425–6436 (2007). Calculation of the Free Energy of Polarization: Quantifying the Effect of Explicitly Treating Electronic Polarization on the Transferability of Force-Field Parameters.

25. G. Lamoureux, E. Harder, I. V. Vorobyov, B. Roux, and A. D. MacKerell, *Chem. Phys. Lett.*, **418**(1-3), 245–249 (2006). A Polarizable Model of Water for Molecular Dynamics Simulations of Biomolecules.

26. G. Lamoureux, A. D. MacKerell, and B. Roux, *J. Chem. Phys.*, **119**(10), 5185–5197 (2003). A Simple Polarizable Model of Water Based on Classical Drude Oscillators.

27. J. L. Banks, G. A. Kaminski, R. H. Zhou, D. T. Mainz, B. J. Berne, and R. A. Friesner, *J. Chem. Phys.*, **110**(2), 741–754 (1999). Parametrizing a Polarizable Force Field from Ab Initio Data I. The Fluctuating Point Charge Model.

28. S. Patel and C. L. Brooks, III, *J. Comput. Chem.*, **25**(1), 1–15 (2004). CHARMM Fluctuating Charge Force Field for Proteins: I. Parameterization and Application to Bulk Organic Liquid Simulations.

29. S. Patel, A. D. MacKerell, and C. L. Brooks, III, *J. Comput. Chem.*, **25**(12), 1504–1514 (2004). CHARMM Fluctuating Charge Force Field for Proteins: II. Protein/Solvent Properties from Molecular Dynamics Simulations Using a Nonadditive Electrostatic Model.

30. A. K. Rappe and W. A. Goddard, *J. Phys. Chem.*, **95**(8), 3358–3363 (1991). Charge Equilibration for Molecular-Dynamics Simulations.

31. L. Song, J. Han, Y. L. Lin, W. Xie, and J. Gao, *J. Phys. Chemistry A*, **113**(43), 11656–11664 (2009). Explicit Polarization (X-Pol) Potential Using Ab Initio Molecular Orbital Theory and Density Functional Theory.

32. W. Xie, M. Orozco, D. G. Truhlar, and J. Gao, *J. Chem. Theory Comput.*, **5**(3), 459–467 (2009). X-Pol Potential: An Electronic Structure-Based Force Field for Molecular Dynamics Simulation of a Solvated Protein in Water.

33. M. J. L. Mills and P. L. A. Popelier, *Theor. Chem. Acc.*, **131**(3) (2012). Polarisable Multipolar Electrostatics from the Machine Learning Method Kriging: An Application to Alanine.

34. P. Cieplak, F. Y. Dupradeau, Y. Duan, and J. M. Wang, *J. Phys. Condense. Mat.*, **21**(33), 333102–333123 (2009). Polarization Effects in Molecular Mechanical Force Fields.

35. T. A. Halgren and W. Damm, *Curr. Opin. Struct. Biol.*, **11**(2), 236–242 (2001). Polarizable Force Fields.

36. P. E. M. Lopes, B. Roux, and A. D. MacKerell, *Theor. Chem. Acc.*, **124**(1-2), 11–28 (2009). Molecular Modeling and Dynamics Studies with Explicit Inclusion of Electronic Polarizability: Theory and Applications.

37. J. W. Ponder, C. J. Wu, P. Y. Ren, V. S. Pande, J. D. Chodera, M. J. Schnieders, I. Haque, D. L. Mobley, D. S. Lambrecht, R. A. DiStasio, M. Head-Gordon, G. N. I. Clark, M. E. Johnson, and T. Head-Gordon, *J. Phys. Chem. B*, **114**(8), 2549–2564 (2010). Current Status of the AMOEBA Polarizable Force Field.

38. A. Warshel, M. Kato, and A. V. Pisliakov, *J. Chem. Theory Comput.*, **3**(6), 2034–2045 (2007). Polarizable Force Fields: History, Test Cases, and Prospects.

39. S. W. Rick and S. J. Stuart, in *Reviews in Computational Chemistry*, K. B. Lipkowitz and D. B. Boyd (Eds.), Wiley-VCH, New York, 2002, Vol. **18**, pp. 89–146, Potentials and Algorithms for Incorporating Polarizability in Computer Simulations.

40. S. Patel, J. E. Davis, and B. A. Bauer, *J. Am. Chem. Soc.*, **131**(39), 13890–13891 (2009). Exploring Ion Permeation Energetics in Gramicidin A Using Polarizable Charge Equilibration Force Fields.

41. J. Applequist, J. R. Carl, and K.-K. Fung, *J. Am. Chem. Soc.*, **94**(9), 2952–2960 (1972). An Atom Dipole Interaction Model for Molecular Polarizability Application to Polyatomic Molecules and Determination of Atom Polarizabilities.

42. C. R. Le Sueur and A. J. Stone, *Mol. Phys.*, **78**(5), 1267–1291 (1993). Practical Schemes for Distributed Polarizabilities.

43. D. R. Garmer and W. J. Stevens, *J. Phys. Chem.*, **93**(25), 8263–8270 (1989). Transferability of Molecular Distributed Polarizabilities from a Simple Localized Orbital Based Method.

44. W. Wang and R. D. Skeel, *J. Chem. Phys.*, **123**(16), 164107 (2005). Fast Evaluation of Polarizable Forces.

45. B. T. Thole, *Chem. Phys.*, **59**(3), 341–350 (1981). Molecular Polarizabilities Calculated with a Modified Dipole Interaction.

46. J. F. Truchon, A. Nicholls, R. I. Iftimie, B. Roux, and C. I. Bayly, *J. Chem. Theory Comput.*, **4**(9), 1480–1493 (2008). Accurate Molecular Polarizabilities Based on Continuum Electrostatics.

47. D. Van Belle, M. Froeyen, G. Lippens, and S. J. Wodak, *Mol. Phys.*, **77**(2), 239–255 (1992). Molecular-Dynamics Simulation of Polarizable Water by an Extended Lagrangian Method.

48. A. J. Stone, *Chem. Phys. Lett.*, **83**(2), 233–239 (1981). Distributed Multipole Analysis, or How to Describe a Molecular Charge Distribution.

49. A. J. Stone, *Mol. Phys.*, **56**(5), 1047–1064 (1985). Distributed Multipole Analysis: Methods and Applications.

50. A. J. Stone, *J. Chem. Theory Comput.*, **1**(6), 1128–1132 (2005). Distributed Multipole Analysis: Stability for Large Basis Sets.

51. A. Grossfield, P. Y. Ren, and J. W. Ponder, *J. Am. Chem. Soc.*, **125**(50), 15671–15682 (2003). Ion Solvation Thermodynamics from Simulation with a Polarizable Force Field.

52. D. Jiao, C. King, A. Grossfield, T. A. Darden, and P. Y. Ren, *J. Phys. Chem. B*, **110**(37), 18553–18559 (2006). Simulation of Ca^{2+} and Mg^{2+} Solvation Using Polarizable Atomic Multipole Potential.

53. J. C. Wu, J. P. Piquemal, R. Chaudret, P. Reinhardt, and P. Y. Ren, *J. Chem. Theory Comput.*, **6**(7), 2059–2070 (2010). Polarizable Molecular Dynamics Simulation of Zn(II) in Water Using the AMOEBA Force Field.

54. P. Ren, C. Wu, and J. W. Ponder, *J. Chem. Theory Comput.*, **7**(10), 3143–3161 (2011). Polarizable Atomic Multipole-Based Molecular Mechanics for Organic Molecules.

55. Y. Shi, C. Wu, J. W. Ponder, and P. Ren, *J. Comput. Chem.*, **32**(5), 967–977 (2011). Multipole Electrostatics in Hydration Free Energy Calculations.

56. J. L. Jiang, Y. B. Wu, Z. X. Wang, and C. Wu, *J. Chem. Theory Comput.*, **6**(4), 1199–1209 (2010). Assessing the Performance of Popular Quantum Mechanics and Molecular Mechanics Methods and Revealing the Sequence-Dependent Energetic Features Using 100 Tetrapeptide Models.

57. D. Jiao, P. A. Golubkov, T. A. Darden, and P. Ren, *Proc. Natl. Acad. Sci. U. S. A.*, **105**(17), 6290–6295 (2008). Calculation of Protein-Ligand Binding Free Energy by Using a Polarizable Potential.

58. J. Zhang, W. Yang, J. P. Piquemal, and P. Ren, *J. Chem. Theory Comput.*, **8**, 1314–1324 (2012). Modeling Structural Coordination and Ligand Binding in Zinc Proteins with a Polarizable Potential.

59. Y. Shi, C. Z. Zhu, S. F. Martin, and P. Ren, *J. Phys. Chem. B*, **116**(5), 1716–27 (2012). Probing the Effect of Conformational Constraint on Phosphorylated Ligand Binding to an SH2 Domain Using Polarizable Force Field Simulations.

60. W. Jiang and B. Roux, *J. Chem. Theory Comput.*, **6**(9), 2559–2565 (2010). Free Energy Perturbation Hamiltonian Replica-Exchange Molecular Dynamics (FEP/H-REMD) for Absolute Ligand Binding Free Energy Calculations.

61. D. Jiao, J. J. Zhang, R. E. Duke, G. H. Li, M. J. Schnieders, and P. Y. Ren, *J. Comput. Chem.*, **30**(11), 1701–1711 (2009). Trypsin-Ligand Binding Free Energies from Explicit and Implicit Solvent Simulations with Polarizable Potential.

62. Y. Shi, D. Jiao, M.J. Schnieders, and P. Ren, Trypsin-Ligand Binding Free Energy Calculation with AMOEBA. Engineering in Medicine and Biology Society (EMBC). EMBC Annual International Conference of the IEEE, 2009: p. 2328-2331.

63. M. J. Schnieders, T. D. Fenn, and V. S. Pande, *J. Chem. Theory Comput.*, **7**(4), 1141–1156 (2011). Polarizable Atomic Multipole X-Ray Refinement: Particle Mesh Ewald Electrostatics for Macromolecular Crystals.

64. M. J. Schnieders, T. D. Fenn, V. S. Pande, and A. T. Brunger, *Acta Crystallogr., Sect. D-Biol. Crystallogr.*, **65**, 952–965 (2009). Polarizable Atomic Multipole X-Ray Refinement: Application to Peptide Crystals.

65. M. J. Schnieders, J. Baltrusaitis, Y. Shi, G. Chattree, L. Zheng, W. Yang, and P. Ren, *J. Chem. Theory Comput.*, **8**(5), 1721–1736 (2012). The Structure, Thermodynamics and Solubility of Organic Crystals from Simulation with a Polarizable Force Field.

66. J. W. Ponder, TINKER: Software Tools for Molecular Design, Washington University, http://dasher.wustl.edu/tinker/, 2012. v6.0.

67. M. S. Friedrichs, P. Eastman, V. Vaidyanathan, M. Houston, S. Legrand, A. L. Beberg, D. L. Ensign, C. M. Bruns, and V. S. Pande, *J. Comput. Chem.*, **30**(6), 864–72 (2009). Accelerating Molecular Dynamic Simulation on Graphics Processing Units.

68. D. A. Case, T. E. Cheatham, 3rd, T. Darden, H. Gohlke, R. Luo, K. M. Merz, Jr., A. Onufriev, C. Simmerling, B. Wang, and R. J. Woods, *J. Comput. Chem.*, **26**(16), 1668–1688 (2005). The AMBER Biomolecular Simulation Programs.

69. M. J. Schnieders, T. D. Fenn, J. Wu, W. Yang, and P. Ren, Force Field X Open Source, Platform Independent Modules for Molecular Biophysics Simulations. http://ffx.kenai.com, 2011.

70. J. W. Ponder, TINKER: Software Tools for Molecular Design. Saint Louis, MO. p. TINKER: Software Tools for Molecular Design. http://dasher.wustl.edu, 2012.

71. M. J. Schnieders, N. A. Baker, P. Y. Ren, and J. W. Ponder, *J. Chem. Phys.*, **126**(12) (2007). Polarizable Atomic Multipole Solutes in a Poisson-Boltzmann Continuum.

72. N. A. Baker, D. Sept, S. Joseph, M. J. Holst, and J. A. McCammon, *Proc. Natl. Acad. Sci. U. S. A.*, **98**(18), 10037–10041 (2001). Electrostatics of Nanosystems: Application to Microtubules and the Ribosome.

73. M. J. Schnieders and J. W. Ponder, *J. Chem. Theory Comput.*, **3**(6), 2083–2097 (2007). Polarizable Atomic Multipole Solutes in a Generalized Kirkwood Continuum.

74. B. R. Brooks, C. L. Brooks, III, A. D. MacKerell, Jr., L. Nilsson, R. J. Petrella, B. Roux, Y. Won, G. Archontis, C. Bartels, S. Boresch, A. Caflisch, L. Caves, Q. Cui, A. R. Dinner, M. Feig, S. Fischer, J. Gao, M. Hodoscek, W. Im, K. Kuczera, T. Lazaridis,

J. Ma, V. Ovchinnikov, E. Paci, R. W. Pastor, C. B. Post, J. Z. Pu, M. Schaefer, B. Tidor, R. M. Venable, H. L. Woodcock, X. Wu, W. Yang, D. M. York, and M. Karplus, *J. Comput. Chem.*, **30**(10), 1545–614 (2009). CHARMM: The Biomolecular Simulation Program.

75. H. L. Woodcock, B. T. Miller, M. Hodoscek, A. Okur, J. D. Larkin, J. W. Ponder, and B. R. Brooks, *J. Chem. Theory Comput.*, **7**(4), 1208–1219 (2011). MSCALE: A General Utility for Multiscale Modeling.

76. L. Zheng, M. Chen, and W. Yang, *Proc. Natl. Acad. Sci. U. S. A.*, **105**(51), 20227–20232 (2008). Random Walk in Orthogonal Space to Achieve Efficient Free-Energy Simulation of Complex Systems.

77. T. D. Fenn and M. J. Schnieders, *Acta Crystallogr. Sect. D*, **67**(11), 957–65 (2011). Polarizable Atomic Multipole X-Ray Refinement: Weighting Schemes for Macromolecular Diffraction.

78. P. Eastman and V. S. Pande, *J. Comput. Chem.*, **31**(6), 1268–1272 (2010). Efficient Nonbonded Interactions for Molecular Dynamics on a Graphics Processing Unit.

79. N. Gresh, P. Claverie, and A. Pullman, *Theor. Chim. Acta*, **66**(1), 1–20 (1984). Theoretical Studies of Molecular Conformation - Derivation of an Additive Procedure for the Computation of Intramolecular Interaction Energies - Comparison with ab Initio SCF Computations.

80. N. Gresh, *J. Comput. Chem.*, **16**(7), 856–882 (1995). Energetics of Zn^{2+} Binding to a Series of Biologically Relevant Ligands - A Molecular Mechanics Investigation Grounded on Ab-Initio Self-Consistent-Field Supermolecular Computations.

81. J. P. Piquemal, B. Williams-Hubbard, N. Fey, R. J. Deeth, N. Gresh, and C. Giessner-Prettre, *J. Comput. Chem.*, **24**(16), 1963–1970 (2003). Inclusion of the Ligand Field Contribution in a Polarizable Molecular Mechanics: SIBFA-LF.

82. J. P. Piquemal, H. Chevreau, and N. Gresh, *J. Chem. Theory Comput.*, **3**(3), 824–837 (2007). Toward a Separate Reproduction of the Contributions to the Hartree-Fock and DFT Intermolecular Interaction Energies by Polarizable Molecular Mechanics with the SIBFA Potential.

83. N. Gresh, G. A. Cisneros, T. A. Darden, and J. P. Piquemal, *J. Chem. Theory Comput.*, **3**(6), 1960–1986 (2007). Anisotropic, Polarizable Molecular Mechanics Studies of Inter- and Intramolecular Interactions and Ligand-Macromolecule Complexes. A Bottom-up Strategy.

84. J. P. Piquemal, N. Gresh, and C. Giessner-Prettre, *J. Phys. Chem. A*, **107**(48), 10353–10359 (2003). Improved Formulas for the Calculation of the Electrostatic Contribution to the Intermolecular Interaction Energy from Multipolar Expansion of the Electronic Distribution.

85. N. Gresh, P. Claverie, and A. Pullman, *Int. J. Quant. Chem.*, **29**(1), 101–118 (1986). Intermolecular Interactions - Elaboration on an Additive Procedure Including an Explicit Charge-Transfer Contribution.

86. N. Gresh, *J. Chim. Phys. Phys. Chim. Biol.*, **94**(7–8), 1365–1416 (1997). Inter- and Intramolecular Interactions. Inception and Refinements of the SIBFA, Molecular Mechanics (SMM) Procedure, A Separable, Polarizable Methodology Grounded on Ab Initio SCF/MP2 Computations. Examples of Applications to Molecular Recognition Problems.

87. J. Antony, J. P. Piquemal, and N. Gresh, *J. Comput. Chem.*, **26**(11), 1131–1147 (2005). Complexes of Thiomandelate and Captopril Mercaptocarboxylate Inhibitors to Metallo-Beta-Lactamase by Polarizable Molecular Mechanics Validation on Model Binding Sites by Quantum Chemistry.

88. C. Roux, N. Gresh, L. E. Perera, J. P. Piquemal, and L. Salmon, *J. Comput. Chem.*, **28**(5), 938–957 (2007). Binding of 5-Phospho-D-Arabinonohydroxamate and 5-Phospho-D-Arabinonate Inhibitors to Zinc Phosphomannose Isomerase from Candida Albicans Studied by Polarizable Molecular Mechanics and Quantum Mechanics.

89. L. M. M. Jenkins, T. Hara, S. R. Durell, R. Hayashi, J. K. Inman, J. P. Piquemal, N. Gresh, and E. Appella, *J. Am. Chem. Soc.*, **129**(36), 11067–11078 (2007). Specificity of Acyl Transfer from 2-Mercaptobenzamide Thioesters to the HIV-1 Nucleocapsid Protein.

90. J. Foret, B. de Courcy, N. Gresh, J. P. Piquemal, and L. Salmon, *Bioorg. Med. Chem.*, **17**(20), 7100–7107 (2009). Synthesis and Evaluation of Non-Hydrolyzable D-Mannose 6-Phosphate Surrogates Reveal 6-Deoxy-6-Dicarboxymethyl-D-Mannose as a New Strong Inhibitor of Phosphomannose Isomerases.

91. N. Gresh, N. Audiffren, J. P. Piquemal, J. de Ruyck, M. Ledecq, and J. Wouters, *J. Phys. Chem. B*, **114**(14), 4884–4895 (2010). Analysis of the Interactions Taking Place in the Recognition Site of a Bimetallic Mg(II)-Zn(II) Enzyme, Isopentenyl Diphosphate Isomerase. A Parallel Quantum-Chemical and Polarizable Molecular Mechanics Study.

92. C. Roux, F. Bhatt, J. Foret, B. de Courcy, N. Gresh, J. P. Piquemal, C. J. Jeffery, and L. Salmon, *Protein. Struct. Funct. Bioinf.*, **79**(1), 203–20 (2011). The Reaction Mechanism of Type I Phosphomannose Isomerases: New Information from Inhibition and Polarizable Molecular Mechanics Studies.

93. M. Ledecq, F. Lebon, F. Durant, C. Giessner-Prettre, A. Marquez, and N. Gresh, *J. Phys. Chem. B*, **107**(38), 10640–10652 (2003). Modeling of Copper(II) Complexes with the SIBFA Polarizable Molecular Mechanics Procedure. Application to a New Class of HIV-1 Protease Inhibitors.

94. A. Marjolin, C. Gourlaouen, C. Clavaguéra, P. Ren, J. Wu, N. Gresh, J.-P. Dognon, and J.-P. Piquemal, *Theor. Chem. Acc.: Theory Comput. Model. (Theor. Chim. Acta)*, **131**(4), 1–14 (2012). Toward Accurate Solvation Dynamics of Lanthanides and Actinides in Water Using Polarizable Force Fields: From Gas-Phase Energetics to Hydration Free Energies.

95. F. Rogalewicz, G. Ohanessian, and N. Gresh, *J. Comput. Chem.*, **21**(11), 963–973 (2000). Interaction of Neutral and Zwitterionic Glycine with Zn^{2+} in Gas Phase: ab Initio and SIBFA Molecular Mechanics Calculations.

96. N. Gresh, G. Tiraboschi, and D. R. Salahub, *Biopolymers*, **45**(6), 405–425 (1998). Conformational Properties of a Model Alanyl Dipeptide and of Alanine-Derived Oligopeptides: Effects of Solvation in Water and in Organic Solvents - A Combined SIBFA/Continuum Reaction Field, ab Initio Self-Consistent Field, and Density Functional Theory Investigation.

97. J. Graf, P. H. Nguyen, G. Stock, and H. Schwalbe, *J. Am. Chem. Soc.*, **129**(5), 1179–1189 (2007). Structure and Dynamics of the Homologous Series of Alanine Peptides: A Joint Molecular Dynamics/NMR Study.

98. G. A. Cisneros, T. A. Darden, N. Gresh, J. Pilme, P. Reinhardt, O. Parisel, and J. P. Piquemal, in *Multi-scale Quantum Models for Biocatalysis*, D. M. York and T. S. Lee (Eds.), Springer Science, 2009, Design of Next Generation Force Fields from ab Initio Computations: Beyond Point Charges Electrostatics.

99. J. P. Piquemal, G. A. Cisneros, P. Reinhardt, N. Gresh, and T. A. Darden, *J. Chem. Phys.*, **124**(10), 104101 (2006). Towards a Force Field Based on Density Fitting.

100. G. A. Cisneros, J. P. Piquemal, and T. A. Darden, *J. Chem. Phys.*, **125**(18), 184101 (2006). Generalization of the Gaussian Electrostatic Model: Extension to Arbitrary Angular Momentum, Distributed Multipoles, and Speedup with Reciprocal Space Methods.

101. S. Brdarski and G. Karlström, *J. Phys. Chem. A*, **102**(42), 8182–8192 (1998). Modeling of the Exchange Repulsion Energy.

102. M. A. Carignano, G. Karlström, and P. Linse, *J. Phys. Chem. B*, **101**(7), 1142–1147 (1997). Polarizable Ions in Polarizable Water: A Molecular Dynamics Study.

103. A. Holt and G. Karlström, *Int. J. Quant. Chem.*, **109**(6), 1255–1266 (2009). Improvement of the NEMO Potential by Inclusion of Intramolecular Polarization.

104. L. Gagliardi, R. Lindh, and G. Karlström, *J. Chem. Phys.*, **121**(10), 4494–4500 (2004). Local Properties of Quantum Chemical Systems: The LoProp Approach.

105. J. M. Hermida-Ramon, S. Brdarski, G. Karlström, and U. Berg, *J. Comput. Chem.*, **24**(2), 161–176 (2003). Inter- and Intramolecular Potential for the N-Formylglycinamide-Water System. A Comparison between Theoretical Modeling and Empirical Force Fields.

106. D. Hagberg, G. Karlström, B. O. Roos, and L. Gagliardi, *J. Am. Chem. Soc.*, **127**(41), 14250–14256 (2005). The Coordination of Uranyl in Water: A Combined Quantum Chemical and Molecular Simulation Study.

107. P. A. Kollman, I. Massova, C. Reyes, B. Kuhn, S. H. Huo, L. Chong, M. Lee, T. Lee, Y. Duan, W. Wang, O. Donini, P. Cieplak, J. Srinivasan, D. A. Case, and T. E. Cheatham, *Acc. Chem. Res.*, **33**(12), 889–897 (2000). Calculating Structures and Free Energies of Complex Molecules: Combining Molecular Mechanics and Continuum Models.

108. P. Drude, C. R. Mann, and R. A. Millikan, *The Theory of Optics*, Longmans, Green, and Co., New York, 1902, pp. 1–572.

109. E. Harder, V. M. Anisimov, T. W. Whitfield, A. D. MacKerell, and B. Roux, *J. Phys. Chem. B*, **112**(11), 3509–3521 (2008). Understanding the Dielectric Properties of Liquid Amides from a Polarizable Force Field.

110. C. M. Baker, P. E. Lopes, X. Zhu, B. Roux, and A. D. MacKerell, Jr., *J. Chem. Theory Comput.*, **6**(4), 1181–1198 (2010). Accurate Calculation of Hydration Free Energies Using Pair-Specific Lennard-Jones Parameters in the CHARMM Drude Polarizable Force Field.

111. J. M. Wang, R. M. Wolf, J. W. Caldwell, P. A. Kollman, and D. A. Case, *J. Comput. Chem.*, **25**(9), 1157–1174 (2004). Development and Testing of a General AMBER Force Field.

112. R. O. Dror, R. M. Dirks, J. P. Grossman, H. Xu, and D. E. Shaw, *Annu. Rev. Biophys.*, **41**(1), 429–452 (2012). Biomolecular Simulation: A Computational Microscope for Molecular Biology.

113. E. M. Myshakin, H. Jiang, and K. D. Jordan, *Abstr. Pap. Am. Chem. Soc.*, **234** (2007). Phys 549-Molecular Dynamics Simulations of Methane Hydrate Decomposition Using a Polarizable Force Field.

114. P. E. M. Lopes, G. Lamoureux, B. Roux, and A. D. MacKerell, *J. Phys. Chem. B*, **111**(11), 2873–2885 (2007). Polarizable Empirical Force Field for Aromatic Compounds Based on the Classical Drude Oscillator.

115. J. E. Davis and S. Patel, *J. Phys. Chem. B*, **113**(27), 9183–9196 (2009). Charge Equilibration Force Fields for Lipid Environments: Applications to Fully Hydrated DPPC Bilayers and DMPC-Embedded Gramicidin A.

116. R. Xiong, X. M. Cai, J. Wei, and P. Y. Ren, *J. Mol. Model.*, 3049–3060 (2012). Some Insights into the Binding Mechanism of Aurora B Kinase Gained by Molecular Dynamics Simulation.

117. T. S. Kaoud, H. Park, S. Mitra, C. Yan, C. C. Tseng, Y. Shi, J. Jose, J. M. Taliaferro, K. Lee, P. Ren, J. Hong, and K. N. Dalby, *ACS Chem. Biol.*, 1873–1883 (2012). Manipulating JNK Signaling with (-)-Zuonin A.

118. T. W. Whitfield, S. Varma, E. Harder, G. Lamoureux, S. B. Rempe, and B. Roux, *J. Chem. Theory Comput.*, **3**(6), 2068–2082 (2007). Theoretical Study of Aqueous Solvation of K(+) Comparing Ab Initio, Polarizable, and Fixed-Charge Models.

119. H. B. Yu, T. W. Whitfield, E. Harder, G. Lamoureux, I. Vorobyov, V. M. Anisimov, A. D. MacKerell, and B. Roux, *J. Chem. Theory Comput.*, **6**(3), 774–786 (2010). Simulating Monovalent and Divalent Ions in Aqueous Solution Using a Drude Polarizable Force Field.

120. Z. Y. Lu and Y. K. Zhang, *J. Chem. Theory Comput.*, **4**(8), 1237–1248 (2008). Interfacing Ab Initio Quantum Mechanical Method with Classical Drude Oscillator Polarizable Model for Molecular Dynamics Simulation of Chemical Reactions.

121. B. R. Brooks, R. E. Bruccoeri, B. D. Olafson, D. J. States, S. Swaminathan, and M. Karplus, *J. Comput. Chem.*, **4**(2), 187–217 (1983). CHARMM: A Program for Macromolecular Energy, Minimization, and Dynamics Calculations.

122. J. C. Phillips, R. Braun, W. Wang, J. Gumbart, E. Tajkhorshid, E. Villa, C. Chipot, R. D. Skeel, L. Kale, and K. Schulten, *J. Comput. Chem.*, **26**(16), 1781–802 (2005). Scalable Molecular Dynamics with NAMD.

123. W. Jiang, D. J. Hardy, J. C. Phillips, A. D. MacKerell, K. Schulten, and B. Roux, *J. Phys. Chem. Lett.*, **2**(2), 87–92 (2011). High-Performance Scalable Molecular Dynamics Simulations of a Polarizable Force Field Based on Classical Drude Oscillators in NAMD.

124. T. S. Kaoud, C. Yan, S. Mitra, C.-C. Tseng, J. Jose, J. M. Taliaferro, M. Tuohetahuntila, A. Devkota, R. Sammons, J. Park, H. Park, Y. Shi, J. Hong, P. Ren, and K. N. Dalby, *ACS*

Med. Chem. Lett., **3**(9), 721–725 (2012). From in Silico Discovery to Intracellular Activity: Targeting JNK-Protein Interactions with Small Molecules.

125. J. D. Chodera, W. C. Swope, F. Noe, J. H. Prinz, M. R. Shirts, and V. S. Pande, *J. Chem. Phys.*, **134**(24), 244107 (2011). Dynamical Reweighting: Improved Estimates of Dynamical Properties from Simulations at Multiple Temperatures.

126. M. R. Shirts, D. L. Mobley, J. D. Chodera, and V. S. Pande, *J. Phys. Chem. B*, **111**(45), 13052–63 (2007). Accurate and Efficient Corrections for Missing Dispersion Interactions in Molecular Simulations.

127. B. A. Bauer, T. R. Lucas, D. J. Meninger, and S. Patel, *Chem. Phys. Lett.*, **508**(4-6), 289–294 (2011). Water Permeation through DMPC Lipid Bilayers Using Polarizable Charge Equilibration Force Fields.

128. B. A. Bauer, G. L. Warren, and S. Patel, *J. Chem. Theory Comput.*, **5**(2), 359–373 (2009). Incorporating Phase-Dependent Polarizability in Nonadditive Electrostatic Models for Molecular Dynamics Simulations of the Aqueous Liquid-Vapor Interface.

129. W. Xie and J. Gao, *J. Chem. Theory Comput.*, **3**(6), 1890–1900 (2007). The Design of a Next Generation Force Field: The X-Pol Potential.

130. J. L. Gao, *J. Chem. Phys.*, **109**(6), 2346–2354 (1998). A Molecular-Orbital Derived Polarization Potential for Liquid Water.

131. S. J. Wierzchowski, D. A. Kofke, and J. L. Gao, *J. Chem. Phys.*, **119**(14), 7365–7371 (2003). Hydrogen Fluoride Phase Behavior and Molecular Structure: A QM/MM Potential Model Approach.

132. W. S. Xie, L. C. Song, D. G. Truhlar, and J. L. Gao, *J. Phys. Chem. B*, **112**(45), 14124–14131 (2008). Incorporation of a QM/MM Buffer Zone in the Variational Double Self-Consistent Field Method.

133. W. S. Xie, L. C. Song, D. G. Truhlar, and J. L. Gao, *J. Chem. Phys.*, **128**(23), 234108 (2008). The Variational Explicit Polarization Potential and Analytical First Derivative of Energy: Towards a Next Generation Force Field.

134. Y. J. Wang, C. P. Sosa, A. Cembran, D. G. Truhlar, and J. L. Gao, *J. Phys. Chem. B*, **116**(23), 6781–6788 (2012). Multilevel X-Pol: A Fragment-Based Method with Mixed Quantum Mechanical Representations of Different Fragments.

135. G. A. Kaminski, R. A. Friesner, and R. H. Zhou, *J. Comput. Chem.*, **24**(3), 267–276 (2003). A Computationally Inexpensive Modification of the Point Dipole Electrostatic Polarization Model for Molecular Simulations.

136. G. A. Kaminski, H. A. Stern, B. J. Berne, and R. A. Friesner, *J. Phys. Chem. A*, **108**(4), 621–627 (2004). Development of an Accurate and Robust Polarizable Molecular Mechanics Force Field from Ab Initio Quantum Chemistry.

137. B. C. Kim, T. Young, E. Harder, R. A. Friesner, and B. J. Berne, *J. Phys. Chem. B*, **109**(34), 16529–16538 (2005). Structure and Dynamics of the Solvation of Bovine Pancreatic Trypsin Inhibitor in Explicit Water: A Comparative Study of the Effects of Solvent and Protein Polarizability.

138. G. A. Kaminski, S. Y. Ponomarev, and A. B. Liu, *J. Chem. Theory Comput.*, **5**(11), 2935–2943 (2009). Polarizable Simulations with Second Order Interaction Model - Force Field and Software for Fast Polarizable Calculations: Parameters for Small Model Systems and Free Energy Calculations.

139. S. Y. Ponomarev and G. A. Kaminski, *J. Chem. Theory Comput.*, **7**(5), 1415–1427 (2011). Polarizable Simulations with Second Order Interaction Model (POSSIM) Force Field: Developing Parameters for Alanine Peptides and Protein Backbone.

140. S. Y. Ponomarev and G. Kaminski, *Abstr. Pap. Am. Chem. Soc.*, **241** (2011). Polarizable Force Field for Protein Simulations POSSIM: Alanine Dipeptide and Tetrapeptide Parameters, and Stability of the Alanine 13 Alpha-Helix in Water.

141. O. Borodin, R. Douglas, G. A. Smith, F. Trouw, and S. Petrucci, *J. Phys. Chem. B*, **107**(28), 6813–6823 (2003). MD Simulations and Experimental Study of Structure, Dynamics, and Thermodynamics of Poly(Ethylene Oxide) and Its Oligomers.

142. O. Borodin and G. D. Smith, *J. Phys. Chem. B*, **107**(28), 6801–6812 (2003). Development of Quantum Chemistry-Based Force Fields for Poly(Ethylene Oxide) with Many-Body Polarization Interactions.

143. O. Borodin, G. D. Smith, and R. Douglas, *J. Phys. Chem. B*, **107**(28), 6824–6837 (2003). Force Field Development and MD Simulations of Poly(Ethylene Oxide)/LIBF$_4$ Polymer Electrolytes.

144. P. Paricaud, M. Predota, A. A. Chialvo, and P. T. Cummings, *J. Chem. Phys.*, **122**(24), 244511 (2005). From Dimer to Condensed Phases at Extreme Conditions: Accurate Predictions of the Properties of Water by a Gaussian Charge Polarizable Model.

145. J. L. Rivera, F. W. Starr, P. Paricaud, and P. T. Cummings, *J. Chem. Phys.*, **125**(9), 094712 (2006). Polarizable Contributions to the Surface Tension of Liquid Water.

146. Z. Tao and P. T. Cummings, *Mol. Simul.*, **33**(15), 1255–1260 (2007). Molecular Dynamics Simulation of Inorganic Ions in POE Aqueous Solution.

147. O. Borodin, *J. Phys. Chem. B*, **113**(33), 11463–11478 (2009). Polarizable Force Field Development and Molecular Dynamics Simulations of Ionic Liquids.

148. F. H. Stillinger and C. W. David, *J. Chem. Phys.*, **69**(4), 1473–1484 (1978). Polarization Model for Water and Its Ionic Dissociation Products.

149. B. Guillot, *J. Mol. Liq.*, **101**(1-3), 219–260 (2002). A Reappraisal of What We Have Learnt During Three Decades of Computer Simulations of Water.

150. P. Ahlström, A. Wallqvist, S. Engström, and B. Jonsson, *Mol. Phys.*, **68**(3), 563–581 (1989). A Molecular-Dynamics Study of Polarizable Water.

151. M. Sprik and M. L. Klein, *J. Chem. Phys.*, **89**(12), 7556–7560 (1988). A Polarizable Model for Water Using Distributed Charge Sites.

152. P. Cieplak, P. Kollman, and T. Lybrand, *J. Chem. Phys.*, **92**(11), 6755–6760 (1990). A New Water Potential Including Polarization - Application to Gas-Phase, Liquid, and Crystal Properties of Water.

153. U. Niesar, G. Corongiu, E. Clementi, G. R. Kneller, and D. K. Bhattacharya, *J. Phys. Chem.*, **94**(20), 7949–7956 (1990). Molecular-Dynamics Simulations of Liquid Water Using the NCC Ab Initio Potential.

154. L. X. Dang, *J. Chem. Phys.*, **96**(9), 6970–6977 (1992). Development of Nonadditive Intermolecular Potentials Using Molecular-Dynamics - Solvation of Li$^+$ and F$^-$ Ions in Polarizable Water.

155. L. X. Dang and T. M. Chang, *J. Chem. Phys.*, **106**(19), 8149–8159 (1997). Molecular Dynamics Study of Water Clusters, Liquid, and Liquid-Vapor Interface of Water with Many-Body Potentials.

156. C. J. Burnham and S. S. Xantheas, *J. Chem. Phys.*, **116**(12), 5115–5124 (2002). Development of Transferable Interaction Models for Water. III. A Flexible, All-Atom Polarizable Potential (TTM2-F) Based on Geometry Dependent Charges Derived from an Ab Initio Monomer Dipole Moment Surface.

157. C. J. Burnham and S. S. Xantheas, *J. Chem. Phys.*, **116**(4), 1479–1492 (2002). Development of Transferable Interaction Models for Water. I. Prominent Features of the Water Dimer Potential Energy Surface.

158. C. J. Burnham and S. S. Xantheas, *J. Chem. Phys.*, **116**(4), 1500–1510 (2002). Development of Transferable Interaction Models for Water. III. Reparametrization of an All-Atom Polarizable Rigid Model (Ttm2-R) from First Principles.

159. S. S. Xantheas, C. J. Burnham, and R. J. Harrison, *J. Chem. Phys.*, **116**(4), 1493–1499 (2002). Development of Transferable Interaction Models for Water. II. Accurate Energetics of the First Few Water Clusters from First Principles.

160. H. B. Yu, T. Hansson, and W. F. van Gunsteren, *J. Chem. Phys.*, **118**(1), 221–234 (2003). Development of a Simple, Self-Consistent Polarizable Model for Liquid Water.

161. H. B. Yu and W. F. van Gunsteren, *J. Chem. Phys.*, **121**(19), 9549–9564 (2004). Charge-on-Spring Polarizable Water Models Revisited: From Water Clusters to Liquid Water to Ice.

162. S. W. Rick, S. J. Stuart, and B. J. Berne, *J. Chem. Phys.*, **101**(7), 6141–6156 (1994). Dynamical Fluctuating Charge Force-Fields - Application to Liquid Water.

163. H. A. Stern, F. Rittner, B. J. Berne, and R. A. Friesner, *J. Chem. Phys.*, **115**(5), 2237–2251 (2001). Combined Fluctuating Charge and Polarizable Dipole Models: Application to a Five-Site Water Potential Function.

164. A. G. Donchev, N. G. Galkin, A. A. Illarionov, O. V. Khoruzhii, M. A. Olevanov, V. D. Ozrin, M. V. Subbotin, and V. I. Tarasov, *Proc. Natl. Acad. Sci. U. S. A.*, **103**(23), 8613–8617 (2006). Water Properties from First Principles: Simulations by a General-Purpose Quantum Mechanical Polarizable Force Field.

165. J. M. Wang, P. Cieplak, and P. A. Kollman, *J. Comput. Chem.*, **21**(12), 1049–1074 (2000). How Well Does a Restrained Electrostatic Potential (RESP) Model Perform in Calculating Conformational Energies of Organic and Biological Molecules?

166. J. Jeon, A. E. Lefohn, and G. A. Voth, *J. Chem. Phys.*, **118**(16), 7504–7518 (2003). An Improved Polarflex Water Model.

167. M. K. Gilson and H. X. Zhou, *Annu. Rev. Biophys. Biomol. Struct.*, **36**, 21–42 (2007). Calculation of Protein-Ligand Binding Affinities.

168. A. J. Lee and S. W. Rick, *J. Chem. Phys.*, **134**(18), 184507 (2011). The Effects of Charge Transfer on the Properties of Liquid Water.

169. E. Muchova, I. Gladich, S. Picaud, P. N. M. Hoang, and M. Roeselova, *J. Phys. Chem. A*, **115**(23), 5973–5982 (2011). The Ice-Vapor Interface and the Melting Point of Ice I-H for the Polarizable POL3 Water Model.

170. B. A. Bauer and S. Patel, *J. Chem. Phys.*, **131**(8), 084709 (2009). Properties of Water Along the Liquid-Vapor Coexistence Curve Via Molecular Dynamics Simulations Using the Polarizable TIP4P-QDP-LJ Water Model.

171. T. P. Lybrand and P. A. Kollman, *J. Chem. Phys.*, **83**(6), 2923–2933 (1985). Water-Water and Water-Ion Potential Functions Including Terms for Many-Body Effects.

172. S. J. Stuart and B. J. Berne, *J. Phys. Chem.*, **100**(29), 11934–11943 (1996). Effects of Polarizability on the Hydration of the Chloride Ion.

173. M. Sprik, M. L. Klein, and K. Watanabe, *J. Phys. Chem.*, **94**(16), 6483–6488 (1990). Solvent Polarization and Hydration of the Chlorine Anion.

174. L. X. Dang, J. E. Rice, J. Caldwell, and P. A. Kollman, *J. Am. Chem. Soc.*, **113**(7), 2481–2486 (1991). Ion Solvation in Polarizable Water - Molecular-Dynamics Simulations.

175. B. Roux, B. Prodhom, and M. Karplus, *Biophys. J.*, **68**(3), 876–892 (1995). Ion-Transport in the Gramicidin Channel - Molecular-Dynamics Study of Single and Double Occupancy.

176. G. L. Warren and S. Patel, *J. Chem. Phys.*, **127**(6), 64509–64528 (2007). Hydration Free Energies of Monovalent Ions in Transferable Intermolecular Potential Four Point Fluctuating Charge Water: An Assessment of Simulation Methodology and Force Field Performance and Transferability.

177. I. S. Joung and T. E. Cheatham, *J. Phys. Chem. B*, **112**(30), 9020–9041 (2008). Determination of Alkali and Halide Monovalent Ion Parameters for Use in Explicitly Solvated Biomolecular Simulations.

178. G. L. Warren and S. Patel, *J. Phys. Chem. B*, **112**(37), 11679–93 (2008). Electrostatic Properties of Aqueous Salt Solution Interfaces: A Comparison of Polarizable and Nonpolarizable Ion Models.

179. M. Masia, M. Probst, and R. Rey, *J. Chem. Phys.*, **121**(15), 7362–7378 (2004). On the Performance of Molecular Polarization Methods. I. Water and Carbon Tetrachloride Close to a Point Charge.

180. M. Masia, M. Probst, and R. Rey, *J. Chem. Phys.*, **123**(16), 164505–164518 (2005). On the Performance of Molecular Polarization Methods. II. Water and Carbon Tetrachloride Close to a Cation.

181. X. Li and Z. Z. Yang, *J. Chem. Phys.*, **122**(8), 84514 (2005). Hydration of Li^+-Ion in Atom-Bond Electronegativity Equalization Method-7P Water: A Molecular Dynamics Simulation Study.

182. Z. Z. Yang and X. Li, *J. Chem. Phys.*, **123**(9), 94507 (2005). Molecular-Dynamics Simulations of Alkaline-Earth Metal Cations in Water by Atom-Bond Electronegativity Equalization Method Fused into Molecular Mechanics.

183. P. Jungwirth and D. J. Tobias, *J. Phys. Chem. A*, **106**(2), 379–383 (2002). Chloride Anion on Aqueous Clusters, at the Air-Water Interface, and in Liquid Water: Solvent Effects on Cl⁻ Polarizability.

184. G. Archontis, E. Leontidis, and G. Andreou, *J. Phys. Chem. B*, **109**(38), 17957–17966 (2005). Attraction of Iodide Ions by the Free Water Surface, Revealed by Simulations with a Polarizable Force Field Based on Drude Oscillators.

185. G. Archontis and E. Leontidis, *Chem. Phys. Lett.*, **420**(1-3), 199–203 (2006). Dissecting the Stabilization of Iodide at the Air-Water Interface into Components: A Free Energy Analysis.

186. M. A. Brown, R. D'Auria, I. F. W. Kuo, M. J. Krisch, D. E. Starr, H. Bluhm, D. J. Tobias, and J. C. Hemminger, *Phys. Chem. Chem. Phys.*, **10**(32), 4778–4784 (2008). Ion Spatial Distributions at the Liquid-Vapor Interface of Aqueous Potassium Fluoride Solutions.

187. X. W. Wang, H. Watanabe, M. Fuji, and M. Takahashi, *Chem. Phys. Lett.*, **458**(1-3), 235–238 (2008). Molecular Dynamics Simulation of NaCl at the Air/Water Interface with Shell Model.

188. A. Grossfield, *J. Chem. Phys.*, **122**(2), 024506 (2005). Dependence of Ion Hydration on the Sign of the Ion's Charge.

189. P. S. Bagus, K. Hermann, and J. C. W. Bauschlicher, *J. Chem. Phys.*, **80**(9), 4378–4386 (1984). A New Analysis of Charge Transfer and Polarization for Ligand--Metal Bonding: Model Studies of Al_4Co and Al_4NH_3.

190. J.-P. Piquemal, L. Perera, G. A. Cisneros, P. Ren, L. G. Pedersen, and T. A. Darden, *J. Chem. Phys.*, **125**(5), 054511–7 (2006). Towards Accurate Solvation Dynamics of Divalent Cations in Water Using the Polarizable AMOEBA Force Field: From Energetics to Structure.

191. M. Devereux, M. C. van Severen, O. Parisel, J. P. Piquemal, and N. Gresh, *J. Chem. Theory Comput.*, **7**(1), 138–147 (2011). Role of Cation Polarization in Holo- and Hemi-Directed $[Pb(H2O)(N)]^{2+}$ Complexes and Development of a Pb^{2+} Polarizable Force Field.

192. T. P. Straatsma and J. A. McCammon, *Chem. Phys. Lett.*, **177**(4-5), 433–440 (1991). Free Energy Evaluation from Molecular Dynamics Simulations Using Force Fields Including Electronics Polarization.

193. D. L. Mobley, C. I. Bayly, M. D. Cooper, M. R. Shirts, and K. A. Dill, *J. Chem. Theory Comput.*, **5**, 350–359 (2009). Small Molecules Hydration Free Energies in Explicit Solvent: An Extensive Test of Fixed-Charge Atomistic Simulations.

194. D. L. Mobley, E. Dumont, J. D. Chodera, and K. A. Dill, *J. Phys. Chem. B*, **111**(9), 2242–2254 (2007). Comparison of Charge Models for Fixed-Charge Force Fields: Small-Molecule Hydration Free Energies in Explicit Solvent.

195. R. C. Rizzo, T. Aynechi, D. A. Case, and I. D. Kuntz, *J. Chem. Theory Comput.*, **2**(1), 128–139 (2006). Estimation of Absolute Free Energies of Hydration Using Continuum Methods: Accuracy of Partial, Charge Models and Optimization of Nonpolar Contributions.

196. M. R. Shirts and V. S. Pande, *J. Chem. Phys.*, **122**(13), 134508 (2005). Solvation Free Energies of Amino Acid Side Chain Analogs for Common Molecular Mechanics Water Models.

197. X. Q. Sun and L. X. Dang, *J. Chem. Phys.*, **130**(21), 124709–124713 (2009). Computational Studies of Aqueous Interfaces of RbBr Salt Solutions.

198. V. M. Anisimov, I. V. Vorobyov, B. Roux, and A. D. MacKerell, *J. Chem. Theory Comput.*, **3**(6), 1927–1946 (2007). Polarizable Empirical Force Field for the Primary and Secondary Alcohol Series Based on the Classical Drude Model.

199. A. Hesselmann and G. Jansen, *Chem. Phys. Lett.*, **357**(5-6), 464–470 (2002). First-Order Intermolecular Interaction Energies from Kohn-Sham Orbitals.

200. J. C. Wu, G. Chattre, and P. Ren, *Theor. Chem. Acc.*, **131**(3), 1138–1148 (2012). Automation of AMOEBA Polarizable Force Field Parameterization for Small Molecules.

201. T. Ogawa, N. Kurita, H. Sekino, O. Kitao, and S. Tanaka, *Chem. Phys. Lett.*, **397**(4-6), 382–387 (2004). Consistent Charge Equilibration (CQEQ) Method: Application to Amino Acids and Crambin Protein.

202. E. Harder, B. C. Kim, R. A. Friesner, and B. J. Berne, *J. Chem. Theory Comput.*, **1**(1), 169–180 (2005). Efficient Simulation Method for Polarizable Protein Force Fields: Application to the Simulation of BPTI in Liquid.

203. P. Llinas, M. Masella, T. Stigbrand, A. Menez, E. A. Stura, and M. H. Le Du, *Protein Sci.*, **15**(7), 1691–1700 (2006). Structural Studies of Human Alkaline Phosphatase in Complex with Strontium: Implication for Its Secondary Effect in Bones.

204. Z. X. Wang, W. Zhang, C. Wu, H. X. Lei, P. Cieplak, and Y. Duan, *J. Comput. Chem.*, **27**(6), 781–790 (2006). Erratum - Strike a Balance: Optimization of Backbone Torsion Parameters of AMBER Polarizable Force Field for Simulations of Proteins and Peptides.

205. Z. Z. Yang and Q. Zhang, *J. Comput. Chem.*, **27**(1), 1–10 (2006). Study of Peptide Conformation in Terms of the ABEEM/MM Method.

206. Q. M. Guan, B. Q. Cui, D. X. Zhao, L. D. Gong, and Z. Z. Yang, *Chin. Sci. Bull.*, **53**(8), 1171–1174 (2008). Molecular Dynamics Study on BPTI Aqueous Solution by ABEEM/MM Fluctuating Charge Model.

207. B. Q. Cui, Q. M. Guan, L. D. Gong, D. X. Zhao, and Z. Z. Yang, *Chem. J. Chin. Universities-Chinese*, **29**(3), 585–590 (2008). Studies on the Heme Prosthetic Group's Geometry by ABEEM/MM Method.

208. B. de Courcy, J. P. Piquemal, C. Garbay, and N. Gresh, *J. Am. Chem. Soc.*, **132**(10), 3312–3320 (2010). Polarizable Water Molecules in Ligand-Macromolecule Recognition Impact on the Relative Affinities of Competing Pyrrolopyrimidine Inhibitors for FAK Kinase.

209. B. de Courcy, L. G. Pedersen, O. Parisel, N. Gresh, B. Silvi, J. Pilme, and J. P. Piquemal, *J. Chem. Theory Comput.*, **6**(4), 1048–1063 (2010). Understanding Selectivity of Hard and Soft Metal Cations within Biological Systems Using the Subvalence Concept. I. Application to Blood Coagulation: Direct Cation-Protein Electronic Effects Vs Indirect Interactions through Water Networks.

210. B. de Courcy, J. P. Dognon, C. Clavaguera, N. Gresh, and J. P. Piquemal, *Int. J. Quant. Chem.*, **111**(6), 1213–1221 (2011). Interactions within the Alcohol Dehydrogenase Zn(II)-Metalloenzyme Active Site: Interplay between Subvalence, Electron Correlation/Dispersion, and Charge Transfer/Induction Effects.

211. A. De La Lande, D. R. Salahub, J. Maddaluno, A. Scemama, J. Pilme, O. Parisel, H. Gerard, M. Caffarel, and J. P. Piquemal, *J. Comput. Chem.*, **32**(6), 1178–1182 (2011). Rapid Communication Spin-Driven Activation of Dioxygen in Various Metalloenzymes and Their Inspired Models.

212. A. G. Donchev, N. G. Galkin, A. A. Illarionov, O. V. Khoruzhii, M. A. Olevanov, V. D. Ozrin, L. B. Pereyaslavets, and V. I. Tarasov, *J. Comput. Chem.*, **29**(8), 1242–1251 (2008). Assessment of Performance of the General Purpose Polarizable Force Field QMPFF3 in Condensed Phase.

213. E. Harder, A. D. MacKerell, and B. Roux, *J. Am. Chem. Soc.*, **131**(8), 2760–2761 (2009). Many-Body Polarization Effects and the Membrane Dipole Potential.

214. J. E. Davis, O. Raharnan, and S. Patel, *Biophys. J.*, **96**(2), 385–402 (2009). Molecular Dynamics Simulations of a DMPC Bilayer Using Nonadditive Interaction Models.

215. B. Roux and T. Simonson, *Biophys. Chem.*, **78**(1-2), 1–20 (1999). Implicit Solvent Models.

216. N. A. Baker, *Curr. Opin. Struct. Biol.*, **15**(2), 137–143 (2005). Improving Implicit Solvent Simulations: A Poisson-Centric View.

217. N. A. Baker, in Reviews in Computational Chemistry, K. B. Lipkowitz, R. Larter and T. Cundari (Eds.), Wiley-VCH Inc., New York, 2005, Vol. **21**, pp 349–379. Biomolecular Applications of Poisson-Boltzmann Methods.

218. N. A. Baker, *Method. Enzymol.*, **383**, 94–118 (2004). Poisson-Boltzmann Methods for Biomolecular Electrostatics.

219. J. Tomasi, B. Mennucci, and R. Cammi, *Chem. Rev.*, **105**(8), 2999–3093 (2005). Quantum Mechanical Continuum Solvation Models.

220. J. Tomasi, *Theor. Chem. Acc.*, **112**(4), 184–203 (2004). Thirty Years of Continuum Solvation Chemistry: A Review, and Prospects for the near Future.

221. A. Klamt and G. Schuurmann, *J. Chem. Soc.: Perkin Trans.*, **2**(5), 799–805 (1993). COSMO - A New Approach to Dielectric Screening in Solvents with Explicit Expressions for the Screening Energy and Its Gradient.

222. C. J. Cramer and D. G. Truhlar, *Chem. Rev.*, **99**(8), 2161–2200 (1999). Implicit Solvation Models: Equilibria, Structure, Spectra, and Dynamics.

223. J. R. Maple, Y. X. Cao, W. G. Damm, T. A. Halgren, G. A. Kaminski, L. Y. Zhang, and R. A. Friesner, *J. Chem. Theory Comput.*, **1**(4), 694–715 (2005). A Polarizable Force Field and Continuum Solvation Methodology for Modeling of Protein-Ligand Interactions.

224. J.-F. Truchon, A. Nicholls, R. I. Iftimie, B. Roux, and C. I. Bayly, *J. Chem. Theory Comput.*, **4**(9), 1480–1493 (2008). Accurate Molecular Polarizabilities Based on Continuum Electrostatics.

225. J.-F. Truchon, A. Nicholls, B. Roux, R. I. Iftimie, and C. I. Bayly, *J. Chem. Theory Comput.*, **5**(7), 1785–1802 (2009). Integrated Continuum Dielectric Approaches to Treat Molecular Polarizability and the Condensed Phase: Refractive Index and Implicit Solvation.

226. T. Yang, J. C. Wu, C. Yan, Y. Wang, R. Luo, M. B. Gonzales, K. N. Dalby, and P. Ren, *Protein. Struct. Funct. Bioinf.*, **79**(6), 1940–1951 (2011). Virtual Screening Using Molecular Simulations.

227. D.-X. Zhao, L. Yu, L.-D. Gong, C. Liu, and Z.-Z. Yang, *J. Chem. Phys.*, **134**(19), 194115 (2011). Calculating Solvation Energies by Means of a Fluctuating Charge Model Combined with Continuum Solvent Model.

228. A. McCoy, *Acta Crystallogr. Sect. D*, **60**(12 Part 1), 2169–2183 (2004). Liking Likelihood.

229. W. I. Weis, A. T. Brunger, J. J. Skehel, and D. C. Wiley, *J. Mol. Biol.*, **212**(4), 737–761 (1990). Refinement of the Influenza-Virus Hemagglutinin by Simulated Annealing.

230. L. Moulinier, D. A. Case, and T. Simonson, *Acta Crystallogr. Sect. D-Biol. Crystallogr.*, **59**, 2094–2103 (2003). Reintroducing Electrostatics into Protein X-Ray Structure Refinement: Bulk Solvent Treated as a Dielectric Continuum.

231. C. Sagui and T. A. Darden, *Annu. Rev. Biophys. Biomol. Struct.*, **28**, 155–179 (1999). Molecular Dynamics Simulations of Biomolecules: Long-Range Electrostatic Effects.

232. P. P. Ewald, *Ann. Phys.*, **369**(3), 253–287 (1921). Die Berechnung Optischer Und Elektrostatischer Gitterpotentiale.

233. W. Smith, *CCP5 Inf. Quart.*, **4**(13) (1982). Point Multipoles in the Ewald Summation (Revisited).

234. P. Ren and J. W. Ponder, *J. Phys. Chem. B*, **107**(24), 5933–5947 (2003). Polarizable Atomic Multipole Water Model for Molecular Mechanics Simulation.

235. T. Darden, D. York, and L. Pedersen, *J. Chem. Phys.*, **98**(12), 10089–10092 (1993). Particle-Mesh Ewald - An N Log(N) Method for Ewald Sums in Large Systems.

236. U. Essmann, L. Perera, M. L. Berkowitz, T. Darden, H. Lee, and L. G. Pedersen, *J. Chem. Phys.*, **103**(19), 8577–8593 (1995). A Smooth Particle-Mesh Ewald Method.

237. C. Sagui, L. G. Pedersen, and T. A. Darden, *J. Chem. Phys.*, **120**(1), 73–87 (2004). Towards an Accurate Representation of Electrostatics in Classical Force Fields: Efficient Implementation of Multipolar Interactions in Biomolecular Simulations.

238. T. D. Fenn, M. J. Schnieders, A. T. Brunger, and V. S. Pande, *Biophys. J.*, **98**(12), 2984–2992 (2010). Polarizable Atomic Multipole X-Ray Refinement: Hydration Geometry and Application to Macromolecules.

239. T. D. Fenn, M. J. Schnieders, M. Mustyakimov, C. Wu, P. Langan, V. S. Pande, and A. T. Brunger, *Structure*, **19**(4), 523–533 (2011). Reintroducing Electrostatics into Macromolecular Crystallographic Refinement: Application to Neutron Crystallography and DNA Hydration.

240. M. J. Schnieders, T. S. Kaoud, C. Yan, K. N. Dalby, and P. Ren, *Curr. Pharm. Des.*, 18(9), 1173–1185 (2012). Computational Insights for the Discovery of Non-ATP Competitive Inhibitors of MAP Kinases.

241. V. B. Chen, W. B. Arendall, J. J. Headd, D. A. Keedy, R. M. Immormino, G. J. Kapral, L. W. Murray, J. S. Richardson, and D. C. Richardson, *Acta Crystallogr. Sect. D*, 66, 12–21 (2009). MolProbity: All-Atom Structure Validation for Macromolecular Crystallography.

242. J. Maddox, *Nature*, 335(6187), 201 (1998). Crystals from First Principles.

243. G. M. Day, T. G. Cooper, A. J. Cruz-Cabeza, K. E. Hejczyk, H. L. Ammon, S. X. M. Boerrigter, J. S. Tan, R. G. Della Valle, E. Venuti, J. Jose, S. R. Gadre, G. R. Desiraju, T. S. Thakur, B. P. van Eijck, J. C. Facelli, V. E. Bazterra, M. B. Ferraro, D. W. M. Hofmann, M. A. Neumann, F. J. J. Leusen, J. Kendrick, S. L. Price, A. J. Misquitta, P. G. Karamertzanis, G. W. A. Welch, H. A. Scheraga, Y. A. Arnautova, M. U. Schmidt, J. van de Streek, A. K. Wolf, and B. Schweizer, *Acta Crystallogr. B*, 65, 107–125 (2009). Significant Progress in Predicting the Crystal Structures of Small Organic Molecules - A Report on the Fourth Blind Test.

244. D. A. Bardwell, C. S. Adjiman, Y. A. Arnautova, E. Bartashevich, S. X. M. Boerrigter, D. E. Braun, A. J. Cruz-Cabeza, G. M. Day, R. G. Della Valle, G. R. Desiraju, B. P. van Eijck, J. C. Facelli, M. B. Ferraro, D. Grillo, M. Habgood, D. W. M. Hofmann, F. Hofmann, K. V. J. Jose, P. G. Karamertzanis, A. V. Kazantsev, J. Kendrick, L. N. Kuleshova, F. J. J. Leusen, A. V. Maleev, A. J. Misquitta, S. Mohamed, R. J. Needs, M. A. Neumann, D. Nikylov, A. M. Orendt, R. Pal, C. C. Pantelides, C. J. Pickard, L. S. Price, S. L. Price, H. A. Scheraga, J. van de Streek, T. S. Thakur, S. Tiwari, E. Venuti, and I. K. Zhitkov, *Acta Crystallogr. B*, 67(6), 535–551 (2011). Towards Crystal Structure Prediction of Complex Organic Compounds - A Report on the Fifth Blind Test.

245. W. Cabri, P. Ghetti, G. Pozzi, and M. Alpegiani, *Org. Process Res. Dev.*, 11(1), 64–72 (2006). Polymorphisms and Patent, Market, and Legal Battles: Cefdinir Case Study.

246. S. L. Price and L. S. Price, in *Solid State Characterization of Pharmaceuticals*, John Wiley & Sons, Ltd, 2011, pp. 427–450, Computational Polymorph Prediction.

247. S. L. Price, *Phys. Chem. Chem. Phys.*, 10(15), 1996–2009 (2008). From Crystal Structure Prediction to Polymorph Prediction: Interpreting the Crystal Energy Landscape.

248. P. Ren, M. Marucho, J. Zhang, and N. A. Baker, *Q. Rev. Biophys.*, 45(4), 427–491 (2011). Biomolecular Electrostatics and Solvation: A Computational Perspective.

249. G. W. A. Welch, P. G. Karamertzanis, A. J. Misquitta, A. J. Stone, and S. L. Price, *J. Chem. Theory Comput.*, 4(3), 522–532 (2008). Is the Induction Energy Important for Modeling Organic Crystals?

250. B. Civalleri, C. M. Zicovich-Wilson, L. Valenzano, and P. Ugliengo, *CrystEngComm*, 10(4), 405–410 (2008). B3LYP Augmented with an Empirical Dispersion Term (B3LYP-D*) as Applied to Molecular Crystals.

251. S. L. Price, *CrystEngComm*, 6(61), 344–353 (2004). Quantifying Intermolecular Interactions and Their Use in Computational Crystal Structure Prediction.

252. A. Gavezzotti, *J. Phys. Chem. B*, 106(16), 4145–4154 (2002). Calculation of Intermolecular Interaction Energies by Direct Numerical Integration over Electron Densities. I. Electrostatic and Polarization Energies in Molecular Crystals.

253. A. Gavezzotti, *J. Phys. Chem. B*, 107(10), 2344–2353 (2003). Calculation of Intermolecular Interaction Energies by Direct Numerical Integration over Electron Densities. 2. An Improved Polarization Model and the Evaluation of Dispersion and Repulsion Energies.

254. W. T. M. Mooij, B. P. van Eijck, and J. Kroon, *J. Phys. Chem. A*, 103(48), 9883–9890 (1999). Transferable Ab Initio Intermolecular Potentials. 2. Validation and Application to Crystal Structure Prediction.

255. P. Ren and J. W. Ponder, *J. Comput. Chem.*, 23(16), 1497–1506 (2002). Consistent Treatment of Inter- and Intramolecular Polarization in Molecular Mechanics Calculations.

256. A. J. Misquitta and A. J. Stone, *J. Chem. Theory Comput.*, **4**(1), 7–18 (2007). Accurate Induction Energies for Small Organic Molecules. 1. Theory.

257. A. J. Misquitta, A. J. Stone, and S. L. Price, *J. Chem. Theory Comput.*, **4**(1), 19–32 (2007). Accurate Induction Energies for Small Organic Molecules. 2. Development and Testing of Distributed Polarizability Models against SAPT(DFT) Energies.

258. C. Ouvrard and S. L. Price, *Cryst. Growth Des.*, **4**(6), 1119–1127 (2004). Toward Crystal Structure Prediction for Conformationally Flexible Molecules: The Headaches Illustrated by Aspirin.

CHAPTER 3

Modeling Protein Folding Pathways

Clare-Louise Towse and Valerie Daggett

Department of Bioengineering, University of Washington, Seattle, WA 98195-5013, United States

INTRODUCTION

Proteins are an important component of all living things. Some proteins exist as autonomously folded structures, while others become structured only when functional or interacting with binding partners.[1] Moreover, there is great diversity within the range of structures, and where similarities do exist, they are not necessarily conserved within their primary sequences.[2–4] For many proteins there are well-defined structure–function relationships; however, an increasing number of proteins have been found that are functional when lacking defined structure.[5] Nevertheless, the common thread is that with a multitude of structural possibilities there still remains some mystery in the accurate prediction of protein structure from the amino acid sequence. Part of solving this mystery requires developing an understanding of how proteins assume their three-dimensional folds from the one-dimensional information encoded in the amino acid sequence: in other words, the challenge of protein folding.[6–8]

To characterize the folding pathway of a protein, one must not focus on a single facet alone but instead identify the native, transition, and denatured states as well as any intermediate states that are present. Although experimental techniques have progressed over the years, with sensitivity improvements and shorter timescale detection, they still lack the ability to provide comprehensive atomistic detail for other than well-folded structures.[9] Molecular dynamics (MD) simulation, developed in 1964,[10] remains a more powerful method for providing atomic-level information at the desired temporal resolutions.

Reviews in Computational Chemistry, Volume 28, First Edition.
Edited by Abby L. Parrill and Kenny B. Lipkowitz.
© 2015 John Wiley & Sons, Inc. Published 2015 by John Wiley & Sons, Inc.

By harnessing Newton's laws of motion with empirically parameterized potential energy functions, MD simulations can compute the changes in conformational energy and, hence, predict atomic motions within a protein. When validated against the available experimental data, these predictive trajectories accrue a level of confidence to both explain empirical observations and guide further experimentation.

The use of MD simulations to gain insight into protein structure and motions started in the 1970s; continuing advancement in this field has more or less paralleled advancements in computer technology. Theoretical calculations of protein folding were actually attempted before a significant amount of development had gone into the simulation of the native state dynamics of a protein structure.[11] The first successful protein folding study of a globular protein was that of bovine pancreatic trypsin inhibitor (BPTI),[12] a small and popular target for both experimental and computational folding studies. Levitt and Warshel investigated the formation of the tertiary structure in BPTI as early as 1975 using a simplified model of the X-ray crystal structure *in vacuo* with the backbone dihedral angles set to that of an extended conformation to represent the unfolded state.[12] This model contained significantly more information than the noncomputational and somewhat restrictive methods that had previously been applied to recapitulate the packing of α-helices in myoglobin;[13] yet the model was still simple enough to be computationally tractable at that time. Although not strictly MD, the method used would become a part of many MD protocols used today: minimization. Further, they managed to "renature" the BPTI protein to within 7 Å of the simplified model built from the experimental tertiary structure.[12]

As computational power advanced and became more accessible, so did the detail incorporated into the protein models. By 1977, work had moved toward united atom representations of proteins and the use of molecular mechanics to characterize the native state dynamics of BPTI *in vacuo* with no water included except for four crystalline waters.[11] The empirical energy function used in this first protein MD simulation forms the basis of today's protein force fields.[14,15] The "extended" or "united" atom representations reduced the computational expense by incorporating hydrogen atoms with their bound heavy atoms and enabled a simulation of 8.8 ps of native state dynamics.[11] An early study of the effect of environment on protein dynamics confirmed how critical realistic description of the environment could be,[16] but water was not effectively incorporated until 1984 when van Gunsteren et al.[17] used a truncated octahedron of 1467 waters surrounding BPTI. The protein model had evolved to explicitly contain all hydrogen atoms except for those bound to carbon, and the water model used was a simple three-point charge model, the single-point charge (SPC) model.[18] This was followed by all-atom MD with all atoms included in the description of the protein model in 1988[19] as well as the use of a flexible three-point charge water model, the F3C model, that confirmed the importance of incorporating solvent into simulations.

Although previously applied to folding studies of short peptides and secondary structure fragments,[20-22] the first use of an explicitly solvated all-atom MD simulation to study protein unfolding was reported in 1992.[23] A number of excellent reviews cover the following period where MD simulations gained a stronghold in the investigation of protein unfolding as a consequence of being able to observe this process at atomic-level detail.[24,25] In 1998, what became one of the most notorious long-timescale simulations was performed, that of the refolding of the 36-residue villin headpiece subdomain, HP-36.[26] The structure was first denatured *in silico* using extremely high temperature and subjected to an extensive equilibration procedure. Then the simulation was continued under physiological conditions for 1 μs, over which time the villin headpiece subdomain was observed to refold to a stable near-native structure. Duan et al.[27] had also made a prior attempt of the same simulation for 200 ns; this was still an order of magnitude longer than any other reported simulations at that time.[28] Despite the 1 μs simulation of Duan et al.,[26] for the most part, the majority of the biological modeling community was restricted to more modest timescales and kept within the submicrosecond limit for some time.[29,30] Only more recently in the new millennium did we start to see an increase in groups modeling 1 μs and beyond.[31-37] Breaking the 1 μs limit was a consequence of improvements in the parallelization and scalability of MD programs, faster processors, efficient usage of multicore processors, and the use of dynamic load balancing.[38-42] A recent contributor to the area of long-timescale MD simulations is D. E. Shaw Research who built Anton, a computer housing processors designed specifically for performing MD calculations. Since the introduction of Anton, they have become a front runner in long-timescale simulations, jumping the millisecond hurdle in 2009.[43-47]

Although computational power has increased, the major limitation that still remains is the computational expense required for calculations to comfortably access the biological timescales that pertain to protein folding, which range from nanoseconds to seconds and beyond.[48] We are at the point where we can observe folding events using conventional MD methods, but often no more than a single folding event. To obtain meaningful statistics to describe protein folding requires significantly more sampling at these timescales than is possible at present.[49] For this reason, a large subfield of protein modeling is the development of advanced sampling methods that enable the time constraints to be circumvented.[50] There are other ceaseless areas of development to address additional known deficiencies: inaccuracies in some of the force fields and water models, determination and inclusion of empirical parameters more relevant to the study of protein folding and characterization of nonnative states. Depending on the available resources, time constraints and protein size, resorting to simplified models of proteins or using protocols that accelerate protein folding/unfolding *in silico*, are still relevant strategies to enable the biological timescales of interest to be accessible to observation via MD simulations.

Outline of this Chapter

We begin this chapter with an introduction to protein simulation methodology aimed at experimentalists and graduate students new to *in silico* investigations. Here, more emphasis is placed on the knowledge needed to select appropriate simulation protocols, leaving theoretical and mathematical depth for other texts to take care of.[51−54] We explain some of the more practical considerations of performing simulations of proteins, in particular, the additional considerations required when studying protein folding where nonnative environments are modeled. Following this introduction to the practicalities of protein folding simulations we move on to a more in-depth review of some of the established methods in the field as it stands today.

PROTEIN SIMULATION METHODOLOGY

Force Fields, Models and Solvation Approaches

One of the steepest learning curves for the beginner in this field is usually not the actual use of the modeling program to perform the calculation, but gaining a sufficient understanding of the methods to enable design of a practical simulation protocol. The freedom in selecting a simulation protocol is restricted by three main factors: the computing resources available, the time constraints and the desired level of accuracy. The accuracy itself depends upon the force field used, the protein and solvent models, and adequate sampling of the timescale accessible in order to quantify the desired biological process being examined. Although we strive to perform calculations with the highest accuracy possible, it is futile to use the most accurate force field and models if you are lacking the computational resources to sample biologically relevant timescales within a reasonable time.

Owing to the size and complexity of protein systems, the dynamics cannot be computed analytically but must be determined using numerical integration of Newton's equations of motion over a large number of timesteps.[55,56] To ensure errors and instabilities are minimized, integration is performed for timesteps where the time interval between them is finite and short, typically femtoseconds. Each timestep coincides with a fixed conformation of the protein with coordinates that reflect the change in atomic positions determined by Newton's equations of motions based on the potential energy and forces acting on the proceeding structure. The negative gradient of the potential energy, V, gives the force, F, used in the equations of motions to propagate the forces and predict motion (Eq. [1]). Hence, a simulation begins with known coordinates for a given protein structure from which a series of sequential conformations will be generated providing a trajectory from which various properties can be determined.

Force Fields and Protein Models

Force fields are composed of a potential energy function that describes all the forces acting upon the atoms within a protein model and the corresponding parameters, such as atomic radii, force constants, and bond lengths. The potential energy function used in MD simulations is a simple summation of the energetic contribution of all the bonding, V_b, and nonbonding, V_{nb}, interactions (Eq. [2]). Equations [3] and [4] depict the bonding and nonbonding portions of a common force field illustrating how the potential energy function, V, is cast as a summation of the bond lengths, b, bond angles, θ, torsion angles, φ, and nonbonded pair interactions based on the distance between i,j atoms pairs, r_{ij}. The commonly used force fields[39,57−60] have very similar potential energy functions based on the original consistent force field.[61] The force field described here is the Levitt et al.[62] force field which is used in ENCAD[63] and our in-house package *in lucem* molecular mechanics (*il*mm).[64]

$$F = -\nabla V \tag{1}$$

$$V = V_b + V_{nb} \tag{2}$$

$$V_b = \overset{\text{bonds}}{\sum_i} K_{b,i}(b_i - b_{0,i})^2 + \overset{\text{angles}}{\sum_i} K_{\theta,i}(\theta_i - \theta_{0,i})^2$$

$$+ \overset{\text{torsions}}{\sum_i} K_{\varphi,i}\{1 - \cos[n_i(\varphi_i - \varphi_{0,i})]\} \tag{3}$$

$$V_{nb} = \sum_{i,j}\left[\varepsilon_{ij}\left(\frac{r_{0,ij}}{r_{ij}}\right)^{12} - 2\varepsilon_{ij}\left(\frac{r_{0,ij}}{r_{ij}}\right)^{6}\right] + 332\sum_{i,j}\left(\frac{q_i q_j}{r_{ij}}\right) \tag{4}$$

The individual components of a force field reflect established physical principles, such as representing bond lengths and angles as harmonic oscillators, and using Lennard-Jones 12-6 and Coulomb-type functions for the van der Waals and electrostatic nonbonding interactions, respectively. However, the force fields are not independent of the protein models used. Each force field has associated with it atomic and molecular definitions and parameters, definitions that will differ between united atom and all-atom representations, for example. Similarly, water models are parameterized for use with certain force fields and are not necessarily interchangeable.[65−68]

Force fields are empirically based functions that use experimentally derived parameters for the bond lengths, angles, and force constants. Parameters are typically obtained from small organic molecules or, where experimental data are undetermined, can be calculated using *ab initio* methods.[69] Hence, force fields have their limitations, and there have been many force field comparisons performed to assess their accuracy and abilities.[68,70−73] Arguably, a stronger assessment of a force field is whether it is able to adequately reproduce experimental observables. It is important to note that many of the force field deficiencies reported are force field specific and, despite some claims, all the cited shortcomings are not necessarily applicable to all force fields. A recent examination of a subset of the commonly used force fields demonstrated their inequality in obtaining the same, or correct, native structure of a β-hairpin.[70]

The speed of an MD simulation can be increased through removing progressively more detail. However, doing so will produce results of increasingly lower accuracy. All-atom MD simulations with explicit water models included are by far the most complete representation of the system and, if executed effectively, the most accurate approach. United atom representations of a protein, where groups of atoms such as methyl groups are represented as a single entity, reduce the degrees of freedom and in consequence some of the computational cost. This is no longer commonly considered a justifiable approximation for protein simulations given the current speed of processors. Switching from an explicit description of the water to an implicit description, such as the EEF1 force field, can reduce the cost even further.[74] Further reduction in detail can be obtained through using coarse-grained (CG) methods such as lattice models.[75,76] Again, such approximations come with a reduction in accuracy, and the relevancy of the results may be diminished. However, such steps may be unavoidable for some applications[77,78] and an appropriate balance between the accuracy and speed possible for modeling a given system size and timescale must be found.

It is reasonable to expect that as computational power increases and methodological improvements are made, there would be a decrease in demand for using CG methods as all-atom MD becomes more feasible for larger and larger protein systems. Contrary to this, there has instead been an increase in the number of researchers using CG methods.[77,78] This increased demand can be linked to the interest in examining larger systems and complexes as more experimental data on these assemblies become available. For these reasons, CG methods warrant a mention here, but we refer the reader to many excellent texts and reviews that provide more detail regarding this subfield of biomolecular modeling.[75,76,78−80]

Solvation Approaches and Solvent Models

The description of the solvent, a crucial element of the force field, falls within two broad categories: implicit and explicit solvation. Again, the approach selected depends on the size of the system and the desired timescale targeted. Within these two broad categories, there are a plethora of more specific options. Despite the multitude of water models that exist, there is some restriction to selection because of the proven compatibility of a given water model with a chosen force field.[65,66] As a general rule, water model usage is restricted to the force fields with which they were parameterized. The most widely used models are the empirical water models, of which there are many, including the transferable intermolecular potential (TIP) models,[81] TIP3P,

TIP4P, TIP5P; SPC models, SPC and SPC/E;[82] rigid Stillinger and Rahman model, ST2;[83] and the flexible three-centered water model, F3C.[84]

Many of these models are rigid geometry representations that have the electrostatics described by three, four, or five interaction sites. The simplest forms, TIP3P and SPC, have just three interaction sites, point charges, positioned on each of the hydrogen and oxygen atoms in the water model.[81,82] More complex models have additional interaction sites. For example, TIP4P has one additional negative charge displaced from the oxygen atom to produce improved water structure and density in agreement with experiment.[81] The TIP5P and ST2 models have five interaction sites. Instead of a single interaction site displaced from the oxygen, as in the TIP4P model, the TIP5P and ST2 models have two sites that correspond to the individual lone pairs on the oxygen atom.[81,85] The F3C model is similar to the TIP3P and SPC models, having just three interaction sites and similar partial charge parameters; however, it has a potential where the bond lengths and angles are not fixed and is more compatible with general protein force fields. Moreover, this flexible model is in better agreement with the structural and dynamic properties of water determined experimentally.[84]

A simplification that one can use instead of treating the solvent explicitly is to incorporate the solvent contribution by modeling it as a continuous dielectric medium. Such implicit solvation methods are widely used to reduce the computational expense of a simulation. A popular choice is the generalized Born method; a review of this method and case studies of its application are contained in Bashford et al.[86]

Simulation Setup

To perform an MD simulation the main steps are as follows. The atomic coordinates for the protein typically come from NMR or X-ray crystal structures in Protein Data Bank (PDB) format. The very first step is to inspect the atomic coordinates to ensure they are all present and determine the existence of any nonstandard residues or other experimental artifacts, such as purification tags, that you might not want to include in the simulation. The desired protonation states should then be selected for the conditions being simulated. For example, under standard conditions at neutral pH, the arginine and lysine residues should be positively charged with glutamic and aspartic acid negatively charged. The histidine side chain has a pK_a of ~6 and is neutral at pH 7 and predominantly present as the neutral τ tautomer, where the Nε2 atom is protonated. However, potential hydrogen bonding with other residues in the vicinity should be considered to determine if the alternate π tautomer, where the Nδ1 atom is protonated, should be selected. The positively charged protonation state, where both the Nδ1 and Nε2 atoms are protonated, is rarer at neutral pH but should not be overlooked as in the case of cytochrome b_5 where experimental pK_a values indicate that one of the histidine residues is positively charged at neutral pH.[87]

When cofactors are present, one should ensure parameters exist for them in the force field being used; parameterization may be needed where the relevant parameters are not available.[88–90] Once these initial aspects of protein structure have been considered, any missing atoms should be added to the protein model. Hydrogen atoms are generally missing from X-ray crystal structures, except for the most highly resolved structures, and many atoms or entire residues in mobile regions may also lack coordinates in the PDB file. Also, one must often select rotameric states for the side-chains, manually if not implicitly executed within the program, where more than one side-chain conformation is observed and included in the PDB file. The model should then be energy minimized to remove bad contacts and, if explicit solvent is being used, should be solvated with an appropriate water model.

The use of explicit solvation models introduces more complexity than just the consideration of the water model accuracy. One such complexity is how to incorporate or remove the need to replicate surface effects arising at the interface with air, such as surface tension and evaporation. Periodic boundary conditions (PBCs) allow a more realistic system to be generated by removing the water–air interface and negating the need to incorporate surface effects. In this approach, the system, the protein, and all solvent molecules are surrounded by mirrored images to reduce edge effects (Figure 1a). If a molecule in the central unit cell leaves through the bottom wall of that cell, it will simultaneously reenter the box from a neighboring cell through the opposite wall, in this case the top wall (Figure 1a). Hence, the number of atoms at any one time remains constant, and surface effects are removed by having all simulated atoms surrounded at all times by neighboring atoms within a cell or in neighboring cells.

To avoid artificial stabilization and dynamics of the protein structure because of finite size effects when using PBC,[91–93] the solvent should be added to generate a unit cell large enough to leave a gap, typically at least 10 Å, between the protein and the edge to prevent periodic contacts between neighboring replicas. PBC also requires the use of congruent unit cells, the cubic box being the simplest of these (Figure 1b). The choice of unit cell geometry does not have a large effect on the simulation outcome.[94] However, to increase the calculation efficiency, the unit cell needs to be kept as small as possible to reduce the number of water molecules required to solvate the protein while maintaining an adequate water layer. Two commonly used unit cells for compact protein structures are the cubic box and the truncated octahedron. The truncated octahedron is more efficient than the cubic box; this is a unit that most closely resembles a sphere enabling an appropriate water layer around the protein with a minimal number of water molecules. However, other unit cells, such as the hexagonal prism or rhombic dodecahedron, may be more appropriate depending on the geometry of the protein structure.[94,95]

Another consideration when using PBC is that calculation of all pairwise nonbonding interactions in such an infinite system would be computationally

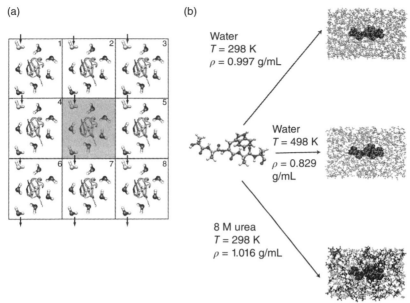

Figure 1 (a) Simplified illustration of periodic boundary conditions. Identical replicas, numbered 1–8, surround the original system of a protein and six water models in the central unit. During a simulation, translation outside the unit cell across the bottom wall is counteracted by an identical copy that enters the cell instantaneously through the top wall. In this manner, the number of atoms remains constant, and no surface effects need to be incorporated into the calculation. (b) Examples of different solvation approaches used for the study of protein folding. A small peptide is shown solvated with correct water density for simulation under native state conditions at 298 K, solvated at reduced density equal to that of liquid water at 498 K, and solvated in an 8 M urea solution: dark gray solvent molecules, urea; light gray solvent molecules, water.

exhaustive. Hence, nonbonding interactions are commonly treated using distance cutoffs, predefined distances beyond which atoms pairs are not considered in the calculation. Typically, the cutoff distance ranges from 8 to 12 Å. In many cases the choice of cutoff distance does not have an impact on the outcome of the simulation;[96] however, this cannot be assumed in all cases.[97] Abrupt termination of the distance over which nonbonding interactions are considered can cause artifacts; a common solution is to employ force-shifted truncation where the force is slowly ramped down to maintain energy conservation.[96] Other approaches often employed for just the electrostatic interactions are Ewald summations, primarily a particle mesh Ewald summation. A comparison of these solutions to determining the energetic contribution of nonbonding interactions with our spherical force-shifted cutoff is discussed in Beck et al.,[96] and artifacts induced by the artificial periodicity of the Ewald method are reviewed.

Once solvated, the system should again be minimized. It is recommended that the solvent and protein are minimized independently while the other is restrained before minimizing them together as a whole system. Once minimization has been complete, the system should be heated to the desired temperature and then equilibrated adequately by simulation at this temperature until stability in the energetics and physical properties are observed. The portion of the simulation after this point is referred to as the production dynamics and should be used in the analysis. To increase the sampling of conformational space, it is common practice to acquire multiple simulations of a system. This can be achieved by using either alternate NMR structures, different random seeds for the assignment of initial velocities, or by some other preparatory method to generate different starting geometries for a set of independent simulations.[98] An often-asked question is: how many simulations are required to adequately sample conformational space? To answer this, we compared the results from 100 independent simulations with those obtained from smaller subsets. It was determined that average properties of states along a folding pathway could be sufficiently captured using as few as just five independent simulations.[98]

Simulation of Unfolding Environments

For the study of some parts of the folding pathway, such as characterizing the native state or trying to establish a transition state using experimental restraints to determine the putative structure, one uses a standard simulation setup with the chosen water model at appropriate temperatures to agree with experiment or observations made at physiological conditions between 298 K and 310 K. However, investigation of folding/unfolding pathways can require simulations being acquired under different conditions and environments. As discussed later, unfolding simulations are often used to study the folding pathway. Common examples of conditions used to encourage unfolding are by mimicking either thermal or chemical denaturation *in silico* by increasing the temperature or adding a cosolvent, such as urea (Figure 1b). Alternatively, other conditions, such as increasing the acidity or pressure, can be used.

For elevated temperature unfolding simulations, canonical (NVT) and microcanonical (NVE) ensembles are commonly used so that system density is maintained. Although the isothermal-isobaric (NPT) ensemble can also be used, the dynamical behavior under increasing temperature is not equivalent to that obtained using the NVT ensemble.[99] If using either ensemble with the same density as that of the native state, $\sim$1 g/mL, then unfolding will be satisfactory, but one should be aware that the unfolding will be exacerbated by both the high temperature and pressure and will likely not correspond with the thermal denaturation probed experimentally. Thus, another more realistic approach is to use reduced densities to replicate the experimental density at a given temperature.[100] As shown in Figure 1b, the system is set at the same volume as that used for a standard simulation at 298 K but at a lower density appropriate for the elevated temperature of 498 K.

Mimicking chemical denaturation requires inclusion of cosolvents into the water box at the correct mole fraction to generate the desired concentration at a given temperature. To do this, one typically proceeds by first generating a box of water molecules at a reduced density equivalent to that required to produce the target density for the final water-denaturant system, for example, 0.98 g/mL of water for a final density of 1.02 g/mL for 8 M urea at 298 K. Individual water models are then randomly removed and replaced with cosolvent molecules at the requisite mole fraction to give the denaturant concentration and density required (Figure 1b).[101] If using low pH to induce unfolding, one only need to address the protonation states of the residues before solvating the protein to be consistent with those at the desired pH. Below pH 6, histidine is positively charged and simulations should be performed with hydrogens added to both the Nδ1 and Nε2 nitrogen atoms. Moving to acidic conditions below pH 5 will also require protonation of the glutamic and aspartic acid residues within the protein as well as the C-terminus.

UNFOLDING: THE REVERSE OF FOLDING

Using MD simulations to investigate protein folding faces two main challenges. The most obvious relates to the timescale of protein folding and the computational expense required for adequate sampling. Protein folding *in vivo* falls between microseconds and minutes[102] with a double-norleucine mutant of the villin headpiece being the fastest folder identified so far.[103] The second challenge is the need for well-defined atomic coordinates of the starting state for any given simulation. It has always been difficult to derive structural coordinates for denatured states experimentally, a challenge compounded by the expected heterogeneous nature and conformational uncertainty of the unfolded proteins. Even now, complete structural characterization of an unfolded protein is intractable and there is much debate over the nature of, and residual structure in, denatured states.[104]

In the past, many researchers have used artificially generated unfolded conformations as starting structures by setting all backbone dihedrals to produce an extended chain.[105] Although objective and unbiased, this approach is far from ideal. An alternative strategy, because of the availability of known experimental coordinates, is to study the protein folding mechanism using solved native state structures and to study the reverse: the unfolding pathway. With protein folding being a nonequilibrium process, the validity of such an approach has been disputed regarding symmetry in the folding and unfolding pathways.[106,107] Consideration of the "principle of microscopic reversibility"[108] in some part addresses this; the postulated principle that the mechanisms or molecular rearrangement that occurs in the forward and reverse pathways for a thermodynamically reversible process must be the same. By extrapolation, we can infer that as protein folding is a thermodynamically reversible process under some conditions, the conformational transitions during unfolding will be the reverse of those made on folding. Proving that this principle applies to protein folding would provide validity in the use of protein unfolding studies as a way to determine folding pathways.

This principle was first shown to hold for the case of chymotrypsin inhibitor 2 (CI2).[109] A comparative study of the unfolding and folding pathways under identical conditions marginally below the melting temperature (T_m) of CI2, where both folded and unfolded conformations should be approximately populated, demonstrated reflected mechanisms that progressed via the same transition state (Figure 2a). Four well-defined states – shown in Figure 2a, the native (NS), near-native (NS′), transition (TS), and denatured (DS) states – were identified through which both the unfolding and folding pathways passed. When simulated at 348 K, close to the experimental T_m, the CI2 crystal structure rapidly expanded to a state more stable at that temperature, the NS′; the correct topology was maintained with the majority of native contacts present, albeit elongated because of the small ($\sim$4%) increase in volume. This NS′ conformation is believed to be the NS for CI2 at the elevated temperature used in this study. Unfolding then proceeded with the maximum conformational change registered at 25.6 ns (Figure 2a) where only nonnative hydrophobic interactions were present. This correlated with the unpacking that disrupted the contacts between the N- and C-termini as well as some loss of helix structure, which until that point had been relatively robust. Folding was then witnessed to be a simple reversal of the unfolding pathway with the expanded native state (NS′) revisited at 200 ns where native and native-like contacts were again present, as well as native interactions that defined the secondary structure packing.[109] This was the first time that microscopic reversibility had been directly observed.

Microscopic reversibility has also been observed for the engrailed homeodomain (EnHD).[110] EnHD is a 61-residue protein with a melting temperature of 325 K; simulations at this approximate temperature (323 K) demonstrated a number of unfolding and refolding events that again proceeded through four well-defined states (Figure 2b). For example, unfolding at 22 ns was followed by refolding between 39.7 and 41 ns as illustrated in Figure 2b. As the unfolding proceeds, EnHD assumes a near-native state, NS′, where helix α3 shifts toward the N-terminus by approximately 10 Å. Unfolding continues, and the denatured state is reached at 39 ns where helix α3 undocks from the core of EnHD. Refolding from this point was rapid, α3 moved back, and the native state was reached after approximately 1 ns (Figure 2b). In the many simulations that were part of this study, the folding and unfolding pathways were observed proceeding through the four states in the same order, providing further evidence of protein folding being microscopically reversible.[110]

Although an unfolding simulation can provide information on the folding pathway and microscopic reversibility has been confirmed for some systems, it should be remembered that this assumption is valid and unfolding is

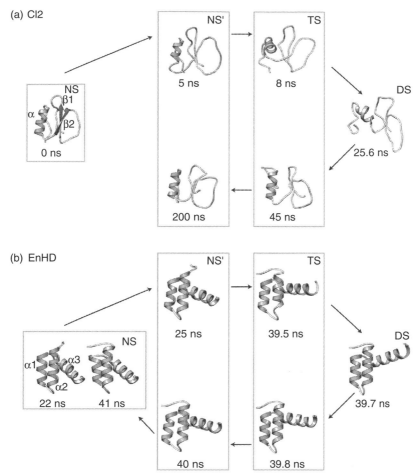

Figure 2 Demonstration of microscopic reversibility for the (a) chymotrypsin inhibitor 2 (CI2) and (b) engrailed homeodomain (EnHD). In both (a) and (b) the top unfolding pathways traverse conformational states that are revisited on the lower folding pathways illustrating the equality of the unfolding and refolding pathways.

a true reversal of the folding pathway only when both pathways are observed under identical conditions. Studies comparing temperature-quenched refolding and high-temperature unfolding have shown, however, that the forward and reverse pathways can be the same at different temperatures.[55,111] Nevertheless, care should be taken when drawing conclusions for the folding pathway under native state conditions when using an unfolding simulation under different conditions.

Approaches taken to induce protein unfolding by MD simulation have involved the use of high temperature, chemical denaturants, or high pressure.[112–120] Commensurate with their common use in experimental methods, both urea[121–133] and guanidinium chloride (GdmCl)[129–131,134] have been used to generate denaturing conditions *in silico*. A few of these studies compared how denaturation was induced by the differential atomic interactions of urea and GdmCl and to elucidate whether the direct interaction mechanism of denaturation was valid. Initial computational work to establish whether direct or indirect effects, or both, were the mechanism behind chemical denaturation with urea demonstrated the unfolding to be rapid in the presence of 8 M urea and the pathway to be conserved across both urea denatured and thermal denatured simulations.[133] Urea induced the unfolding through direct interaction with the protein and indirectly via changes in the solvent structure. The urea molecules both competed with the water–protein interactions disrupting native hydrogen bonds and stabilizing nonnative conformations and weakened water–water interactions encouraging solvation of hydrophobic groups.[133] The contrast in the unfolding pathways in GdmCl and urea was examined demonstrating residual secondary structure propensities that were denaturant dependent.[130,131] Progress made in understanding the denaturation mechanisms of urea and GdmCl owes much to the details elucidated from MD simulations, consideration of which, along with a discussion of the current contentions, was the subject of a recent review.[135] More recently, the effect of using a mixture of these two denaturants was studied showing that although the denaturants on their own induce unfolding, as a mixture, they promote hydrophobic collapse, with the protein observed to undergo hydrophobic collapse of the denatured state driven by nonnative interactions between hydrophobic residues.[136] Even though this result confirmed previous predictions, the behavior of combined denaturants remains to be verified experimentally.

Although not successful in all cases, high-pressure MD simulations have been used to investigate pressure-induced unfolding. In the first attempt at high-pressure simulations of BPTI at 10 kbar, unfolding was not observed.[112] This was likely because of the trajectory being limited to 100 ps as pressure-induced unfolding was observed both experimentally[137] and from another MD study on BPTI at 10–15 kbar for 800 ps.[138] High pressure often has relevance to functional activations such as cooperative water binding within protein cavities without a large change in the structure being observed.[117]

The most widely used approach by far, however, is elevated temperature unfolding simulations, which are discussed in more detail below.

ELEVATED TEMPERATURE UNFOLDING SIMULATIONS

The use of elevated temperature to expedite the unfolding of a protein on a timescale accessible by computational methods appeared in the early

1990s.[20,23,139] The most successful early study was of reduced BPTI where high-temperature simulations from the native state were able to generate and characterize what was believed to be the molten globule state within 250 ps.[23] This first success was quickly followed by other attempts.[140−142] This approach has since grown to be one of the most popular approaches for studying unfolding with MD simulations having been useful for direct comparison with experimental unfolding techniques that use thermal, pH, or chemical denaturation to induce unfolding.

Unlike chemical and pressure-induced unfolding simulations that are typically compared with experimental investigations under the same conditions, high-temperature simulations are often compared with a wider range of unfolding and folding experiments. This includes comparison with data acquired under physiological conditions rather than a restriction to comparisons made only against thermal denaturing experiments. This has led to some criticism due to the potential of elevated temperatures to change the folding landscape and because it was not known, until more recently, if the principle of microscopic reversibility could truly be applied.[109,110] As comparisons are typically drawn between unfolding pathways observed in high-temperature simulations and data obtained at different temperatures, the inequality of these unfolding and folding pathways at different temperatures is possible. Even so, the good agreement observed between simulation and experiment when comparing a variety of different conditions suggests that both are probing the same process.

Some have argued that the energy landscape is altered by temperature and may differ significantly from what occurs at physiological temperature, with the presence of intermediate or misfolded states lost as the temperature is increased.[107,143−145] In these cases, although there is reported loss of detail, the overall pathways are retained.[107,144] And, if a temperature dependence is observed, it is often weak[145] indicating that changes to the protein landscape are not detrimental enough to circumvent the value of using high-temperature approaches to accelerate unfolding. Observed temperature variations of the folding pathway may often be attributed to using inappropriately high temperatures.[146] While the use of high temperature is valuable, the existence of proteins with unfolding that does not adhere to Arrhenius-type kinetics means temperature selection should be wisely considered and one should establish that the temperature ranges used provide consistency in the unfolding pathways.[146]

The most successful use of high-temperature MD is in combination with experimental validation to confirm the states and pathways identified are valid. The unfolding pathway has been shown to generate experimentally comparable results for many proteins, including CI2,[147,148] EnHD,[149−151] dihydrofolate reductase,[152] $\beta2$ microglobulin,[153] and protein G.[154] Many in-depth studies have addressed the concerns regarding the validation and use of high-temperature simulations to infer the folding pathway by showing the results of high and low temperature to be comparable.[147,149,150,155] Two of the most rigorously studied proteins are CI2[147,155] and EnHD.[149,150] Repeated simulations of CI2 at seven different temperatures spanning 298–498 K demonstrated the unfolding pathway to be unchanged on increasing the temperature.[147] The timescale on which the transition state (TS) appeared did change, varying from 0.3 to 19.9 ns, but the TS ensembles themselves were comparable both with each other and with experiment (Figure 3). At all temperatures the trajectories were observed to share the same events along essentially the same pathway, with $\beta1$ and the active site loop between $\beta1$ and $\beta2$ moving after disruption of the hydrophobic core packing prior to the TS, followed by solvent exposure of the core and collapsing of the active site to give the unfolded state (Figure 3). In addition to the movement of the TS, the only other notable differences are the order of contact loss that occurs. However, it was believed that, being of similar magnitude, these contacts had an equivalence such that the order of loss did not distort the unfolding

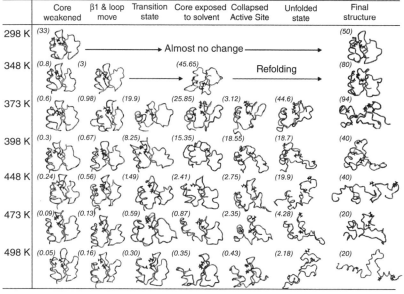

Figure 3 Unfolding pathways of chymotrypsin inhibitor 2 (CI2) acquired at different temperatures: the simulation times corresponding to the individual structures are given in nanoseconds within the brackets. The topmost native state simulation shows the structure to be stable after 50 ns; no unfolding events were witnessed. At the higher temperatures, 348–498 K, unfolding events are witnessed with unfolding occurring at shorter time lengths as the temperature is increased. Close to the native state, at 348 K, only partial unfolding is obtained before refolding back to the native state structure is observed. Above 348 K, more complete unfolding occurs, and all unfolding pathways traverse the same conformational changes irrespective of the temperature used in the unfolding simulations. Reproduced with permission from Day et al.[147]

pathway.[147] Because of the known mild sensitivity of CI2 to experimental denaturing conditions, further simulations were performed and compared with simulations also performed in urea.[155] Again, although increasing the temperature shifted the position of the TS toward the native state, the TS ensembles were in quantitative agreement with experiment.[155] The same adherence to the unfolding pathway irrespective of the temperature used in the unfolding simulations was also observed for EnHD.[149]

Another criticism of MD simulations is that by simulation of a single molecule in a water box, comparable to infinite dilution, the true dynamics reflective of the crowding and protein–protein interactions present *in vivo* or even in an experimental sample for *in vitro* studies are not captured. A recent study addressed this and demonstrated that the unfolding pathway of EnHD was unaffected by the presence of multiple copies of this protein within the same simulation box. Elevated temperature simulations were performed on a multimolecule system of 32 EnHD molecules, equivalent to a concentration of 18 mM, referred to as a test-tube simulation.[156] This work showed that ensemble simulations were not necessary to capture the unfolding pathway, the only measurable difference being that unfolding progressed at a marginally slower rate in the test-tube simulations than had been observed for the single-molecule simulations. This slower unfolding rate was perhaps a consequence of the protein aggregation witnessed in the test-tube simulations; nevertheless, this did not change the NS, TS, intermediate, and DS, all of which agreed with the experimental data.[156]

Although there is substantially more evidence supporting the suitability of high-temperature simulations to elucidate unfolding pathways relevant to the physiological mechanism,[150,151,155] a few reports have revealed folding landscapes altered by temperature; however, these studies have employed nonatomistic models in the simulation or analysis methods, which can affect the results.[107,144] Nevertheless, the suitability of high-temperature simulations should not be assumed for any given protein without either comparison with experimental data, if the data exist, or rigorous testing of the unfolding pathway over a range of plausible temperatures.

BIOLOGICAL RELEVANCE OF FORCED UNFOLDING

Forced unfolding simulations complement the experimental approaches of atomic force microscopy (AFM), light optical tweezers (LOT), and single-molecule force spectroscopy (SMFS) that employ mechanical means to investigate the forces within biological molecules.[157–160] Such approaches manifested from the discovery of the mechanical proteins within biological systems that are involved in maintaining the structural integrity of cells and where exposure to mechanical stress is a facet of their function.[158]

The first reported use of AFM to analyze protein folding was in a study of the mechanical behavior of titin.[161] Titin is a protein component of the filaments that form the contractile units of vertebrate muscle cells and is a large, elastic 3 MDa protein with a multidomain structure. This multidomain structure is predominantly composed of a super-repeating pattern of 11 alternating fibronectin III (Fn3)-like and immunoglobulin (Ig)-like domains in what is termed the A-band and two segments of tandem highly homologous Ig-like domains that bracket a flexible PEVK region rich in proline, glutamate, valine, and lysine amino acids in the more elastic I-band.[162–164] The AFM-probed unfolding was performed on both individual titin molecules and two segments containing four or eight sequentially connected Ig-like domains isolated from the mechanically active I-band.[161] The samples were immobilized onto a gold surface through one of the free termini and adsorption onto the AFM tip of the other termini through which a pulling force was exerted as the AFM tip was retracted away from the gold surface.[161] Unfolding of the full titin protein under force produced an initial gradual increase in force as the protein was extended followed by a saw-tooth force-extension profile with periodic high force peaks. The initial force response was attributed to the extension of the less structured PEVK regions (Figure 4a). Repeated observation of the saw-tooth

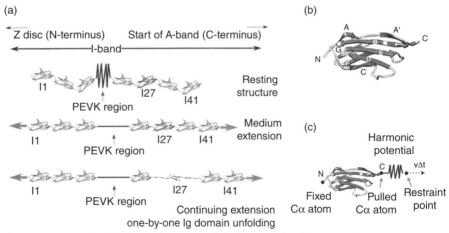

Figure 4 Forced unfolding of immunoglobulin (Ig)-like domains of titin. (a) Cartoon of the titin I-band function reproduced with permission from Lu et al.[165] Sequential Ig-like domains are shown in ribbon representation, labeled to reflect the larger number of tandem Ig-like domains present in this region (I1–I41), bracketing the less structured PEVK region. This cartoon illustrates the three stages in the mechanical unfolding of the I-band as it functions in muscle extension, from at rest, with both Ig-like domain regions and PEVK regions contracted, under small forces where PEVK extends, followed by the unfolding of Ig-like domains one by one at higher forces. (b) Structure of the I27 Ig-like domain β-sandwich fold and (c) schematic of the method used to perform forced unfolding simulation by use of a harmonic potential applied between the atom to be pulled and a restraint point.

force-extension profiles for the forced unfolding of the four and eight Ig-like domain segments confirmed that the high force peaks witnessed for the full titin protein coincided with each of the individual Ig-like domains unfolding in turn (Figure 4a). Complete refolding of titin and the Ig-like domains was incident with the removal of all force. Although a mechanism for the mechanical behavior of the I-band was inferred from this study, as depicted in Figure 4a, this lacked atomic-level information of the conformational changes that accompanied the detected unfolding, an opportunity amenable to MD approaches that are able to mimic similar mechanical processes and provide the missing atomic detail.

There have been two main approaches for the study of mechanical unfolding *in silico*. This first approach is steered molecular dynamics (SMD),[166,167] where force is directly applied to atoms within the simulation to reproduce the forces applied experimentally. The second method – referred to as biased, targeted, or restrained MD – employs a biasing potential to encourage a protein structure to reside in a targeted region of a chosen reaction coordinate by making movement outside of that region less favorable.[168] This latter approach for performing forced unfolding simulations emerged from concerns with the artificially high pulling speeds used in SMD and associated overestimates of hydrogen bond interactions between the β-strands in titin[165] and will be discussed later.

Steered MD[166,167] originated from an attempt to investigate the atomic-level detail of the binding forces within the streptavidin–biotin complex by imitating the AFM experiments.[169] This method was soon applied to understand the mechanical unfolding of titin; the first study being of the I27 and I28 Ig-like domains of titin using an all-atom force field and explicit solvation model.[165] The structure of I27 is illustrated in Figure 4b to demonstrate the β-sandwich fold shared by I28 and other homologous Ig-like domains from titin. In replication of the AFM experiments,[161] a Cα atom at one terminus of the Ig-like domain was constrained and a time-dependent force applied to a Cα atom at the other (the pulled atom) (Figure 4c). The force, in the form of a harmonic potential (Eq. [5]), was used to restrain the atom being pulled to a second point, the restraint point, which was moved at near-constant velocity over 100 fs time intervals to effect stretching of the domain (Figure 4c).

$$F = k(vt - x)$$ [5]

Pulling was applied in a single direction in line with the N- and C-termini and away from the center of mass of the Ig-like domain. The force exerted on the pulled atom, F, is governed by the spring constant, k, where x is the displacement of the pulled atom from its original position at velocity, v (Eq. [5]).

From the experimental data acquired for titin unfolding, two force-dependent stages were inferred (Figure 4a).[165] At low force, the tandem Ig domains align followed by the extension of the PEVK region as the force increases. On reaching higher forces, above ~100 pN, there is one-by-one unfolding of the tandem Ig domains, similar to that seen in AFM experiments where the tandem unfolding produced a saw-tooth profile.[161] Although being on different timescales to the AFM experiments and applied to a lone Ig domain, the MD simulations of Lu et al.[165] were able to provide atomic-level information that revealed the existence of two barriers to forced unfolding of these individual Ig domains. The barriers correlated with the number of hydrogen bonds between the strands: a lower energy barrier for the breaking of the two hydrogen bonds between strands A and B and a higher energy barrier for the disruption of the six hydrogen bonds between the A' and G strands (Figure 4b). The unfolding mechanism began gradually with the sliding of the sheets and then strands relative to one another. However, once these two events occur, unfolding of the rest of the domain is expeditious with the structure coincident with strands E and D being lost last and the domain reaching 260 Å once completely unfolded. These asymmetric barriers were confirmed later by other experiments.[158] Further, the lower energy event of hydrogen bond disruption between A and B strands was confirmed to explain a 7 Å domain extension experimentally observed prior to the main unfolding event.[170]

This approach mimics not only a mechanical process, but also enables the process to be accelerated through the application of force and stretching speeds significantly higher than AFM and LOT experiments.[165] However, the force used or the speed at which pulling is performed can affect the outcome. In the example of Lu et al.[165] constant velocity SMD was used at two pulling speeds (0.5 and 1.0 Å/ps); both speeds presented similar unfolding pathways and comparable force-extension profiles. In both cases, the pulling speeds resulted in the overestimation of the unfolding force compared with experiment because of the accelerated unfolding.[165] This work was repeated also using constant force SMD over forces ranging from 750 to 1200 pN for comparison with the constant velocity approach.[171,172] The constant force approach was deemed more appropriate, and an intermediate involving water-mediated hydrogen bond disruption was observed in detail.[167,171] Constant velocity SMD, on the contrary, had resulted in this intermediate being passed too quickly and produced an altered pathway.[171]

In experiments using tandem domain chains, the AFM tip is indiscriminate in its selection of where it adheres to.[161] Hence, the effect of pulling position and direction is not only one of general curiosity in understanding the mechanical capabilities of proteins, but is also a valid factor when comparing simulation to experiment.[165,171,173] The effect of pulling direction, when applied to different termini, and speed can provide similar results.[165,171,172] However, more extensive investigations that involved movement of the anchor point along the Cα of sequential amino acids showed that direction impacts the unfolding rate in a manner related to the protein topology.[173]

Forced unfolding simulations have also been applied to globular proteins to determine whether the unfolding response mimics that observed for mechanical protein domains and if the unfolding pathways reveal some proteins to have a function-specific response to mechanical stress. This approach has been used on a number of proteins: barnase,[174] ubiquitin,[175] ubiquitin-like,[176] protein G, and protein L.[177] None of these proteins has functions under which they are exposed to mechanical stress. The first pulling studies of a "normal" globular protein were on barnase, and it unfolded at lower forces than had been observed for mechanical proteins, despite sharing comparable solution unfolding rates with the titin I27 domain.[174] Neither was the periodic saw-tooth profile coincident with unfolding. High resistance to force, typical of mechanical proteins, observed experimentally, confirmed a divergence did exist in the reactions of globular and mechanical proteins to force.[174] Moreover, the corresponding simulations revealed a different unfolding pathway to that observed under thermal denaturing conditions (Figure 5).[178]

Force-induced unfolding of barnase was initiated through extension and gradual detachment of the termini in the pulling directions; the unfolding skewed toward the two termini throughout the pathway with helix α1 separated from the core protein at increased pulling speeds (Figure 5). In contrast, elevated temperature simulations produced a more uniform disruption across the structure as unfolding proceeded (Figure 5).[178] In both instances, the simulations agreed with the experimental observations, and simulation of the different pulling speeds showed slightly different mechanisms, a phenomenon also observed at both low and high force for other proteins.[179] A force-dependent switch between unfolding mechanisms identified for protein

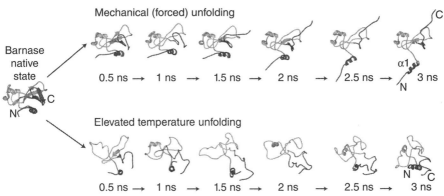

Figure 5 Comparison of the forced unfolding and elevated unfolding trajectories obtained for barnase, adapted from Best et al.[174] The mechanical unfolding pathway is skewed toward the termini on which force was applied and the core of the fold remains compact. Except for the termini pulling away from the rest of the fold, barnase retains native-like topology throughout the mechanical unfolding pathway. In contrast, the elevated temperature unfolding pathway has unfolding of barnase proceeding in a more uniform manner with conformational changes distributed across the fold rather than localized toward the termini, as observed for the forced unfolding pathway. With permission by Elsevier, *Biophys. J.*, 2001, Vol 81.

L[180] led to the conclusion that the switch was a symptom of multiple pathways detectable at lower forces converging at high forces.[179]

More recently, an extensive survey of the correlation between protein size and the maximum force required to induce unfolding (F_{max}) was performed for a set of 7510 proteins of nonfragmented, uncomplexed single domains from the PDB ranging in size from 40 to 150 residues.[181] F_{max} varied up to ~342 pN with most proteins having F_{max} values clustered around ~94 pN and a relationship between F_{max} and protein length implied. A subset of 134 proteins with the highest F_{max} was designated to be "strong" proteins.[181,182] Interestingly, I27 of titin was not a member of this subset but I28 was, a disparity consistent with the greater force resistance of I28 observed experimentally. None of the members in this "strong" subset has all-α topologies; the "strong" proteins fall into the all-β and $\alpha+\beta$ topologies, specifically the α/β complex, four-layer sandwich, three-layer $\alpha\beta\alpha$ sandwich, 2-layer $\alpha\beta$ sandwich, $\alpha\beta$-roll, and β-barrel architectures and predominantly in the immunoglobulin-like and ubiquitin-like fold families.

Forced unfolding simulations are highly relevant and invaluable in characterizing proteins naturally exposed to mechanical stress as a component of their biological function. In this chapter, we have illustrated this utility by discussing research that has been done primarily on the giant muscle protein titin. Although mechanical unfolding is not directly relevant to the biological function of globular proteins, it is plausible that even nonmechanical proteins are subjected to pulling forces during some cellular activities, such as translocation across membranes and through the degradation channel of the proteasome.[176,183,184] Hence, the use of forced unfolding simulations to provide insights into the response of nonmechanical proteins to such stress and impact on the unfolding pathway should not be overlooked. From work in this area to date, it appears that globular proteins react differently under the application of force than do mechanical proteins, supporting the notion that mechanical proteins have been selected by evolution for their ability to withstand higher forces than globular proteins.

BIASED OR RESTRAINED MD

There are many approaches to biasing the potential energy function or restraining the system to a specific region of a defined reaction coordinate to produce conformations consistent with a given state along the folding pathway.[38,185−189] Previously mentioned was an alternative approach to SMD; in this method the potential is perturbed to encourage sampling along a defined reaction coordinate toward a "target" conformation producing a biased molecular dynamics (BMD) simulation.[168]

The perturbation applied to the potential energy function has the harmonic form given in Eq. [6], where W is the biasing force applied on the reaction coordinate at time t, and α is a coupling constant, with $\rho_{(t)}$ and ρ_a values along the reaction coordinate. For a chosen reaction coordinate, $\rho_{(t)}$ is the value of the reaction coordinate at a given time step and ρ_a a reference value that determines whether or not the biasing potential is applied. At the start of a simulation, ρ_a will have the value of the starting structure, for example, the number of native contacts. To bias the simulation toward a chosen "target" conformation on a reaction coordinate, the position of a conformation on the reaction coordinate at a given time point of the MD trajectory, $\rho_{(t)}$, is compared with this reference value. If $\rho_{(t)}$ is further from the "target" value than ρ_a, the perturbation comes into effect making fluctuations that move the trajectory away from the "target" conformation less favorable. If the fluctuations of the trajectory move along the reaction coordinate in the desired direction, no perturbation is applied, $W = 0$, and the reference value, ρ_a, is then updated with the value from the preceding time step such that both the reference value and the trajectory increasingly approach the target. When the perturbation is applied, the coupling constant, α, determines the strength of the biasing potential applied and consequently how rapidly the energy barriers are surmounted through determining the magnitude of allowed fluctuations in the reverse direction along the reaction coordinate; smaller values of α produce more realistic results but require longer simulation times.

$$W(r, t) = \frac{\alpha}{2}(\rho - \rho_a)^2 \qquad [6]$$

For forced unfolding, the reaction coordinate typically selected is the square of the N-to-C distance of the protein structure (Eq. [6]). The external bias is placed on this chosen reaction coordinate to affect mechanical unfolding by encouraging sampling at increasingly greater end-to-end distances between the N and C-termini of the protein. Conversely, any fluctuations that decrease the distance are energetically disfavored. Hence, rather than apply a direct force, extension of the protein is achieved indirectly through preferential selection of the natural fluctuations the protein undergoes during the MD simulation that increase the N-to-C distance. If the fluctuations in the C-to-N terminus continue to increase, then no perturbation is applied. However, if the fluctuations decrease the C-to-N distance then the biasing potential is enforced.

The first instance of using a biased potential to generate a "target" conformation on the basis of experimental input was by Paci and Karplus.[168] They attempted to unfold two fibronectin type 3 domains, [9]Fn3 and [10]Fn3, from the titin protein. By not having an external force directly applied to a specific atom or atoms, and instead functioning by a process previously labeled as a "conformation fluctuation mechanism,"[168] this gentler alternative to SMD led to a dependence on the initial conformation; progression along a given reaction coordinate did not always proceed at the same rate.[168]

However, BMD is more versatile than SMD as it can be applied to any reaction coordinate, such as radius of gyration,[190] native contacts,[191] or the distance between a protein and a binding partner.[192] One valuable application has been the generation of transition state ensembles (TSEs) using experimental Φ-values.[191,193–195] Experimental Φ-values are obtained through protein engineering to determine the effect of a mutation and are a measure of the structural impact of a mutation and the change in the relative stability of the transition states and the native state.[196] These Φ-values can range from 0 to 1; a value of 0 implies the mutation has had little or no impact on the transition state, and a value of 1 indicates instability in the transition state is equal to that introduced into the native state by the mutation. For comparison with MD trajectories, Φ-values are approximate to the fraction of native contacts present in a conformation. Hence, the reaction coordinate can be defined by the change in the fraction of native contacts, where Φ_i^{exp} is the experimental Φ-value of residue i and Φ_i^{calc} the calculated value from the MD simulation, using the relations in Eq. [7] and [8] for a set, E, of NΦ experimental values.

By applying the same biasing potential, as in Eq. [6], depending on whether the simulation moves toward satisfaction of the target Φ-values or not, can produce conformations with the fraction of native contacts for each residue in correspondence with the Φ-values.

$$\rho(t) = \frac{1}{N_\Phi} \sum_{i \in E} (\Phi_i^{calc}(t) - \Phi_i^{exp}(t))^2 \qquad [7]$$

$$\Phi_i^{calc} = \frac{N_i(t)}{N_i^{nat}} \qquad [8]$$

A recent example of such Φ-restrained simulations concerned determination of the putative structures of two sequential on-pathway transition states (TS1 and TS2) of the Phox and Bem1p (PB1) domain from the Next-to-Breast Cancer Gene 1 (NBR1) protein.[194] Two sets of experimental Φ-values pertaining to TS1 and TS2 were obtained using mutagenesis and kinetic experiments. The positions of the two transition states were determined from Tanford β (β_T) values that are calculated from the denaturant m values; TS1 and TS2 had β_T of 0.71 and 0.93, respectively (Figure 6a). The β_T values reflect the compactness of the transition states and can be related to changes in the solvent accessible surface areas (SASAs) from that of the native state, with TS1 having approximately 22% more solvent exposure than TS2 (Figure 6).[197,198] The two sets of Φ-values were used in individual sets of BMD simulations to generate putative conformational ensembles for each of the transition states. The relationship between SASA and the β_T values was in turn used to validate the conformations predicted through the MD simulations and selection of the TSEs. A range of elevated temperatures was also used in conjunction with the biasing potential to expand the sampling of the conformational space coincident with the Φ-values.

(a)

(b)

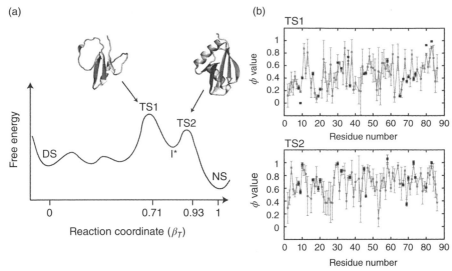

Figure 6 (a) Schematic of potential folding pathway for the Phox and Bem1p (PB1) domain of the NBR1 protein showing the relative position of the two on-pathway transition states in terms of their Tanford β (β_T) values. (b) Predicted Φ-values obtained from biased MD simulations; experimental Φ-values used to restrain the trajectories are shaded black and demonstrate the good agreement between the Φ-values from the generated TSEs and experiment.

The earlier of the two transition states, TS1, was indicative of initial hydrophobic collapse whereas TS2 was more compact and had more native-like structure. The predicted structures are shown in Figure 6a along with their corresponding predicted Φ-values (Figure 6b); Φ-values were predicted for all residues from the simulations. Those to which the simulations were biased are in close agreement with the experimental values, highlighted in Figure 6b; the others provide a basis against which further mutations can be used to test the validity of these putative TSEs. As expected from the position of TS1 and TS2 in relation to each other, the majority of predicted Φ-values from the TS2 simulations are higher than those in TS1 consistent with the consolidation of more secondary structure close to the NS.

CHARACTERIZING DIFFERENT STATES

Although there are exceptions, many of the smaller proteins studied to date fold by a two-state mechanism with a single rate-limiting transition state.[199] The transition state is a crucial but unstable stage along a folding pathway; traversal back to the native state has equal probability to that of the system progressing to the unfolded state. Transition states are often a less compact version of the native state, with native-like secondary structure still present. However, movement out of the transition state toward the unfolded states should be accompanied with rapid structural changes. Large-scale analysis of representative structures of 67% of all known protein folds indicated TSEs share global features, comparable across all topologies.[200] In general, there was prevalence for residues initially in α-helices in the native state to remain structured in the TSE, whereas residues in β-strands were less likely to be structured. Although short-range contacts were similar in both native and transition states, a significant reduction of long-range charged and hydrophobic contacts occurred, consistent with an expansion of the native state.[200]

Restrained or biased simulations, mentioned previously, can be used to generate ensembles of structures consistent with transition states using experimental data as restraints. In these cases, identification of the transition states is a straightforward outcome of the simulation. Conformations that have reached the restrained target experimental values, such as Φ-values, are selected as members of the TSE.[168,194,201] When using other approaches, such as elevated temperature simulations, the dynamics are allowed to more freely sample the conformational space, and successful identification of the TSE requires careful selection of an appropriate reaction coordinate. Many have used native contacts and radius of gyration as reaction coordinates for unfolding trajectories;[202] however, these are not always sufficiently discriminative.[203]

The transition state is briefly populated early on along an unfolding pathway and is expected to appear just as the trajectory exits the native state but before any large structural changes associated with extensive unfolding are observed. Hence, one way to identify the transition state is to perform temporal-dependent conformational clustering, as used to identify the TSE of CI2.[178,204,205] Using just the RMSD between each conformation within a trajectory and a reference or starting structure can be misleading because two widely different structures can have similar RMS deviations in comparison with a single reference structure. Hence, conformational clustering analysis where the RMS deviations between each conformation in a trajectory are taken into account can provide an improved measure of structural similarity and has been used to some success in isolating TSEs.[155,178,200,204–206] Figure 7a illustrates the use of conformational clustering in identification of a TS for the tyrosine kinase Fyn SH3 domain.[203] Calculation of the Cα RMSD between each conformation within the unfolding trajectory provides an RMSD matrix, which through multidimensional scaling can be reduced into three-dimensional space (Figure 7a). This allows one to identify a subset of structures corresponding to the TSE on the basis of the proximity both in time and in 3D Cα RMSD space as shown here for cold shock protein B (Figure 7a).

More sophisticated approaches use principal component analyses or property space analysis. The merits of these methods were discussed

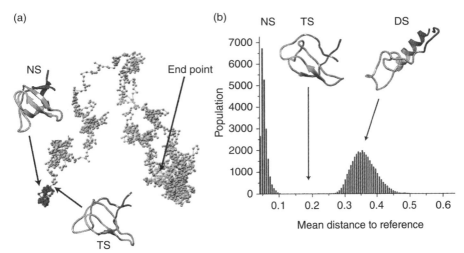

Figure 7 Transition state determination from an unfolding trajectory of the Fyn tyrosine kinase SH3 domain adapted from Toofanny et al.[203] (a) Conformational clustering of Cα RMSD all-by-all comparison of structures within the unfolding trajectory projected in 3D using multidimensional scaling. The shaded cluster contains structures of similar Cα RMSD pertaining to the native state (NS) with the transition state (TS) identified at the exit of this native state cluster. (b) One-dimensional reaction coordinate representation of multidimensional property space for each conformation in the same unfolding trajectory. The different states are labeled showing the separation between the NS and DS populations in property space. The region pertaining to the TS is indicated with a corresponding snapshot of the SH3 TS. With permission by Elsevier, *Biophys. J.*, 2010, Vol 98.

in Toofanny et al.[203] where we introduced a multidimensional embedded one-dimensional reaction coordinate that reflected multivariate descriptors for the different states along unfolding pathways. Examples of its usage are shown in Figures 7b and 8b where it enabled the refolded state of EnHD to be successfully identified and was able to efficiently discriminate the TSE from the native and denatured states of the Fyn SH3 domain.

The one-dimensional reaction coordinate was constructed by compilation of 15 properties for each conformation within a trajectory; chosen properties were native and nonnative contacts, radius of gyration, end-to-end distance, fraction of α-helix and β-sheet content, and various SASA measures.[203] The properties are normalized and then used to calculate a distance in property space between each individual conformation in an unfolding trajectory, i, and that of a structure from a reference ensemble, j, such as the native state ensemble. The distances in the 15-dimensional property space, d_{prop}, were calculated using

$$d_{prop} = \sqrt{\frac{(a_i - a_j)^2 + (b_i - b_j)^2 + \cdots + (N_i - N_j)^2}{N_p}} \qquad [9]$$

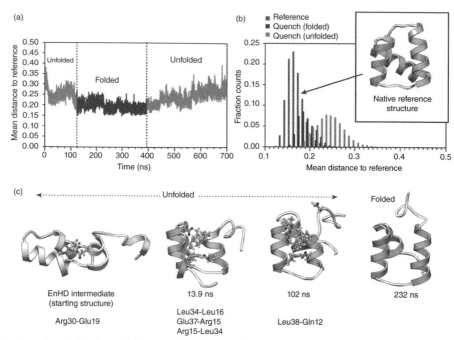

Figure 8 Refolding of the engrailed homeodomain (EnHD) using the quench method adapted from McCully et al.[111] (a) and (b) show the results of property space analysis in identifying the refolded, folded, and unfolded conformations and the relative points in the trajectory. (c) Snapshots taken from the trajectory illustrating the refolding and highlighting the gradual formation of five contacts identified as critical in the folding of EnHD. With permission by Elsevier, *Biophys. J.*, 2010, Vol 99.

where *a*, *b*, and so on represent each of the chosen normalized properties, and N_p is the number of properties being considered.

Hence, a mean distance is determined for an individual conformation as an average to all the distances determined to the reference ensemble; histograms of the mean distances provide a simple discrimination between the different states (Figure 7b). For the tyrosine kinase Fyn SH3 domain, the native and denatured states are well separated in this one-dimensional representation of the property space (Figure 7b). The TSE is a subset of conformations continuous in time within the trajectory that are positioned within the region between the NS and DS populations, with no overlap with the NS evident.

Protein folding intermediates are another metastable state that are challenging to isolate from MD trajectories. The existence of these intermediates is generally expected where the protein size surpasses 100 residues, which encompasses the majority of proteins *in vivo*.[207] However, extensive research has established that small or single-domain proteins are not prohibited from having intermediates.[207] Plausible intermediate conformations can be generated by inclusion of experimental restraints into MD simulations as discussed previously for TSEs. One noteworthy example of this approach for the elucidating of intermediate state ensembles is that of the bacterial immunity protein Im7 using protection factors as restraints.[208] Again, multidimensional approaches similar to the conformational clustering and property space analyses illustrated in Figure 7 are also powerful ways to isolate intermediates; conformational clustering was successful in identifying intermediate structures for barnase[178] and EnHD.[149] In fact, in the case of EnHD the intermediate was predicted by MD[149] and five years later was confirmed through NMR structure determination.[209] Property space analysis using 32 different properties was also able to identify successfully all states from the folding pathway of EnHD (Figure 8), a population corresponding to the intermediate was easily established from those of the native, transition, and denatured states.[210]

PROTEIN FOLDING AND REFOLDING

Thus far we have discussed the investigation of the unfolding pathway from experimentally solved structures to derive conclusions for the reverse folding pathway. This methodology was driven by the requirement that we have a well-defined starting structure to instigate simulations; with the abundance of experimentally solved native structures being greater than those obtained for nonnative states, access to the study of the unfolding pathway has had a much reduced barrier. Early reluctance to commit resources to the study of folding was compounded by the belief that exploration of early stage protein folding would remain inaccessible to MD simulations for some time, until the increase in computational power allowed such calculations to be feasible.[211] Despite these limitations, there were early attempts to capture the folding pathway using simplified models or extended structures.[12,212,213] Here we summarize the key research where native state coordinates did not limit the study to just the unfolding pathways.

Many folding simulations have used artificially extended structures for the initial coordinates.[12,212–215] In one of the earliest studies, Levitt and Warshel[12] were able to fold BPTI to within 7 Å of the native structure using a combination of minimization and heating from a simplified model with all backbone dihedral angles set to 180°. This method of structural minimization from an extended structure was shortly reapplied to the α-helical protein, carp myogen, with all α-helices fixed to assess their potential to be nucleation sites.[213] Despite failing to fold from the fully extended structure, folding to native-like structures did succeed from extended structures of two and four helical fragments of carp myogen, giving credence to the hypothesis that full folding occurs only after some of the helical regions initially form in isolation.[213] More extensive approaches using all-atom representations and a range of solvation environments have been applied since these initial efforts.

The most appropriate contenders for folding simulations are fast folders, such as the designed Trp cage miniprotein,[105] WW domain,[31] λ-repressor protein, and other small proteins and peptides.[216,217] Simulations starting from an extended structure of the Trp cage miniprotein, performed in an attempt to predict the native state structure, were able to isolate a low-energy conformation.[105] A duplicate simulation showed convergence to this same low-energy state.[105] This predicted native state conformation was later confirmed by NMR spectroscopy;[218] the predicted conformation was within a heavy atom RMSD of 1.1 Å from the experimental structure and revealed that folding likely proceeded through the formation of a polyproline (PPII) helix.[105] Similarly, a β-hairpin extracted from the staphylococcal nuclease structure, in which dihedrals were either set to fully extended or PPII conformations, successfully refolded and provided agreement with experimental H/D protection factors.[214]

These approaches to study folding used extended structures often generated by building structures with dihedral angles set to 180°.[148,214,219] Although this simplified approach allowed insights into protein folding, it is given to artificiality on the basis of the probability that denatured states are not as simple as a fully extended polypeptide. Although we may refer to denatured states as "random coil," there is still much discussion over what constitutes the denatured state.[220–223]

More recently, the increasing availability of structural data for nonnative states has allowed researchers to explore the folding pathway through refolding simulations; those simulations use more reliable, and often experimentally verified, nonnative starting structures, rather than being restricted to artificially generated extended conformations. Many refolding studies employ a quench methodology where the protein model is first exposed to denaturing conditions to induce unfolding and then placed under conditions where refolding can be observed.[224] Others use experimental restraints to generate TSs from which refolding can be followed.[225]

As a way of reducing the computational expense, many early unfolding efforts had focused on peptides and fragments taken from full proteins to allow more manageable key components to be used to build up a picture of unfolding. Similarly, the first reported use of the more sophisticated quench methodology was in refolding studies of a fragment of barnase.[226] The fragment was refolded from an unfolded conformation obtained through elevated temperature simulations to understand two aspects, the role of a potential initiation site for folding and β-sheet formation, which at the time were poorly understood. In agreement with experimental data, they concluded that the fragment was a plausible initiation site for folding. Although some nonnative contacts persisted, refolding of the unfolded conformations into a β-hairpin was observed with native contacts present as in the full native state barnase. It was further established that simultaneous formation of side-chain interactions and the backbone hydrogen bonding network between the strands stabilize the folded conformation.[226] Two other β-hairpins that were studied using this method are the designed miniprotein CLN025[227] and a hairpin excised from protein G.[154] The first, CLN025, was refolded from a partially unfolded conformation obtained through forced unfolding simulations, and the second refolded from putative TS structures generated by elevated temperature simulations. Although the β-hairpin from protein G successfully refolded through an initial hydrophobic collapse, that of CLN025 misfolded to an unstable hairpin-like state.[227]

Attempts to refold a full protein began with ubiquitin: the first using changes in solvent to drive the denaturation and refolding where movement from alcoholic conditions (60% methanol) to pure water induced refolding of ubiquitin.[228] The hydrophobic collapse to a near-native fold was also observed by refolding thermally denatured ubiquitin conformations at lower temperature. [229] Two notable case studies, in terms of length as well as being substantial refolding attempts, are the 200 ns and 1 μs refolding simulation of the 36 residue villin headpiece subdomain (HP-36) performed by Duan et al.[26,27] An extremely high-temperature simulation (1000 K) was used to generate an unfolded conformation of the villin headpiece and the initial stages of folding investigated through inducing refolding at the lower temperature of 300 K. Refolding was successful from the unfolded structure, which contained no helical structure and only 3% of native contacts. However, folding did not reach completion; a metastable state persisted for ~150 ns where all but the side-chains failed to reach their native state positions. As the 1 μs simulation is 10-fold shorter than the estimated folding rate for HP-36, it was believed that this metastable state was an intermediate. Together with other refolding simulations of the villin headpiece,[32,216,230,231] a consistent picture of the early folding stages of the HP-36 was obtained with subsequent refolding attempts reaching the native state. Common to all trajectories was an initial rapid hydrophobic collapse, where appearance of secondary structure was driven by the burial of the hydrophobic groups, followed by a slower conformational rearrangement about the hydrophobic core concurrent with expulsion of trapped water molecules.[26,230–233] In contrast, the later stages of the folding pathway are still under debate, some of which has been attributed to the use of unrealistic denatured states.[232] This conflict is not surprising given that none of these refolding attempts had experimentally verified nonnative starting structures.

A more stringent study preceded the experimental verification of the EnHD folding intermediate structure by NMR spectroscopy,[209] a structure that had already been predicted from elevated temperature unfolding simulations,[149,151] as mentioned earlier. To probe the later stages of the EnHD folding pathway, 46 independent simulations were started from the intermediate structure and refolding induced by quenching the temperature to one that was favorable for folding to occur (319 K).[110] The average simulation time was 326 ns to ensure the probability that, on the basis of the experimental half-life of 15 µs, at least one simulation would go to completion and full refolding would be observed.

Refolding was assessed through property space calculations using multiple properties determined as being indicative of correct refolding, with data from 1249 ns of native state simulation used as a reference (Figures 8a and b).[111] The property space consisted of 35 properties that allowed the unfolded and folded conformational ensembles to be distinguished from the native state reference (Figure 8b). Consequently, a segment of the trajectory over which folding to the native state had been attained could be identified (Figure 8a). The folding pathway of EnHD is given in Figure 8c showing the progression from the intermediate starting structure, through gradual refolding conformations until folding had completed, as shown at 232 ns. As folding proceeded native contacts gradually appeared, as highlighted for the five native contacts between H1 and H2 crucial to successful refolding as discussed in McCully et al.[111] (Figure 8c). Interestingly, the order of the contacts formed were the reverse of those lost in the elevated temperature unfolding of EnHD and in the multimolecule test-tube simulations.[156]

FOLDING IN FAMILIES

As an increasing number of protein structures were solved, one question became pertinent: do proteins or domains with similar fold topologies share protein folding mechanisms? It appears there is still much disagreement and contradiction in attempts to determine the level of protein folding conservation within protein families; it remains unknown whether the transition states, intermediate states, or even residues within the folding nuclei are conserved across all families. A number of MD simulations have been performed on related folds, such as the 9th and 10th fibronectin domains from titin,[168] to determine if the folding pathways share any similarities.[155,198,234,235] In some cases, there appears to be growing evidence that there is conservation in the pathways, folding nuclei, transition states, and presence of intermediate states within fold families.[168,200,236,237]

The largest endeavor so far has been our Dynameomics initiative.[200,238–241] By first consolidating a view of the protein fold universe, by generating a consensus domain dictionary of all known folds,[242,243] we selected representative structures as targets to be studied from each fold. In total, 807 targets were simulated under native state conditions and the unfolding pathways examined through elevated temperature simulations.[200,238,241,244] Targets selected to represent the five most populated folds are shown in Figure 9a. These folds are the Ig-like β-sandwich, the flavodoxin-like, TIM β/α-barrel, and the ferrodoxin-like and 3-α-helical bundle folds. By focusing on representatives from a consolidated view of protein fold space, our work provides

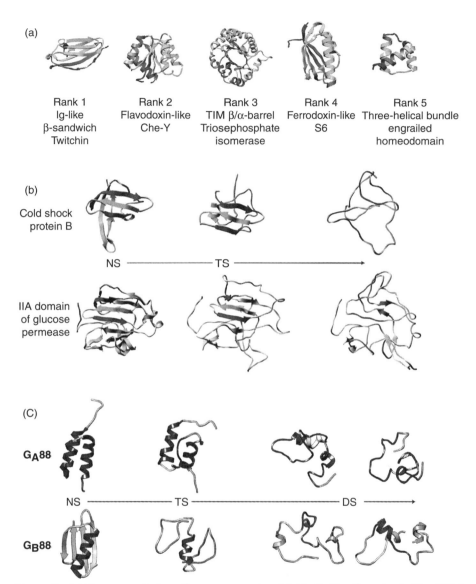

Figure 9 (a) Dynameomics' top five most populated folds: Ig-like, flavodoxin-like, TIM β/α barrel, ferrodoxin-like, and three-helical bundle. (b) Unfolding of the β-hairpin motif within similar β-sheet contexts. (c) G_A88 and G_B88 unfolding pathways showing that the different final conformations for these similar sequences (88% identity) are committed to by the time the transition state (TS) is reached and that the departure between the two folding pathways begins early on within the denatured states (DS).

atomic-level detail of the unfolding pathways for nearly all known folds (97%).[244,245]

We have been systematically analyzing both the native state dynamics and folding pathways of these representative targets.[238,239,241,246,247] The A and B domains of the cold shock protein, CspA and CspB, have the same folds and were shown to share similar unfolding pathways.[247] Further, the role of this β-hairpin motif in unfolding was examined both within the context of different folds and in isolation: FBP28 WW domain, CspA, CspB, and glucose permease IIA domain.[241] The WW domain has a simple motif composed of 3 β-strands in a double hairpin formation, a motif that is repeated in CspA, CspB, and the glucose permease IIA domain. The unfolding of this hairpin motif in CspA and CspB matches that observed for the WW domain; however, within the larger IIA domain the unfolding of the hairpin differs. The differences between the unfolding pathways of CspB and the IIA domain are shown in Figure 9b. Although the nature of the double hairpin in the transition states appears similar, the strands are seen to pull apart in different manners from this point onward (Figure 9b). The effect of the surrounding IIA domain structure on changing the unfolding of the hairpin is compounded by a comparison of the double hairpin motifs extracted from the four structures. In this case, the hairpins unfolded in a similar manner regardless of the different amino acid sequences that formed them, including that for the IIA domain.[247]

Clearly, there is a matter of context in folding pathways. Where some proteins that share a fold demonstrate different unfolding mechanisms, despite isolated components from within them exhibiting similarities, other similar folds do share unfolding mechanisms despite differences in the amino acid sequences. So, just how similar must two sequences be to fold the same? In 2007, Alexander et al.[248–250] designed two protein domains, named $G_A 88$ and $G_B 88$, that had 88% sequence identity yet different folds building on their prior success in generating the same pair with 59% sequence identity. Both domains originate from protein G: the all-α G_A binding domain and the β+α G_B binding domain. Through mutation they were able to bring both domains to 88% sequence identity while still retaining their original wild-type folds (Figure 9c). Even at 95% and 98% sequence similarity, these two protein domains retain their different all-α and β+α wild-type folds.[249,251] While experiment can only tell us that the structure of these two proteins is indeed different and what critical mutation may result in the transformation from one fold into the other, it lacks in providing the atomic-level detail explaining how and why they fold as they do. As MD simulations can provide this missing detail we addressed this by performing numerous MD simulations of these proteins and found that residual structure in the denatured state and contacts in the very early stages of folding determine which fold is adopted.[252,253] The predictions were recently confirmed by experiment.[253] To investigate the 88% identity domains, two slightly different approaches, both using MD simulations, were recently taken in an attempt to resolve the puzzle surrounding these different folds with high sequence identity.[253,254]

Elevated temperature unfolding simulations showed that both proteins have similar helical content in the denatured state; in the case of $G_B 88$ this means that the helical content of the denatured state is higher than in the native state (Figure 9c). However, the simulations demonstrated a commitment to one of the two different topologies that occurred very early in the folding process. By the time the transition state was reached the two proteins had assumed their general topologies with hairpins being formed between the residues involved in the β-sheets in the native state of $G_B 88$ (Figure 9c).[253] Allison et al.[254] also simulated the $G_A 88$ and $G_B 88$ structures, albeit for significantly shorter time lengths. They compared the dynamics of the 88% structures with the more recent $G_A 95$ and $G_B 95$ structures and performed *in silico* mutations to form "crossover" structures where the $G_A 88/95$ structures had the $G_B 88/95$ sequences and *vice versa*. This chapter underscored the utility of studying things *in silico* that cannot be broached via experiment; in this case it was the homology modeling of sequences onto different structures and calculation of the energy penalties involved that would result in one structure being favored over the other for a given sequence. Although minimal structural changes were observed in these simulations, likely a result of the relatively short production dynamics, both these studies concluded that it is local interactions that determine the fold topology outcome.

CONCLUSIONS AND OUTLOOK

The longest simulations to date have been the result of specially designed hardware, with the most recent achievement being a 1 ms simulation of BPTI.[43,47] The built-for-purpose designed hardware is a parallel supercomputer, called Anton, for which the architecture of the chips was specially designed to be efficient for the calculations required for MD simulation.[255] Using built-for-purpose hardware makes the timescales required to monitor protein folding more accessible. Even where access to such specialized architectures is restricted, simulations are now sampling timescales more relevant to the study of protein folding than in past years, albeit at the lower end of the protein folding range suitable for the study of fast folders, such as the WW domain, mutants of which fold in less than 15 μs.[31] Hence, sufficient coverage of the timescales required to monitor the protein folding of all but the smaller sized proteins are still out of comfortable reach and the nonequilibrium, denaturing methods introduced here still have their place.

Although special-purpose parallelized designs are rare, other technological advances are still being made to improve the efficiency of MD simulations. Effective parallelization of widely available MD packages is an ongoing effort, with repeated demonstrations of improved performance evident; a recent review places the current capability of these packages to be between 100 and 500 ns a day on the basis of simulations of dihydrofolate reductase.[255] Another area gathering momentum is the use of graphics processing units (GPUs) for MD simulations.[49,256,257] The movement toward using GPUs over CPUs is based on their differing architectures; GPUs were engineered to perform single-chip-based parallel computations to enable efficient rendering of high-quality graphics. Although the computation per chip may be highly efficient, the current drawback with using GPUs is the latency times for communication between chips, which is slower than the parallelization possible across many CPUs on current HPC machines.[256]

As noted by Vendruscolo and Dobson,[258] at the current rate of advancement in this field it is likely that we will see the second barrier being broken in the 2020s and extending past the minute timescale as we enter the 2030s. Given that protein folding occurs on the timescale of nanoseconds to seconds,[48] it is exciting that within 10 years we may be able to simulate long enough to observe folding events under equilibrium conditions without resorting to advanced sampling and reduced models. Until then, the power of data mining collective results from MD simulations, such as that pioneered through the Dynameomics initiative, should not be ignored.[238–243,245,246,259–261] The improved sampling provided by performing multiple independent simulations, even of restricted length, can together elucidate valuable relationships relevant to solving the protein folding problem.

ACKNOWLEDGMENT

This chapter is based in part on the work supported by NIH grant GM 50789 and through computational time and resources provided by the National Energy Research Supercomputing Center, supported by the Office of Science of the US Department of Energy under contract No. DE-AC02-05CH11231, for which we are grateful.

REFERENCES

1. P. E. Wright and H. J. Dyson, *J. Mol. Biol.*, **293**, 321 (1999). Intrinsically Unstructured Proteins: Re-assessing the Protein Structure-Function Paradigm.

2. P. Cossio, A. Trovato, F. Pietrucci, F. Seno, A. Maritan, and A. Laio, *PLoS Comput. Biol.*, **6**, e1000957 (2010). Exploring the Universe of Protein Structures Beyond the Protein Data Bank.

3. J. Skolnick, A. K. Arakaki, S. Y. Lee, and M. Brylinski, *Proc. Natl. Acad. Sci. U.S.A.*, **106**, 15690 (2009). The Continuity of Protein Structure Space is an Intrinsic Property of Proteins.

4. M. Vendruscolo and C. M. Dobson, *Proc. Natl. Acad. Sci. U.S.A.*, **102**, 5641 (2005). A Glimpse at the Organization of the Protein Universe.

5. H. J. Dyson and P. E. Wright, *Nat. Rev. Mol. Cell Bio.*, **6**, 197 (2005). Intrinsically Unstructured Proteins and their Functions.

6. C. B. Anfinsen, *Les Prix Nobel en*, **1972**, 103 (1973). Studies on the Principles that Govern the Folding of Protein Chains.

7. R. Zwanzig, A. Szabo, and B. Bagchi, *Proc. Natl. Acad. Sci. U.S.A.*, **89**, 20 (1992). Levinthal's Paradox.

8. C. Levinthal, *J. Med. Phys.*, **65**, 44 (1968). Are there Pathways for Protein Folding?

9. A. I. Bartlett and S. E. Radford, *Nat. Struct. Mol. Biol.*, **16**, 582 (2009). An Expanding Arsenal of Experimental Methods Yields an Explosion of Insights into Protein Folding Mechanisms.

10. A. Rahman, *Phys. Rev. A*, **136**, A405 (1964). Correlations in Motion of Atoms in Liquid Argon.

11. J. McCammon, B. Gelin, and M. Karplus, *Nature*, **267**, 585 (1977). Dynamics of Folded Proteins.

12. M. Levitt and A. Warshel, *Nature*, **253**, 694 (1975). Computer Simulation of Protein Folding.

13. O. B. Ptitsyn and A. A. Rashin, *Biophys. Chem.*, **3**, 1 (1975). A Model of Myoglobin Self-Organization.

14. M. Levitt and S. Lifson, *J. Mol. Biol.*, **46**, 269 (1969). Refinement of Protein Conformations using a Macromolecular Energy Minimization Procedure.

15. B. R. Gelin and M. Karplus, *Proc. Natl. Acad. Sci. U.S.A.*, **72**, 2002 (1975). Sidechain Torsional Potentials and Motion of Amino Acids in Proteins: Bovine Pancreatic Trypsin Inhibitor.

16. W. F. van Gunsteren and M. Karplus, *Biochemistry*, **21**, 2259 (1982). Protein Dynamics in Solution and in a Crystalline Environment: A Molecular Dynamics Study.

17. W. F. van Gunsteren and H. J. Berendsen, *J. Mol. Biol.*, **176**, 559 (1984). Computer Simulation as a Tool for Tracing the Conformational Differences between Proteins in Solution and in the Crystalline State.

18. W. F. van Gunsteren, H. J. Berendsen, J. Hermans, W. G. Hol, and J. P. Postma, *Proc. Natl. Acad. Sci. U.S.A.*, **80**, 4315 (1983). Computer Simulation of the Dynamics of Hydrated Protein Crystals and its Comparison with X-Ray Data.

19. M. Levitt and R. Sharon, *Proc. Natl. Acad. Sci. U.S.A.*, **85**, 7557 (1988). Accurate Simulation of Protein Dynamics in Solution.

20. V. Daggett and M. Levitt, *J. Mol. Biol.*, **223**, 1121 (1992). Molecular Dynamics Simulations of Helix Denaturation.

21. V. Daggett, P. A. Kollman, and I. D. Kuntz, *Biopolymers*, **31**, 1115 (1991). A Molecular Dynamics Simulation of Polyalanine: An Analysis of Equilibrium Motions and Helix-Coil Transitions.

22. D. J. Tobias, J. E. Mertz, and C. L. Brooks, III, *Biochemistry*, **30**, 6054 (1991). Nanosecond Time Scale Folding Dynamics of a Pentapeptide in Water.

23. V. Daggett and M. Levitt, *Proc. Natl. Acad. Sci. U.S.A.*, **89**, 5142 (1992). A Model of the Molten Globule State from Molecular Dynamics Simulations.

24. V. Daggett and M. Levitt, *Curr. Opin. Struct. Biol.*, **4**, 291 (1994). Protein Folding↔Unfolding Dynamics.

25. M. Karplus and A. Sali, *Curr. Opin. Struct. Biol.*, **5**, 58 (1995). Theoretical Studies of Protein Folding and Unfolding.

26. Y. Duan and P. A. Kollman, *Science*, **282**, 740 (1998). Pathways to a Protein Folding Intermediate Observed in a 1-Microsecond Simulation in Aqueous Solution.

27. Y. Duan, L. Wang, and P. A. Kollman, *Proc. Natl. Acad. Sci. U.S.A.*, **95**, 9897 (1998). The Early Stage of Folding of Villin Headpiece Subdomain Observed in a 200-Nanosecond Fully Solvated Molecular Dynamics Simulation.

28. A. Li and V. Daggett, *Protein Eng.*, **8**, 1117 (1995). Investigation of the Solution Structure of Chymotrypsin Inhibitor 2 using Molecular Dynamics: Comparison to X-ray Crystallographic and NMR Data.

29. V. Daggett, *Curr. Opin. Struct. Biol.*, **10**, 160 (2000). Long Timescale Simulations.

30. V. Wong and D. A. Case, *J. Phys. Chem. B*, **112**, 6013 (2008). Evaluating Rotational Diffusion from Protein MD Simulations.

31. P. L. Freddolino, F. Liu, M. Gruebele, and K. Schulten, *Biophys. J.*, **94**, L75 (2008). Ten-Microsecond Molecular Dynamics Simulation of a Fast-Folding WW Domain.

32. D. L. Ensign, P. M. Kasson, and V. S. Pande, *J. Mol. Biol.*, **374**, 806 (2007). Heterogeneity Even at the Speed Limit of Folding: Large-Scale Molecular Dynamics Study of a Fast-Folding Variant of the Villin Headpiece.

33. M. M. Seibert, A. Patriksson, B. Hess, and D. van der Spoel, *J. Mol. Biol.*, **354**, 173 (2005). Reproducible Polypeptide Folding and Structure Prediction using Molecular Dynamics Simulations.

34. P. Maragakis, K. Lindorff-Larsen, M. P. Eastwood, R. O. Dror, J. L. Klepeis, I. T. Arkin, M. Ø. Jensen, H. Xu, N. Trbovic, R. A. Friesner, A. G. Palmer, III,, and D. E. Shaw, *J. Phys. Chem. B*, **112**, 6155 (2008). Microsecond Molecular Dynamics Simulation Shows Effect of Slow Loop Dynamics on Backbone Amide Order Parameters of Proteins.

35. K. Martínez-Mayorga, M. C. Pitman, A. Grossfield, S. E. Feller, and M. F. Brown, *J. Am. Chem. Soc.*, **128**, 16502 (2006). Retinal Counterion Switch Mechanism in Vision Evaluated by Molecular Simulations.

36. A. Grossfield, M. C. Pitman, S. E. Feller, O. Soubias, and K. Gawrisch, *J. Mol. Biol.*, **381**, 478 (2008). Internal Hydration Increases During Activation of the G-Protein-Coupled Receptor Rhodopsin.

37. R. O. Dror, D. H. Arlow, D. W. Borhani, M. Ø. Jensen, S. Piana, and D. E. Shaw, *Proc. Natl. Acad. Sci. U.S.A.*, **106**, 4689 (2009). Identification of Two Distinct Inactive Conformations of the β2-Adrenergic Receptor Reconciles Structural and Biochemical Observations.

38. J. L. Klepeis, K. Lindorff-Larsen, R. O. Dror, and D. E. Shaw, *Curr. Opin. Struct. Biol.*, **19**, 120 (2009). Long-Timescale Molecular Dynamics Simulations of Protein Structure and Function.

39. J. Phillips, R. Braun, W. Wang, J. Gumbart, E. Tajkhorshid, E. Villa, C. Chipot, R. Skeel, L. Kale, and K. Schulten, *J. Comput. Chem.*, **26**, 1781 (2005). Scalable Molecular Dynamics with NAMD.

40. J. C. Phillips, G. Zheng, S. Kumar, and L. V. Kale. NAMD: Biomolecular Simulation on Thousands of Processors, *Supercomputing, ACM/IEEE 2002 Conference*, 36 (2002).

41. K. Y. Sanbonmatsu and C.-S. Tung, *J. Struct. Biol.*, **157**, 470 (2007). High Performance Computing in Biology: Multimillion Atom Simulations of Nanoscale Systems.

42. A. Bhatelé, L. V. Kalé, and S. Kumar. Dynamic Topology Aware Load Balancing Algorithms for Molecular Dynamics Applications, *Proceedings of the 23rd International Conference on Supercomputing (ICS'09)*, 110 (2009).

43. D. E. Shaw, P. Maragakis, K. Lindorff-Larsen, S. Piana, R. O. Dror, M. P. Eastwood, J. A. Bank, J. M. Jumper, J. K. Salmon, Y. Shan, and W. Wriggers, *Science*, **330**, 341 (2010). Atomic-Level Characterization of the Structural Dynamics of Proteins.

44. S. Piana, K. Lindorff-Larsen, and D. E. Shaw, *Proc. Natl. Acad. Sci. U.S.A.* (2012). Protein Folding Kinetics and Thermodynamics from Atomistic Simulation.

45. K. Lindorff-Larsen, S. Piana, R. O. Dror, and D. E. Shaw, *Science*, **334**, 517 (2011). How Fast-Folding Proteins Fold.

46. K. Lindorff-Larsen, N. Trbovic, P. Maragakis, S. Piana, and D. E. Shaw, *J. Am. Chem. Soc.*, **134**, 3787 (2012). Structure and Dynamics of an Unfolded Protein Examined by Molecular Dynamics Simulation.

47. D. E. Shaw, K. J. Bowers, E. Chow, M. P. Eastwood, D. J. Ierardi, J. L. Klepeis, J. S. Kuskin, R. H. Larson, K. Lindorff-Larsen, P. Maragakis, M. A. Moraes, R. O. Dror, S. Piana, Y. Shan, B. Towles, J. K. Salmon, J. P. Grossman, K. M. Mackenzie, J. A. Bank, C. Young, M. M. Deneroff, and B. Batson, 1 (2009). Millisecond-Scale Molecular Dynamics Simulations on Anton, *Proceedings of the Conference on High Performance Computing Networking, Storage and Analysis (SC'09)*.

48. S. E. Radford, *Trends Biochem. Sci*, **25**, 611 (2000). Protein Folding: Progress Made and Promises Ahead.

49. P. L. Freddolino, C. B. Harrison, Y. Liu, and K. Schulten, *Nat. Phys.*, **6**, 751 (2010). Challenges in Protein-Folding Simulations.

50. K. Klenin, B. Strodel, D. J. Wales, and W. Wenzel, *Biochim. Biophys. Acta*, **1814**, 977 (2011). Modelling Proteins: Conformational Sampling and Reconstruction of Folding Kinetics.

51. F. Jensen, *Introduction to Computational Chemistry*, John Wiley & Sons, Ltd., Chichester, England; Hoboken, NJ, 2007.

52. A. R. Leach, *Molecular Modelling: Principles and Applications*, Prentice Hall, Harlow, 2001.

53. T. Schlick, R. Collepardo-Guevara, L. A. Halvorsen, S. Jung, and X. Xiao, *Q. Rev. Biophys.*, **44**, 191 (2011). Biomolecular Modeling and Simulation: A Field Coming of Age.

54. M. P. Allen, in *Computational Soft Matthew: From Synthetic Polymers to Proteins*, N. Attig, K. Binder, H. Grubmüller, and K. Kremer (Eds.), John von Neumann fur Computing, NIC Series Volume, 2004, pp. 1–28, Introduction to Molecular Dynamics Simulation.

55. D. A. C. Beck and V. Daggett, *Methods*, **34**, 112 (2004). Methods for Molecular Dynamics Simulations of Protein Folding/Unfolding in Solution.

56. H. A. Scheraga, M. Khalili, and A. Liwo, *Annu. Rev. Phys. Chem.*, **58**, 57 (2012). Protein-Folding Dynamics: Overview of Molecular Simulation Techniques.

57. B. R. Brooks, C. L. Brooks, III, A. D. MacKerell, Jr., L. Nilsson, R. Petrella, B. Roux, Y. Won, G. Archontis, C. Bartels, and S. Boresch, *J. Comput. Chem.*, **30**, 1545 (2009). CHARMM: The Biomolecular Simulation Program.

58. D. A. Case, T. E. Cheatham, III, T. Darden, H. Gohlke, R. Luo, K. M. Merz, A. Onufriev, C. Simmerling, B. Wang, and R. J. Woods, *J. Comput. Chem.*, **26**, 1668 (2005). The Amber Biomolecular Simulations Programs.

59. M. Christen, P. H. Hünenberger, D. Bakowies, R. Baron, R. Bürgi, D. P. Geerke, T. N. Heinz, M. A. Kastenholz, V. Kräutler, C. Oostenbrink, C. Peter, D. Trzesniak, and W. F. van Gunsteren, *J. Comput. Chem.*, **26**, 1719 (2005). The GROMOS Software for Biomolecular Simulation: GROMOS05.

60. W. L. Jorgensen and J. Tirado-Rives, *J. Am. Chem. Soc.*, **110**, 1657 (1988). The OPLS Potential Functions for Proteins, Energy Minimizations for Crystals of Cyclic Peptides and Crambin.

61. M. Levitt, *Nat. Struct. Biol.*, **8**, 392 (2001). The Birth of Computational Structural Biology.

62. M. Levitt, M. Hirshberg, R. Sharon, and V. Daggett, *Comput. Phys. Commun.*, **91**, 215 (1995). Potential Energy Function and Parameters for Simulations of the Molecular Dynamics of Proteins and Nucleic Acids in Solution.

63. M. Levitt, *J. Mol. Biol.*, **168**, 595 (1983). Molecular Dynamics of Native Protein. I. Computer Simulation of Trajectories.

64. D. A. C. Beck, M. E. McCully, D. O. V. Alonso, and V. Daggett, In Lucem *Molecular Mechanics (ilmm)*, University of Washington, 2000.

65. D. R. Nutt and J. C. Smith, *J. Chem. Theory Comput.*, **3**, 1550 (2007). Molecular Dynamics Simulations of Proteins: Can the Explicit Water Model be Varied?

66. B. Hess and N. F. A. van der Vegt, *J. Phys. Chem. B*, **110**, 17616 (2006). Hydration Thermodynamic Properties of Amino Acid Analogues: A Systematic Comparison of Biomolecular Force Fields and Water Models.

67. D. J. Price and C. L. Brooks, III, *J. Chem. Phys.*, **121**, 10096 (2004). A Modified TIP3P Water Potential for Simulation with Ewald Summation.

68. K. A. Beauchamp, Y. S. Lin, and R. Das, *J. Chem. Theory Comput.* (2012). Are Protein Force Fields Getting Better? A Systematic Benchmark on 524 Diverse NMR Measurements.

69. S. J. Weiner, P. A. Kollman, D. A. Case, U. C. Singh, C. Ghio, G. Alagona, S. Profeta, Jr., and P. Weiner, *J. Am. Chem. Soc.*, **106**, 765 (1984). A New Force Field for Molecular Mechanical Simulation of Nucleic Acids and Proteins.

70. E. A. Cino, W.-Y. Choy, and M. Karttunen, *J. Chem. Theory Comput.*, **8**, 2725 (2012). Comparison of Secondary Structure Formation using 10 Different Force Fields in Microsecond Molecular Dynamics Simulations.

71. V. Hornak, R. Abel, A. Okur, B. Strockbine, A. Roitberg, and C. Simmerling, *Proteins: Struct., Funct., Bioinf.*, **65**, 712 (2006). Comparison of Multiple AMBER Force Fields and Development of Improved Protein Backbone Parameters.

72. R. B. Best, N.-V. Buchete, and G. Hummer, *Biophys. J.*, **95**, L07 (2008). Are Current Molecular Dynamics Force Fields too Helical?

73. K. Lindorff-Larsen, P. Maragakis, S. Piana, M. P. Eastwood, R. O. Dror, and D. E. Shaw, *PLoS One*, **7**, e32131 (2012). Systematic Validation of Protein Force Fields Against Experimental Data.

74. T. Lazaridis and M. Karplus, *Proteins Struct., Funct., Genet.*, **35**(133) (1999). Effective Energy Function for Proteins in Solution.

75. J.-E. Shea, M. R. Friedel, and A. Baumketner, in *Reviews in Computational Chemistry*, K. B Lipkowitz, T. R. Cundari, and V. J. Gillet (Eds.), Wiley-VCH Inc., New York, 2006, Vol. 22, pp. 169–228, Simulations of Protein Folding.

76. M. G. Saunders and G. A. Voth, *Curr. Opin. Struct. Biol.*, **22**, 144 (2012). Coarse-Graining of Multiprotein Assemblies.

77. A. V. Sinitskiy, M. G. Saunders, and G. A. Voth, *J. Phys. Chem. B*, **116**, 8363 (2012). Optimal Number of Coarse-Grained Sites in Different Components of Large Biomolecular Complexes.

78. S. Takada, *Curr. Opin. Struct. Biol.*, **22**, 130 (2012). Coarse-Grained Molecular Simulations of Large Biomolecules.

79. J. Zhang, W. Li, J. Wang, M. Qin, L. Wu, Z. Yan, W. Xu, G. Zuo, and W. Wang, *IUBMB Life*, **61**, 627 (2009). Protein Folding Simulations: From Coarse-Grained Model to All-Atom Model.

80. C. Chen, P. Depa, V. G. Sakai, J. K. Maranas, J. W. Lynn, I. Peral, and J. R. D. Copley, *J. Chem. Phys.*, **124**, 234901 (2006). A Comparison of United Atom, Explicit Atom, and Coarse-Grained Simulation Models for Poly(ethylene oxide).

81. W. L. Jorgensen, J. Chandrasekhar, J. D. Madura, R. W. Impey, and M. L. Klein, *J. Chem. Phys.*, **79**, 926 (1983). Comparison of Simple Potential Functions for Simulating Liquid Water.

82. H. Berendsen, J. Postma, W. F. van Gunsteren, and J. Hermans, *Intermol. Forces*, **11**, 331 (1981). Interaction Models for Water in Relation to Protein Hydration.

83. F. H. Stillinger and A. Rahman, *J. Chem. Phys.*, **60**, 1545 (1974). Improved Simulation of Liquid Water by Molecular Dynamics.

84. M. Levitt, M. Hirshberg, R. Sharon, K. E. Laidig, and V. Daggett, *J. Phys. Chem. B*, **101**, 5051 (1997). Calibration and Testing of a Water Model for Simulation of the Molecular Dynamics of Proteins and Nucleic Acids in Solution.

85. M. W. Mahoney and W. L. Jorgensen, *J. Chem. Phys.*, **112**, 8910 (2000). A Five-Site Model for Liquid Water and the Reproduction of the Density Anomaly by Rigid, Nonpolarizable Potential Functions.

86. D. Bashford and D. A. Case, *Annu. Rev. Phys. Chem.*, **51**, 129 (2000). Generalized Born Models of Macromolecular Solvation Effects.

87. E. M. Storch and V. Daggett, *Biochemistry*, **34**, 9682 (1995). Molecular Dynamics Simulation of Cytochrome b5: Implications for Protein-Protein Recognition.

88. P. Comba and R. Remenyi, *Coord. Chem. Rev.*, **238**, 9 (2003). Inorganic and Bioinorganic Molecular Mechanics Modeling – The Problem of the Force Field Parameterization.

89. E. Project, E. Nachliel, and M. Gutman, *J. Comput. Chem.*, **29**, 1163 (2008). Parameterization of Ca+2–Protein Interactions for Molecular Dynamics Simulations.

90. A. Hansson, P. C. T. Souza, R. L. Silveira, L. Martínez, and M. S. Skaf, *Int. J. Quantum Chem.*, **111**, 1346 (2011). CHARMM Force Field Parameterization of Rosiglitazone.

91. O. N. de Souza and R. L. Ornstein, *Biophys. J.*, **72**, 2395 (1997). Effect of Periodic Box Size on Aqueous Molecular Dynamics Simulation of a DNA Dodecamer with Particle-Mesh Ewald Method.

92. K. Takemura and A. Kitao, *J. Phys. Chem. B*, **111**, 11870 (2007). Effects of Water Model and Simulation Box Size on Protein Diffusional Motions.

93. J. Higo, H. Kono, N. Nakajima, H. Shirai, H. Nakamura, and A. Sarai, *Chem. Phys. Lett.*, **306**, 395 (1999). Molecular Dynamics Study on Mobility and Dipole Ordering of Solvent Around Proteins: Effects of Periodic-Box Size and Protein Charge.

94. T. A. Wassenaar and A. E. Mark, *J. Comput. Chem.*, **27**, 316 (2006). The Effect of Box Shape on the Dynamic Properties of Proteins Simulated Under Periodic Boundary Conditions.

95. H. Bekker, *J. Comput. Chem.*, **18**, 1930 (1997). Unification of Box Shapes in Molecular Simulations.

96. D. A. C. Beck, R. S. Armen, and V. Daggett, *Biochemistry*, **44**, 609 (2005). Cutoff Size Need Not Strongly Influence Molecular Dynamics Results for Solvated Polypeptides.

97. H. Schreiber and O. Steinhauser, *Biochemistry*, **31**, 5856 (1992). Cutoff Size Does Strongly Influence Molecular Dynamics Results on Solvated Polypeptides.

98. R. Day and V. Daggett, *Proc. Natl. Acad. Sci. U.S.A.*, **102**, 13445 (2005). Ensemble Versus Single-Molecule Protein Unfolding.

99. R. Walser, A. E. Mark, and W. F. van Gunsteren, *Biophys. J.*, **78**, 2752 (2000). On the Temperature and Pressure Dependence of a Range of Properties of a Type of Water Model Commonly Used in High-Temperature Protein Unfolding Simulations.

100. G. S. Kell, *J. Chem. Eng. Data*, **12**, 66 (1967). Precise Representation of Volume Properties of Water at One Atmosphere.

101. Q. Zou, B. J. Bennion, V. Daggett, and K. P. Murphy, *J. Am. Chem. Soc.*, **124**, 1192 (2002). The Molecular Mechanism of Stabilization of Proteins by TMAO and its Ability to Counteract the Effects of Urea.

102. F. U. Hartl and M. Hayer-Hartl, *Nat. Struct. Mol. Biol.*, **16**, 574 (2009). Converging Concepts of Protein Folding *In Vitro* and *In Vivo*.

103. T. Cellmer, M. Buscaglia, E. R. Henry, J. Hofrichter, and W. A. Eaton, *Proc. Natl. Acad. Sci. U.S.A.*, **108**, 6103 (2011). Making Connections Between Ultrafast Protein Folding Kinetics and Molecular Dynamics Simulations.

104. T. R. Sosnick and D. Barrick, *Curr. Opin. Struct. Biol.*, **21**, 12 (2011). The Folding of Single Domain Proteins - Have We Reached a Consensus?

105. C. Simmerling, B. Strockbine, and A. E. Roitberg, *J. Am. Chem. Soc.*, **124**, 11258 (2002). All-Atom Structure Prediction and Folding Simulations of a Stable Protein.

106. D. Bhatt and D. M. Zuckerman, *J. Chem. Theory Comput.*, **7**, 2520 (2011). Beyond Microscopic Reversibility: Are Observable Non-Equilibrium Processes Precisely Reversible?

107. A. R. Dinner and M. Karplus, *J. Mol. Biol.*, **292**, 403 (1999). Is Protein Unfolding the Reverse of Protein Folding? A Lattice Simulation Analysis.

108. R. C. Tolman, *Proc. Natl. Acad. Sci. U.S.A.*, **11**, 436 (1925). The Principle of Microscopic Reversibility.

109. R. Day and V. Daggett, *J. Mol. Biol.*, **366**, 677 (2007). Direct Observation of Microscopic Reversibility in Single-Molecule Protein Folding.

110. M. E. McCully, D. A. C. Beck, and V. Daggett, *Biochemistry*, **47**, 7079 (2008). Microscopic Reversibility of Protein Folding in Molecular Dynamics Simulations of the Engrailed Homeodomain.

111. M. E. McCully, D. A. C. Beck, A. R. Fersht, and V. Daggett, *Biophys. J.*, **99**, 1628 (2010). Refolding the Engrailed Homeodomain: Structural Basis for the Accumulation of a Folding Intermediate.

112. D. B. Kitchen, L. H. Reed, and R. M. Levy, *Biochemistry*, **31**, 10083 (1992). Molecular Dynamics Simulation of Solvated Protein at High Pressure.

113. D. Paschek and A. E. García, *Phys. Rev. Lett.*, **93**, 238105 (2004). Reversible Temperature and Pressure Denaturation of a Protein Fragment: A Replica Exchange Molecular Dynamics Simulation Study.

114. N. Smolin and R. Winter, *Biochim. Biophys. Acta*, **1764**, 522 (2006). A Molecular Dynamics Simulation of SNase and its Hydration Shell at High Temperature and High Pressure.

115. D. Trzesniak, R. D. Lins, and W. F. van Gunsteren, *Proteins: Struct., Funct., Bioinf.*, **65**, 136 (2006). Protein Under Pressure: Molecular Dynamics Simulation of the Arc Repressor.

116. O. Chara, J. R. Grigera, and A. N. McCarthy, *J. Biol. Phys.*, **33**, 515 (2007). Studying the Unfolding Kinetics of Proteins Under Pressure Using Long Molecular Dynamics Simulation Runs.

117. M. D. Collins, G. Hummer, M. L. Quillin, B. W. Matthews, and S. M. Gruner, *Proc. Natl. Acad. Sci. U.S.A.*, **102**, 16668 (2005). Cooperative Water Filling of a Nonpolar Protein Cavity Observed by High-Pressure Crystallography and Simulation.

118. A. N. McCarthy and J. R. Grigera, *Biochim. Biophys. Acta*, **1764**, 506 (2006). Pressure Denaturation of Apomyoglobin: A Molecular Dynamics Simulation Study.

119. E. Paci, *Biochim. Biophys. Acta*, **1595**, 185 (2002). High Pressure Simulations of Biomolecules.

120. S. Sarupria, T. Ghosh, A. E. García, and S. Garde, *Proteins: Struct., Funct., Bioinf.*, **78**, 1641 (2010). Studying Pressure Denaturation of a Protein by Molecular Dynamics Simulations.

121. D. R. Canchi and A. E. García, *Biophys. J.*, **100**, 1526 (2011). Backbone and Side-Chain Contributions in Protein Denaturation by Urea.

122. D. R. Canchi, D. Paschek, and A. E. García, *J. Am. Chem. Soc.*, **132**, 2338 (2010). Equilibrium Study of Protein Denaturation by Urea.

123. L. J. Smith, R. M. Jones, and W. F. van Gunsteren, *Proteins: Struct., Funct., Bioinf.*, **58**, 439 (2004). Characterization of the Denaturation of Human α-Lactalbumin in Urea by Molecular Dynamics Simulations.

124. A. Caballero-Herrera, K. Nordstrand, K. D. Berndt, and L. Nilsson, *Biophys. J.*, **89**, 842 (2005). Effect of Urea on Peptide Conformation in Water: Molecular Dynamics and Experimental Characterization.

125. A. G. Rocco, L. Mollica, P. Ricchiuto, A. M. Baptista, E. Gianazza, and I. Eberini, *Biophys. J.*, **94**, 2241 (2008). Characterization of the Protein Unfolding Processes Induced by Urea and Temperature.

126. M. C. Stumpe and H. Grubmüller, *PLoS Comput. Biol.*, **4**, e1000221 (2008). Polar or Apolar--The Role of Polarity for Urea-Induced Protein Denaturation.

127. L. Hua, R. Zhou, D. Thirumalai, and B. J. Berne, *Proc. Natl. Acad. Sci. U.S.A.*, **105**, 16928 (2008). Urea Denaturation by Stronger Dispersion Interactions with Proteins than Water Implies a 2-Stage Unfolding.

128. M. C. Stumpe and H. Grubmüller, *Biophys. J.*, **96**, 3744 (2009). Urea Impedes the Hydrophobic Collapse of Partially Unfolded Proteins.

129. C. Camilloni, L. Sutto, D. Provasi, G. Tiana, and R. A. Broglia, *Protein Sci.*, **17**, 1424 (2008). Early Events in Protein Folding: Is There Something More Than Hydrophobic Burst?

130. C. Camilloni, A. Guerini Rocco, I. Eberini, E. Gianazza, R. A. Broglia, and G. Tiana, *Biophys. J.*, **94**, 4654 (2008). Urea and Guanidinium Chloride Denature Protein L in Different Ways in Molecular Dynamics Simulations.

131. E. P. O'Brien, R. I. Dima, B. R. Brooks, and D. Thirumalai, *J. Am. Chem. Soc.*, **129**, 7346 (2007). Interactions Between Hydrophobic and Ionic Solutes in Aqueous Guanidinium Chloride and Urea Solutions: Lessons for Protein Denaturation Mechanism.

132. J. Tirado-Rives, M. Orozco, and W. L. Jorgensen, *Biochemistry*, **36**, 7313 (1997). Molecular Dynamics Simulations of the Unfolding of Barnase in Water and 8 M Aqueous Urea.

133. B. J. Bennion and V. Daggett, *Proc. Natl. Acad. Sci. U.S.A.*, **100**, 5142 (2003). The Molecular Basis for the Chemical Denaturation of Proteins by Urea.

134. P. E. Mason, C. E. Dempsey, L. Vrbka, J. Heyda, J. W. Brady, and P. Jungwirth, *J. Phys. Chem. B*, **113**, 3227 (2009). Specificity of Ion-Protein Interactions: Complementary and Competitive Effects of Tetrapropylammonium, Guanidinium, Sulfate, and Chloride Ions.

135. J. L. England and G. Haran, *Annu. Rev. Phys. Chem.*, **62**, 257 (2011). Role of Solvation Effects in Protein Denaturation: From Thermodynamics to Single Molecules and Back.

136. Z. Xia, P. Das, E. I. Shakhnovich, and R. Zhou, *J. Am. Chem. Soc.*, **134**, 18266 (2012). Collapse of Unfolded Proteins in a Mixture of Denaturants.

137. K. Goossens, L. Smeller, J. Frank, and K. Heremans, *Eur. J. Biochem.*, **236**, 254 (1996). Pressure-Tuning the Conformation of Bovine Pancreatic Trypsin Inhibitor Studied by Fourier-Transform Infrared Spectroscopy.

138. B. Wroblowski, J. F. Díaz, K. Heremans, and Y. Engelborghs, *Proteins: Struct., Funct., Bioinf.*, **25**, 446 (1996). Molecular Mechanisms of Pressure Induced Conformational Changes in BPTI.

139. P. Fan, D. Kominos, D. B. Kitchen, R. M. Levy, and J. Baum, *Chem. Phys.*, **158**, 295 (1991). Stabilization of α-Helical Secondary Structure During High-Temperature Molecular-Dynamics Simulations of α-Lactalbumin.

140. A. E. Mark and W. F. van Gunsteren, *Biochemistry*, **31**, 7745 (1992). Simulation of the Thermal Denaturation of Hen Egg White Lysozyme: Trapping the Molten Globule State.

141. V. Daggett and M. Levitt, *J. Mol. Biol.*, **232**, 600 (1993). Protein Unfolding Pathways Explored Through Molecular Dynamics Simulations.

142. A. Caflisch and M. Karplus, *Proc. Natl. Acad. Sci. U.S.A.*, **91**, 1746 (1994). Molecular Dynamics Simulation of Protein Denaturation: Solvation of the Hydrophobic Cores and Secondary Structure of Barnase.

143. A. V. Finkelstein, *Protein Eng.*, **10**, 843 (1997). Can Protein Unfolding Simulate Protein Folding?

144. W. Guo, S. Lampoudi, and J.-E. Shea, *Proteins*, **55**, 395 (2004). Temperature Dependence of the Free Energy Landscape of the Src-SH3 Protein Domain.

145. A. Cavalli, P. Ferrara, and A. Caflisch, *Proteins: Struct., Funct., Bioinf.*, **47**, 305 (2002). Weak Temperature Dependence of the Free Energy Surface and Folding Pathways of Structured Peptides.

146. T. Wang and R. C. Wade, *J. Chem. Theory Comput.*, **3**, 1476 (2007). On the Use of Elevated Temperature in Simulations to Study Protein Unfolding Mechanisms.

147. R. Day, B. J. Bennion, S. Ham, and V. Daggett, *J. Mol. Biol.*, **322**, 189 (2002). Increasing Temperature Accelerates Protein Unfolding Without Changing the Pathway of Unfolding.

148. X. Daura, B. Jaun, D. Seebach, W. F. van Gunsteren, and A. E. Mark, *J. Mol. Biol.*, **280**, 925 (1998). Reversible Peptide Folding in Solution by Molecular Dynamics Simulation.

149. U. Mayor, C. M. Johnson, V. Daggett, and A. R. Fersht, *Proc. Natl. Acad. Sci. U.S.A.*, **97**, 13518 (2000). Protein Folding and Unfolding in Microseconds to Nanoseconds by Experiment and Simulation.

150. U. Mayor, N. R. Guydosh, C. M. Johnson, J. G. Grossmann, S. Sato, G. S. Jas, S. M. V. Freund, D. O. V. Alonso, V. Daggett, and A. R. Fersht, *Nature*, **421**, 863 (2003). The Complete Folding Pathway of a Protein from Nanoseconds to Microseconds.

151. M. L. Demarco, D. O. V. Alonso, and V. Daggett, *J. Mol. Biol.*, **341**, 1109 (2004). Diffusing and Colliding: The Atomic Level Folding/Unfolding Pathway of a Small Helical Protein.

152. Y. Y. Sham, B. Ma, C.-J. Tsai, and R. Nussinov, *Proteins: Struct., Funct., Bioinf.*, **46**, 308 (2002). Thermal Unfolding Molecular Dynamics Simulation of Escherichia Coli Dihydrofolate Reductase: Thermal Stability of Protein Domains and Unfolding Pathway.

153. B. Ma and R. Nussinov, *Protein Eng.*, **16**, 561 (2003). Molecular Dynamics Simulations of the Unfolding of β_2-Microglobulin and its Variants.

154. V. S. Pande and D. S. Rokhsar, *Proc. Natl. Acad. Sci. U.S.A.*, **96**, 9062 (1999). Molecular Dynamics Simulations of Unfolding and Refolding of a β-Hairpin Fragment of Protein G.

155. R. Day and V. Daggett, *Protein Sci.*, **14**, 1242 (2005). Sensitivity of the Folding/Unfolding Transition State Ensemble of Chymotrypsin Inhibitor 2 to Changes in Temperature and Solvent.

156. M. E. McCully, D. A. C. Beck, and V. Daggett, *Proc. Natl. Acad. Sci. U.S.A.*, **109**, 17851 (2012). Multimolecule Test-Tube Simulations of Protein Unfolding and Aggregation.

157. J. Forman and J. Clarke, *Curr. Opin. Struct. Biol.*, **17**, 58 (2007). Mechanical Unfolding of Proteins: Insights into Biology, Structure and Folding.

158. A. F. Oberhauser and M. Carrión-Vázquez, *J. Biol. Chem.*, **283**, 6617 (2008). Mechanical Biochemistry of Proteins One Molecule at a Time.

159. E. M. Puchner and H. E. Gaub, *Curr. Opin. Struct. Biol.*, **19**, 605 (2009). Force and Function: Probing Proteins with AFM-Based Force Spectroscopy.

160. A. Galera-Prat, A. Gómez-Sicilia, A. F. Oberhauser, M. Cieplak, and M. Carrión-Vázquez, *Curr. Opin. Struct. Biol.*, **20**, 63 (2010). Understanding Biology by Stretching Proteins: Recent Progress.

161. M. Rief, M. Gautel, F. Oesterhelt, J. M. Fernandez, and H. E. Gaub, *Science*, **276**, 1109 (1997). Reversible Unfolding of Individual Titin Immunoglobulin Domains by AFM.

162. J. Hsin, J. Strümpfer, E. H. Lee, and K. Schulten, *Annu. Rev. Biophys.*, **40**, 187 (2011). Molecular Origin of the Hierarchical Elasticity of Titin: Simulation, Experiment, and Theory.

163. S. Labeit and B. Kolmerer, *Science*, **270**, 293 (1995). Titins: Giant Proteins in Charge of Muscle Ultrastructure and Elasticity.

164. L. Tskhovrebova, M. L. Walker, J. G. Grossmann, G. N. Khan, A. Baron, and J. Trinick, *J. Mol. Biol.*, **397**, 1092 (2010). Shape and Flexibility in the Titin 11-Domain Super-Repeat.

165. H. Lu, B. Isralewitz, A. Krammer, V. Vogel, and K. Schulten, *Biophys. J.*, **75**, 662 (1998). Unfolding of Titin Immunoglobulin Domains by Steered Molecular Dynamics Simulation.

166. S. Izrailev, S. Stepaniants, B. Isralewitz, D. Kosztin, H. Lu, F. Molnar, W. Wriggers, and K. Schulten, in *Computational Molecular Dynamics: Challenges, Methods, Ideas*, P. Deuflhard, J. Hermans, B. Leimkuhler, A. E. Mark, S. Reich, and R. D. Skeel (Eds.), Springer-Verlag, Berlin, 1999, pp. 39–65, Steered Molecular Dynamics.

167. B. Isralewitz, M. Gao, and K. Schulten, *Curr. Opin. Struct. Biol.*, **11**, 224 (2001). Steered Molecular Dynamics and Mechanical Functions of Proteins.

168. E. Paci and M. Karplus, *J. Mol. Biol.*, **288**, 441 (1999). Forced Unfolding of Fibronectin Type 3 Modules: An Analysis by Biased Molecular Dynamics Simulations.

169. H. Grubmuller, B. Heymann, and P. Tavan, *Science*, **271**, 997 (1996). Ligand Binding: Molecular Mechanics Calculation of the Streptavidin-Biotin Rupture Force.

170. J. M. Fernandez, P. E. Marszalek, H. Lu, H. Li, M. Carrión-Vázquez, A. F. Oberhauser, and K. Schulten, *Nature*, **402**, 100 (1999). Mechanical Unfolding Intermediates in Titin Modules.

171. H. Lu and K. Schulten, *Biophys. J.*, **79**, 51 (2000). The Key Event in Force-Induced Unfolding of Titin's Immunoglobulin Domains.

172. H. Lu and K. Schulten, *Chem. Phys.*, **247**, 141 (1999). Steered Molecular Dynamics Simulation of Conformational Changes of Immunoglobulin Domain I27 Interpret Atomic Force Microscopy Observations.

173. R. Toofanny and P. M. Williams, *J. Mol. Graph. Model.*, **24**, 396 (2006). Simulations of Multi-Directional Forced Unfolding of Titin I27.

174. R. B. Best, B. Li, A. Steward, V. Daggett, and J. Clarke, *Biophys. J.*, **81**, 2344 (2001). Can Non-Mechanical Proteins Withstand Force? Stretching Barnase by Atomic Force Microscopy and Molecular Dynamics Simulation.

175. D. J. Brockwell, G. S. Beddard, E. Paci, D. K. West, P. D. Olmsted, D. A. Smith, and S. E. Radford, *Biophys. J.*, **89**, 506 (2005). Mechanically Unfolding the Small, Topologically Simple Protein L.

176. A. Das and C. Mukhopadhyay, *Proteins: Struct., Funct., Bioinf.*, **75**, 1024 (2009). Mechanical Unfolding Pathway and Origin of Mechanical Stability of Proteins of Ubiquitin Family: An Investigation by Steered Molecular Dynamics Simulation.

177. A. V. Glyakina, N. K. Balabaev, and O. V. Galzitskaya, *J. Chem. Phys.*, **131**, 045102 (2009). Mechanical Unfolding of Proteins L and G with Constant Force: Similarities and Differences.

178. A. Li and V. Daggett, *J. Mol. Biol.*, **275**, 677 (1998). Molecular Dynamics Simulation of the Unfolding of Barnase: Characterization of the Major Intermediate.

179. A. V. Glyakina, N. K. Balabaev, and O. V. Galzitskaya, *Biochemistry (Moscow)*, **74**(316) (2009). Comparison of Transition States Obtained Upon Modeling of Unfolding of Immunoglobulin-Binding Domains of Proteins L and G Caused by External Action with Transition States Obtained in the Absence of Force Probed by Experiments.

180. D. K. West, E. Paci, and P. D. Olmsted, *Phys. Rev. E*, **74**, 061912 (2006). Internal Protein Dynamics Shifts the Distance to the Mechanical Transition State.

181. J. I. Sukowska and M. Cieplak, *Biophys. J.*, **94**, 6 (2008). Stretching to Understand Proteins - A Survey of the Protein Data Bank.

182. M. Sikora, J. I. Sukowska, and M. Cieplak, *PLoS Comput. Biol.*, **5**, e1000547 (2009). Mechanical Strength of 17,134 Model Proteins and Cysteine Slipknots.

183. S. Prakash and A. Matouschek, *Trends Biochem. Sci*, **29**, 593 (2004). Protein Unfolding in the Cell.

184. C. Bustamante, Y. R. Chemla, N. R. Forde, and D. Izhaky, *Annu. Rev. Biochem.*, **73**, 705 (2004). Mechanical Processes in Biochemistry.

185. A. Caflisch and E. Paci, in *Protein Folding Handbook*, J. Buchner and T. Kiefhaber (Eds.), Wiley-VCH Verlag GmbH, 2008, pp. 1143–1169, Molecular Dynamics Simulations to Study Protein Folding and Unfolding.

186. J.-E. Shea and C. L. Brooks, III, *Annu. Rev. Phys. Chem.*, **52**, 499 (2001). From Folding Theories to Folding Proteins: A Review and Assessment of Simulation Studies of Protein Folding and Unfolding.

187. I. Daidone, A. Amadei, D. Roccatano, and A. D. Nola, *Biophys. J.*, **85**, 2865 (2003). Molecular Dynamics Simulation of Protein Folding by Essential Dynamics Sampling: Folding Landscape of Horse Heart Cytochrome C.

188. H. Lei and Y. Duan, *Curr. Opin. Struct. Biol.*, **17**, 187 (2007). Improved Sampling Methods for Molecular Simulation.

189. M. C. Zwier and L. T. Chong, *Curr. Opin. Pharmacol.*, **10**, 745 (2010). Reaching Biological Timescales with All-Atom Molecular Dynamics Simulations.

190. E. Paci, L. J. Smith, C. M. Dobson, and M. Karplus, *J. Mol. Biol.*, **306**, 329 (2001). Exploration of Partially Unfolded States of Human α-Lactalbumin by Molecular Dynamics Simulation.

191. E. Paci, M. Vendruscolo, C. M. Dobson, and M. Karplus, *J. Mol. Biol.*, **324**, 151 (2002). Determination of a Transition State at Atomic Resolution from Protein Engineering Data.

192. E. Paci, A. Caflisch, A. Pluckthun, and M. Karplus, *J. Mol. Biol.*, **314**, 589 (2001). Forces and Energetics of Hapten-Antibody Dissociation: A Biased Molecular Dynamics Simulation Study.

193. C. D. Geierhaas, R. B. Best, E. Paci, M. Vendruscolo, and J. Clarke, *Biophys. J.*, **91**, 263 (2006). Structural Comparison of the Two Alternative Transition States for Folding of TI I27.

194. P. Chen, C.-L. Evans, J. D. Hirst, and M. S. Searle, *Biochemistry*, **50**, 125 (2011). Structural Insights into the Two Sequential Folding Transition States of the PB1 Domain of NBR1 from Φ Value Analysis and Biased Molecular Dynamics Simulations.

195. L. R. Allen and E. Paci, *J. Phys.: Condens. Matter*, **19**, 285211 (2007). Transition States for Protein Folding using Molecular Dynamics and Experimental Restraints.

196. A. R. Fersht, A. Matouschek, and L. Serrano, *J. Mol. Biol.*, **224**, 771 (1992). The Folding of an Enzyme. I. Theory of Protein Engineering Analysis of Stability and Pathway of Protein Folding.

197. J. K. Myers, N. C. Pace, and J. Martin Scholtz, *Protein Sci.*, **4**, 2138 (1995). Denaturant *m* Values and Heat Capacity Changes: Relation to Changes in Accessible Surface Areas of Protein Unfolding.

198. C. D. Geierhaas, E. Paci, M. Vendruscolo, and J. Clarke, *J. Mol. Biol.*, **343**, 1111 (2004). Comparison of the Transition States for Folding of Two Ig-Like Proteins from Different Superfamilies.

199. A. Fersht, *Proc. Natl Acad. Sci. U.S.A.*, **974**, 1525 (2000). Transition-State Structure as a Unifying Basis in Protein-Folding Mechanisms: Contact Order, Chain Topology, Stability, and the Extended Nucleus Mechanism.

200. A. L. Jonsson, K. A. Scott, and V. Daggett, *Biophys. J.*, **97**, 2958 (2009). Dynameomics: A Consensus View of the Protein Unfolding/Folding Transition State Ensemble Across a Diverse Set of Protein Folds.

201. F. Ding, J. J. LaRocque, and N. V. Dokholyan, *J. Biol. Chem.*, **280**, 40235 (2005). Direct Observation of Protein Folding, Aggregation, and a Prion-Like Conformational Conversion.

202. F. B. Sheinerman and C. L. Brooks, III, *J. Mol. Biol.*, **278**, 439 (1998). Calculations on Folding of Segment B1 of Streptococcal Protein G.

203. R. Toofanny, A. L. Jonsson, and V. Daggett, *Biophys. J.*, **98**, 2671 (2010). A Comprehensive Multidimensional-Embedded, One-Dimensional Reaction Coordinate for Protein Unfolding/Folding.

204. A. Li and V. Daggett, *Proc. Natl. Acad. Sci. U.S.A.*, **91**, 10430 (1994). Characterization of the Transition State of Protein Unfolding by use of Molecular Dynamics: Chymotrypsin Inhibitor 2.

205. A. Li and V. Daggett, *J. Mol. Biol.*, **257**, 412 (1996). Identification and Characterization of the Unfolding Transition State of Chymotrypsin Inhibitor 2 by Molecular Dynamics Simulations.

206. M. Petrovich, A. L. Jonsson, N. Ferguson, V. Daggett, and A. R. Fersht, *J. Mol. Biol.*, **360**, 865 (2006). Φ-Analysis at the Experimental Limits: Mechanism of β-Hairpin Formation.

207. D. Brockwell and S. Radford, *Curr. Opin. Struct. Biol.*, **17**, 30 (2007). Intermediates: Ubiquitous Species on Folding Energy Landscapes?

208. J. Gsponer, H. Hopearuoho, S. Whittaker, G. Spence, G. Moore, E. Paci, S. Radford, and M. Vendruscolo, *Proc. Natl. Acad. Sci. U.S.A.*, **103**, 99 (2006). Determination of an Ensemble of Structures Representing the Intermediate State of the Bacterial Immunity Protein Im7.

209. T. L. Religa, J. S. Markson, U. Mayor, S. M. V. Freund, and A. R. Fersht, *Nature*, **437**, 1053 (2005). Solution Structure of a Protein Denatured State and Folding Intermediate.

210. D. A. C. Beck and V. Daggett, *Biophys. J.*, **93**, 3382 (2007). A One-Dimensional Reaction Coordinate for Identification of Transition States from Explicit Solvent P(fold)-Like Calculations.

211. E. I. Shakhnovich, *Curr. Opin. Struct. Biol.*, **7**, 29 (1997). Theoretical Studies of Protein-Folding Thermodynamics and Kinetics.

212. M. Levitt, *J. Mol. Biol.*, **104**, 59 (1976). A Simplified Representation of Protein Conformations for Rapid Simulation of Protein Folding.

213. A. Warshel and M. Levitt, *J. Mol. Biol.*, **106**, 421 (1976). Folding and Stability of Helical Proteins: Carp Myogen.

214. S. Patel, P. Sista, P. V. Balaji, and Y. U. Sasidhar, *J. Mol. Graph. Model.*, **25**, 103 (2006). β-Hairpins with Native-Like and Non-Native Hydrogen Bonding Patterns Could Form During the Refolding of Staphylococcal Nuclease.

215. H. Wang, J. Varady, L. Ng, and S. S. Sung, *Proteins: Struct., Funct., Genet.*, **37**(325) (1999). Molecular Dynamics Simulations of β-Hairpin Folding.

216. B. Zagrovic, C. D. Snow, M. R. Shirts, and V. S. Pande, *J. Mol. Biol.*, **323**, 927 (2002). Simulation of Folding of a Small α-Helical Protein in Atomistic Detail Using Worldwide-Distributed Computing.

217. H. Lei and Y. Duan, *J. Phys. Chem. B*, **111**, 5458 (2007). *Ab Initio* Folding of Albumin Binding Domain from All-Atom Molecular Dynamics Simulation.

218. J. W. Neidigh, R. M. Fesinmeyer, and N. H. Andersen, *Nat. Struct. Biol.*, **9**, 425 (2002). Designing a 20-Residue Protein.

219. E. Demchuk, D. Bashford, and D. A. Case, *Folding Des.*, 2, 35 (1997). Dynamics of a Type VI Reverse Turn in a Linear Peptide in Aqueous Solution.

220. K.-I. Oh, Y.-S. Jung, G.-S. Hwang, and M. Cho, *J. Biomol. NMR*, 53, 25 (2012). Conformational Distributions of Denatured and Unstructured Proteins are Similar to those of 20 × 20 Blocked Dipeptides.

221. D. A. C. Beck, D. O. V. Alonso, D. Inoyama, and V. Daggett, *Proc. Natl. Acad. Sci. U.S.A.*, 105, 12259 (2008). The Intrinsic Conformational Propensities of the 20 Naturally Occurring Amino Acids and Reflection of these Propensities in Proteins.

222. E. R. McCarney, J. E. Kohn, and K. W. Plaxco, *Crit. Rev. Biochem. Mol. Biol.*, 40, 181 (2005). Is There or Isn't There? The Case For (and Against) Residual Structure in Chemically Denatured Proteins.

223. Z. Shi, K. Chen, Z. Liu, and N. R. Kallenbach, *Chem. Rev.*, 106, 1877 (2006). Conformation of the Backbone in Unfolded Proteins.

224. C. Hyeon, G. Morrison, D. L. Pincus, and D. Thirumalai, *Proc. Natl. Acad. Sci. U.S.A.*, 106, 20288 (2009). Refolding Dynamics of Stretched Biopolymers Upon Force Quench.

225. X. Periole, M. Vendruscolo, and A. E. Mark, *Proteins: Struct., Funct., Bioinf.*, 69, 536 (2007). Molecular Dynamics Simulations from Putative Transition States of α-Spectrin SH3 Domain.

226. M. Prévost and I. Ortmans, *Proteins: Struct., Funct., Bioinf.*, 29, 212 (1997). Refolding Simulations of an Isolated Fragment of Barnase into a Native-Like β Hairpin: Evidence for Compactness and Hydrogen Bonding as Concurrent Stabilizing Factors.

227. G.-J. Zhao and C.-L. Cheng, *Amino Acids*, 43, 557 (2012). Molecular Dynamics Simulation Exploration of Unfolding and Refolding of a Ten-Amino Acid Miniprotein.

228. D. O. V. Alonso and V. Daggett, *J. Mol. Biol.*, 247, 501 (1995). Molecular Dynamics Simulations of Protein Unfolding and Limited Refolding: Characterization of Partially Unfolded States of Ubiquitin in 60% Methanol and in Water.

229. D. O. V. Alonso and V. Daggett, *Protein Sci.*, 7, 860 (1998). Molecular Dynamics Simulations of Hydrophobic Collapse of Ubiquitin.

230. M.-Y. Shen and K. F. Freed, *Proteins: Struct., Funct., Bioinf.*, 49, 439 (2002). All-Atom Fast Protein Folding Simulations: The Villin Headpiece.

231. H. Lei and Y. Duan, *J. Mol. Biol.*, 370, 196 (2007). Two-Stage Folding of HP-35 from *Ab Initio* Simulations.

232. P. L. Freddolino and K. Schulten, *Biophys. J.*, 97, 2338 (2009). Common Structural Transitions in Explicit-Solvent Simulations of Villin Headpiece Folding.

233. H. Lei, C. Wu, H. Liu, and Y. Duan, *Proc. Natl. Acad. Sci. U.S.A.*, 104, 4925 (2007). Folding Free-Energy Landscape of Villin Headpiece Subdomain from Molecular Dynamics Simulations.

234. E. Paci, C. Friel, K. Lindorff-Larsen, S. Radford, M. Karplus, and M. Vendruscolo, *Proteins: Struct., Funct., Bioinf.*, 54, 513 (2004). Comparison of the Transition State Ensembles for Folding of Im7 and Im9 Determined using All-Atom Molecular Dynamics Simulations with Φ Value Restraints.

235. S. Gianni, N. R. Guydosh, F. Khan, T. D. Caldas, U. Mayor, G. W. N. White, M. L. Demarco, V. Daggett, and A. R. Fersht, *Proc. Natl. Acad. Sci. U.S.A.*, 100, 13286 (2003). Unifying Features in Protein-Folding Mechanisms.

236. K. Gunasekaran, S. J. Eyles, A. T. Hagler, and L. M. Gierasch, *Curr. Opin. Struct. Biol.*, 11, 83 (2001). Keeping it in the Family: Folding Studies of Related Proteins.

237. O. B. Ptitsyn and K. L. Ting, *J. Mol. Biol.*, 291, 671 (1999). Non-Functional Conserved Residues in Globins and their Possible Role as a Folding Nucleus.

238. M. W. van der Kamp, R. D. Schaeffer, A. L. Jonsson, A. D. Scouras, A. M. Simms, R. Toofanny, N. C. Benson, P. C. Anderson, E. D. Merkley, S. Rysavy, D. Bromley, D. A. C. Beck, and V. Daggett, *Structure*, 18, 423 (2010). Dynameomics: A Comprehensive Database of Protein Dynamics.

239. N. C. Benson and V. Daggett, *Protein Sci.*, **17**, 2038 (2008). Dynameomics: Large-Scale Assessment of Native Protein Flexibility.

240. D. A. C. Beck, A. L. Jonsson, R. D. Schaeffer, K. A. Scott, R. Day, R. Toofanny, D. O. V. Alonso, and V. Daggett, *Protein Eng. Des. Sel.*, **21**, 353 (2008). Dynameomics: Mass Annotation of Protein Dynamics and Unfolding in Water by High-Throughput Atomistic Molecular Dynamics Simulations.

241. A. L. Jonsson, R. D. Schaeffer, M. W. van der Kamp, and V. Daggett, *BioMol. Concepts*, **335** (2011). Dynameomics: Protein Dynamics and Unfolding Across Fold Space.

242. R. Day, D. A. C. Beck, R. S. Armen, and V. Daggett, *Protein Sci.*, **12**, 2150 (2003). A Consensus View of Fold Space: Combining SCOP, CATH, and the Dali Domain Dictionary.

243. R. D. Schaeffer, A. L. Jonsson, A. M. Simms, and V. Daggett, *Bioinformatics*, **27**, 46 (2011). Generation of a Consensus Protein Domain Dictionary.

244. R. D. Schaeffer and V. Daggett, *Protein Eng. Des. Sel.*, **24**, 11 (2011). Protein Folds and Protein Folding.

245. C.-L. Towse and V. Daggett, *BioEssays*, **34**, 1060 (2012). When a Domain is Not a Domain, and Why it is Important to Properly Filter Proteins in Databases.

246. A. D. Scouras and V. Daggett, *Protein Sci.*, **20**, 341 (2011). The Dynameomics Rotamer Library: Amino Acid Side Chain Conformations and Dynamics from Comprehensive Molecular Dynamics Simulations in Water.

247. A. L. Jonsson and V. Daggett, *J. Struct. Biol.*, **176**, 143 (2011). The Effect of Context on the Folding of β-Hairpins.

248. P. A. Alexander, Y. He, Y. Chen, J. Orban, and P. N. Bryan, *Proc. Natl. Acad. Sci. U.S.A.*, **104**, 11963 (2007). The Design and Characterization of Two Proteins with 88% Sequence Identity but Different Structure and Function.

249. Y. He, Y. Chen, P. Alexander, P. N. Bryan, and J. Orban, *Proc. Natl. Acad. Sci. U.S.A.*, **105**, 14412 (2008). NMR Structures of Two Designed Proteins with High Sequence Identity but Different Fold and Function.

250. Y. He, D. C. Yeh, P. Alexander, P. N. Bryan, and J. Orban, *Biochemistry*, **44**, 14055 (2005). Solution NMR Structures of IgG Binding Domains with Artificially Evolved High Levels of Sequence Identity but Different Folds.

251. Y. He, Y. Chen, P. A. Alexander, P. N. Bryan, and J. Orban, *Structure*, **20**, 283 (2012). Mutational Tipping Points for Switching Protein Folds and Functions.

252. K. A. Scott and V. Daggett, *Biochemistry*, **46**, 1545 (2007). Folding Mechanisms of Proteins with High Sequence Identity but Different Folds.

253. A. Morrone, M. E. McCully, P. N. Bryan, M. Brunori, V. Daggett, S. Gianni, and C. Travaglini-Allocatelli, *J. Biol. Chem.*, **286**, 3863 (2011). The Denatured State Dictates the Topology of Two Proteins with Almost Identical Sequence but Different Native Structure and Function.

254. J. R. Allison, M. Bergeler, N. Hansen, and W. F. van Gunsteren, *Biochemistry*, **50**, 10965 (2011). Current Computer Modeling Cannot Explain why Two Highly Similar Sequences Fold into Different Structures.

255. R. O. Dror, R. M. Dirks, J. P. Grossman, H. Xu, and D. E. Shaw, *Biophysics*, **41**, 429 (2012). Biomolecular Simulation: A Computational Microscope for Molecular Biology.

256. J. A. Baker and J. D. Hirst, *Mol. Inf.*, **30**, 498 (2011). Molecular Dynamics Simulations using Graphics Processing Units.

257. M. S. Friedrichs, P. Eastman, V. Vaidyanathan, M. Houston, S. Legrand, A. L. Beberg, D. L. Ensign, C. M. Bruns, and V. S. Pande, *J. Comput. Chem.*, **30**, 864 (2009). Accelerating Molecular Dynamic Simulation on Graphics Processing Units.

258. M. Vendruscolo and C. M. Dobson, *Curr. Biol.*, **21**, R68 (2011). Protein Dynamics: Moore's Law in Molecular Biology.

259. C. Kehl, A. M. Simms, R. Toofanny, and V. Daggett, *Protein Eng. Des. Sel.*, **21**, 379

(2008). Dynameomics: A Multi-Dimensional Analysis-Optimized Database for Dynamic Protein Data.

260. A. M. Simms, R. Toofanny, C. Kehl, N. C. Benson, and V. Daggett, *Protein Eng. Des. Sel.*, **21**, 369 (2008). Dynameomics: Design of a Computational Lab Workflow and Scientific Data Repository for Protein Simulations.

261. K. A. Scott, D. O. V. Alonso, S. Sato, A. R. Fersht, and V. Daggett, *Proc. Natl. Acad. Sci. U.S.A.*, **104**, 2661 (2007). Conformational Entropy of Alanine Versus Glycine in Protein Denatured States.

## CHAPTER 4

Assessing Structural Predictions of Protein–Protein Recognition: The CAPRI Experiment

Joël Janin,[a] Shoshana J. Wodak,[b] Marc F. Lensink,[c] and Sameer Velankar[d]

[a]IBBMC, Université Paris-Sud, Orsay 91405, France
[b]VIB Structural Biology Research Center, VUB Building E Pleinlaan 2 1050, Brussel, Belgium
[c]UGSF, CNRS UMR8576, University Lille North of France, Villeneuve d'Ascq 59658, France
[d]European Bioinformatics Institute (EMBL-EBI), European Molecular Biology Laboratory, Wellcome Trust Genome Campus, Hinxton, Cambridgeshire CB10 1SD, UK

INTRODUCTION

Macromolecular recognition is ubiquitous in living organisms, and it plays a central role in all biological processes. Proteins specifically interact with other proteins and nucleic acids to form a wide variety of assemblies, from binary complexes to the elaborate multicomponent machines that perform many of the cellular functions.[1,2] The interaction can be transient, which is the rule in processes such as catalysis or signal transduction, or it can be permanent and build stable assemblies. In either case, it depends on two macromolecules associating to form an interface held by noncovalent forces similar to the forces that stabilize the conformation of the component macromolecules.

Reviews in Computational Chemistry, Volume 28, First Edition.
Edited by Abby L. Parrill and Kenny B. Lipkowitz.
© 2015 John Wiley & Sons, Inc. Published 2015 by John Wiley & Sons, Inc.

A binary complex contains one such interface, a larger assembly contains several, but they are of the same nature.

The rules that govern the formation of specific interfaces are only partly understood, and much of what we know about them derives from the data acquired over the years by structural biologists, and stored in the Protein Data Bank (PDB).[3] The PDB contained over 96,000 entries at the end of 2013, and more than 9000 new entries have been deposited that year. These entries report the structure of many binary protein–protein complexes and oligomeric (multisubunit) proteins and of a few large assemblies such as ATP synthase[4] or the ribosome.[5–7] The PDB is a rich source of information on protein–protein interaction, and the protein–protein complexes present in it have been subjected to a number of generic studies over the years.[8–13] These studies stressed the diversity of the protein–protein interfaces, but they also pointed out some common properties. Features such as the interface size, geometry, chemical composition, and conservation in evolution can be used in modeling protein–protein interaction, and they have been the basis of a bevy of computational approaches in recent years.[14]

A well-established computational approach is docking, which simulates the association of two molecular structures. Small molecule docking has been a major tool in drug design for over 20 years. Protein–protein docking goes back even earlier,[15] but most of the current procedures date from the 1990s. They perform well on rigid molecules that have only six degrees of freedom, but with real proteins, the conformation of the components may change during the association reaction, and more degrees of freedom have to be explored. To be of any practical use, a docking procedure must be able to assemble proteins in their free conformation, not just reassemble a complex of known structure. Its value is best assessed in a blind prediction that starts from the known structure of the free components and produces models that are compared to an unpublished experimental structure. The CAPRI (Critical Assessment of PRedicted Interactions) community-wide experiment has been designed to do just that. The experiment, which has been running since 2001, has much in common with CASP (Critical Assessment of protein Structure Prediction), which started earlier and from which it was inspired. We describe here how CAPRI is organized, summarize the results of the predictions, and show how it has fostered progress of the field during the past 12 years.

PROTEIN–PROTEIN DOCKING

A Short History of Protein–Protein Docking

Introducing Docking in the 1970s

Protein–protein docking may be defined as the search for stable modes of association between two preformed protein structures. The concept of docking, first introduced by Wodak and Janin,[15] was extended a few years later to protein–ligand (small molecules) associations,[16] and small molecule docking has become one of the most active areas of computational drug discovery today.

Docking procedures start from the atomic coordinates of the individual molecules, generate models of putative complexes, give the models a score, and rank them. With small molecules, internal degrees of freedom can be taken into account, although at a significant computational expense. But accounting for protein flexibility either in protein–protein or protein–ligand docking remains a challenge. The main difficulty lies with modeling the flexibility of the protein backbone, as is discussed in this chapter. A critical test of a docking procedure is therefore its capacity to successfully operate on "unbound" conformations, those based on atomic coordinates derived from the crystal structures of the free component proteins (Figure 1). Operating on the "bound" conformations taken from the crystal structure of the protein–protein or protein–ligand complex itself has little predictive value, given that these conformations are invariably biased toward the native structure in the complex.

In the rigid-body approximation, where internal degrees of freedom are ignored, the docking problem has only six degrees of freedom; yet, it maintains a considerable degree of complexity. Assuming that each protein is a sphere of 15 Å radius on the surface of which atom scale surface features are drawn on a 1 Å grid, a systematic search requires probing approximately 10^9 distinct association modes.[17] Even with modern-day computers, this is feasible only for a limited set of protein pairs. Thus, most current docking procedures resort to various means of data reduction, by introducing biological or biochemical data to constrain the search space, using simplified representations of the protein, or both. Hence, *ab initio* genome-scale docking calculations still remain out of reach. A detailed review of the early protein docking algorithms that formed the basis of today's methods can be found in Wodak and Janin.[18] The earliest algorithm[15,19,20] mapped the rigid-body search space using a polar coordinate system developed in the laboratory of Pr. Cyrus Levinthal at Columbia University, New York.[21] The docking poses it generated were given a score on the basis of the surface area buried between the two molecules in contact. This area was computed using an analytical approximation,[22] on the basis of a simplified protein model where each amino acid residue was approximated by a sphere with an appropriate radius.[23] A soft repulsive residue-pair potential[24] was used to limit interpenetration of the spheres. The algorithm ignored the chemical nature of the protein surfaces that are brought in contact and searched only for large surface patches with complementary shapes, a strategy that was later used by many groups. Its application to docking the pancreatic trypsin inhibitor onto the active site of trypsin yielded a native-like mode of association together with 11 other solutions that had similar scores.[15] In a second study, an improved version of the same algorithm was used to generate a reaction path between the two allosteric R and T forms of the hemoglobin tetramer by systematically probing the interaction of two $\alpha\beta$ dimers.[19] Six years later, time-resolved absorption spectroscopy data led to a similar reaction path,[25] thereby validating the results of the docking calculations.

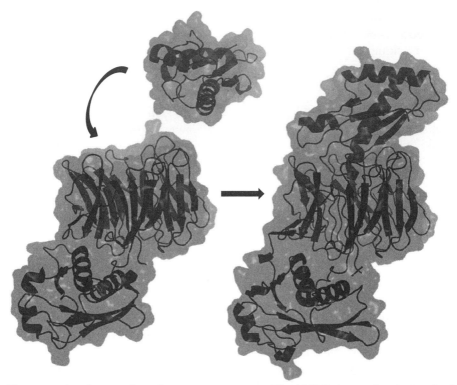

Figure 1 *Docking unbound protein structures.* The TolB/Pal complex is involved in maintaining the integrity of the outer membrane of Gram-negative bacteria. The "unbound" structures on the left represent Pal (PDB entry 1OAP, on top) and TolB (entry 1CRZ). The complex can be modeled by rotating and translating the smaller Pal protein, so as to bring it in contact with TolB. The X-ray structure of the complex[161] (entry 2HQS) is shown on the right. Local conformation changes affect both components as they form the complex; side chains rotate, and surface loops move in the Pal-binding site of TolB. As a result, interface backbone atoms are displaced by 0.6 Å RMS, side-chain atoms, by 1.7 Å RMS, between the unbound and the bound structures. The TolB/Pal complex was target T26 of CAPRI. Its prediction yielded 29 medium-quality models, which were submitted by seven different predictor groups and four scorer groups. Six other groups submitted models of acceptable quality.[105] See Figure 3 and Section "The CAPRI Evaluation Procedure" for the quality criteria used to assess models in CAPRI.

Developing Algorithms in the 1990s

The importance of protein–protein interactions in biology became obvious in the early 1990s, and access to more powerful computers attracted more groups to the docking problem. Thus, Cherfils, Duquerroy, and Janin[26] combined a rigid-body search using the Wodak–Janin algorithm with a refinement step carried out on detailed atomic models of the protein to identify native-like antibody–lysozyme complexes. Other docking procedures based on shape complementarity soon followed. Wang[27] used a modification of an approach pioneered by Connolly,[17,28] which identifies "knobs" and "holes" by triangulating the surface of the two component molecules and searches for quartets of complementary knob/hole pairs. In the procedure of Shoichet et al.[29,30] developed for docking small ligands into a known binding site on a receptor protein, candidate solutions were generated by matching a reduced representation of the ligand to a "negative image" of the receptor binding site. A version of the algorithm was adapted to docking protein inhibitors onto an enzyme or receptor recognition site with encouraging results.

The knobs and holes complementarity search was updated and made efficient in docking algorithms on the basis of geometric hashing.[31–33] In these algorithms the molecular surfaces are processed in order to identify triplets of critical points. The triplets are indexed by the distance between each pair in the triplet and stored in hash tables. The relative orientation and translation of the interacting molecules is then obtained by matching cliques of surface features described by the critical point triplets. Surface complementarity could be defined and tested in several ways. The one proposed by Fischer et al.[34] performed well on bound molecules, but significantly less so when the calculation was applied to the unbound species, indicating that the method was highly sensitive to small changes in surface features.

Another breakthrough occurred when it was realized that the translation portion of the surface complementarity search could be sped up significantly by representing each protein as a set of weighted points on a cubic grid[35] and calculating a correlation between the two sets with the Fast Fourier Transform algorithm (FFT).[36] In addition to speed, the FFT approach enables the implementation of more elaborate scoring functions, which incorporate information on surface properties such as hydrophobicity, electrostatic, and van der Waals interactions.[37–40] The addition of a simple electrostatic term in the FTDOCK program[41] was shown to improve the rank of correct solutions when docking 10 enzyme–inhibitor or antigen–antibody complexes. A further advantage of the FFT method is that one can easily vary the resolution at which the features of the interacting molecules are considered. Low-resolution searches are very fast and may be sufficient for detecting the regions with complementary shape.[42,43] They also tend to blur the effects of small conformational changes, always present when docking "unbound" molecules.[41] But high-resolution searches, representing much heavier calculations, are required for deriving accurate atomic models of a complex, prompting various strategies to speed up the calculations.[44,45]

Major Current Algorithms

Docking and Scoring

Today's major docking procedures use a multipronged approach (reviewed in Refs. 46–48). They typically start with a rigid-body search that employs a coarse-grained representation of the proteins and allows for some degree of overlap between their surfaces. Subsets of the highest scoring solutions (often in the thousands) are then subjected to one or more refinement steps using detailed atomic models and increasingly sophisticated scoring functions.

Packages such as HEX,[49] ZDOCK,[50] CLUSPRO,[51] PIPER,[52] Patch-Dock[53] implement efficient rigid-body search algorithms. They score docking poses on the basis of shape complementarity, augmented with terms representing electrostatic interactions, solvent contributions, and/or knowledge-based interaction potentials.[54,55] The docking landscape sampled during the global search may inform subsequent steps in the docking pipeline. The docking poses are clustered, and only those representing the most densely populated clusters are kept for further processing.[56] Some procedures also compile statistics on how frequently individual residues appear in the interfaces of cluster members and use this information to select candidate docking solutions.[57]

Once the space of likely docking solutions has been sufficiently reduced, various refinement strategies are applied to zero in on near-native poses. They range from removing a few side-chain clashes[58] to the fully fledged optimization of both side-chain and backbone conformations using molecular dynamics simulations or Monte-Carlo/simulated annealing techniques in RosettaDock[59] and HADDOCK.[60] These calculations are usually carried out on detailed atomic models, and they may be complemented with molecular dynamics simulations in the presence of explicit water molecules in order to test the robustness of the identified optimal solutions.[60]

A major stumbling block in both the refinement step and the final ranking of the solutions is the limited reliability of the scoring functions used for singling out near-native models. The best performing docking procedures implement a variety of scoring functions, which may include residue- or atom-pair potentials (e.g., Refs. 54 and 55), as well as a range of structural and physical–chemical features derived from known protein interfaces.[61] Scoring may use the observation that the docking energy landscape is funnel like, and the functions can be optimized to single out funnels leading to the native solution.[62] Overall, however, while progress has been achieved, the results of the CAPRI evaluations described below[63–65] – confirmed by an independent assessment[66] – indicate that none of the scoring schemes can be trusted to correctly rank docking poses. A lingering issue has been the lack of consistent experimental data on binding affinities of protein–protein associations that could be used to benchmark scoring schemes (see Section CAPRI & the Seattle challenge).

Modeling Flexibility

Accurately modeling backbone and loop rearrangements also remains work in progress for most docking procedures, although there have been tangible advances. Effective strategies include modeling large-scale motions governed by hinge deformations (FlexDock[67]) or performing many-body multistage docking of individual protein domains.[68] Docking ensembles of conformers generated for the individual components is also quite popular. Such ensembles may be derived by molecular dynamics or related methods, by Monte-Carlo sampling, by simulating motion along normal modes,[69–71] or by using the structural homologues that the PDB may contain.[72]

With increasing computer speed, handling of side-chain and loop flexibility during the rigid-body sampling step has become tractable and shown to yield encouraging results (e.g., see Refs. 60 and 73). Methods for modeling the association of protein subunits coupled to their folding have also emerged. One such method uses the Rosetta package, and it has produced very promising results in modeling the simultaneous folding and docking of symmetric interleaved homo-oligomers.[74]

Making Use of External Information and Other Approaches

Successful docking predictions more often than not exploit a range of supporting information in order to bias the global search or to filter solutions. Typically, this information pertains to data on residues known or predicted to be at the interface, structural data on related or similar complexes, or distance constraints derived from NMR or cross-linking experiments, when these are available. Data-driven docking procedures systematically integrate restraints on the basis of such data. Thus, the HADDOCK package[60,75] and program ATTRACT[73,76] integrate a wide range of data types, whereas program MultiFit uses low-resolution maps from cryo-electron microscopy as restraints in docking atomic resolution models.[77] These methods are quite powerful when the information they rely on is correct, but also particularly vulnerable when it is incorrect or ambiguous.

Two recent approaches that differ substantially from earlier docking methods are worth mentioning. In the procedure of Li, Moal, and Bates,[78] a simulation framework is used to search for encounter complexes by simulating collisions between multiple copies of the component proteins in a crowded environment. The frequency of collisions and the retention time (the time interval any two proteins remain in contact during the simulations) are then employed to discriminate between specific and nonspecific complexes, with promising results.

A second approach, termed template-based docking, in fact dispenses with docking altogether. It relies on identifying in the PDB known structures that can serve as templates for the query assembly of two proteins. The templates may be found by homology modeling[79,80] or by threading,[81] two methods based on the amino acid sequences, but structural alignment is definitely more effective when structures are available for the query proteins.[82,83] Template-based methods are increasingly used in large-scale predictions.[84,85] Recent studies suggest that they can be as reliable as predictions based on classical docking or on nonstructural evidence, and that the relative paucity of known protein complexes that may serve as templates is not a limiting factor if both the sequence and the structural similarity relationships to known structures are more thoroughly explored.[86,87]

THE CAPRI EXPERIMENT

Why Do Blind Predictions?

As described in the previous section, several major algorithms for protein–protein docking were developed during the 1990s, and by the year 2001, the field had become mature. The procedures had been extensively tested on a number of binary protein–protein complexes of known X-ray structure. However, nearly all the tests involved target complexes of the same two types: enzymes (mostly proteases) binding a protein inhibitor; a monoclonal antibody binding the antigen hen lysozyme. Those were realistic targets, because "unbound" structures were available for the enzymes, the inhibitors, lysozyme, and at least one of the antilysozyme antibodies. But the sample was small, and very far from illustrating the diversity of protein–protein interaction in biology. In practice, it represented the state of protein crystallography a decade earlier: the first antibody–lysozyme structure is dated 1986.[88] In the late 1990s, protein crystallographers had begun analyzing many other aspects of protein–protein interaction, and its role in signal transduction was illustrated by some remarkable X-ray structures. The transducin $G\alpha$-$G\beta\gamma$ complex,[89] part of the visual system in the mammalian retina, is an early example. Transducin, like many of the other new complexes, displays large conformation changes and other features not observed in earlier structures. How docking would perform on them was a crucial question, the answer of which determined the value of the method as a predictive tool. It was also timely, because in 2001, a new field of application had opened to docking. Following the completion of the human genome sequence, structural genomics programs were going to determine the structure of thousands of new proteins in high-throughput X-ray and NMR studies. Their targets included the components of many binary or larger assemblies, and docking could in principle build models of all these assemblies from the component structures.

The role of structural modeling and docking in the postgenomic era was the subject of a meeting on "Modeling Protein Interactions in Genomes" organized in Charleston, South Carolina, by S. Vajda and I. Vakser, in June 2001. A conclusion of the meeting was that docking procedures should be subjected to unbiased tests in a blind prediction experiment.[90] This had already been attempted before on a few occasions, and the prediction had met with success on an enzyme–inhibitor complex,[91] but it failed completely on two antigen–antibody complexes.[92,93] The new experiment would be similar to the CASP (Critical Assessment of Structural Predictions) community-wide experiment. Since 1994, CASP had been assessing methods to predict a protein fold on the basis of its amino acid sequence, by comparing to unpublished X-ray or NMR structures of the target proteins, the models that predictors submitted to an independent assessors' body (http://www.predictioncenter.org).[94] The new experiment was named CAPRI. In CAPRI, the starting point is the experimental structure of two proteins: participants model their complex, and the models are assessed by comparison with an unpublished X-ray structure of the complex.[95] As in CASP, both the prediction and the assessment are done blindly. The predictors have no access to the experimental structure of the targets, and the identity of the groups submitting models is hidden from the assessors, who measure on each model a set of objective parameters that define the quality of the prediction.

Organizing CAPRI

The Docking and Scoring Experiments

Both CASP and CAPRI depend for targets on the cooperation of structural biologists, willing to communicate data before they are published, but this dependence is more decisive in the case of CAPRI. CASP runs prediction rounds every 2 years, and the recent rounds had more than 100 targets, offered mostly by structural genomics centers who determine thousands of new protein structures every year. In comparison, new structures of protein–protein complexes are rare, and CAPRI has far fewer targets than does CASP. In the early years of the experiment, no more than a dozen new structures were reported per year, often in high-profile journals. Those that involved two proteins for which an unbound structure was available were even rarer. As a consequence, the paucity and value of its targets determine the way CAPRI is organized. A prediction round is started each time a target is offered by experimentalists, and the round is kept short to fit their publication schedule. Over the years, CAPRI rounds have had one to four targets, and lasted two to eight weeks depending on the number of targets. With two to four rounds each year, CAPRI was in its twenty-seventh round in 2012, while CASP was holding its tenth.

The scoring experiment, introduced in 2006, adds a few days to this schedule. It is designed to assess procedures to score and rank docking models independently of the algorithms that produce the models. Groups that develop such procedures may register as scorers in a CAPRI round. Near the end of the round, the predictor groups are invited to upload a large set of their models in parallel with submitting their 10 best models for the docking assessment. Merging the uploaded models yields an ensemble of 1000–2000 models for each target, originating from different groups. This ensemble is communicated to the registered scorers, who have a couple of days to rank the models and submit 10 that score the highest. The CAPRI assessors evaluate these submissions with the same criteria as the predictors' submissions.

Confidentiality

Because CAPRI relies on unpublished information, confidentiality has been a major concern from the start. X-ray structures of protein–protein complexes are highly valuable items, and all relevant data must be kept confidential until the authors' publication comes out and the PDB releases the atomic coordinates. In practice, the structural information that CAPRI participants need to build their models concerns the unbound components, not the complex, and it is generally public. But this information may be sufficient to reveal the nature of the target, which must be protected, and therefore, we request participants to register and sign a confidentiality agreement at each prediction round. When the round opens, registered participants are given a password to access the CAPRI website (http://pdbe.org/capri) hosted by the European Bioinformatics Institute (EBI-EMBL, Hinxton, UK). A few weeks later, they submit their models to the same site.

In most cases, the target information provided to the predictors comes from the PDB in the form of atomic coordinates for the two components, or for homologue proteins from which the components can be modeled, in which case the amino acid sequence of the target is also provided. Predictors are free to use other PDB entries, and all the data that is publicly available from mutant studies or other sources. In several cases, the unbound structure was new, and therefore confidential, and it came from an X-ray study of the free protein performed in parallel with that of the complex. In early rounds of CAPRI, we also had what we called "bound/unbound" targets: due to the lack of an unbound structure, the coordinates of one of the components were taken from those of the complex itself. Such targets proved easier to model than with both components unbound, probably because conformation changes could affect only one side, but the prediction was not realistic, because the answer had to be known in advance. "Bound" targets with both components taken from the complex have never been accepted in CAPRI, and recent rounds have had only unbound targets.

Leakage of target information did occur, and it perturbed the experiment although it did not come from the CAPRI side. In at least two cases, images of the complex, or even its atomic coordinates, had been left on the Web unprotected by the authors of the X-ray structure or their collaborators. A simple Google search with the protein names as keywords retrieved them in a matter of seconds, and we had to cancel the target.

The CAPRI Targets

Protein–Protein and Protein–RNA Complexes

The 26 CAPRI prediction rounds performed from 2001 to 2012 have had a total of 56 targets. With the exception of an NMR structure and of the "affinity prediction" targets described below, the targets have come from X-ray studies, and most have been offered through personal connections with the crystallographers. In the early years of the experiment, a majority of the targets were still antigen–antibody complexes (with the antibody component bound), and enzyme–inhibitor complexes. But that changed quickly, and of the 44 protein–protein complexes that have been subjected to prediction, only eight were antigen–antibody, and seven were enzyme–inhibitor complexes. The remainder represents a wide variety of biological processes that includes

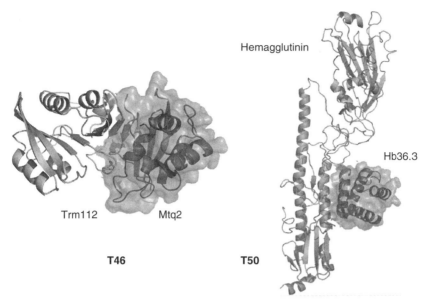

Figure 2 *Two targets for CAPRI predictions.* T46, a target of CAPRI Round 22 (June 2010), is a complex of the Mtq2 methyltransferase with the Trm112 activator protein of *E. cuniculi* (PDB entry 3Q87).[96] Predictors had to model-build Mtq2 from other methyltransferases and Trm112 from the Zn-binding domain of a yeast homologue. The low level of sequence identity between the homologues made for a difficult target and the prediction yielded only models of "acceptable" quality.[64] T50, a target of Round 24 (March 2011), is a engineered complex between the flu virus hemagglutinin and HB36.3, a small protein construct designed to bind the conserved stem region of the viral antigen (PDB entry 3R2X).[101] Starting from X-ray structures of the hemagglutinin and a precursor of the designed ligand, ten predictor and six scorer groups were able to submit medium-quality models.[64] Variants of the same complex were later offered as targets for affinity prediction[149] (see Section CAPRI & the Seattle challenge Engineering Interactions and Predicting Affinity).

the cell cycle, signal transduction, membrane trafficking, and so on. The list of the targets with their authors, the relevant literature references and the PDB entries, is available at http://pdbe.org/capri/pdb_ids.html. Figure 2 illustrates two examples. Target T46 is the Mtq2/Trm112 methyltransferase involved in the maturation of translation termination factors;[96] T50 is an engineered complex discussed below. The components of T46 had no previously known structures, but they had homologues in the PDB that could be used for model-building prior to docking.

In addition to protein–protein complexes, one CAPRI target was a protein–RNA complex, and three were homo-oligomers. Target T33, the protein–RNA complex, contained a segment of ribosomal RNA, and the predictors were given coordinates taken from an X-ray structure of the ribosome. Because the RNA of the target complex has a different sequence and a very different conformation, docking predictions were expected to fail. They did, but in a second step we took the same complex to make up target T34, this time with the RNA component in its bound form, and its prediction produced good models. Offering the same complex twice for prediction, first unbound, and then bound/unbound after completing the unbound prediction, is a trick that we used in three other cases where one component required homology modeling with a low degree of identity. In each case, the prediction produced much better models with the bound/unbound than the unbound version of the target, again indicating that conformation changes play a crucial role in docking.

Homo-Oligomeric Proteins

Homo-oligomers are proteins made of several chains with the same sequence. They are abundant in nature, and in the PDB, a majority of the proteins are homo-oligomers. Such proteins can serve as targets of docking procedures, as they can in principle be modeled by assembling models of the subunits. Many CASP targets are homo-oligomers, but the participants are asked to predict only the subunit fold. A few predictors nevertheless attempted to model the assembly, but the results were poor.[97] In CAPRI, a homo-oligomer can be a target only if it exists in several different states of assembly, one state serving as a starting point to model the others; reassembling subunits dissociated *in silico* would be bound docking of no predictive value. Proteins that change quaternary structure may not be uncommon in nature, but they are very few in the PDB. A rare example is the envelope protein of the tick-borne encephalitis virus. It is a dimer in the virus particle, but it rearranges into a trimer during the infection process, and the trimer drives the fusion of the viral and cell membranes.[98] The envelope protein was target T10 of CAPRI, the known structure of the dimer serving to predict that of the trimer before it was published. Large changes within the subunits accompany the quaternary structure change, and this target, like all those of the same type, proved challenging, in spite of the constraints imposed by the symmetry of the trimer.

Designed Complexes

In the past few years, experimentalists have offered a new type of target to CAPRI: designed complexes. Protein engineering groups evolve novel interactions by mutating surface residues on a scaffold protein and selecting variants capable of binding a target ligand protein. The design involves a modeling step followed by a selection procedure carried out either *in vitro* or *in vivo*.[99] Typically, a dozen amino acid substitutions allow the variant to achieve nanomolar affinity for the target. A number of novel protein–protein complexes have been produced in this way, and several have been crystallized. An example is the complex of protein HB36.3 with the hemagglutinin, which is the main antigen of the influenza virus. HB36.3 was designed *in silico* with the Rosetta procedure developed by the Baker lab at the University of Washington (Seattle),[100] taking as scaffold a small bacterial protein of unknown function. The designed variant bears 11 mutations and binds hemagglutinin with a dissociation constant (K_d)

of 200 nM. In a second step, a round of *in vitro* selection was performed on a DNA library. It introduced three more mutations and brought K_d down to 4 nM. An X-ray structure of the complex is shown in Figure 2. It confirms that the mode of binding predicted by Rosetta is correct, and that the conformation of HB36.3 remains close to that of the scaffold in spite of all the mutations.[101]

The HB36.3-hemagglutinin complex was CAPRI target T50. Given atomic coordinates for the unbound hemagglutinin and the scaffold protein, plus the amino acid sequence of the variant, several groups returned good-quality models of the complex. This success was repeated on two other targets of the same type. It indicates that the rules governing the design of artificial complexes are consistent with those of the docking procedures and suggests that these rules capture major features of protein–protein interaction in biology. However, the results of the affinity prediction experiment described in Section CAPRI & the Seattle challenge indicate that they are still far from accurate.

Creating a Community

Some 50 groups worldwide have participated in CAPRI as predictors and 20 as scorers. The experiment has created strong links between them directly or through the intermediary of the management committee and the Web site. Groups exchange information on methods and results between prediction rounds, and they often share data on targets during the rounds. The scoring experiment is an outcome of this cooperation: it relies on the willingness of predictor groups to make the models they have built available to the scorers.

Moreover, participants meet in person on a regular basis at Evaluation Meetings. Six have been held, the last two in Barcelona, Spain, in December 2009, and Utrecht, The Netherlands, in April 2013. At each meeting, the CAPRI assessors first present a survey of their evaluation; then a selection of the predictor groups report their results and describe their procedures, with an emphasis on new methods. The same groups are invited to contribute papers to a special issue of the journal *Proteins: Structure, Function, and Bioinformatics*, which has been supporting CASP and CAPRI from the start. CAPRI special issues have appeared in 2003, 2005, 2007, 2010, and 2013. The journal was founded by Cyrus Levinthal (1922–1990) of Columbia University, New York. Cy, who was the mentor of two of us (JJ and SJW), is one of the founders of computational structural biology, including computer graphics and the first major protein modeling package.[21]

The Evaluation Meetings also offer an opportunity for all to discuss the schedule and organization details of the experiment. The model quality criteria used by the assessors (Section "Evaluation Criteria") were an outcome of these discussions. Another outcome was the docking benchmark elaborated by the group of Z. Weng (Boston University and University of Massachusetts Medical School, Worcester, MA). The benchmark is designed for testing docking procedures in unbound docking, and it contains atomic coordinates for both the protein–protein complexes and their unbound components. It is regularly updated as new structures are deposited in the PDB. Version 1, presented in 2003, contained 59 test cases, mostly enzyme–inhibitor and antigen–antibody complexes;[102] version 4 has 176, and these two categories are now a minority.[103] As the unbound component structures are present in the benchmark, the extent of conformation changes in the complexes can be estimated, and the benchmark classifies the test cases as "rigid-body," "medium difficulty," and "difficult," depending on the value of parameter I_rms, similar to the one used in CAPRI assessment (Section "Evaluation Criteria"). I_rms is the root-mean-square distance between the C-alpha atoms of the interface residue in the bound and unbound structures. Standard docking procedures produce near-native models of most of the "rigid-body" cases (I_rms < 1.5 Å), which make up 70% of the benchmark. They generally fail on the "difficult" cases (I_rms > 2.2 Å), which make up 15%, and may differ widely in their performance on the "medium difficulty" cases.

ASSESSING DOCKING PREDICTIONS

The CAPRI Evaluation Procedure

Evaluation Criteria

In a prediction round, the participant groups are invited to submit up to 10 docking models of each target complex to the CAPRI website. Objectively and fairly evaluating the quality of these models is central to the experiment. To this end, the submissions are made anonymous (the identity of the submitting groups is withheld), and they are evaluated by an independent team against the atomic coordinates of the target structure deposited by their authors with the CAPRI management team on a confidential basis.

A set of parameters is computed providing insights into the useful or problematic aspects of the models. For example, to design high-affinity inhibitors on the basis of a docking model, residue–residue contacts at the interface or preferably atomic contacts must be accurately defined, whereas merely identifying the residues that make up the interface may be sufficient for designing mutants that interfere with the interaction. On the other hand, fitting an atomic model into a low-resolution electron density map derived from electron microscopy[104] is not sensitive to the exact nature of the interacting residues but rather to the relative orientation and position of the receptor and ligand molecules.

On the basis of a consensus reached with the CAPRI community, a total of eight parameters are evaluated.[63,64,105] Two assess residue–residue contacts between the docked proteins. f_nat is the fraction of intersubunit residue–residue contacts present in the X-ray structure of the target that are reproduced (recalled) in the model (residues are deemed in contact if any of their atoms are within 5 Å); f_non-nat is the fraction of residue contacts in the model that are not present in the target. In a perfect model, f_nat and f_non-nat should equal 1 and 0, respectively.

It was realized early on that the fraction of native contacts f_nat can be made artificially high if the subunits interpenetrate, and it was decided that models with more than a threshold number of atomic clashes (atom pairs separated by less than 3 Å) should be disqualified. This threshold is derived from the distribution of the number of clashes observed in the ensemble of submitted models and equal to the average number of clashes plus twice the standard deviation of this distribution. Typically, 40–150 clashes are allowed in a model, depending on the size of the proteins and the interfaces.

The extent to which a model identifies the correct regions of the protein surfaces that interact may be assessed independently of the contacts between these regions. For this purpose, two parameters f_IR and f_OP are computed that represent respectively the fraction of interface residues in the target that is reproduced in the model and the fraction of interface residues in the model that does not match those in the target. Both are evaluated separately for each component of the complex. Interface residues are defined as those that lose accessible surface area when the two proteins associate, computed from the difference in solvent accessible surface area of the complex and the individual components taken in the bound form.

Three more parameters assess the global geometry of the model: L_rms, θ_L, and d_L, which measure the fit between the ligands in the model and the target after the receptors have been superimposed.[106] By convention, "ligand" and "receptor" refer to the smaller and the larger of the two docked proteins, independent of their biological function. L_rms is the root-mean-square displacement (RMSD) of the backbone atoms of the ligand in the model versus the target complex. θ_L and d_L are, respectively, the residual rigid-body rotation angle and the residual translation vector required to superimpose the ligand molecules once the receptors have been superimposed.

However, a global fit may not provide a good picture of the fit at the interface, especially when the ligand is a large protein. This is assessed by parameter I_rms, the RMSD of the backbone atoms of the interface residues only, which includes contributions from rigid-body motion as well as local conformational changes. To this end, interface residues are defined as those having at least one atom within 10 Å of an atom on the other molecule in the target structure, a broader definition of the interface than for other assessment parameters. As the quality of the side-chain conformations in the docking models is also of interest, an RMSD is also computed over side-chain atoms of the same interface residues.

The final ranking of the models submitted for each target is based on a subset of these criteria, namely, f_nat, L_rms, and I_rms. Their values define a model as being either incorrect, or of acceptable, medium, or high quality. A high-quality model must have f_nat above 0.5, while both L_rms and I_rms are below 1.0 Å, which in practice requires that it approach the quality of an

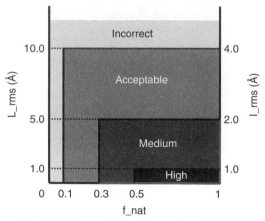

Figure 3 *Quality of the docking models submitted to CAPRI.* Quality criteria f_nat, L_rms, and I_rms are estimated by comparing a model with the target X-ray structure. As described in Section "Evaluation Criteria", f_nat measures the recall of the native residue–residue contacts at the interface; L_rms and I_rms are root-mean-square displacement of the backbone atoms calculated for the ligand protein after superposing the receptor (L_rms) or for the interface residues only (I_rms). The values of all three parameters define the quality of the model as high, medium, acceptable, or incorrect, as represented in the figure. See Refs. 63 and 105 for the exact application of these criteria.

X-ray structure. A graphical representation of the assessment criteria and how they translate into the CAPRI ranking can be found in Figure 3.

To ensure a fair and consistent evaluation, all the assessment parameters must be calculated on the same number of residues and atoms across all submitted models. The submissions are therefore filtered to exclude flexible residues that adopt different conformations in the bound and unbound protein components, or not present across all models. These residues are defined using structural alignments of the individual ligand and receptor entities[107] and identifying those with largest deviations. Typically, residues located in (short) loops are excluded from the RMSD fits and subsequent analysis.

An additional filter is applied to ensure that every residue used in the various calculations is in fact present in the model. It uses pairwise alignments of the sequences in the predicted model onto the target sequences and disqualifies models that do not meet a minimum threshold of sequence identity.

Assessing Interface Predictions

All published information on the targets may be used in the prediction. Thus, residues may be known to be part of the interface from experiment or predicted to be so. In addition to enhancing the performance of docking procedure, a reliable interface prediction may provide insight into protein function, guide mutagenesis studies, or inform drug design applications. While many prediction methods have been proposed (reviewed in Refs. 108–111), their performance remains poor, especially when applied to complexes between proteins that are stable on their own. In the CAPRI assessment procedure, the f_IR and f_OP parameters monitor whether a model correctly identifies interface residues of the binding partners. Lensink and Wodak[112] extended this analysis by comparing 46 interfaces present in the X-ray structures of 20 CAPRI targets to those of the models submitted by 76 participant groups. The comparison showed that the best performing methods predict interfaces with average recall and precision of about 60% for a 60% majority of the analyzed interfaces. This is significantly better than the published performance of most interface prediction methods on nonobligate complexes. Moreover, a sizable fraction (24%) of the models that rank as incorrect in the CAPRI assessment also contain an interface that has a recall and precision greater than 50% and can be counted as correctly predicted albeit not correctly oriented and positioned. Overall, 70% of the interface predictions are correct in CAPRI submissions, suggesting that docking methods predict the interfaces better than the geometry of the complexes and the residue–residue contacts.

Prediction of Interfacial Water Positions

Protein–protein interfaces often contain water molecules that can be located in an X-ray structure.[113,114] An example is the barnase/barstar complex, which is of very high affinity, yet contains 18 water molecules at its interface.[115] This has recently led to expanding the CAPRI experiment by asking participants to predict water positions at the interface. The assessment procedure[116] considers water-mediated receptor–ligand contacts, in which both the ligand and the receptor have atoms within 3.5 Å of the same water molecule and computes the fraction f_wnat of the water-mediated contacts present in the target and recalled in the model. Its value places a given model into one of five categories, the top two having f_wnat above 0.5 and 0.8, respectively. Models containing too many water molecules are disqualified by allowing only one water–water clash shorter than 2.5 Å per water present in the target.

Up to now, only target T47 has been specifically analyzed for the interface water prediction in CAPRI. T47 is a complex between the DNase domain of colicin E2 and the cognate immunity protein Im2.[117] Because it is very similar to a previously published E9 DNase/Im2 complex, which was the target T41 shown in Figure 7, T47 made for an easy target, and many high- or medium-quality models were submitted. However, the main goal was to predict the position of interface waters, and that proved difficult: of 204 high- or medium-quality models, 11 had f_wnat above 0.5, but none above 0.8. On the other hand, 28% of all submitted models had f_wnat above 0.3, and the quality of the water predictions correlated with that of the side-chain conformations at the interface. Rather than modeling the solvent a posteriori into docked models, water prediction now becomes an integral part of docking procedures,[118,119] and its assessment should become an integral part of the CAPRI evaluation.

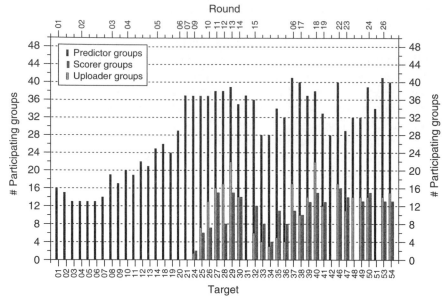

Figure 4   *The participants in each CAPRI prediction round.* Numbers on the vertical axis count the groups that submitted predictions for each of the CAPRI targets covered in this review. Numbers on the bottom and top horizontal axes refer to the targets and the corresponding CAPRI rounds, respectively.

A Survey of the Results of 12 Years of Blind Predictions on 45 Targets

During the 12 years since CAPRI started, there have been 26 prediction rounds, with a total of 56 targets. The number of participants in the CAPRI experiments is summarized in Figure 4. The number of groups submitting docking predictions climbed steadily during the first five rounds from an average of about 15 to stabilize at 35 and thereafter, each submitting 10 models per target. The scoring experiment, launched in round 9 with target T24, and routinely performed thereafter, involves from 3 to 20 predictor groups who upload 100 models each of a given target, and 10–12 scorer groups who rerank these models to make their own ten-model submission.

Predicting the Structure of Protein–Protein Complexes: Docking Results

Figure 5 illustrates the collective performance of CAPRI predictors on the 43 docking targets that have been assessed; affinity prediction targets are excluded. A 70% majority of the docking targets obtained at least one medium- or high-quality model. There were often many such models, submitted by several groups applying different docking procedures. In the ten-model submissions made by each group, incorrect models were usually in a majority, and those of acceptable or higher quality rarely ranked first.[63,64,105] Still, the

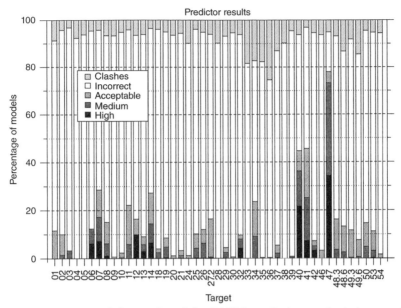

Figure 5 *A summary of the results of the CAPRI predictions: submissions made by predictor groups.* For each target offered for docking prediction since CAPRI started in 2001, bars represent the cumulative percentages of submitted models evaluated as high-, medium-, or acceptable quality. The remainder comprises incorrect models and models disqualified because of an excess of clashes or low-sequence identity. With targets T27 and T48-49, two possible modes of assembly had to be considered when assessing the models (see Figure 7). Detailed results of the assessment are reported in Refs. 63, 64, 105, 162, and 163.

predictions were highly enriched with the correct solution, and, therefore, they would have been useful to experimentalists.

A very welcome development has been the steady increase in docking Web servers. Table 1 lists those currently operating, all of which have participated in CAPRI, and Figure 6 summarizes their performance on individual CAPRI targets. Servers operate completely automatically and offer wide access to docking procedures by nonexperts. In CAPRI, servers are allowed only 1–3 days to make a submission, much less than human predictors, so as to prevent manual interventions. Because servers usually implement older but more extensively tested versions of the docking and scoring methods, their performance is understandably not on par with that of human participants, yet several do respectably well. ClusPro and HADDOCK, for instance, have a high success rate and are easy to use, which makes them very popular.

Success and Failure in Docking Predictions
The targets where the components undergo small conformational changes on association have been well predicted with few exceptions. This includes by definition the targets that had a bound component and also those where an accurate

Table 1   Docking Servers

Server	URL	References
3D-Garden	http://www.sbg.bio.ic.ac.uk/~3dgarden/	151
ClusPro	http://cluspro.bu.edu/login.php	152
GRAMM-X	http://vakser.bioinformatics.ku.edu/ resources/gramm/grammx	153
HADDOCK	http://www.nmr.chem.uu.nl/haddock/	154
HEX SERVER	http://hexserver.loria.fr/	155
FiberDock	http://bioinfo3d.cs.tau.ac.il/FiberDock/	156
FireDock	http://bioinfo3d.cs.tau.ac.il/FireDock/	58
FlexDock	http://bioinfo3d.cs.tau.ac.il/FlexDock/	67
PatchDock	http://bioinfo3d.cs.tau.ac.il/PatchDock/	53
RosettaDock	http://rosettadock.graylab.jhu.edu/	157
SKE-DOCK	http://www.pharm.kitasato-u.ac.jp/bmd/ files/SKE_DOCK.html	158
SwarmDock	http://bmm.cancerresearchuk.org/~Swarm Dock/	159
SymmDock	http://bioinfo3d.cs.tau.ac.il/SymmDock/	53
ZDock	http://zdock.umassmed.edu/	160

A list of the Web servers that participated in CAPRI. For each server are listed its acronym, the URL link, and the literature reference where details on the server can be found. The list is alphabetically ordered.

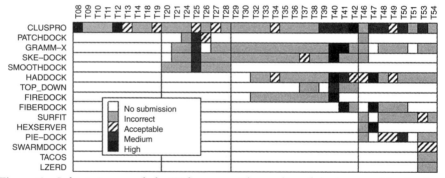

Figure 6   *Submissions made by Web servers.* The quality of the best model submitted by each server for a given target is represented by the indicated code. See Table 1 for a description of the servers. Detailed results of the assessment are reported in Refs. 63 and 64.

model could be built by homology from the amino acid sequence. Good models were also obtained when nonstructural information was available on specific residues or a region of the protein surface that contributes to the interaction, or when related structures were known. This was, for example, the case of target T41 (Figure 7), a complex of the colicin E9 DNase domain with the Im2 immunity protein.[120] Both components were given in the unbound form for docking prediction, and the conformational changes relative to the bound components

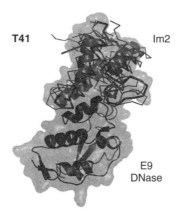

Figure 7 *Target T41 and its CAPRI prediction.* Target T41 is a complex of the DNase domain of colicin E9 with the Im2 immunity protein (PDB entry 2WPT).[120] The bound and unbound conformations differ by 1.5 Å RMSD for the DNase domain, and 2.0 Å for Im2. 327 docking models were submitted for this target; 67 were of acceptable, 58 of medium, and 24 of high quality.[63] The target X-ray structure is drawn as a ribbon with a molecular surface. The thick line represents Im2 in one of the high-quality models; thinner lines represent two medium-quality models.

were significant. However, most predictors assumed that the complex involved the same interface as in related colicin/immunity protein complexes published previously.[121] This was correct, and the prediction had a high success rate: 22 predictor groups and four docking servers submitted a total 24 high- and 58 medium-quality models, plus many acceptable ones. A few of these models are illustrated in Figure 7 along with the X-ray structure of the target.

The more poorly predicted targets were in general those displaying large conformational changes, those having unexpected features, and those for which little information was available on regions involved in the interface. In such cases, CAPRI predictions yielded at best "acceptable" models, which recall only a fraction of the native interactions. Only five targets (T04/T05, T28, T33, T38) yielded no correct model at all. In three of those, the failure of the prediction could be attributed to large conformational changes. In other cases, it could be blamed to misleading or incorrect biochemical information.

Target T27 offers a particularly interesting example of how CAPRI predictions can be instrumental in interpreting experimental data. T27 is a complex between the SUMO-conjugating enzyme UBC9 and the ubiquitin-conjugating enzyme E2-25 K (HIP2), a substrate of UBC9. In the crystal structure (PDB code 2O25; Walker et al. unpublished), the asymmetric unit contains two UBC9 and two HIP2 molecules that can be paired in at least two different ways (Figure 8). None of the 38 CAPRI predictor groups predicted the first binding mode, initially preferred by the authors of the structure. Instead, they submitted many models displaying the second binding mode,[105] which involves a lysine that reacts with SUMO. The biologically

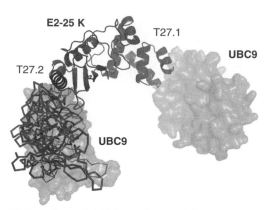

Figure 8 *Target T27 and its CAPRI prediction.* T27 is a complex between the SUMO-conjugating enzyme UBC9 and the ubiquitin-conjugating enzyme E2-25 K (HIP2), a substrate of UBC9 (PDB code 2O25; J. R. Walker et al. unpublished). The figure shows the contacts observed in the crystal between the E2-25 K molecule drawn as a ribbon and two symmetry-related UBC9 molecules drawn as molecular surfaces. Because the biologically relevant contact was uncertain, two alternative solutions T27.1 and T27.2 were considered when assessing the predictions.[63] The submissions contained two medium-quality and 55 acceptable models of T27.2, but none of T27.1. Thick lines represent UBC9 in the medium-quality models; a thinner line represents an acceptable model.

relevant contact is still uncertain, and in reality, both may be relevant, as UBC9 and HIP2 are part of a multicomponent protein degradation system, and they may interact with other components through several regions of their surface.

Finally, it is important to mention that even incorrect docking models may bring useful information, as many models that poorly recall the native residue–residue contacts nevertheless predict interface residues quite well (see Section "Assessing Interface Predictions").

Ranking Docking Poses: Results of the Scoring Experiment

The CAPRI scoring experiment was designed to help method developers test scoring function independently from the docking calculation. For a given target, scorers are given access to an ensemble of models contributed on a voluntary basis by "uploader" groups, participants in the CAPRI docking experiment who provide on the order of 100 models each to the scoring experiment. The scorers select models from the uploaded ensemble, and they submit a set of 10 to be assessed using the standard CAPRI evaluation criteria.

The scoring experiment started with target T24, and results are now available for a total of 24 targets. The scoring submissions are usually significantly enriched in correct models relative to the docking submissions.

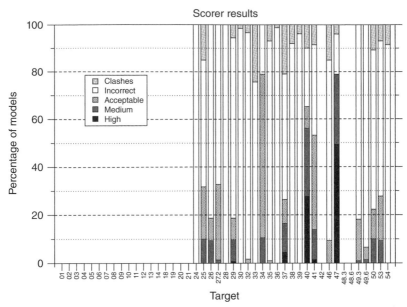

Figure 9 *A summary of the results of the scoring experiment.* Starting with T24, most of the docking targets of CAPRI were also proposed for scoring by registered scorer groups, who had access to a large ensemble of models that predictor groups had uploaded after making their own submission. For each target, bars represent the cumulative percentages of submitted models of high-, medium, or acceptable quality. The remainder represents incorrect models or models disqualified because of clashes or low-sequence identity. Detailed results of the assessment are reported in Refs. 63 and 64.

For six targets, correct models represent over 30% of the scorers' submissions (Figure 9), whereas the docking submissions achieved that success rate on 3 targets only out of 43 (compare with Figure 5). Although predictor groups may select their best models to be uploaded, so that part of the enrichment is already present in the uploaded ensemble, this result proves that some of the scorer groups are able to identify good models that were ranked low by the predictor groups who produced them.[63,64]

An independent test of the quality of the scoring functions is their ability to rank the models in the list of 10 that predictors and scorers submit for each target. The rank of a model should reflect a degree of confidence, with high-confidence models appearing at the top of the list. Yet, no obvious correlation has been found so far between the rank of models and their accuracy as determined by the CAPRI criteria. Thus, designing effective scoring functions capable of reliably discriminating between correct and incorrect docking models remains an outstanding problem.

RECENT DEVELOPMENTS IN MODELING PROTEIN–PROTEIN INTERACTION

Modeling Multicomponent Assemblies. The Multiscale Approach

Genome-wide studies performed in recent years have shown the prominence of protein–protein interaction. In single-cell organisms, more than half of the proteins are part of protein complexes, which may contain tens of subunits. Examples are complexes involved in DNA replication, gene transcription, protein synthesis, protein folding, protein degradation, and protein transport. The structural characterization of these multicomponent assemblies and of the relevant protein–protein interaction networks provides us with insights into their function and is fundamental to our understanding of their role in life processes.[1,2,122]

Macromolecular complexes can be subjected to biophysical techniques such as isothermal calorimetry, mass spectrometry, or chemical cross-linking, to characterize the interaction between their components. At the atomic level, nuclear magnetic resonance spectroscopy is not suitable for large complexes, but X-ray crystallography has successfully determined the atomic structure of such large macromolecular machines as ATP synthase and the ribosome.[4–7] Still, obtaining suitable crystals and diffraction data remains a challenge, and the number of large macromolecular assemblies that can be studied in this way is limited. On the other hand, cryo-electron microscopy is well adapted to that purpose, and its latest advances that include electron tomography yield information at a quasi-atomic level on macromolecular complexes.

Given the limitations of each technique and the difficulty of obtaining high-resolution information on large assemblies, the community has developed multiscale "hybrid" approaches that integrate data from diverse origins.[123] This requires sophisticated computational methods that allow the integration of structural and biophysical data with data coming from experiments such as mutagenesis, yeast two-hybrid, affinity purification, chemical cross-linking, mass spectrometry, small-angle X-ray scattering, or fluorescence resonance energy transfer spectroscopy.[124–126] Thus, a model of the nuclear pore complex containing 456 proteins could be assembled from a combination of structural and nonstructural data.[127–131] Assessing the quality of such a model remains a major challenge facing the structural biology community, but the study of the component proteins and subassemblies by X-ray crystallography and cryo-electron microscopy should give an answer. Figure 10 shows structures taken from the PDB and the Electron Microscopy Data Bank (EMDB, http://www.ebi.ac.uk/pdbe/emdb/).[132] They represent some of the components or subassemblies of the nuclear pore complex that can be used in modeling the assembly.

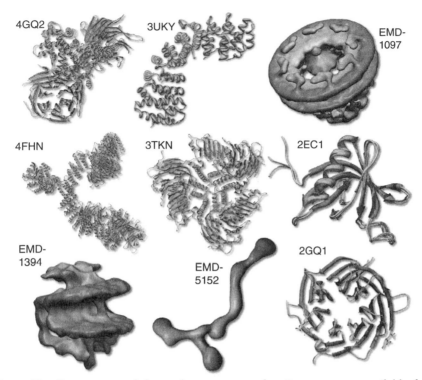

Figure 10 *Components of the nuclear pore complex.* Structures are available from the PDB and EMDB for some of the components of the nuclear core complex. The individual proteins and small subassemblies shown in ribbon representation are from six PDB entries that provide atomic-level information from X-ray (4GQ2, 3UKY, 4FHN, 3TKN, 4GQ1) or nuclear magnetic resonance spectroscopy (2EC1) studies. The three larger subassemblies drawn as surfaces have been analyzed by cryo-electron microscopy (EMD-5152, EMD-1097) or cryo-electron tomography (EMD-1394).

The development of protein–protein docking methods and their assessment in CAPRI contribute to the progress of the hybrid approaches. CAPRI targets have included not only protein–protein complexes, but also targets with nonprotein components such as a carbohydrate or a nucleic acid, and small-angle scattering data have been provided for a recent target. The scoring experiment may also contribute by testing methods to identify the correct docking solution from a number of models generated during the first stage of modeling.

Genome-Wide Modeling of Protein–Protein Interaction

A large amount of proteomic data on protein complexes is now available for several model organisms.[133–138] Their analysis alongside the information available from phylogenetic and sequence homology data has been the focus of research for more than a decade with the aim of building protein–protein interaction maps and understanding biological processes.[139] Recent developments in the field of modeling of protein interactions have been reviewed with an emphasis on methods that are applied to complete proteomes to build a "3D-interactome."[140,141] One of the challenges predictive methods must meet is the subtle changes at the interface that confer specificity to the protein–protein interaction.

For predictive methods that depend on the availability of structural data, the limited set of experimental information on large assemblies is an issue. The structures of complexes deposited to PDB may already be in sufficient number to greatly improve the accuracy and coverage of prediction of protein–protein interactions on a genomic scale.[86,87] Efforts are made to archive predictions alongside the experimental data and make them available via a user interface. An example is the Genome-WIde Docking Database (GWIDD) that archives genome-wide protein interaction data from experimentally determined complex structures alongside the predicted models. The data is made available for 771 organisms including data for human proteome.[142,143]

Multicomponent protein complexes are involved in a number of diseases and of great interest to drug design.[144] Many have emerged as novel therapeutic targets, and structural information helps designing molecules that inhibit their assembly. Residues or regions in the interface that are critical for the specificity of the interaction can be identified, and off-target binding resulting in side effects may be reduced by targeting those.

Engineering Interactions and Predicting Affinity

Engineering Novel Interactions

The procedures that have been developed to model natural protein–protein interactions can also be tools to create novel interactions. In the past decade, a number of laboratories have been engineering proteins to make high-affinity ligands for a target protein. Usually, a scaffold protein is subjected to cycles of site-directed mutagenesis followed by the selection of mutants best able to bind the target. Many mutants must be tested, which in most cases is done *in vivo* or *in vitro* on a DNA library.[99] The group of D. Baker (University of Washington, Seattle) does it first *in silico*, relying on the Rosetta force field that they initially developed for protein folding and then successfully applied to docking.[100] Their design procedure begins with a coarse-grained docking search that identifies candidate complexes, which are computationally mutated, and energy refined to get an estimate of their binding free energy. Then, the top-scoring candidates are expressed on the surface of yeast cells, and the cells that bind a fluorescent derivative of the target are selected on a cell sorter. Finally, the successful designs are purified and tested *in vitro*. In a proof-of-concept experiment, the scaffold protein was an ankyrin repeat and the targets a set of structurally diverse proteins. The experiment yielded an engineered complex that could be crystallized and formed the expected interaction, except for the relative orientation of the two components.[145] Another experiment aimed to mimic the way a neutralizing antibody binds the hemagglutinin of the flu virus. Two binders were identified, and the crystal structures of the complexes were very close to the designs.[101] Moreover, these constructs have themselves been subjected to further engineering experiments that enhanced their affinity and specificity.[146]

CAPRI and the Seattle Challenge

These experiments demonstrated that rational design can create novel functional interactions, but the efficiency was low. In the flu hemagglutinin experiment, millions of mutants had been tested computationally, and 88 had been predicted to have a high affinity for the target. Yet, only two actually bound when expressed on the yeast surface. Could other force fields do better than Rosetta? The CAPRI community was representative of the current state of the art, and D. Baker and S. Fleishman challenged it in a novel type of prediction round, which took place in January 2010. Instead of predicting a structure, the participants were given atomic models generated by Rosetta for 87 designed complexes, only one of which actually formed in the experiment, and they were asked to rank the models in terms of affinity relative to 120 natural complexes taken from the docking benchmark. Twenty-eight predictor groups made submissions, but when the rankings were assessed, all very poorly distinguished the designs from the natural complexes, and none could identify the successful design.[65]

Thus, the scoring functions used by the participants did not perform significantly better than that of the Seattle group. Nevertheless, a detailed analysis of the results showed that the electrostatics and solvation terms present in some of the functions could partially distinguish the designs from the natural complexes. Moreover, the designed binding surfaces tended to be less polar and less embedded in the protein structure than the natural ones and to have a less rigid backbone.[65] Such properties can be taken into account in computational design, and appropriate terms can be introduced in Rosetta and other force fields to improve their performance.[147] This is one of the ways in which a community-wide experiment such as CAPRI can help methods to progress. The publication that reports the challenge and its output is a fitting witness to the cooperative spirit that CAPRI promotes: it is cosigned by the Seattle group and all the participants who provide details of their respective scoring procedures.[65]

A second round of affinity prediction, performed at the end of 2011, was based on a large set of mutant data issued from the study of Whitehead et al.,[146] also by the Baker group. Two of the designed flu hemagglutinin binders were systematically mutated to introduce all 20 amino acid types at each position of their sequence in turn, yielding 1007 variants. Yeast clones displaying the variants on their surface were subjected to large-scale DNA sequencing, and the variant affinity for the target estimated from its enrichment in the yeast population selected on a cell sorter.[148] CAPRI participants were asked to predict the changes in affinity. Twenty-one groups made submissions, and again, the correlation with the experiment was poor. As the method was entirely novel and the affinity test was indirect, the round was extended and the participants given access to a subset of the experimental data on which they could train their procedures. Training generally improved the quality of the submissions, but not in a way that could be generalized to other systems.[149]

Modeling and Predicting Affinity

The failure to predict whether a designed complex is stable in solution demonstrates that the force fields and scoring functions used in docking do not correctly represent the binding free energy. In recent years, several procedures have been published to estimate the binding free energy of a protein–protein complex from its structure. Kastritis and Bonvin[66] ran them on 46 complexes taken from the docking benchmark and found very little correlation with the experimental values. Force fields and scoring functions employed in protein engineering are optimized for folding or docking, not to represent affinity. Developing appropriate force fields for affinity prediction is now an active field of research. New procedures may be benchmarked against a set of experimental data collated by Kastritis et al.[150] to form a structure-affinity benchmark. The benchmark contains values of K_d and binding free energies taken from the biochemical literature for 144 protein–protein complexes, together with the method and conditions under which they have been measured. The complexes are also part of the docking benchmark, and, therefore, structures are available for their components. Previous test sets contained only the structure of the complex, and, therefore, all the procedures based on these sets neglect conformation changes, which in many cases make a major contribution to the binding free energy. They also ignore the effect of experimental conditions, especially pH, which can be taken into account with the new benchmark.

CONCLUSION

Protein–protein docking is a lively field of research, and the CAPRI experiment plays a central role in fostering its development. Initially designed to assess the quality of predicted structures of protein–protein complexes, CAPRI has taken on the evaluation of scoring functions and more recently that of affinity predictions. Its evolution has been driven by the demand of experimentalists as much as by the interests of the computational biologists who participate in predictions. The CAPRI community has responded to new challenges arising from exciting new protein design endeavors. Several of its members are actively involved in the development of multiscale integrative procedures enabling structural biologists to build meaningful models of large macromolecular assemblies such as the nuclear pore complex by combining information from many different sources.

Genome-wide analyses are currently in the process of charting the landscape of protein–protein interactions in many organisms including human, revealing many interactions that are involved in disease. This landscape may be extended to three dimensions by building reliable structural models of the interacting proteome. This is a powerful incentive for developing procedures that integrate docking with other approaches that make the best use of all available information from biochemical, structural, or sequence studies. The performance of these procedures needs to be assessed, and we trust that the 12-year experience acquired in CAPRI will guarantee that this is done with the same high standard and objectivity as for structural predictions of protein–protein complexes.

ACKNOWLEDGMENTS

The authors acknowledge the valuable contribution to the management of CAPRI by J. Moult (Rockville, MD), M.J. Sternberg (London, UK), L. Ten Eyck (La Jolla, CA), S. Vajda (Boston, MA), and I. Vakser (Lawrence, KS), and they convey the gratitude and appreciation of the entire CAPRI community to all the experimentalists who have provided targets. We thank G. van Ginkel (Hinxton, UK) for their help with figures.

REFERENCES

1. B. Alberts , *Cell*, **92**, 291–294 (1998). The Cell as a Collection of Protein Machines: Preparing the Next Generation of Molecular Biologists.

2. C. V. Robinson, A. Sali, and W. Baumeister, *Nature*, **450**, 973–982 (2007). The Molecular Sociology of the Cell.

3. H. Berman, K. Henrick, H. Nakamura, and J. L. Markley, *Nucleic Acids Res.*, **35**, D301–D303 (2007). The Worldwide Protein Data Bank (wwPDB): Ensuring a Single, Uniform Archive of PDB Data.

4. D. Stock, A. G. Leslie, and J. E. Walker, *Science*, **286**, 1700–1705 (1999). Molecular Architecture of the Rotary Motor in ATP Synthase.

5. F. Schluenzen, A. Tocilj, R. Zarivach, J. Harms, M. Gluehmann, D. Janell, A. Bashan, H. Bartels, I. Agmon, F. Franceschi, et al., *Cell*, **102**, 615–623 (2000). Structure of Functionally Activated Small Ribosomal Subunit at 3.3 Angstroms Resolution.

6. N. Ban, P. Nissen, J. Hansen, P. B. Moore, and T. A. Steitz, *Science*, **289**, 905–920 (2000). The Complete Atomic Structure of the Large Ribosomal Subunit at 2.4 Å Resolution.

7. B. T. Wimberly, D. E. Brodersen, W. M. Clemons, Jr.,, R. J. Morgan-Warren, A. P. Carter, C. Vonrhein, T. Hartsch, and V. Ramakrishnan, *Nature*, **407**, 327–339 (2000). Structure of the 30S Ribosomal Subunit.

8. C. Chothia and J. Janin, *Nature*, **256**, 705–708 (1975). Principles of Protein-Protein Recognition.

9. J. Janin and C. Chothia, *J. Biol. Chem.*, **265**, 16027–16030 (1990). The Structure of Protein-Protein Recognition Sites.

10. S. Jones and J. M. Thornton, *Proc. Natl. Acad. Sci. U. S. A.*, **93**, 13–20 (1996). Principles of Protein-Protein Interactions.

11. L. Lo Conte, C. Chothia, and J. Janin, *J. Mol. Biol.*, **285**, 2177–2198 (1999). The Atomic Structure of Protein-Protein recognition Sites.

12. I. M. Nooren and J. M. Thornton, *EMBO J.*, **22**, 3486–3492 (2003). Diversity of Protein-Protein Interactions.

13. J. Janin, R. P. Bahadur, and P. Chakrabarti, *Q. Rev. Biophys.*, **41**, 133–180 (2008). Protein-Protein Interaction and Quaternary Structure.

14. R. Nussinov and G. Schreiber (Eds.), *Computational Protein-Protein Interaction*, CRC Press, Boca Raton, FL, 2009.

15. S. J. Wodak and J. Janin, *J. Mol. Biol.*, **124**, 323–342 (1978). Computer Analysis of Protein-Protein Interaction.

16. I. D. Kuntz, J. M. Blaney, S. J. Oatley, R. Langridge, and T. E. Ferrin, *J. Mol. Biol.*, **161**, 269–288 (1982). A Geometric Approach to Macromolecule-Ligand Interactions.

17. M. L. Connolly, *Biopolymers*, **25**, 1229–1247 (1986). Shape Complementarity at the Hemoglobin alpha 1 beta 1 Subunit Interface.

18. S. J. Wodak and J. Janin, *Adv. Protein Chem.*, **61**, 9–73 (2002). Structural Basis of Macromolecular Recognition.

19. J. Janin and S. J. Wodak, *Biopolymers*, **24**, 509–526 (1985). Reaction Pathway for the Quaternary Structure Change in Hemoglobin.

20. J. Janin and S. J. Wodak, *Adv. Protein Chem.*, **61**, 1–8 (2002). Protein Modules and Protein-Protein Interaction. Introduction.

21. C. Levinthal, S. J. Wodak, P. Kahn, and A. K. Dadivanian, *Proc. Natl. Acad. Sci. U. S. A.*, **72**, 1330–1334 (1975). Hemoglobin Interaction in Sickle Cell Fibers. I: Theoretical Approaches to the Molecular Contacts.

22. S. J. Wodak and J. Janin, *Proc. Natl. Acad. Sci. U. S. A.*, **77**, 1736–1740 (1980). Analytical Approximation to the Accessible Surface Area of Proteins.

23. M. Levitt, *J. Mol. Biol.*, **104**, 59–107 (1976). A Simplified Representation of Protein Conformations for Rapid Simulation of Protein Folding.

24. J. A. McCammon, B. R. Gelin, M. Karplus, and P. G. Wolynes, *Nature*, **262**, 325–326 (1976). The Hinge-Bending Mode in Lysozyme.

25. W. A. Eaton, E. R. Henry, and J. Hofrichter, *Proc. Natl. Acad. Sci. U. S. A.*, **88**, 4472–4475 (1991). Application of Linear Free Energy Relations to Protein Conformational Changes: The Quaternary Structural Change of Hemoglobin.

26. J. Cherfils, S. Duquerroy, and J. Janin, *Proteins*, **11**, 271–280 (1991). Protein-Protein Recognition Analyzed by Docking Simulation.

27. H. Wang, *J. Comput. Chem.*, **12**, 746–750 (1991). Grid-Search Molecular Accessible Surface Algorithm for Solving the Protein Docking Problem.

28. M. L. Connolly, *J. Appl. Cryst.*, **18**, 499–505 (1985). Molecular Surface Triangulation.

29. B. K. Shoichet, D. L. Bodian, et al., *J. Comput. Chem.*, **13**, 380–397 (1992). Molecular Docking Using Shape Descriptors.

30. B. K. Shoichet and I. D. Kuntz, *J. Mol. Biol.*, **221**, 327–346 (1991). Protein Docking and Complementarity.

31. D. Fischer, R. Norel, H. J. Wolfson, and R. Nussinov, *Proteins*, **16**, 278–292 (1993). Surface Motifs by a Computer Vision Technique: Searches, Detection, and Implications for Protein-Ligand Recognition.

32. S. L. Lin, R. Nussinov, D. Fischer, and H. J. Wolfson, *Proteins*, **18**, 94–101 (1994). Molecular Surface Representations by Sparse Critical Points.

33. R. Norel, D. Fischer, H. J. Wolfson, and R. Nussinov, *Protein Eng.*, **7**, 39–46 (1994). Molecular Surface Recognition by a Computer Vision-based Technique.

34. D. Fischer, S. L. Lin, H. J. Wolfson, and R. Nussinov, *J. Mol. Biol.*, **248**, 459–477 (1995). A Geometry-Based Suite of Molecular Docking Processes.

35. F. Jiang and S. H. Kim, *J. Mol. Biol.*, **19**, 79–102 (1991). "Soft Docking": Matching of Molecular Surface Cubes.

36. E. Katchalski-Katzir, I. Shariv, M. Eisenstein, A. A. Friesem, C. Aflalo, and I. A. Vakser, *Proc. Natl. Acad. Sci. U. S. A.*, **89**, 2195–2199 (1992). Molecular Surface Recognition: Determination of Geometric Fit Between Proteins and Their Ligands by Correlation Techniques.

37. I. A. Vakser and C. Aflalo, *Proteins*, **20**, 320–329 (1994). Hydrophobic Docking: A Proposed Enhancement to Molecular Recognition Techniques.

38. R. W. Harrison, I. V. Kourinov, and L. C. Andrews, *Protein Eng.*, **7**, 359–369 (1994). The Fourier-Green's Function and the Rapid Evaluation of Molecular Potentials.

39. N. S. Blom and J. Sygusch, *Proteins*, **27**, 493–506 (1997). High Resolution Fast Quantitative Docking Using Fourier Domain Correlation Techniques.

40. J. G. Mandell, V. A. Roberts, M. E. Pique, V. Kotlovyi, J. C. Mitchell, E. Nelson, I. Tsigelny, and L. F. Ten Eyck, *Protein Eng.*, **14**(2), 105–113 (2001). Protein Docking Using Continuum Electrostatics and Geometric Fit.

41. H. A. Gabb, R. M. Jackson, and M. J. Sternberg, *J. Mol. Biol.*, **272**, 106–120 (1997). Modelling Protein Docking Using Shape Complementarity, Electrostatics and Biochemical Information.

42. I. A. Vakser, *Protein Eng.*, **8**, 371–377 (1995). Protein Docking for Low-Resolution Structures.

43. I. A. Vakser, *Biopolymers*, **39**, 455–464 (1996). Low-Resolution Docking: Prediction of Complexes for Underdetermined Structures.

44. M. Meyer, P. Wilson, and D. Schomburg, *J. Mol. Biol.*, **264**, 199–210 (1996). Hydrogen Bonding and Molecular Surface Shape Complementarity as a Basis for Protein Docking.

45. F. Ackermann, G. Herrmann, S. Posch, and G. Sagerer, *Bioinformatics*, **14**, 196–205 (1998). Estimation and Filtering of Potential Protein-Protein Docking Positions.

46. D. W. Ritchie, *Curr. Protein Pept. Sci.*, **9**, 1–15 (2008). Recent Progress and Future Directions in Protein-Protein Docking.

47. S. Vajda and D. Kozakov, *Curr. Opin. Struct. Biol.*, **19**, 164–170 (2009). Convergence and Combination of Methods in Protein-Protein Docking.

48. M. Zacharias, *Curr. Opin. Struct. Biol.*, **20**, 180–186 (2010). Accounting for Conformational Changes During Protein-Protein Docking.

49. D. W. Ritchie, D. Kozakov, and S. Vajda, *Bioinformatics*, **24**, 1865–1873 (2008). Accelerating and Focusing Protein-Protein Docking Correlations Using Multi-Dimensional Rotational FFT Generating Functions.

50. K. Wiehe, B. Pierce, W. W. Tong, H. Hwang, J. Mintseris, and Z. Weng, *Proteins*, **69**, 719–725 (2007). The Performance of ZDOCK and ZRANK in Rounds 6-11 of CAPRI.

51. S. R. Comeau, D. Kozakov, R. Brenke, Y. Shen, D. Beglov, and S. Vajda, *Proteins*, **69**, 781–785 (2007). ClusPro: Performance in CAPRI Rounds 6-11 and the New Server.

52. Y. Shen, R. Brenke, D. Kozakov, S. R. Comeau, D. Beglov, and S. Vajda, *Proteins*, **69**, 734–742 (2007). Docking with PIPER and Refinement with SDU in Rounds 6-11 of CAPRI.

53. D. Schneidman-Duhovny, Y. Inbar, R. Nussinov, and H. J. Wolfson, *Nucleic Acids Res.*, **33**, W363–W367 (2005). PatchDock and SymmDock: Servers for Rigid and Symmetric Docking.

54. D. Kozakov, R. Brenke, S. R. Comeau, and S. Vajda, *Proteins*, **65**, 392–406 (2006). PIPER: An FFT-Based Protein Docking Program with Pairwise Potentials.

55. J. Mintseris, B. Pierce, K. Wiehe, R. Anderson, R. Chen, and Z. Weng, *Proteins*, **69**, 511–520 (2007). Integrating Statistical Pair Potentials into Protein Complex Prediction.

56. D. Kozakov, K. H. Clodfelter, S. Vajda, and C. J. Camacho, *Biophys. J.*, **89**, 867–875 (2005). Optimal Clustering for Detecting Near-native Conformations in Protein Docking.

57. S. Qin and H. X. Zhou, *Proteins*, **78**, 3166–3173 (2010). Selection of Near-Native Poses in CAPRI Rounds 13-19.

58. E. Mashiach, D. Schneidman-Duhovny, N. Andrusier, R. Nussinov, and H. J. Wolfson, *Nucleic Acids Res.*, **36**, W229–W232 (2008). FireDock: A Web Server for Fast Interaction Refinement in Molecular Docking.

59. J. J. Gray, S. Moughon, C. Wang, O. Schueler-Forman, B. Kuhlman, C. A. Rohl, and D. Baker, *J. Mol. Biol.*, **331**, 281–299 (2003). Protein-Protein Docking with Simultaneous Optimization of Rigid-Body Displacement and Side-Chain Conformations.

60. C. Dominguez, R. Boelens, and A. M. Bonvin, *J. Am. Chem. Soc.*, **125**, 1731–1737 (2003). HADDOCK: A Protein-Protein Docking Approach Based on Biochemical or Biophysical Information.

61. W. Muller and H. Sticht, *Proteins*, **67**, 98–111 (2007). A Protein-Specifically Adapted Scoring Function for the Reranking of Docking Solutions.

62. N. London and O. Schueler-Furman, *Biochem. Soc. Trans.*, **36**, 1418–1421 (2008). FunHunt: Model Selection Based on Energy Landscape Characteristics.

63. M. F. Lensink and S. J. Wodak, *Proteins*, **78**, 3073–3084 (2010). Docking and Scoring Protein Interactions: CAPRI 2009.

64. M. F. Lensink and S. J. Wodak, *Proteins*, **81**, 2082–2095 (2013). Docking, Scoring, and Affinity Prediction in CAPRI.

65. S. J. Fleishman, T. A. Whitehead, S. J. Wodak, J. Janin, J. D. Baker, et al., *J. Mol. Biol.*, **414**, 289–302 (2011). Community-wide Assessment of Protein-interface Modeling Suggests Improvements to Design Methodology.

66. P. L. Kastritis and A. M. Bonvin, *J. Proteome Res.*, **9**, 2216–2225 (2010). Are Scoring Functions in Protein-Protein Docking Ready to Predict Interactomes? Clues From a Novel Binding Affinity Benchmark.

67. D. Schneidman-Duhovny, R. Nussinov, and H. J. Wolfson, *Proteins*, **69**, 764–773 (2007). Automatic Prediction of Protein Interactions with Large Scale Motion.

68. E. Ben-Zeev, N. Kowalsman, A. Ben-Shimon, D. Segal, T. Atarot, O. Noivit, T. Shay, and M. Eisenstein, *Proteins*, **60**, 195–201 (2005). Docking to Single-Domain and Multiple-Domain Proteins: Old and New Challenges.

69. D. Mustard and D. W. Ritchie, *Proteins*, **60**, 269–274 (2005). Docking Essential Dynamics Eigenstructures.

70. M. Krol, R. A. Chaleil, A. L. Tournier, and P. A. Bates, *Proteins*, **69**, 750–757 (2007). Implicit Flexibility in Protein Docking: Cross-Docking and Local Refinement.

71. S. Chaudhury and J. J. Gray, *J. Mol. Biol.*, **381**, 1068–1087 (2008). Conformer Selection and Induced Fit in Flexible Backbone Protein-Protein Docking using Computational and NMR Ensembles.

72. O. N. Demerdash, A. Buyan, and J. C. Mitchell, *Proteins*, **78**, 3156–3165 (2010). Replic Opter: A Replicate Optimizer for Flexible Docking.

73. A. May and M. Zacharias, *Proteins*, **69**, 774–780 (2007). Protein-Protein Docking in CAPRI Using ATTRACT to Account for Global and Local Flexibility.

74. R. Das, I. Andre, Y. Shen, Y. Wu, A. Lemak, S. Bansal, C. H. Arrowsmith, T. Szyperski, and D. Baker, *Proc. Natl. Acad. Sci. U. S. A.*, **106**, 18978–18983 (2009). Simultaneous Prediction of Protein Folding and Docking at High Resolution.

75. E. Karaca, A. S. Melquiond, S. J. de Vries, P. L. Kastritis, and A. M. Bonvin, *Mol. Cell. Proteomics*, **9**, 1784–1794 (2010). Building Macromolecular Assemblies by Information-Driven Docking: Introducing the HADDOCK Multibody Docking Server.

76. S. Fiorucci and M. Zacharias, *Proteins*, **78**, 3131–3139 (2010). Binding Site Prediction and Improved Scoring During Flexible Protein-Protein Docking with ATTRACT.

77. K. Lasker, A. Sali, and H. J. Wolfson, *Proteins*, **78**, 3205–3211 (2010). Determining Macromolecular Assembly Structures by Molecular Docking and Fitting into an Electron Density Map.

78. X. Li, I. H. Moal, and P. A. Bates, *Proteins*, **78**, 3189–3196 (2010). Detection and Refinement of Encounter Complexes for Protein-Protein Docking: Taking Account of Macromolecular Crowding.

79. A. Fiser and A. Sali, *Methods Enzymol.*, **374**, 461–491 (2003). Modeller: Generation and Refinement of Homology-Based Protein Structure Models.

80. L. Bordoli, F. Kiefer, K. Arnold, P. Benkert, J. Battey, and T. Schwede, *Nat. Protoc.*, **4**, 1–13 (2009). Protein Structure Homology Modeling using SWISS-MODEL Workspace.

81. L. Lu, H. Lu, and J. Skolnick, *Proteins*, **49**, 350–364 (2002). MULTIPROSPECTOR: An Algorithm for the Prediction of Protein-Protein Interactions by Multimeric Threading.

82. O. Keskin, A. S. Aytuna, R. Nussinov, and A. Gursoy, *Nucleic Acids Res.*, **33**, W331–W336 (2005). PRISM: Protein Interactions by Structural Matching.

83. Q. C. Zhang, D. Petrey, R. Norel, and B. Honig, *Proc. Natl. Acad. Sci. U. S. A.*, **107**, 10896–10901 (2010). Protein Interface Conservation Across Structure Space.

84. L. Lu, A. K. Arakaki, H. Lu, and J. Skolnick, *Genome Res.*, **13**, 1146–1154 (2003). Multimeric Threading-based Prediction of Protein-Protein Interactions on a Genomic Scale: Application to the Saccharomyces cerevisiae Proteome.

85. Z. Zhu, A. Tovchigrechko, T. Baronova, Y. Gao, D. Douguet, N. O'Toole, and I. A. Vakser, *J. Bioinform. Comput. Biol.*, **6**, 789–810 (2008). Large-Scale Structural Modeling of Protein Complexes at Low Resolution.

86. Q. C. Zhang, D. Petrey, L. Deng, L. Qiang, Y. Shi, C. A. Thu, B. Bisikirska, C. Lefebvre , D. Accili, T. Hunter, T. Maniatis, A. Califano, and B. Honig, *Nature*, **490**, 556–560 (2012). Structure-Based Prediction of Protein-Protein Interactions on a Genome-Wide Scale.

87. P. J. Kundrotas, Z. Zhu, J. Janin, and I. Vakser, *Proc. Natl. Acad. Sci. U. S. A.*, **109**, 9438–9441 (2012). Templates are Available to Model Nearly All Complexes of Structurally Characterized Proteins.

88. A. G. Amit, R. A. Mariuzza, S. E. Phillips, and R. J. Poljak, *Science*, **233**, 747–753 (1986). Three-Dimensional Structure of an Antigen-Antibody Complex at 2.8 Å Resolution.

89. D. G. Lambright, J. Sondek, A. Bohm, N. P. Skiba, H. E. Hamm, and P. B. Sigler, *Nature*, **379**, 311–319 (1996). The 2.0 Å Crystal Structure of a Heterotrimeric G Protein.

90. S. Vajda, I. A. Vakser, M. J. Sternberg, and J. Janin, *Proteins*, **47**, 444–446 (2002). Modeling of Protein Interactions in Genomes.

91. N. C. Strynadka, M. Eisenstein, E. Katchalski-Katzir, B. K. Shoichet, I. D. Kuntz, R. Abagyan, M. Totrov, J. Janin, J. Cherfils, F. Zimmerman, A. Olson, B. Duncan, M. Rao, R. Jackson, M. Sternberg, and M. N. James, *Nat. Struct. Biol.*, **3**, 233–239 (1996). Molecular Docking Programs Successfully Predict the Binding of a Beta-Lactamase Inhibitory Protein to TEM-1 Beta-Lactamase.

92. J. Cherfils, T. Bizebard, M. Knossow, and J. Janin, *Proteins*, **18**, 8–18 (1994). Rigid-Body Docking with Mutant Constraints of Influenza Hemagglutinin with Antibody HC19.

93. J. S. Dixon, *Proteins*, **29**(Suppl. 1), 198–204 (1997). Evaluation of the CASP2 Docking Section.

94. J. Moult, J. T. Pedersen, R. Judson, and K. Fidelis, *Proteins*, **23**(3), 2–5 (1995). A Large-Scale Experiment to Assess Protein Structure Prediction Methods.

95. J. Janin, K. Henrick, J. Moult, L. TenEyck, M. J. Sternberg, S. Vajda, I. A. Vakser, and S. J. Wodak, *Proteins*, **52**, 2–9 (2003). Critical Assessment of Predicted Interactions.

96. D. Liger, L. Mora, N. Lazar, S. Figaro, J. Henri, N. Scrima, R. H. Buckingham, H. van Tilbeurgh, V. Heurgué-Hamard, and M. Graille, *Nucleic Acids Res.*, **39**, 6249–6259 (2011). Mechanism of Activation of Methyltransferases Involved in Translation by the Trm112 'Hub' Protein.

97. V. Mariani, F. Kiefer, T. Schmidt, J. Haas, and T. Schwede, *Proteins*, **79**(S10), 37–58 (2011). Assessment of Template Based Protein Structure Predictions in CASP9.

98. S. Bressanelli, K. Stiasny, S. L. Allison, E. A. Stura, S. Duquerroy, J. Lescar, F. X. Heinz, and F. A. Rey, *EMBO J.*, **23**, 728–738 (2004). Structure of a Flavivirus Envelope Glycoprotein in Its Low-pH-induced Membrane Fusion Conformation.

99. Y. L. Boersma and A. Plückthun, *Curr. Opin. Biotechnol.*, **22**, 849–857 (2011). DARPins and Other Repeat Protein Scaffolds: Advances in Engineering and Applications.

100. O. Schueler-Furman, C. Wang, P. Bradley, K. Misura, and D. Baker, *Science*, 310, 638–642 (2005). Progress in Modeling of Protein Structures and Interactions.

101. S. J. Fleishman, T. A. Whitehead, D. C. Ekiert, C. Dreyfus, J. E. Corn, E. M. Strauch, I. A. Wilson, and D. Baker, *Science*, 332, 816–821 (2011). Computational Design of Proteins Targeting the Conserved Region of Influenza Hemagglutinin.

102. R. Chen, J. Mintseris, J. Janin, and Z. Weng, *Proteins*, 52, 88–91 (2003). A Protein-Protein Docking Benchmark.

103. H. Hwang, T. Vreven, J. Janin, and Z. Weng, *Proteins*, 78, 3111–3114 (2010). Protein-Protein Docking Benchmark Version 4.0.

104. W. Wriggers, R. A. Milligan, and J. A. McCammon, *J. Struct. Biol.*, 125, 185–195 (1999). Situs: A Package for Docking Crystal Structures into Low-Resolution Maps from Electron Microscopy.

105. M. F. Lensink, R. Mendez, and S. J. Wodak, *Proteins*, 69, 704–718 (2007). Docking and Scoring Protein Complexes: CAPRI 3rd Edition.

106. N. S. Boutonnet, M. J. Rooman, M. E. Ochagavia, J. Richelle, and S. J. Wodak, *Protein Eng.*, 8, 647–662 (1995). Optimal Protein Structure Alignments by Multiple Linkage Clustering: Application to Distantly Related Proteins.

107. N. S. Boutonnet, M. J. Rooman, and S. J. Wodak, *J. Mol. Biol.*, 253, 633–647 (1995). Automatic Analysis of Protein Conformational Changes by Multiple Linkage Clustering.

108. H. X. Zhou and S. Qin, *Bioinformatics*, 23, 2203–2209 (2007). Interaction-Site Prediction for Protein Complexes: a Critical Assessment.

109. S. J. de Vries and A. M. Bonvin, *Curr. Protein Pept. Sci.*, 9, 394–406 (2008). How Proteins Get in Touch: Interface Prediction in the Study of Biomolecular Complexes.

110. I. Ezkurdia, L. Bartoli, P. Fariselli, R. Casadio, A. Valencia, and M. L. Tress, *Brief. Bioinform.*, 10, 233–246 (2009). Progress and Challenges in Predicting Protein-Protein Interaction Sites.

111. I. Roterman-Konieczna (Ed.), *Identification of Ligand Binding Site and Protein-Protein Interaction Area*, Springer, Dordrecht, 2013.

112. M. F. Lensink and S. J. Wodak, *Proteins*, 78, 3085–3095 (2010). Blind Predictions of Protein Interfaces by Docking Calculations in CAPRI.

113. J. Janin, *Structure*, 7, R277–R279 (1999). Wet and Dry Interfaces: The Role of Solvent in Protein-Protein and Protein-DNA Recognition.

114. F. Rodier, R. P. Bahadur, P. Chakrabarti, and J. Janin, *Proteins*, 60, 36–45 (2005). Hydration of Protein-Protein Interfaces.

115. A. M. Buckle, G. Schreiber, and A. R. Fersht, *Biochemistry*, 33, 8878–8889 (1994). Protein-Protein Recognition: Crystal Structural Analysis of a Barnase-Barstar Complex at 2.0-Å Resolution.

116. M. F. Lensink, I. H. Moal, P. A. Bates, P. L. Kastritis, et al., *Proteins*, 82, 620–632 (2014). Blind Prediction of Interfacial Water Positions in CAPRI.

117. J. A. Wojdyla, S. J. Fleishman, D. Baker, and C. Kleanthous, *J. Mol. Biol.*, 417, 79–94 (2012). Structure of the Ultra-high-affinity Colicin E2 DNase--Im2 Complex.

118. A. D. van Dijk and A. M. Bonvin, *Bioinformatics*, 22, 2340–2347 (2006). Solvated Docking: Introducing Water into the Modelling of Biomolecular Complexes.

119. P. L. Kastritis, A. D. van Dijk, and A. M. Bonvin, *Methods Mol. Biol.*, 819, 355–374 (2012). Explicit Treatment of Water Molecules in Data-driven Protein-Protein Docking: the Solvated HADDOCKing Approach.

120. N. A. Meenan, A. Sharma, S. J. Fleishman, C. J. Macdonald, B. Morel, R. Boetzel, G. R. Moore, D. Baker, and C. Kleanthous, *Proc. Natl. Acad. Sci. U. S. A.*, 107, 10080–10085 (2010). The Structural and Energetic Basis for High Selectivity in a High-Affinity Protein-Protein Interaction.

121. U. C. Kühlmann, A. J. Pommer, G. R. Moore, R. James, and C. Kleanthous, *J. Mol. Biol.*, **301**, 1163–1178 (2000). Specificity in Protein-Protein Interactions: The Structural Basis for Dual Recognition in Endonuclease Colicin-Immunity Protein Complexes.

122. R. B. Russell, F. Alber, P. Aloy, F. P. Davis, D. Korkin, M. Pichaud, M. Topf, and A. Sali, *Curr. Opin. Struct. Biol.*, **14**, 313–324 (2004). A Structural Perspective on Protein–Protein Interactions.

123. A. C. Steven and W. Baumeister, *J. Struct. Biol.*, **163**, 186–195 (2008). The Future is Hybrid.

124. F. Alber, F. Forster, D. Korkin, M. Topf, and A. Sali, *Annu. Rev. Biochem.*, **77**, 443–477 (2008). Integrating Diverse Data for Structure Determination of Macromolecular Assemblies.

125. D. Russel, K. Lasker, B. Webb, J. Velázquez-Muriel, E. Tjioe, et al., *PLoS Biol.*, **10**, e1001244 (2012). Putting the Pieces Together: Integrative Modeling Platform Software for Structure Determination of Macromolecular Assemblies.

126. D. Schneidman-Duhovny, A. Rossi, A. Avila-Sakar, S. J. Kim, J. Velázquez-Muriel, P. Strop, H. Liang, K. A. Krukenberg, M. Liao, H. M. Kim, S. Sobhanifar, V. Dötsch, A. Rajpal, J. Pons, D. A. Agard, Y. Cheng, and A. Sali, *Bioinformatics*, **28**, 3282–3289 (2012). A Method for Integrative Structure Determination of Protein-Protein Complexes.

127. F. Alber, S. Dokudovskaya, L. M. Veenhoff, W. Zhang, J. Kipper, D. Devos, A. Suprapto, O. Karni-Schmidt, R. Williams, B. T. Chait, A. Sali, and M. P. Rout, *Nature*, **450**, 695–701 (2007). The Molecular Architecture of the Nuclear Pore Complex.

128. F. Alber, S. Dokudovskaya, L. M. Veenhoff, W. Zhang, J. Kipper, D. Devos, A. Suprapto, O. Karni-Schmidt, R. Williams, B. T. Chait, M. P. Rout, and A. Sali, *Nature*, **450**, 683–694 (2007). Determining the Architectures of Macromolecular Assemblies.

129. M. Beck, V. Lucić, F. Forster, W. Baumeister, and O. Medalia, *Nature*, **449**, 611–615 (2007). Snapshots of Nuclear Pore Complexes in Action Captured by Cryo-Electron Tomography.

130. D. Frenkiel-Krispin, B. Maco, U. Aebi, and O. Medalia, *J. Mol. Biol.*, **395**, 578–586 (2010). Structural Analysis of a Metazoan Nuclear Pore Complex Reveals a Fused Concentric Ring Architecture.

131. A. Hoelz, E. W. Debler, and G. Blobel, *Annu. Rev. Biochem.*, **80**, 613–643 (2011). The Structure of the Nuclear Pore Complex.

132. C. L. Lawson, M. L. Baker, C. Best, C. Bi, M. Dougherty, P. Feng, G. van Ginkel, B. Devkota, I. Lagerstedt, R. H. Newman, T. J. Oldfield, I. Rees, G. Sahni, R. Sala, S. Velankar, J. Warren, J. D. Westbrook, K. Henrick, G. J. Kleywegt, H. M. Berman, and W. Chiu, *Nucleic Acids Res.*, **39**, D456–D464 (2011). EMDataBank.org: Unified Data Resource for CryoEM.

133. A. C. Gavin, P. Aloy, et al., *Nature*, **440**, 631–636 (2006). Proteome Survey Reveals Modularity of the Yeast Cell Machinery.

134. S. J. Wodak, J. Vlasblom, and S. Pu, *Methods Mol. Biol.*, **759**, 381–406 (2011). High-Throughput Analyses and Curation of Protein Interactions in Yeast.

135. K. G. Guruharsha, J.-F. Rual, et al., *Cell*, **147**, 690–703 (2011). A Protein Complex Network of Drosophila melanogaster.

136. Arabidopsis Interactome Mapping Consortium, *Science*, **333**, 601–607 (2011). Evidence for Network Evolution in an *Arabidopsis* Interactome Map.

137. J.-F. Rual, K. Venkatesan, et al., *Nature*, **437**, 1173–1178 (2005). Towards a Proteome-scale Map of the Human Protein–protein Interaction Network.

138. P. C. Havugimana, G. T. Hart, et al., *Cell*, **150**, 1068–1081 (2012). A Census of Human Soluble Protein Complexes.

139. A. J. M. Walhout and M. Vidal, *Nat. Rev. Mol. Cell Biol.*, **2**, 55–63 (2001). Protein Interaction Maps for Model Organisms.

140. P. Aloy and R. B. Russell, *Nat. Rev. Mol. Cell Biol.*, **7**, 188–197 (2006). Structural Systems Biology: Modelling Protein Interactions.

141. A. Stein, R. Mosca, and P. Aloy, *Curr. Opin. Struct. Biol.*, **21**, 200–208 (2011). Three-Dimensional Modeling of Protein Interactions and Complexes is Going 'Omics.

142. P. J. Kundrotas, Z. Zhu, and I. A. Vakser, *Nucleic Acids Res.*, 38, D513–D517 (2010). GWIDD: Genome-Wide Protein Docking Database.

143. P. J. Kundrotas, Z. Zhu, and I. A. Vakser, *Hum. Genomics*, 6, 7–10 (2012). GWIDD: A Comprehensive Resource for Genome-wide Structural Modeling of Protein-Protein Interactions.

144. H. Jubb, A. P. Higueruelo, A. Winter, and T. L. Blundell, *Trends Pharmocol Sci.*, 33, 241–248 (2012). Structural Biology and Drug Discovery for Protein-Protein Interactions.

145. J. Karanicolas, J. E. Corn, I. Chen, L. A. Joachimiak, O. Dym, S. H. Peck, S. Albeck, T. Unger, W. Hu, G. Liu, S. Delbecq, G. T. Montelione, C. P. Spiegel, D. R. Liu, and D. Baker, *Mol. Cell*, 42, 250–260 (2011). A De Novo Protein Binding Pair by Computational Design and Directed Evolution.

146. T. A. Whitehead, A. Chevalier, Y. Song, C. Dreyfus, S. J. Fleishman, C. De Mattos, C. A. Myers, H. Kamisetty, P. Blair, I. A. Wilson, and D. Baker, *Nat. Biotechnol.*, 30, 543–548 (2012). Optimization of Affinity, Specificity and Function of Designed Influenza Inhibitors Using Deep Sequencing.

147. P. B. Stranges and B. Kuhlman, *Protein Sci.*, 22, 74–82 (2013). A Comparison of Successful and Failed Protein Interface Designs Highlights The Challenges of Designing Buried Hydrogen Bonds.

148. D. M. Fowler, C. L. Araya, S. J. Fleishman, E. H. Kellogg, J. J. Stephany, D. Baker, and S. Fields, *Nat. Methods*, 7, 741–746 (2010). High-Resolution Mapping of Protein Sequence-function Relationships.

149. R. Moretti, S. J. Fleishman, R. Agius, M. Torchala, P. A. Bates, et al., S. Velankar, J. Janin, S. J. Wodak, and D. Baker, *Proteins*, 81, 1980–1987 (2013). Community-Wide Evaluation of Methods for Predicting the Effect of Mutations on Protein-Protein Interactions.

150. P. L. Kastritis, I. H. Moal, H. Hwang, Z. Weng, P. A. Bates, A. M. Bonvin, and J. Janin, *Protein Sci.*, 20, 482–491 (2011). A Structure-Based Benchmark for Protein-Protein Binding Affinity.

151. V. I. Lesk and M. J. Sternberg, *Bioinformatics*, 24, 1137–1144 (2008). 3D-Garden: A System for Modelling Protein-Protein Complexes Based on Conformational Refinement of Ensembles Generated with the Marching Cubes Algorithm.

152. D. Kozakov, D. R. Hall, D. Beglov, R. Brenke, S. R. Comeau, Y. Shen, K. Li, J. Zheng, P. Vakili, I. Paschalidis, and S. Vajda, *Proteins*, 78, 3124–3130 (2010). Achieving Reliability and High Accuracy in Automated Protein Docking: ClusPro, PIPER, SDU, and Stability Analysis in CAPRI Rounds 13-19.

153. A. Tovchigrechko and I. A. Vakser, *Nucleic Acids Res.*, 34, W310–W314 (2006). GRAMM-X Public Web Server for Protein-Protein Docking.

154. S. J. de Vries, A. D. van Dijk, M. Krzeminski, M. van Dijk, A. Thureau, V. Hsu, T. Wassenaar, and A. M. Bonvin, *Proteins*, 69, 726–733 (2007). HADDOCK versus HADDOCK: New Features and Performance of HADDOCK2.0 on the CAPRI Targets.

155. G. Macindoe, L. Mavridis, V. Venkatraman, M. D. Desvignes, and D. W. Ritchie, *Nucleic Acids Res.*, 38(Web Server issue), W445–W449 (2010). HexServer: An FFT-Based Protein Docking Server Powered by Graphics Processors.

156. E. Mashiach, R. Nussinov, and H. J. Wolfson, *Nucleic Acids Res.*, 38, W457–W461 (2010). FiberDock: A Web Server for Flexible Induced-Fit Backbone Refinement in Molecular Docking.

157. S. Lyskov and J. J. Gray, *Nucleic Acids Res.*, 36, W233–W238 (2008). The RosettaDock Server for Local Protein-Protein Docking.

158. G. Terashi, M. Takeda-Shitaka, D. Takaya, K. Komatsu, and H. Umeyama, *Proteins*, 60, 289–295 (2005). Searching for Protein-Protein Interaction Sites and Docking by the Methods of Molecular Dynamics, Grid Scoring, and the Pairwise Interaction Potential of Amino Acid Residues.

159. I. H. Moal and P. A. Bates, *Int. J. Mol. Sci.*, 11, 3623–3648 (2010). SwarmDock and the Use of Normal Modes in Protein-Protein Docking.

160. B. G. Pierce, Y. Hourai, and Z. Weng, *PLoS One*, **6**, e24657 (2011). Accelerating Protein Docking in ZDOCK Using an Advanced 3D Convolution Library.

161. D. A. Bonsor, I. Grishkovskaya, E. J. Dodson, and C. Kleanthous, *J. Am. Chem. Soc.*, **129**, 4800–4807 (2007). Molecular Mimicry Enables Competitive Recruitment by a Natively Disordered Protein.

162. R. Mendez, R. Leplae, L. De Maria, and S. J. Wodak, *Proteins*, **52**, 51–67 (2003). Assessment of Blind Predictions of Protein-Protein Interactions: Current Status of Docking Methods.

163. R. Mendez, R. Leplae, M. F. Lensink, and S. J. Wodak, *Proteins*, **60**, 150–169 (2005). Assessment of CAPRI Predictions in Rounds 3-5 Shows Progress in Docking Procedures.

CHAPTER 5

Kinetic Monte Carlo Simulation of Electrochemical Systems

C. Heath Turner,[a] Zhongtao Zhang,[a] Lev D. Gelb,[b] and Brett I. Dunlap[c]

[a]Department of Chemical and Biological Engineering, The University of Alabama, Tuscaloosa, AL 35487-0203, USA
[b]Department of Materials Science and Engineering, University of Texas at Dallas, Richardson, TX 75080, USA
[c]Chemistry Division, US Naval Research Laboratory, Washington, DC 20375-5342, USA

BACKGROUND

Over the past decade or so, there has been an accelerated emphasis on providing devices for electrical energy storage and electrochemical conversion (batteries, capacitors, fuel cells, etc.). Although these basic technologies have been in development for over a century, their importance has grown substantially because of the increased electrification of the transportation sector and the energy storage requirements associated with intermittent renewable energy sources such as wind and solar energy.[1] Many aspects of our future energy portfolio will be influenced by the efficiency, economics, and stability of electrochemical conversion and storage devices. Research progress in electrochemical conversion and storage can drive the commercialization of technologies ranging from large-scale power-grid applications all the way down to portable electronic devices, which rely heavily on rechargeable Li-ion batteries.[2] To bring these technologies to market, significant research effort has

Reviews in Computational Chemistry, Volume 28, First Edition.
Edited by Abby L. Parrill and Kenny B. Lipkowitz.
© 2015 John Wiley & Sons, Inc. Published 2015 by John Wiley & Sons, Inc.

focused on the development of new materials, especially their compositions and structures. Many of the recent advancements in this field have pushed the feature size of the materials down to the micrometer to nanometer length scale. At the nanometer length scale, many of the traditional engineering and design relationships are beyond the limits of continuum models. In parallel, experimental characterization of the detailed electrochemical events and material behavior becomes more elusive at the atomic scale, as the complex molecular-level processes are more challenging to capture.

To make meaningful progress toward the design of next-generation electrochemical systems and devices, more information is needed about how the underlying atomic-scale phenomena contribute to the overall electrochemical performance. While experimental investigations and advanced synthesis techniques play a major role in this field, we focus on the insight that can be gained by applying atomic-scale modeling tools to contemporary electrochemical systems.

One of the primary characteristics of any modeling technique is the inherent trade-off between computational speed and model resolution (both timescale and length scale). Thus, one of the initial considerations in choosing a modeling hierarchy is the resolution needed to capture or predict the properties of interest, which can be classified by the most relevant length scales and timescales. System attributes range from localized lattice defects and dopant distributions occurring at the nanometer length scale, all the way up to grain boundaries, and interconnect interfacial structures at the micron and millimeter length scales.

INTRODUCTION TO KINETIC MONTE CARLO

The kinetic Monte Carlo (KMC) method has found applications in several broad areas of science and engineering, with many modeling studies of surface deposition processes,[3–9] heterogeneous catalysis,[10–16] and now in electrochemical systems. In electrochemical systems, it is particularly difficult to isolate and identify the fundamental cause–effect relationships that exist because of the many highly correlated relationships between the structures, chemical compositions, voltages, thermal gradients, and so on that are driven by the chemical reactions that ultimately move ionic charge. At meso- and macroscopic scales, these simultaneous influences can be approximated with finite element approaches or other mesh-based continuum methods. At an atomic resolution with traditional molecular dynamics methods,[17,18] however, it is computationally prohibitive to model all of these thermodynamic and kinetic effects on the timescales needed for an ion to move an appreciable distance – even though the atomic-level details contain critical information about the chemical reactions needed for moving these technologies forward. Here we describe the KMC simulation method, which can mitigate that problem and play an important role in the modeling hierarchy of electrochemical systems.

In terms of applicable length scales and timescales, the KMC method generally ranks between molecular dynamics methods and mesoscale simulation techniques (Figure 1). Subramanian and his coworkers have recently reviewed modeling and simulations in lithium-ion battery research,[19] including the importance of KMC in describing detailed electrochemical events, such as the growth of the passive solid electrolyte interphase (SEI) layer.

The KMC method emerged from the very early contributions of Young and Elcock,[20] Cox and Miller,[21] and Gillespie.[22] In later years, Fichthorn and Weinberg[23] provided an in-depth analysis of the method, and others[24,25] have since provided important descriptions and reviews of the method. These previous works provide an important methodological foundation, so we refer the reader to these earlier examples for a more rigorous analysis of KMC. Here we spend more effort discussing the implementation of this technique and highlight current examples in the literature that have used this method to accelerate the understanding and development within the field of electrochemical systems and processes.

Although our focus is on electrochemical systems, KMC is a general approach for modeling the overall behavior of many different systems that can be decomposed into a collection of discrete events and their associated rate constants. A KMC simulation relies on the division of the phase space of the system into a set of discrete states and *a priori* knowledge of a set of transition rates characterizing the transitions between these states. For instance, as related to the topic at hand, the states may be different arrangements of ions in a solid electrolyte, and the events are implemented as jumps of individual

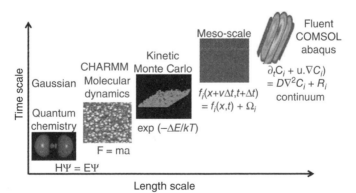

Figure 1 Computational methods roadmap, which illustrates the length scales and timescales accessible with KMC modeling. (Reproduced with permission from Ref. 19. Copyright (2012), The Electrochemical Society.) (For a color version of this figure, please see plate 2 in color plate section.)

ions between positions. Other possible events are adsorption of solution-phase molecules and chemical reactions at a catalyst surface. The events are assumed to obey Poisson statistics, representative of random, uncoordinated rare-event (rate-limited) processes. Although one of the primary outcomes of a KMC simulation is a prediction of the time evolution of a system, the thermodynamic averages can also be obtained under equilibrium conditions.

Due to the discrete state nature of a KMC simulation, as well as the nature of many of its applications, the systems are often defined with respect to an underlying lattice. Although this reduces configurational flexibility, a lattice can still provide a reasonable representation of the system in many cases. Furthermore, comparisons can sometimes be made with other lattice-based models that are exactly solvable, thereby providing insight into phase transitions and thermodynamic scaling behavior. Thus, we will assume here a system where atoms occupy discrete lattice sites. Consider as a simple example a simulation of an electrodeposition process. We might initially begin from a planar metal substrate, with atoms assigned to individual lattice points (intended to accommodate the known crystal structure of the bulk metal) and having open vacancy positions located above the substrate, which may become occupied as the simulation progresses.

Once an appropriate lattice and an initial configuration have been chosen, a catalogue of events (and event rates) must be established. Examples of "on-the-fly" rate assessment exist, such as the work of Henkelman to accelerate on-the-fly molecular dynamics.[26,27] Here we will consider only information that is provided *a priori*. It is often challenging to identify all of the possible events that could occur in the real (experimental) system. However, in many cases a complete catalogue of all possible events/rates is unnecessary; instead, a reasonable subset of the dominant events/rates can provide an adequate representation of the system. These rate constants (Γ_n) are typically supplied in an Arrhenius form (Eq. [1]), consisting of a preexponential factor (k_n^0) and an activation barrier (E_n), where k_B is Boltzmann's constant and the subscript n denotes the event number. Furthermore, these rates may be altered, on the basis of the global operating conditions (such as the temperature) and/or the local site-site interactions in the system (which may change the activation barriers or the preexponential factors). The rates are particularly sensitive to the value of the activation barrier, since the activation barriers appear in the exponential term.

$$\Gamma_{n,\text{site}}(x, y, z) = k_{n,\text{site}}^0 \exp\left(-\frac{E_{n,\text{site}}}{k_B T}\right) \qquad [1]$$

For example, if we assume that atoms occupy sites on a cubic lattice, we identify each lattice site as a unique set of integer coordinates (x, y, z). In a surface deposition process, some of the sites would be labeled as occupied and others vacant. In such a system, important events would include (1) adsorption/desorption from the gas to the surface; (2) diffusion on the surface; and (3) movement of an atom in/out of a surface layer, creating or forming a vacancy site. The associated rate constants may be obtained from experimental measurements, first-principles density functional theory (DFT) calculations, or extrapolated from other known similar systems. With respect to the system configuration, each site (x, y, z) will be able to participate in various events (n), depending on the local environment. For each of these possible events, there exists an associated rate constant $(\Gamma_{n,\text{site}})$, so that an evaluation of the system configuration will provide rate information about all possible transitions at all locations in the model. Thus, the total rate of a particular event (Γ_n) is:

$$\Gamma_n = \sum_x \sum_y \sum_z \Gamma_{n,\text{site}}(x, y, z) \qquad [2]$$

Also, the total rate of all events (Γ_{total}) can be calculated by simply summing all of the individual event rates. This constant bookkeeping in the system is critical for propagating the system through time, according to the instantaneous rate information. Fortunately, the computational cost for this accounting process is very low because the rate information typically requires updating only in the proximity (neighbor or possibly next-nearest-neighbor sites) of where the most recent event occurred. Note that Γ_{total} is dominated by its largest terms, which correspond to events with the fastest rates.

After the system configuration has been established and the initial event rates have been calculated, the time evolution of the system is defined by advancing the system clock before each event, according to Eq. [3] where Δt is the time step and RN is a random number, distributed evenly between 0 and 1.

$$\Delta t = -\frac{\ln(RN)}{\Gamma_{\text{total}}} \qquad [3]$$

Once the clock has been incremented $(t_{\text{new}} = t_{\text{old}} + \Delta t)$, the system configuration is then advanced by stochastically selecting an event to occur, according to the probability of Eq. [4] (based on the relative rates in the system):

$$P_{n,\text{site}} = \frac{\Gamma_{n,\text{site}}}{\Gamma_{\text{total}}} \qquad [4]$$

Because the event selection probability $(P_{n,\text{site}})$ is proportional to the relative rates of the individual events, on average, the event rates will dictate the relative probability (i.e., frequency) that different events are chosen to occur. Once an event is identified, the system configuration is updated, and the list of event rates is likewise updated (according to the new configuration). The algorithm ensures that at each time step, an event is always performed. This is a fundamentally different procedure compared to traditional Monte Carlo calculations, which typically propagate a system by performing *trial* moves, and either accepting or rejecting these moves. The KMC simulation approach can also span timescales that are orders of magnitude larger than traditional molecular dynamics methods, because the timescale of a KMC simulation is inversely proportional to the rate of the fastest processes included in the model. This allows systems with activated processes to be modeled efficiently because each step in the dynamics leading up to each event has a faster rate than the event as a whole. The molecular vibrations are implicitly averaged within the Arrhenius preexponential factor in Eq. [1], so that the focus is on the much rarer event in which the activation barrier is overcome. This computational efficiency is obviously considerable and important. As the simulation progresses, it cycles through many iterations of: (1) incrementing time; (2) selecting a system event; (3) updating the configuration; and (4) updating rates. Concurrently, dynamic properties can be assessed, such as the rate of deposition onto a surface, the average flux of ionic species through an electrolyte, the concentrations and lifetimes of different surface species, and so on. Slight variations to the KMC approach described above exist[24] (such as null-event KMC), and there are a variety of approaches for updating the clock and efficiently searching for the next event to occur.

Although several approximations are made when mapping an experimental system onto a KMC model, one of the primary challenges is creating a catalog of all possible events (or at least the dominant ones), along with accurate event rates. If a reaction or other event is not included in the event list, then it will never occur during the course of the simulation. If that omitted event is important, the KMC results will be unreliable. To address this issue, experimental benchmarking is critical, so that omissions and other weaknesses in the model can be identified and corrected.

ELECTROCHEMICAL RELATIONSHIPS

One of the fundamental challenges in applying KMC to electrochemical systems is the treatment of charges, including the calculation of dynamic electric fields (arising from the local atomic arrangement and possibly from an applied oscillatory voltage), which affect the motion of ions in a system. At an interface between a metal and semiconductor, electrons can spill over from the metal to the semiconductor forming a Schottky barrier. The unbalanced charge forms an electric dipole that results in no long-range electric fields, but a change in electrical potential. The self-consistent response of such interfaces to applied fields is the foundation of the semiconductor industry. Similarly, at a metal-electrolyte interface, not only do the electrons of the metal respond to the interface and applied potential but also the ions in the electrolyte polarize in response to the redistribution of the metallic electrons. Again, there is charge neutrality, a dipole layer, electrical potential drop, and no long-ranged electric field if no ionic current flows through the electrolyte. This is called the electrochemical double layer, which is a critical aspect of electrochemical devices.

The Butler–Volmer expression provides rate information for events that involve the transfer of charge and is widely used in electrochemistry.[28,29] Using a kinetic model of the electrostatic forces on a metal ion crossing the double layer, Butler derived an exponential relationship between the current and electrical potential drop across an electrochemical interface.[30] Erdey-Grúz[31] and Volmer independently noted that electrochemical processes are dominated by slow charge transfer processes at both the anode and cathode due to high kinetic barriers that allow the buildup of overpotentials at both the cathode and anode. The charge transfer coefficients those researchers introduced are extremely useful. The Butler–Volmer expression pertains to ionic species moving through an electrolyte or to charge transfer processes occurring at an anode or cathode. For an electrochemical process involving two electrodes (with anode and cathode charge transfer coefficients of α_a and α_c, respectively), the Butler–Volmer equation relates the electrical current (i) to the electrode potential (E) on the basis of its contribution to the activation overpotential ($\eta = E - E_{eq}$):

$$i = i_0 \cdot \left\{ \exp\left[\frac{\alpha_a nF\eta}{RT} \right] - \exp\left[-\frac{\alpha_c nF\eta}{RT} \right] \right\} \qquad [5]$$

In this expression, i represents the electrode current density, i_0 is the exchange current density, R is the gas constant, T is the temperature, the anode and cathode charge transfer coefficients are often related by $\alpha_a + \alpha_c = 1$, n is the number of electrons transferred, and F is the Faraday constant. In KMC simulations the Butler–Volmer expression is decomposed and applied to individual redox events, such that individual charge transfer processes can be modeled within the KMC protocol. For instance, Hin[32] used KMC to model Li-ion movement at a solution/solid interface, so that the individual Li ion insertion/deletion steps at the interface are dictated by the following rates (rewritten in our notation):

$$\Gamma_{insertion} = \nu_0 \cdot \exp\left[\frac{\alpha_a nF\eta}{RT} \right] \qquad [6]$$

$$\Gamma_{removal} = \nu_0 \cdot \exp\left[\frac{-\alpha_c nF\eta}{RT} \right] \qquad [7]$$

Here, ν_0 is the exchange rate constant, in the absence of an overpotential, and $\Gamma_{insertion}$ and $\Gamma_{removal}$ are the insertion and removal rates of the Li ion across the interface, respectively. This is the same general approach taken by others in the literature. For instance, Pornprasertruk et al.[33] used KMC to study the frequency response of a yttrium-stabilized zirconia (YSZ) solid oxide fuel cell (SOFC) electrolyte, by simulation of the movements of oxygen ion vacancies (though neglecting the interfacial reactions at the electrodes). This movement is affected by the local electrical potential, which varies with the arrangement of ions. To accelerate these simulations, the electrolyte model was subdivided into atomic layers of smeared, uniform net charge per unit area (simplifying the calculation of the electrical potential by reducing the system to a quasi-one-dimensional model). The change in the electric field above and below layer k, $\pm E_k$ is proportional to its net charge density (Gauss' law). The potential energy change because of moving charge q from plane Z_0 to plane Z_i is qV_{sc}^i, where

$$V_{sc}^i = V_{sc}^0 + \sum_{k=0}^{i-1} a \left(\sum_{h=k+1}^{N} E_h - \sum_{g=1}^{k} E_g \right) \quad (1 \leq i \leq N) \qquad [8]$$

and a represents the lattice spacing of the electrolyte material, and V_{sc} is the space-charge electrical potential arising from the distribution of ions within the electrolyte. The electric field must be updated (as the ionic species move) by calculating the instantaneous charge density within each sheet and then summing these individual contributions along the length of the electrolyte. Once

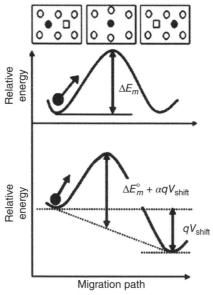

Figure 2 Illustration of the change in the migration barrier under applied potential (q is the charge of the migrating species), where the top and bottom graphs correspond to a system in the absence and the presence of an applied potential, respectively. The circles represent occupied lattice sites (the filled circle is used to identify the migrating atom), while the square represents a vacancy. (Reprinted from Ref. 33. Copyright (2007), with permission from Elsevier.)

the electric field is updated, the electric work is calculated and used to bias the motion of the ions (Figure 2) according to a Butler–Volmer relationship (Eqs. [6] and [7]).

These examples illustrate several key features of KMC simulations of electrochemical systems. First, the rate expressions for the events in the model must be obtained. Second, if any of these rates involves the transfer of ionic species, the rate expressions must then incorporate the influence of the electrical potential on the event rates (the motion of the electrons is not explicitly modeled because the electron dynamics occur on a much faster timescale). Finally, the simulation code must be able to calculate the electrical potential at different points in the system, so that the local event rates can be evaluated correctly.

There are some general implementation challenges and guidelines with KMC codes that should be recognized. First, the memory requirements for the codes can become prohibitively large, if the system arrays are not efficiently designed. With a large three-dimensional system, many lattice points are needed, and each lattice point contains information that describes the type of atom at that site, the type of events that can occur at a particular site, and the electrical potential at the site. These features can expand the system datasets into large multidimensional arrays, which can often exceed the memory capacities of modern computers. Designing compact arrays (with minimal dimensions) can allow the KMC code to address large systems with minimal memory requirements, however. Second, the rates of the different events may span several orders of magnitude. Thus, even though a simulation may run for several million KMC steps, some of the low-probability events might not have an opportunity to occur. Accordingly, the occurrence of each event type should be analyzed after a simulation is completed. If an event is assumed to be in a dynamic equilibrium, then the number of forward and reverse events should be the same (approximately). As very general guidance, about 1000 occurrences or more of the slowest event should take place so that reasonable system statistics can be obtained. To satisfy these requirements, the simulations can be run for extended periods of time, or the codes can be modified to run on parallel computers or optimized with advanced sampling techniques.

The simulation programs used in most prior KMC studies were typically written to accommodate only a narrow range of materials and conditions. However, there are some general KMC codes available that might be modified to account for electrochemical steps. The open-source SPPARKS code,[34] developed at Sandia National Laboratories, has been applied to model many material properties and surface growth processes. The NASCAM code (NAnoSCAle Modeling)[35] can be obtained freely from Moskovkin, Bera, and Lucas at the University of Namur, but the code is designed primarily to model surface deposition processes. Another software package, developed by the Henkelman and Jonsson groups, is EON,[36] which can perform different types of simulations including parallel replica dynamics, hyperdynamics, basin hopping, and KMC. Commercial options for KMC modeling are also available from Kintech Lab[37] and the Kinetix package from Accelrys.[38]

APPLICATIONS

Transport in Li-ion Batteries

Li-ion batteries have drawn increasing attention from both academic researchers and industry because high energy-density electrical storage has become a critical technology for a wide range of emerging products. Applications of Li-ion batteries range from everyday use in portable consumer products to automotive, medical, and aerospace fields.[19] A typical Li-ion battery comprises a metal oxide-based cathode, a carbonaceous anode, a polymer-based electrolyte, and a separator diaphragm. During the charging cycle, Li ions diffuse toward the anode and become intercalated within the porous carbon electrode (usually graphite), and during discharge, the Li ions migrate in the opposite direction toward the cathode (often $LiFePO_4$, $LiCoO_2$, or a spinel). Many of the technical performance challenges facing Li-ion batteries can be traced to the atomistic-level materials characteristics that govern the power density, safety, and performance degradation. Although these are complex systems, many important kinetic properties of rechargeable Li-ion batteries can be captured with efficient KMC simulations. In particular, KMC can provide important insights into the kinetic mechanisms of the microscopic Li diffusion process. It has been shown[39−42] that KMC simulations of Li-ion diffusion in electrode materials provide good qualitative agreement with experimental work. However, two primary challenges prevent quantitative agreement with experiment: (a) the large variation in experimental diffusion coefficient measurements (values from different sources varying by several orders of magnitude); and (b) the difficulty of obtaining accurate activation barriers for the KMC event rates. Therefore, KMC modeling can be improved with more definitive experimental results or from improved rate determination by electronic structure calculations. As the rate information becomes less ambiguous, the quantitative predictions from KMC are greatly enhanced.

One of the first KMC simulations of Li-ion diffusion was reported in 1994 by Deppe et al.[43] who used both a KMC and a lattice gas (LG) model. Their system consisted of a two-dimensional InSe cathode and a Li-doped borated glass separator. The diffusion behavior from both approaches (KMC and LG) led to the same conclusion that the charge interactions at the interface play a crucial role in the ionic diffusion across the interface. However, similar to other applications of KMC, the quantitative results were found to be sensitive to the reaction energy barriers used in the model. Unfortunately, no direct comparisons were made with experimental systems, so it was challenging to identify accurate values for the kinetic parameters, limiting the predictive capacity of the model.

In 1999, a more sophisticated KMC model for Li diffusion in $Li_yMn_2O_4$ was demonstrated by Darling and Newman.[39] Using activation barriers derived from experimental measurements,[44] the authors calculated the Fickian diffusion coefficients (D) versus fractional occupancies (θ) from 0 to 1 in spinel $Li_yMn_2O_4$ lattices, with the fraction of pinned Li ions varying from 0% to 40%. These results were consistent with experiments and mean field simulations by Gao et al.,[44] confirming Gao et al.'s original assumption that pinning Li-ions by defects in the $Li_yMn_2O_4$ host structure will reduce the diffusion coefficients. This finding, based on a comparison of the experimental, KMC, and mean field results, challenged the common assumption that D has a constant value.

Shortly thereafter, Van der Ven and Ceder[45] reported a multiscale KMC model of Li diffusion in layered Li_xCoO_2, using activation barriers estimated by first-principles DFT calculations. Activation energies of two Li migration pathways were characterized using local density approximation (LDA) DFT: (a) oxygen dumbbell hop (ODH) and (b) tetrahedral site hop (TSH). The KMC simulations were conducted at both $T = 300\,K$ and $400\,K$. For lower lithium concentration (x), the predicted diffusion coefficients versus Li ratio (x) reproduced experimental trends successfully.[46] However, the KMC simulations failed to reproduce the experimentally observed sharp drop of D when the value of x is within the range $0.95-1$.[46] According to the authors, this drop can be attributed to the thermodynamic driving force (chemical potential), which deviates from ideality at the higher lithium concentrations. Although it was not attempted, the authors could have extended this study to capture such deviations by including the changes in the diffusion activation barrier (by accounting for Li–Li interactions), so that the predictions would be more robust over the range of operating conditions.

Later, Van der Ven et al.[40] also developed a multiscale KMC model that combined DFT calculations, cluster expansion techniques, and conventional MC and KMC algorithms to study Li diffusion in cathode materials, such as layered Li_xTiS_2,[40] spinel $Li_{1+x}Ti_2O_4$,[41] and graphite anodes.[42] First-principle energies were used to parameterize a cluster expansion calculation,[47–49] so that an energy database could be mapped to different possible local configurations. On the basis of cluster expansions, equilibrium properties were then calculated by conventional MC, and Li diffusion coefficients were simulated by KMC. For the layered Li_xTiS_2 materials,[40] the KMC-calculated dependence of Li diffusion coefficients on the Li concentration (at $300\,K$) was found to be qualitatively similar to the experimental results.[50] Li exhibited low-diffusion coefficients at low and high concentrations, with higher diffusion coefficients at intermediate concentrations (maximum near $x = 0.5$). For spinel $Li_{1+x}Ti_2O_4$ materials,[41] the calculated Li diffusion coefficients varied by several orders of magnitude for negative x. For values of the Li fraction ($1 + x$) ranging from 1 to 2, the Li diffusion coefficient values were similar. In addition, the KMC-calculated intraplanar diffusion of Li within Li_xC_6 decreases smoothly and slowly with increasing Li concentration.[42] Also, the high Li diffusivity found in bulk graphite suggests that the rate limitations in this system likely originate from the surface and/or crystallinity effects.

In 2006, Jung and Pyun[51] developed a KMC-based approach for simulating a potentiostatic current transient voltammogram (current vs time, at fixed potential) and a linear sweep voltammogram (potential vs time) of Li transport through a $Li_{1-x}Mn_2O_4$ film electrode, with a partially inactive fractal surface. In the simulations, the analysis was performed by simply counting the ions transported across the interfaces as a function of time. The potentiostatic current transient and linear sweep voltammograms were calculated under both diffusion-controlled and cell-impedance-controlled boundary conditions and for different partially inactive fractal dimensions. Although no direct comparisons were made with experiments, this work provided a useful technique for comparing and interpreting future experimental studies of related systems.

Hin[32] presented a combined continuum and KMC method for simulating the kinetics of Li intercalation and structural changes, as well as the morphological evolution of the Li-rich/Li-poor phase boundary in Li_xFePO_4 electrode particles. The KMC model was coupled with a finite difference continuum model to treat the Li-ion diffusion flux within the electrolyte. Also, the local particle adsorption was coupled to concentration fields by Butler–Volmer kinetics. The KMC-simulated galvanostatic discharge process was performed at room temperature, and a comparison of the computational and experimental results[52] is shown in Figure 3.

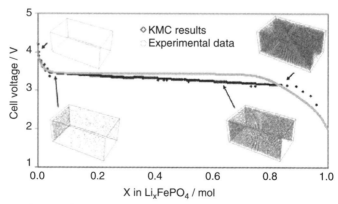

Figure 3 Simulation of the galvanostatic discharge process of Li_xFePO_4 electrode particles at room temperature. Cell voltage is plotted as a function of Li concentration (mol) in the active material. The gray points represent the lithium atoms in the active particle. (Reprinted with permission from Ref. 32. Copyright (2011), Wiley-VCH.)

These simulation results were found to be in general agreement with the experiments. The simulations successfully reproduced the cell voltage variation as a function of lithium concentration (a plateau at around 3.42 V in the experiments with a slowly decreasing curve), as well as the miscibility gap. The relationship between the cell voltage and the direction of far-field Li-ion flux was also discussed.

In addition, several KMC simulation models have been developed to study in more detail the diffusion processes in Li-ion batteries, since the distribution and kinetics of the Li-ion motion have direct consequences in terms of the battery performance and stability. These models include work[53] on Li-ion hopping in polymer electrolytes, parameterized by polarization energy calculations, and a KMC study[54] of ambipolar Li-ion and electron-polaron (e$^-$) diffusion in nanostructured TiO_2 (which investigated the simultaneous diffusion of both Li$^+$ and e$^-$ in the electrode).

Solid Electrolyte Interphase (SEI) Passive Layer Formation

When a battery is cycled for the first time, a passive layer forms around the carbon electrode because of Li-ion intercalation and other chemical reactions (in particular electrolyte decomposition). This is commonly referred to as the SEI layer, and it plays a vital role in the safety, capacity fade, and the life cycle of Li-ion batteries.[55] The development of fundamental models for understanding SEI-layer formation is one of the critical needs in Li-ion battery research.[19] Models developed to simulate the SEI layer formation and capacity fade of Li-ion batteries include continuum models,[56,57] single particle models,[58] and first-principle models.[59] Among these techniques, KMC's variable timescale is well-suited to studying relatively slow processes such as SEI layer formation and capacity fade, unlike more traditional molecular simulation techniques that are bound to fixed timescales. Although very limited atomistic-level simulation work has been reported in this area, Ramadesigan et al.[19] proposed in a recent review that atomistic and mesoscale models could be efficiently combined to predict and understand the real-time performance of batteries, leading to fundamental design improvements.

Methekar et al.[55] reported a KMC simulation of SEI layer formation within a Li-ion battery, in the tangential direction of Li-ion intercalation into a graphite anode. Unlike continuum models, this simulation explicitly addressed the surface heterogeneity, enabling a detailed analysis of the passive SEI layer formation around the electrode particles. The simulations showed that the active surface coverage decreases slowly in the initial stages of battery operation, followed by a rapid decrease (as seen in Figures 4a and b), in good agreement with experiments.[60]

Methekar et al. investigated the effects of different operating parameters including the exchange current density (ECD), charging voltage, and temperature on the SEI layer formation. As expected, the predicted passive SEI layer

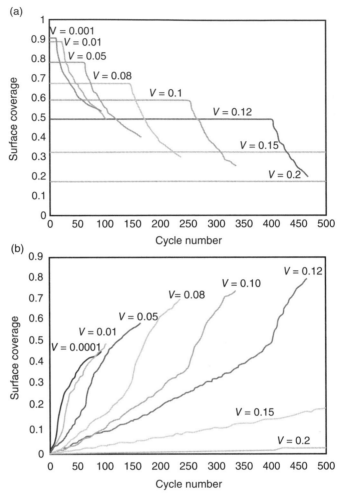

Figure 4 KMC simulations of the (a) active and (b) passive surface coverage of a graphite anode corresponding to various charging potentials. (Reproduced with permission from Ref. 55. Copyright (2012), The Electrochemical Society.) (For a color version of this figure, please see plate 4 in color plate section.)

coverage was found to be low at lower ECDs. Also, the temperature plays a complex role in these systems, since material degradation, SEI formation, and transport properties are all strongly correlated with and significantly affected by the temperature. It was also shown that a lower charging potential will slow the growth of the passive SEI layer. However, such low charging potentials may be unrealistic in practical applications (since the battery capacity would be heavily underutilized). Methekar et al. concluded by proposing a dynamically coupled KMC and two-dimensional continuum model that could be used to simulate the electrolyte and the solid phase efficiently. If developed, such a comprehensive model of SEI layer formation could provide crucial atomistic-level information about operational improvements or material advances to minimize performance fade.

Analysis of Impedance Spectra

Electrochemical impedance spectroscopy (EIS) is widely used for characterizing electrochemical systems.[61] By measuring the impedance as a function of frequency, EIS provides a powerful tool for analyzing the performance losses in batteries and fuel cells. For example, Adler et al.[62] have used EIS to identify the causes of fuel cell inefficiencies over a range of experimental operating conditions. KMC simulations can be used to better interpret experimental EIS observations and trace their atomic origins. Diffusion coefficients, electrode resistance, and reaction rates of elementary reactions can be identified with KMC simulations under various applied frequencies, and this information can be used to clarify the connections between EIS peaks or frequencies and the underlying reaction mechanisms.

In 2002, Kim and Pyun[63] reported a method for using KMC simulations to analyze experimental EIS data. Using experimentally determined activation energies, a KMC simulation of Li-ion diffusion coefficients within a $LiMn_2O_4$ electrode film was obtained, which included information about the resistances versus the electrode potential. By comparing the experimental EIS spectra with the KMC results, both the diffusion of Li-ions at the electrolyte-electrode interface and the Li-ion diffusion within the electrode were found to be rate-controlling steps, depending on the operating conditions.

KMC was also used to interpret the EIS measured frequencies in a SOFC model.[64] At low temperatures, an experimental 8 mol% YSZ SOFC, with Pt and Au electrodes, was analyzed using EIS. The frequency-response data were fit to an equivalent circuit model to extract resistance and capacitance data, which can be used to further analyze the system and aid in engineering designs at larger scales. The low-frequency loop showed a strong correlation to the bias voltage, indicating that the cathode reactions were dominant. Then, with a KMC model, the relative frequencies of each elementary reaction were classified and compared with the different sections of Nyquist plots, generated from the experimental EIS results. Although the temperatures of the experiments and the KMC simulations differ, the relative reaction frequencies were in general agreement with each other, as was the qualitative behavior of the SOFC.

Electrochemical Dealloying

Selective electrochemical dealloying is an important method for preparing nanoporous metal materials. These processes are strongly dictated by the kinetics and thermodynamic driving forces at the atomic scale, thus providing excellent opportunities for KMC-based investigations. Erlebacher and coworkers presented an atomistic KMC routine for modeling the porosity evolution, critical potential, and rate-limiting behavior of electrochemical dealloying within an Au/Ag system.[65,66] The results from these KMC simulations were also compared to those of others who used multiscale modeling.[67] The current-potential behavior for varying Ag/Au compositions and the critical potentials compare well with experimental results (Figure 5). Moreover, the morphological evolution of the pores was elucidated by the KMC simulation, as shown in Figure 6. In addition, the author proposed a continuum model of alloy dissolution that is consistent with the experiments and with the atomistic KMC calculations.

After the Ag/Au simulation study, Erlebacher presented a different KMC model for atomistically modeling the dealloying porosity evolution, critical potential, and rate-limiting behavior. This model is much more general in scope, in that no specific materials were targeted. Instead, the simulation consisted of a two-component system, composed of a "more noble" (MN) metal and a "less noble" (LN) metal, though energetic parameters for Ag and Cu were used.[66] The selective dissolution behavior and porosity formation were reproduced correctly, including the composition and geometric restrictions during the dealloying, a composition-dependent critical potential, a passivation regime, and a regime of steady-state dissolution flux. Furthermore, an intrinsic critical potential was predicted in those simulations (shown in Figure 7), although being lower in value than the experimentally measured result. In addition, the KMC simulations revealed that the less noble atoms, stripped from small terraces at the alloy/electrolyte interface, were responsible for the rate-limiting step in the dissolution process.

In other work, Artymowicz et al.[67] studied the parting limit or so-called dealloying threshold for electrolytic dissolution from an Au/Ag fcc binary alloy using both KMC and a geometric percolation model. The high-density percolation threshold predicted by the geometric model was slightly higher than from the KMC simulation, because some of the atoms in the KMC simulations meeting the percolation requirements did not necessarily lead to sustained dealloying. The same group also used KMC simulations to explore an inverse Gibbs-Thompson effect in dealloyed nanoparticles.[68] In these nanoparticles, the electrochemical potential for selective electrochemical dissolution increased empirically with inverse particle radius $1/r$.

Electrochemical Cells

Several recent KMC studies of electrode reactions and electrode/electrolyte interfaces of electrochemical cells have been reported. Peterson et al.[69] combined KMC with MD simulations to model the behavior of polarizable metallic electrodes held at a constant potential and separated by an electrolyte. Marcus theory[70,71] was used to calculate the electron transport rates. This

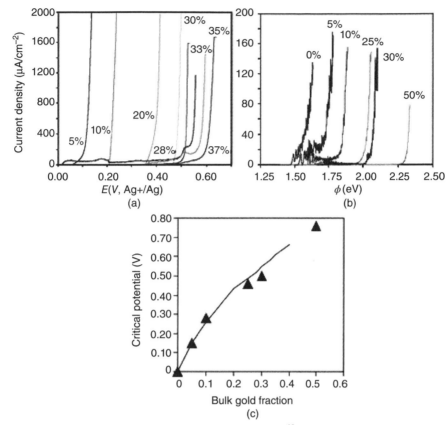

Figure 5 Comparison of experimental and simulated[65] current-potential behavior of Ag/Au dealloying: (a) experimental current-potential behavior for varying Ag/Au alloy compositions (atom% Au) within 0.1 M HClO₄ +0.1 M Ag⁺ (reference electrode 0.1 M Ag⁺/Ag); (b) simulated current-potential behavior of Ag/Au alloys; (c) comparison of experimental (line) and simulated (triangles) critical potentials; the zero of overpotential has been set equal to the onset of dissolution of pure silver both in simulation and in experiment. (Reprinted with permission from Ref. 65. Copyright (2001), Macmillan Publishers, Ltd.)

study also introduced a technique for addressing the polarization of the electrodes by the electrolyte, based on contained explicit image-charges of the electrolytes and a constant uniform charge at the electrodes. In the KMC-MD method, the oxidation/reduction steps were performed with KMC acceptance rules when close to the electrode walls, and molecular dynamics methods were used elsewhere to model the transport behavior.

KMC simulations have also been combined with differential electro-chemical mass spectrometry (DEMS) to give new insights into the mechanisms and kinetics of adsorbed CO electro-oxidation on a platinum electrode.[72] On the basis of DEMS experimental observations, the authors proposed a

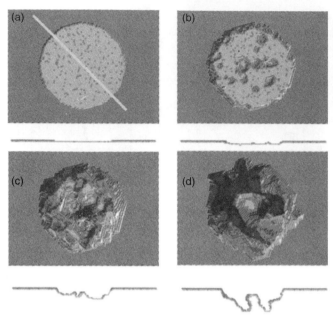

Figure 6   Simulated evolution of an artificial pit in $Au_{10\%}Ag_{90\%}$ (atom%), $\varphi = 1.8$ eV. Cross-sections along the $(1\,1\,\bar{1})$ plane defined by the light line in (a) are shown below each view: (a) the initial condition is a surface fully passivated with gold except within a circular region (the "artificial pit"); (b) after 1 s, the pit has penetrated a few monolayers into the bulk; (c) after 10 s, a gold cluster has nucleated in the center of the pit; and (d) at 100 s, the pit has split into multiple pits. (Reprinted with permission from Ref. 65. Copyright (2001), Macmillan Publishers, Ltd.) (For a color version of this figure, please see plate 3 in color plate section.)

mechanistic model for the electro-oxidation of adsorbed CO and simulated it using KMC. The current transients of the adsorbed CO electro-oxidation process were reproduced and corroborated by the DEMS experiments. Analysis of the combination of DEMS and KMC results suggested that, on 3–4 nm Pt nanoparticles, the adsorbed CO electro-oxidation in the main peak of the current transients can be described by a simple Langmuir–Hinshelwood mechanism (rather than a nucleation and growth mechanism). This suggested mechanism is also in agreement with predictions from mean field theory.[73]

Very recently, Viswanathan et al.[74] developed a multiscale model for simulating linear sweep voltammetry of electrochemical solid–liquid interfaces of H_2O on $Pt(1\,1\,1)$ and on $Pt_3Ni(1\,1\,1)$ facets. In the model, DFT was used to parameterize the reaction kinetics; KMC was used to capture the kinetic steps of the electrochemical oxidation, and conventional MC was used to equilibrate the surface between kinetic steps. The calculated cyclic voltammograms are in good agreement with experimental CV[75,76] and the experimental XPS results[77] (Figure 8).

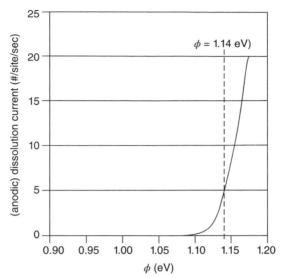

Figure 7 Simulated polarization curve showing the intrinsic critical potential of dealloying. Experimentally, the critical potential for this curve is placed at 1.14 eV, whereas the intrinsic critical potential predicted from KMC simulations is below 0.95 eV. (Reproduced with permission from Ref. 66. Copyright (2012), The Electrochemical Society.)

Also, in 2012, Suzuki et al.[78] used KMC simulations to evaluate the sintering and voltage drops in high-temperature proton exchange membrane fuel cells (HT-PEMFCs). A three-dimensional KMC model, based on DFT calculations, was compared with experimental results at 150 °C, 170 °C, and 190 °C (at nonhumid conditions). The KMC-simulated agglomeration of electrocatalysts over time was quantitatively confirmed by experimental transmission electron microscopy (TEM).

Solid Oxide Fuel Cells

Solid oxide fuel cells are promising electrochemical devices that can achieve high fuel conversion efficiencies (near 70%) with minimal greenhouse gas emissions, and they can operate on a wide range of fuel sources to generate electrical power. Because the diffusion of the ions within the SOFC electrolytes typically occurs on the microsecond timescale,[79] it is difficult for most atomistic models to capture these events. Therefore, KMC simulations have been used to model the ion diffusion within solid oxide electrolytes,[80–88] in SOFC cathodes and anodes,[33,79,89–94] as well as SOFC electrode sintering,[95] and glass seals for SOFCs.[96] Turner and coworkers[94] recently reviewed many examples of KMC models of SOFCs, including the cathode and anode reactions, along with transport within the electrolyte.

To study the ion diffusion within solid oxide electrolytes, Krishnamurthy et al.[80] performed DFT calculations and KMC simulations of oxygen diffusivity in a YSZ electrolyte. Also, Modak and Lusk[79] simulated the open-circuit voltage of a one-dimensional YSZ electrolyte model and then compared the predicted voltages and concentration profiles with an analytical Guoy–Chapman solution. Within the KMC model, many properties were predicted including the oxygen concentration distribution, the voltage profile, the local electric field, and the effects of the temperature and the relative permittivity.

In 2007, Pornprasertsuk et al.[33] performed an EIS simulation analysis of a YSZ electrolyte by combining DFT-predicted activation energy values within a KMC simulation model. The EIS of the YSZ electrolyte was calculated and plotted in Nyquist and Bode plots, allowing for a direct comparison of the simulation and experimental results. The Nyquist plot (Figure 9) of the simulated EIS data (at 400 K) and the experimentally measured EIS of YSZ (at 336 °C) were consistent in the higher frequency range, but deviated at lower frequencies. As mentioned previously, this information can be used to extract capacitance and resistance data for larger-scale system engineering analysis and design.

Lee et al.[83] recently proposed an approach for computing the impedance of oxygen vacancy diffusion in a YSZ electrode, on the basis of KMC simulations. By feeding the results from a single equilibrium KMC simulation into a fluctuation-dissipation theorem (FDT) model, the electrical impedance at all frequencies can be obtained. This allowed the authors to conduct a systematic

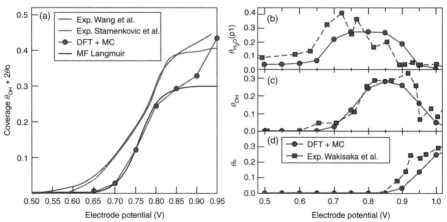

Figure 8   (a) Simulated integrated cyclic voltammogram of the electrochemical oxidation of H_2O on Pt (1 1 1). The electrode potential is measured against RHE. The experimental results are from Wang et al.[75] and Stamenkovic et al.[76] (b)–(d) $\theta_{H2O(pl)}$, θ_{OH}, and θ_O calculated from DFT+MC simulations as compared to *ex-situ* EC-XPS experiments by Wakisaka et al.[77] (Reprinted with permission from Ref. 74. Copyright (2012), American Chemical Society.) (For a color version of this figure, please see plate 5 in color plate section.)

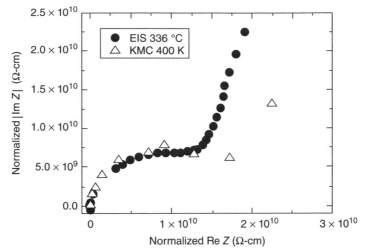

Figure 9   Nyquist plot of the normalized impedance results obtained from KMC simulations at 400 K, compared with the results from the EIS measurements on a pulsed layer deposited polycrystalline thin film YSZ at 336 °C. (Reprinted with permission from Ref. 33. Copyright (2007), Elsevier.)

investigation of the influence of dopant concentration and distribution on the impedance of the electrolyte. The same group later performed a follow-up study,[84] on the basis of DFT, KMC, and a cluster expansion approach, in which they optimized the bulk YSZ ionic conductivity by tailoring the dopant arrangement atomistically. Alignment of the Y cations along the [1 0 0] rectangular superlattice lines was found to maximize the ionic conductivity. The same group conducted another KMC study[85] of the effect of the doping concentration and distribution on the ionic conductivity of a YSZ material. That work also included the temperature dependency of ion conductivity. Using a multiscale KMC model similarly to their previous paper,[84] the authors predicted an optimal dopant concentration of 8 mol%, and an effective energy barrier to diffusion of 0.74–0.81 eV. The optimized arrangements of the Y cations were also predicted in this work.

In a search for alternative electrolyte materials (other than YSZ), Dholabhai et al.[86−88] calculated the ion conductivities of ceria doped with varying fractions of praseodymium, gadolinium, and samarium. Their results were in reasonable agreement with experiments and provide a useful guideline for the future design of ceria-based SOFCs.

A cathode-only KMC SOFC model, intended to model real SOFC operation, was developed by Lau et al.[90,91] and used by Wang et al. in an extensive investigation of the frequency response of the system.[92] The response of the fuel cell at frequencies ranging from 10^4 to 10^9 Hz (and temperatures ranging from 600 to 1000 °C) was calculated by KMC simulations (shown in Figure 10).

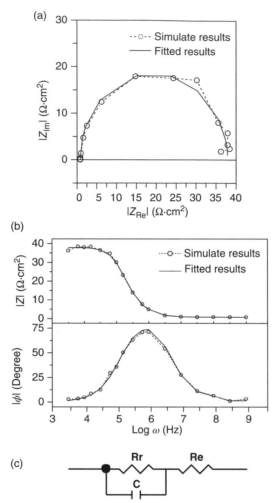

Figure 10 (a) Nyquist and (b) Bode plots of the impedance results obtained from the KMC simulations at $T = 1073\,\mathrm{K}$, where the dashed and solid curves correspond to the simulated data and their fitted results, respectively. (c) The equivalent circuit model with two resistances and one capacitance used to fit the frequency-response data obtained from the KMC simulations. (Reproduced with permission from Ref. 92. Copyright (2010), The Electrochemical Society.)

However, the simulated frequency response was substantially offset from the experimental values ($10^{-2}-10^6\,\mathrm{Hz}$).[97] This discrepancy might be attributed to the shortcomings in elementary reactions used by the KMC model (like missing anode-side reactions or missing details of electrode morphology). However, fitting the KMC results to the equivalent circuit model still exhibited results qualitatively consistent with experiments.

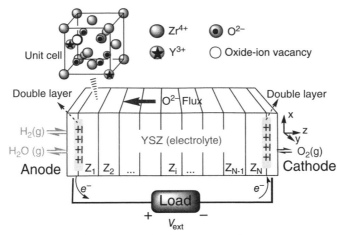

Figure 11 Illustration of KMC model of H_2-powered YSZ SOFC. (Reprinted with permission from Ref. 93. Copyright (2010), Elsevier.) (For a color version of this figure, please see plate 6 in color plate section.)

Soon after the cathode-only model was developed, two KMC models of complete SOFCs were reported.[89,93] A complete (anode and cathode) SOFC model with two different cathode catalyst structures (Pt-strap and Pt-island) was developed by Pornprasertsuk et al.[89] using a DFT + KMC approach. Several simulation results (e.g., different rate-limiting behavior corresponding to different operation conditions and Pt loads) are consonant with experimental investigations. A related complete-cell model (Figure 11) was developed by Wang et al.[93] This model was used to help interpret experimental EIS data by Holme et al.,[64] as discussed earlier. Both material-dependent and material-independent parameters were studied and the results showed consistent trends with early simulations and experiments. Sensitivity analyses of kinetic parameters were also performed.

Other Electrochemical Systems

Electrodeposition of copper on trenches of microchip interconnects is an important process in modern microelectronics.[98] Thus, KMC models[8,99−101] and KMC-based multiscale models[9,102−104] have been used to investigate the nucleation, surface chemistry, and roughness evolution in this process. However, most KMC models of electrodeposition have not incorporated electrochemical influences and so will not be discussed here. Nonetheless, due to the success of KMC with other electrochemical systems, excellent future opportunities exist for applying KMC to electrodeposition processes.

In other work, Pan et al.[105] investigated electrochemical metallization (ECM) resistive switching memory (RRAM) using KMC. The filament forming stage of two systems, Cu/H_2O and Ag/Ag_2S, were modeled and compared with experimental results. The filament forming time, topographies, and formation voltages were simulated to elucidate the so-called "voltage-time dilemma" and to provide a path to optimizing future RRAM design.

CONCLUSIONS AND FUTURE OUTLOOK

Because electrochemical devices will play a growing role in transportation, renewable energy storage, memory devices, portable electronics, and other sectors, there is a need for accurate, atomistic modeling techniques that can be used to support experimental research efforts. The KMC approach reviewed here can provide a great deal of fundamental insights into a variety of processes, but in many cases the validity of the predictions is limited by the availability of accurate mechanistic details. Thus, if KMC simulations are to play more of a role in electrochemical system modeling (as well as in areas such as catalysis and surface deposition), concerted efforts are needed to acquire better reaction rate data, either experimentally or computationally. Ideally, multiscale approaches can compute rates "on-the-fly,"[26,27] and when computationally feasible, these techniques might provide a more comprehensive modeling solution that is less reliant on experimental inputs.

Integration of KMC simulations with larger-scale models such as finite element continuum approaches can enhance the impact of a KMC simulation. In many cases, certain components of an electrochemical device can be effectively approximated by a bulk continuum treatment (e.g., ionic diffusion through the bulk of an electrolyte) while other components of a device that involve interfacial transport require discrete treatments like either Butler–Volmer empiricism or an atomistic treatment to capture important details related to the distribution of charge in the double layer.

As with any computational technique, there is a persistent need for experimental validation of the results from a KMC simulation, as well as appreciating the limits of the model's parameterization. As stressed in the introduction of this chapter, most KMC simulations are not able to predict system behavior beyond the scope of a presupplied database of rates. Thus, predicting behavior far from the original parameterization conditions could be dangerous, and new experimental benchmarks may be necessary. Fortunately, deviations from experimental benchmarks can highlight interesting behavior because the parameters, rates, and events in a KMC model are all well grounded in the system's physical and chemical properties. There are few arbitrarily-fit parameters, so if the KMC model performs poorly, critical information about shortcomings in the assumed mechanism or rate estimation can be identified, and sensitivity analysis may provide an initial indication of the most critical event rates in a model.[91,93]

There have been many excellent contributions to the development and application of KMC for modeling electrochemical systems over the past decade.

However, to move these KMC modeling tools from the hands of experts to a more general audience, there needs to be a more concerted and focused effort in software development. Many of the scientific contributions mentioned in this review are based on individual efforts that rely on in-house codes. The impact of these efforts could be expanded if some flexible, benchmarked codes were readily available. Unfortunately, the variety of electrochemical systems encountered creates limits to simple transferability. Finally, as DFT-based parameter estimation becomes more robust and cost-effective, and as additional KMC simulations of electrochemical systems are demonstrated, we expect that KMC will play an important role in connecting atomistic-level information to experimentally observable phenomena, aiding in the design of many next-generation electrochemical devices.

ACKNOWLEDGMENTS

C. H. T. acknowledges partial funding provided by the National Science Foundation CAREER Award (CTS-0747690), and L. D. G. acknowledges financial support from the National Science Foundation (CHE-0626008). This work was supported by the Office of Naval Research, both directly and through the Naval Research Laboratory.

REFERENCES

1. J. M. Carrasco, L. G. Franquelo, J. T. Bialasiewicz, E. Galvan, R. C. P. Guisado, A. M. Prats, J. I. Leon, and N. Moreno-Alfonso, *IEEE T. Ind. Electron.*, 53, 1002 (2006). Power-Electronic Systems for the Grid Integration of Renewable Energy Sources: A Survey.

2. D. R. Rolison and L. F. Nazar, *MRS Bull.*, 36, 486 (2011). Electrochemical Energy Storage to Power the 21st Century.

3. H. N. G. Wadley, A. X. Zhou, R. A. Johnson, and M. Neurock, *Prog. Mater. Sci.*, 46, 329 (2001). Mechanisms, Models and Methods of Vapor Deposition.

4. C. C. Battaile, D. J. Srolovitz, and J. E. Butler, *J. Appl. Phys.*, 82, 6293 (1997). A Kinetic Monte Carlo Method for the Atomic-Scale Simulation of Chemical Vapor Deposition: Application to Diamond.

5. C. C. Battaile, D. J. Srolovitz, and J. E. Butler, *Diam. Relat. Mater.*, 6, 1198 (1997). Morphologies of Diamond Films from Atomic-Scale Simulations of Chemical Vapor Deposition.

6. C. C. Battaile and D. J. Srolovitz, *Annu. Rev. Mater. Res.*, 32, 297 (2002). Kinetic Monte Carlo Simulation of Chemical Vapor Deposition.

7. Y. M. Lou and P. D. Christofides, *Comput. Chem. Eng.*, 29, 225 (2004). Feedback Control of Surface Roughness of GaAs (001) Thin Films Using Kinetic Monte Carlo Models.

8. T. J. Pricer, M. J. Kushner, and R. C. Alkire, *J. Electrochem. Soc.*, 149, C396 (2002). Monte Carlo Simulation of the Electrodeposition of Copper. I. Additive-Free Acidic Sulfate Solution.

9. E. Rusli, T. O. Drews, D. L. Ma, R. C. Alkire, and R. D. Braatz, *J. Process Contr.*, 16, 409 (2006). Robust Nonlinear Feedback-Feedforward Control of a Coupled Kinetic Monte Carlo-Finite Difference Simulation.

10. E. W. Hansen and M. Neurock, *J. Catal.*, 196, 241 (2000). First-Principles-Based Monte Carlo Simulation of Ethylene Hydrogenation Kinetics on Pd.

11. L. D. Kieken, M. Neurock, and D. H. Mei, *J. Phys. Chem. B*, **109**, 2234 (2005). Screening by Kinetic Monte Carlo Simulation of Pt-Au(100) Surfaces for the Steady-State Decomposition of Nitric Oxide in Excess Dioxygen.

12. D. H. Mei, E. W. Hansen, and M. Neurock, *J. Phys. Chem. B*, **107**, 798 (2003). Ethylene Hydrogenation over Bimetallic Pd/Au(111) Surfaces: Application of Quantum Chemical Results and Dynamic Monte Carlo Simulation.

13. D. Mei, P. A. Sheth, M. Neurock, and C. M. Smith, *J. Catal.*, **242**, 1 (2006). First-Principles-Based Kinetic Monte Carlo Simulation of the Selective Hydrogenation of Acetylene over Pd(111).

14. D. Mei, J. Du, and M. Neurock, *Ind. Eng. Chem. Res.*, **49**, 10364 (2010). First-Principles-Based Kinetic Monte Carlo Simulation of Nitric Oxide Reduction over Platinum Nanoparticles under Lean-Burn Conditions.

15. M. Saeys, M. F. Reyniers, J. W. Thybaut, M. Neurock, and G. B. Marin, *J. Catal.*, **236**, 129 (2005). First-Principles Based Kinetic Model for the Hydrogenation of Toluene.

16. N. K. Sinha and M. Neurock, *J. Catal.*, **295**, 31 (2012). A First Principles Analysis of the Hydrogenation of C-1-C-4 Aldehydes and Ketones over Ru(0001).

17. M. P. Allen and D. J. Tildesley, *Computer Simulation of Liquids*, Oxford University Press, Oxford, 1989.

18. D. Frenkel and B. Smit, *Understanding Molecular Simulations: From Algorithms to Applications*, Academic Press, San Diego, 2002.

19. V. Ramadesigan, P. W. C. Northrop, S. De, S. Santhanagopalan, R. D. Braatz, and V. R. Subramanian, *J. Electrochem. Soc.*, **159**, R31 (2012). Modeling and Simulation of Lithium-Ion Batteries from a Systems Engineering Perspective.

20. W. M. Young and E. W. Elcock, *Proc. Phys. Soc. London*, **89**, 735 (1966). Monte Carlo Studies of Vacancy Migration in Binary Ordered Alloys: I.

21. D. R. Cox and H. D. Miller, *The Theory of Stochastic Processes*, Methuen, London, 1965.

22. D. T. Gillespie, *J. Comput. Phys.*, **22**, 403 (1976). A General Method for Numerically Simulating Stochastic Time Evolution of Coupled Chemical Reactions.

23. K. A. Fichthorn and W. H. Weinberg, *J. Chem. Phys.*, **95**, 1090 (1991). Theoretical Foundations of Dynamic Monte Carlo Simulations.

24. A. Chatterjee and D. G. Vlachos, *J. Comput. Aid. Mater. Des.*, **14**, 253 (2007). An Overview of Spatial Microscopic and Accelerated Kinetic Monte Carlo Methods.

25. A. F. Voter, *Radiation Effects in Solids*, Springer, NATO Publishing Unit, Dortrecht, 2005.

26. G. Henkelman and H. Jonsson, *J. Chem. Phys.*, **115**, 9657 (2001). Long Time Scale Kinetic Monte Carlo Simulations without Lattice Approximation and Predefined Event Table.

27. L. Xu and G. Henkelman, *J. Chem. Phys.*, **129**, 114104 (2008). Adaptive Kinetic Monte Carlo for First-Principles Accelerated Dynamics.

28. P. J. Gellings and H. J. M. Bouwmeester (Eds.), *The CRC Handbook of Solid State Electrochemistry*, CRC Press, Boca Raton, 1997.

29. G. Inzelt, *J. Solid State Electr.*, **15**, 1373 (2011). Milestones of the Development of Kinetics of Electrode Reactions.

30. J. A. V. Butler, *Trans. Faraday Soc.*, **19**, 729 (1924). Studies in Heterogeneous Equilibria. Part II. The Kinetic Interpretation of the Nernst Theory of Electromotive Force.

31. T. Erdey-Grúz and V. M. Zur, *Z. Phys. Chem.*, **150**(A), 203 (1930). Theorie Der Wasserstoffüberspannung.

32. C. Hin, *Adv. Funct. Mater.*, **21**, 2477 (2011). Kinetic Monte Carlo Simulations of Anisotropic Lithium Intercalation into Li_xFePo_4 Electrode Nanocrystals.

33. R. Pornprasertsuk, J. Cheng, H. Huang, and F. B. Prinz, *Solid State Ion.*, **178**, 195 (2007). Electrochemical Impedance Analysis of Solid Oxide Fuel Cell Electrolyte Using Kinetic Monte Carlo Technique.

34. spparks.sandia.gov.

35. www.fundp.ac.be/sciences/physique/pmr/telechargement/logiciels/nascam.

36. theory.cm.utexas.edu/eon/.

37. www.kintechlab.com.

38. accelrys.com.

39. R. Darling and J. Newman, *J. Electrochem. Soc.*, **146**, 3765 (1999). Dynamic Monte Carlo Simulations of Diffusion in $Li_yMn_2O_4$.

40. A. Van der Ven, J. C. Thomas, Q. C. Xu, B. Swoboda, and D. Morgan, *Phys. Rev. B*, **78**, 12 (2008). Nondilute Diffusion from First Principles: Li Diffusion in $Li_{(x)}TiS_{(2)}$.

41. J. Bhattacharya and A. Van der Ven, *Phys. Rev. B*, **81**, 12 (2010). Phase Stability and Nondilute Li Diffusion in Spinel $Li_{1+X}Ti_2O_4$.

42. K. Persson, Y. Hinuma, Y. S. Meng, A. Van der Ven, and G. Ceder, *Phys. Rev. B*, **82**, 9 (2010). Thermodynamic and Kinetic Properties of the Li-Graphite System from First-Principles Calculations.

43. J. Deppe, R. F. Wallis, I. Nachev, and M. Balkanski, *J. Phys. Chem. Solids*, **55**, 759 (1994). 2-Dimensional Hopping Diffusion across Material Interfaces.

44. Y. Gao, J. N. Reimers, and J. R. Dahn, *Phys. Rev. B*, **54**, 3878 (1996). Changes in the Voltage Profile of $Li/Li_{1+X}Mn_{2-X}O_4$ Cells as a Function of X.

45. A. Van der Ven and G. Ceder, *Electrochem. Solid St.*, **3**, 301 (2000). Lithium Diffusion in Layered $LixCoO_2$.

46. J. M. McGraw, C. S. Bahn, P. A. Parilla, J. D. Perkins, D. W. Readey, and D. S. Ginley, *Electrochim. Acta*, **45**, 187 (1999). Li Ion Diffusion Measurements in V_2O_5 and $Li(Co_{1-X}Al_x)O^{-2}$ Thin-Film Battery Cathodes.

47. J. M. Sanchez, F. Ducastelle, and D. Gratias, *Physica A*, **128**, 334 (1984). Generalized Cluster Description of Multicomponent Systems.

48. D. deFontaine, *Solid State Phys.*, **47**, 33 (1994). Cluster Approach to Order–Disorder Transformations in Alloys.

49. D. B. Laks, L. G. Ferreira, S. Froyen, and A. Zunger, *Phys. Rev. B*, **46**, 12587 (1992). Efficient Cluster Expansion for Substitutional Systems.

50. K. Kanehori, F. Kirino, T. Kudo, and K. Miyauchi, *J. Electrochem. Soc.*, **138**, 2216 (1991). Chemical Diffusion Coefficient of Lithium in Titanium Disulfide Single-Crystals.

51. K. N. Jung and S. I. Pyun, *Electrochim. Acta*, **52**, 2009 (2007). Theoretical Approach to Cell-Impedance-Controlled Lithium Transport through $Li_{1-\delta}Mn_2O_4$ Film Electrode with Partially Inactive Fractal Surface by Analyses of Potentiostatic Current Transient and Linear Sweep Voltammogram.

52. N. Meethong, H.-Y. S. Huang, W. C. Carter, and Y.-M. Chiang, *Electrochem. Solid St.*, **10**, A134 (2007). Size-Dependent Lithium Miscibility Gap in Nanoscale $Li_{1-X}FePO_4$.

53. S. Scarle, M. Sterzel, A. Eilmes, and R. W. Munn, *J. Chem. Phys.*, **123**, 13 (2005). Monte Carlo Simulation of Li^+ Motion in Polyethylene Based on Polarization Energy Calculations and Informed by Data Compression Analysis.

54. J. G. Yu, M. L. Sushko, S. Kerisit, K. M. Rosso, and J. Liu, *J. Phys. Chem. Lett.*, **3**, 2076 (2012). Kinetic Monte Carlo Study of Ambipolar Lithium Ion and Electron-Polaron Diffusion into Nanostructured TiO_2.

55. R. N. Methekar, P. W. C. Northrop, K. J. Chen, R. D. Braatz, and V. R. Subramanian, *J. Electrochem. Soc.*, **158**, A363 (2011). Kinetic Monte Carlo Simulation of Surface Heterogeneity in Graphite Anodes for Lithium-Ion Batteries: Passive Layer Formation.

56. H. J. Ploehn, P. Ramadass, and R. E. White, *J. Electrochem. Soc.*, **151**, A456 (2004). Solvent Diffusion Model for Aging of Lithium-Ion Battery Cells.

57. S. Santhanagopalan, Q. Z. Guo, P. Ramadass, and R. E. White, *J. Power Sources*, **156**, 620 (2006). Review of Models for Predicting the Cycling Performance of Lithium Ion Batteries.

58. Q. Zhang and R. E. White, *J. Power Sources*, **179**, 793 (2008). Capacity Fade Analysis of a Lithium Ion Cell.

59. P. Ramadass, B. Haran, P. M. Gomadam, R. White, and B. N. Popov, *J. Electrochem. Soc.*, **151**, A196 (2004). Development of First Principles Capacity Fade Model for Li-Ion Cells.

60. A. T. Stamps, C. E. Holland, R. E. White, and E. P. Gatzke, *J. Power Sources*, **150**, 229 (2005). Analysis of Capacity Fade in a Lithium Ion Battery.

61. D. D. Macdonald, *Electrochim. Acta*, **51**, 1376 (2006). Reflections on the History of Electrochemical Impedance Spectroscopy.

62. S. B. Adler, *Chem. Rev.*, **104**, 4791 (2004). Factors Governing Oxygen Reduction in Solid Oxide Fuel Cell Cathodes.

63. S. W. Kim and S. I. Pyun, *J. Electroanal. Chem.*, **528**, 114 (2002). Analysis of Cell Impedance Measured on the $LiMn_2O_4$ Film Electrode by PITT and EIS with Monte Carlo Simulation.

64. T. P. Holme, R. Pornprasertsuk, and F. B. Prinz, *J. Electrochem. Soc.*, **157**, B64 (2010). Interpretation of Low Temperature Solid Oxide Fuel Cell Electrochemical Impedance Spectra.

65. J. Erlebacher, M. J. Aziz, A. Karma, N. Dimitrov, and K. Sieradzki, *Nature*, **410**, 450 (2001). Evolution of Nanoporosity in Dealloying.

66. J. Erlebacher, *J. Electrochem. Soc.*, **151**, C614 (2004). An Atomistic Description of Dealloying - Porosity Evolution, the Critical Potential, and Rate-Limiting Behavior.

67. D. M. Artymowicz, J. Erlebacher, and R. C. Newman, *Philos. Mag.*, **89**, 1663 (2009). Relationship Between the Parting Limit for De-Alloying and a Particular Geometric High-Density Site Percolation Threshold.

68. I. McCue, J. Snyder, X. Li, Q. Chen, K. Sieradzki, and J. Erlebacher, *Phys. Rev. Lett.*, **108**, 5 (2012). Apparent Inverse Gibbs-Thomson Effect in Dealloyed Nanoporous Nanoparticles.

69. M. K. Petersen, R. Kumar, H. S. White, and G. A. Voth, *J. Phys. Chem. C*, **116**, 4903 (2012). A Computationally Efficient Treatment of Polarizable Electrochemical Cells Held at a Constant Potential.

70. R. A. Marcus, *J. Chem. Phys.*, **24**, 966 (1956). On the Theory of Oxidation-Reduction Reactions Involving Electron Transfer.1.

71. R. A. Marcus and N. Sutin, *Biochim. Biophys. Acta*, **811**, 265 (1985). Electron Transfers in Chemistry and Biology.

72. H. S. Wang, Z. Jusys, R. J. Behm, and H. D. Abruna, *J. Phys. Chem. C*, **116**, 11040 (2012). New Insights into the Mechanism and Kinetics of Adsorbed CO Electrooxidation on Platinum: Online Mass Spectrometry and Kinetic Monte Carlo Simulation Studies.

73. N. P. Lebedeva, M. T. M. Koper, J. M. Feliu, and R. A. van Santen, *J. Phys. Chem. B*, **106**, 12938 (2002). Role of Crystalline Defects in Electrocatalysis: Mechanism and Kinetics of CO Adlayer Oxidation on Stepped Platinum Electrodes.

74. V. Viswanathan, H. A. Hansen, J. Rossmeisl, T. F. Jaramillo, H. Pitsch, and J. K. Norskov, *J. Phys. Chem. C*, **116**, 4698 (2012). Simulating Linear Sweep Voltammetry from First-Principles: Application to Electrochemical Oxidation of Water on Pt(111) and $Pt_3Ni(111)$.

75. J. X. Wang, N. M. Markovic, and R. R. Adzic, *J. Phys. Chem. B*, **108**, 4127 (2004). Kinetic Analysis of Oxygen Reduction on Pt(111) in Acid Solutions: Intrinsic Kinetic Parameters and Anion Adsorption Effects.

76. V. R. Stamenkovic, B. Fowler, B. S. Mun, G. Wang, P. N. Ross, C. A. Lucas, and N. M. Markovic, *Science*, **315**, 493 (2007). Improved Oxygen Reduction Activity on $Pt_3Ni(111)$ Via Increased Surface Site Availability.

77. M. Wakisaka, H. Suzuki, S. Mitsui, H. Uchida, and M. Watanabe, *Langmuir*, **25**, 1897 (2009). Identification and Quantification of Oxygen Species Adsorbed on Pt(111) Single-Crystal and Polycrystalline Pt Electrodes by Photoelectron Spectroscopy.

78. A. Suzuki, Y. Oono, M. C. Williams, R. Miura, K. Inaba, N. Hatakeyama, H. Takaba, M. Hori, and A. Miyamoto, *Int. J. Hydrogen Energy.*, **37**, 18272 (2012). Evaluation for Sintering of Electrocatalysts and Its Effect on Voltage Drops in High-Temperature Proton Exchange Membrane Fuel Cells (HT-PEMFC).

79. A. U. Modak and M. T. Lusk, *Solid State Ion.*, **176**, 2181 (2005). Kinetic Monte Carlo Simulation of a Solid-Oxide Fuel Cell: I. Open-Circuit Voltage and Double Layer Structure.

80. R. Krishnamurthy, Y. G. Yoon, D. J. Srolovitz, and R. Car, *J. Am. Ceram. Soc.*, **87**, 1821 (2004). Oxygen Diffusion in Yttria-Stabilized Zirconia: A New Simulation Model.

81. R. Pornprasertsuk, P. Ramanarayanan, C. B. Musgrave, and F. B. Prinz, *J. Appl. Phys.*, **98**, 8 (2005). Predicting Ionic Conductivity of Solid Oxide Fuel Cell Electrolyte from First Principles.

82. A. Kushima and B. Yildiz, *J. Mater. Chem.*, **20**, 4809 (2010). Oxygen Ion Diffusivity in Strained Yttria Stabilized Zirconia: Where Is the Fastest Strain?

83. E. Lee, F. B. Prinz, and W. Cai, *Electrochem. Commun.*, **12**, 223 (2010). Kinetic Monte Carlo Simulations of Oxygen Vacancy Diffusion in a Solid Electrolyte: Computing the Electrical Impedance Using the Fluctuation-Dissipation Theorem.

84. E. Lee, F. B. Prinz, and W. Cai, *Phys. Rev. B*, **83**, 4 (2011). Enhancing Ionic Conductivity of Bulk Single-Crystal Yttria-Stabilized Zirconia by Tailoring Dopant Distribution.

85. E. Lee, F. B. Prinz, and W. Cai, *Model. Simul. Mater. Sc.*, **20**, 21 (2012). Ab Initio Kinetic Monte Carlo Model of Ionic Conduction in Bulk Yttria-Stabilized Zirconia.

86. P. P. Dholabhai, S. Anwar, J. B. Adams, P. Crozier, and R. Sharma, *J. Solid State Chem.*, **184**, 811 (2011). Kinetic Lattice Monte Carlo Model for Oxygen Vacancy Diffusion in Praseodymium Doped Ceria: Applications to Materials Design.

87. P. P. Dholabhai and J. B. Adams, *J. Mater. Sci.*, **47**, 7530 (2012). A Blend of First-Principles and Kinetic Lattice Monte Carlo Computation to Optimize Samarium-Doped Ceria.

88. P. P. Dholabhai, S. Anwar, J. B. Adams, P. A. Crozier, and R. Sharma, *Model. Simul. Mater. Sci.*, **20**, 13 (2012). Predicting the Optimal Dopant Concentration in Gadolinium Doped Ceria: A Kinetic Lattice Monte Carlo Approach.

89. R. Pornprasertsuk, T. Holme, and F. B. Prinz, *J. Electrochem. Soc.*, **156**, B1406 (2009). Kinetic Monte Carlo Simulations of Solid Oxide Fuel Cell.

90. K. C. Lau, C. H. Turner, and B. I. Dunlap, *Solid State Ion.*, **179**, 1912 (2008). Kinetic Monte Carlo Simulation of the Yttria Stabilized Zirconia (YSZ) Fuel Cell Cathode.

91. K. C. Lau, C. H. Turner, and B. I. Dunlap, *Chem. Phys. Lett.*, **471**, 326 (2009). Kinetic Monte Carlo Simulation of O_2- Incorporation in the Yttria Stabilized Zirconia (YSZ) Fuel Cell.

92. X. Wang, K. C. Lau, C. H. Turner, and B. I. Dunlap, *J. Electrochem. Soc.*, **157**, B90 (2010). Kinetic Monte Carlo Simulation of AC Impedance on the Cathode Side of a Solid Oxide Fuel Cell.

93. X. Wang, K. C. Lau, C. H. Turner, and B. I. Dunlap, *J. Power Sources*, **195**, 4177 (2010). Kinetic Monte Carlo Simulation of the Elementary Electrochemistry in a Hydrogen-Powered Solid Oxide Fuel Cell.

94. C. H. Turner, W. An, B. I. Dunlap, K.-C. Lau, and X. Wang, *Ann. Rep. Comput. Chem.*, **6**, 201 (2010). Atomistic Modeling of Solid-Oxide Fuel Cells.

95. Y. X. Zhang, C. R. Xia, and M. Ni, *Int. J. Hydrogen Energ.*, **37**, 3392 (2012). Simulation of Sintering Kinetics and Microstructure Evolution of Composite Solid Oxide Fuel Cells Electrodes.

96. W. Xu, X. Sun, E. Stephens, I. Mastorakos, M. A. Khaleel, and H. Zbib, *J. Power Sources*, **218**, 445 (2012). A Mechanistic-Based Healing Model for Self-Healing Glass Seals Used in Solid Oxide Fuel Cells.

97. Q.-A. Huang, R. Hui, B. Wang, and H. Zhang, *Electrochim. Acta*, **52**, 8144 (2007). A Review of AC Impedance Modeling and Validation in SOFC Diagnosis.

98. R. D. Braatz, R. C. Alkire, E. Seebauer, E. Rusli, R. Gunawan, T. O. Drews, X. Li, and Y. He, *J. Process Contr.*, **16**, 193 (2006). Perspectives on the Design and Control of Multiscale Systems.

99. T. O. Drews, R. D. Braatz, and R. C. Alkire, *J. Electrochem. Soc.*, **150**, C807 (2003). Parameter Sensitivity Analysis of Monte Carlo Simulations of Copper Electrodeposition with Multiple Additives.

100. T. O. Drews, R. D. Braatz, and R. C. Alkire, *Int. J. Multiscale Comput.*, **2**, 313 (2004). Coarse-Grained Kinetic Monte Carlo Simulation of Copper Electrodeposition with Additives.

101. T. O. Drews, R. Aleksandar, J. Erlebacher, R. D. Braatz, P. C. Searson, and R. C. Alkire, *J. Electrochem. Soc.*, **153**, C434 (2006). Stochastic Simulation of the Early Stages of Kinetically Limited Electrodeposition.

102. T. O. Drews, E. G. Webb, D. L. Ma, J. Alameda, R. D. Braatz, and R. C. Alkire, *AICHE J.*, **50**, 226 (2004). Coupled Mesoscale – Continuum Simulations of Copper Electrodeposition in a Trench.

103. T. O. Drews, S. Krishnan, J. C. Alameda, D. Gannon, R. D. Braatz, and R. C. Alkire, *IBM J. Res. Dev.*, **49**, 49 (2005). Multiscale Simulations of Copper Electrodeposition onto a Resistive Substrate.

104. X. H. Li, T. O. Drews, E. Rusli, F. Xue, Y. He, R. Braatz, and R. Alkire, *J. Electrochem. Soc.*, **154**, D230 (2007). Effect of Additives on Shape Evolution During Electrodeposition. I. Multiscale Simulation with Dynamically Coupled Kinetic Monte Carlo and Moving-Boundary Finite-Volume Codes.

105. F. Pan, S. Yin, and V. Subramanian, *IEEE Electr. Device L.*, **32**, 949 (2011). A Detailed Study of the Forming Stage of an Electrochemical Resistive Switching Memory by KMC Simulation.

CHAPTER 6

Reactivity and Dynamics at Liquid Interfaces

Ilan Benjamin

Department of Chemistry and Biochemistry, University of California, Santa Cruz, CA 95064, USA

INTRODUCTION

Many phenomena of interest in science and technology take place at the interface between a liquid and a second phase. Corrosion, the operation of solar cells, and the water splitting reaction are examples of chemical processes that take place at the liquid/solid interface.[1,2] Electron transfer, ion transfer, and proton transfer reactions at the interface between two immiscible liquids are important for understanding processes such as ion extraction,[3,4] phase transfer catalysis,[5,6] drug delivery,[7] and ion channel dynamics in membrane biophysics.[8] The study of reactions at the water liquid/vapor interface is of crucial importance in atmospheric chemistry.[9,10] Understanding the behavior of solute molecules adsorbed at these interfaces and their reactivity is also of fundamental theoretical interest. The surface region is an inhomogeneous environment where the asymmetry in the intermolecular forces may produce unique behavior.

Because of its importance, it is not surprising that the study of the neat liquid surface, as well as of solute adsorption, spectroscopy, and reactivity, goes back many years. However, up until the last decade of the twentieth century most of the experimental studies involved the measurement of macroscopic properties such as surface tension and surface potential,[11,12] and generally speaking, the spectroscopic techniques employed lacked the specificity and

Reviews in Computational Chemistry, Volume 28, First Edition.
Edited by Abby L. Parrill and Kenny B. Lipkowitz.
© 2015 John Wiley & Sons, Inc. Published 2015 by John Wiley & Sons, Inc.

sensitivity required to probe the surface region. Although these techniques contribute significantly to our knowledge, they lack the ability to provide a detailed understanding at the molecular level.

In recent years, advances in a number of new experimental methods have provided unprecedented sensitivity and selectivity in the measurement of liquid interfacial phenomena. Prominent among these are nonlinear spectroscopic techniques such as Second Harmonic Generation (SHG) and Sum Frequency Generation (SFG), which probe the surface region selectively.[13-15] These techniques have been used to explore the liquid/vapor interface[16-18] and buried interfaces, such as liquid/liquid[19-24] and liquid/solid interfaces,[25-29] as well as biological interfaces.[30,31] Other techniques that have been used in recent years to study liquid surfaces and interfaces include light scattering,[32-36] X-ray and neutron scattering,[34,37-44] atomic scattering,[45] fluorescence anisotropy decay,[46,47] scanning electrochemical microscopy,[48,49] infrared spectroscopy in a total reflection geometry,[50,51] and X-ray absorption spectroscopy.[52,53]

In parallel to these experimental approaches, much progress has been made in theoretical studies of liquid surfaces. Advances in the statistical mechanics of inhomogeneous fluids[54-57] have contributed significantly to our understanding of the molecular structure of liquid/solid, liquid/liquid, and liquid/vapor interfaces. However, the mathematical complexity because of losing the spherical symmetry of the bulk has limited the application to mainly calculating a small number of properties (such as density profile, surface tension, and molecular orientation) of neat inhomogeneous liquids.[58]

A major breakthrough in the theoretical understanding of the structure and dynamics of neat liquid interfaces, especially the behavior of reactive and nonreactive solutes adsorbed at these interfaces, has occurred over the past two decades thanks to advances in computer simulation methodology and the availability of high-speed computers. Computer simulations were initially used to test the validity of statistical mechanical approximations to calculating bulk and interfacial liquid molecular structure. However, their main contribution, together with experiments, has been to demonstrate the crucial role that the molecular structure of the liquid plays in understanding the spectroscopy, energy relaxation, and reactivity of solute molecules dissolved in the liquid.

While the long-standing picture of the solvent as a structureless medium has been very useful for offering a qualitative understanding of the solvent's effect on structure and dynamics, computer simulations and experiments clearly suggest that a microscopic molecular description of the solvent is necessary. This is particularly so for interfacial phenomena, because the interfacial region itself is only a few molecular diameters thick.

The purpose of this chapter is to discuss the computational tools that were developed to specifically address liquid interfacial systems and to summarize the microscopic insight gained about the structure and dynamics of neat liquid interfaces and the behavior of solute molecules adsorbed at these interfaces. Because most of these computational tools are based on molecular dynamics and Monte Carlo simulations of liquids, about which many excellent review articles and books exist, we will refer the reader to these sources when needed. As far as the new physical insight that these tools provide, our focus will be on presenting unifying concepts rather than on results that are specific for a given system. Because the subject of liquid–solid interface simulations has already received good coverage in this series,[59,60] we limit our discussion of these surfaces to presenting a contrasting view with the liquid/vapor and liquid/liquid interfaces.

We begin this chapter with a brief summary of the simulation methodology developed to deal specifically with liquid surfaces and interfaces. We then describe the application of this methodology to the neat interface. The focus is on molecular-level information that in recent years has been compared directly with experiments. This provides the necessary background for discussing the methodology and general insights that computer simulations provided for solute adsorption, transport, relaxation, and reactivity at liquid interfaces. The emphasis is on presenting general concepts that underlie different phenomena and focusing on the unique effect of the interface region in contrast with bulk behavior.

SIMULATION METHODOLOGY FOR LIQUID INTERFACES

Most simulation techniques applied to date to liquid interfaces are based on classical molecular dynamics and Monte Carlo methods. With few exceptions (discussed below), these techniques can be used straightforwardly to simulate the neat interface between a liquid and a second phase and to investigate the thermodynamics and dynamics of solute adsorption and reaction. There are several excellent books on the fundamentals of these techniques[61–64] as well as free software available on the internet.[65] (For a library of free software, see http://www.ccp5.ac.uk/librar.shtml.)

Force Fields for Molecular Simulations of Liquid Interfaces

In the classical molecular dynamics method, the dynamics of a system composed of N particles are followed by numerically solving the $3N$-coupled Newton equations of motion:

$$m_i \frac{d^2 \mathbf{r}_i}{dt^2} = -\nabla_i U(\mathbf{r}_1, \mathbf{r}_2, \ldots, \mathbf{r}_N) \qquad [1]$$

where $\mathbf{r}_i$ is the vector position of particle i with mass m_i. These particles are usually identified as the individual atoms in the molecules, but in many applications a particle can represent a group of atoms ("united" atom) or a fictitious mass for modeling coupling to an external bath[66] or for describing fluctuating charges in a molecule (see below).

The key ingredient in a molecular dynamics (and a Monte Carlo) simulation is the potential energy function $U(r_1, r_2, \ldots, r_N)$ describing the interactions between all the particles. This is sometimes called the "force field" of the system. In principle, this function can be determined by solving the Schrodinger equation for the ground state energy as a function of all the particle positions. Because this is practical only for a system of a few atoms, the approach taken for simulating a condensed phase system composed of thousands of particles is an empirical representation of U in terms of simple functions that are determined by fitting experimental data and utilizing solutions of the Schrodinger equation to small parts of the system (e.g., by looking at a small cluster of molecules). There exists an extensive literature and many databases for empirical force fields used in simulations of condensed phase systems;[67–70] here we give only a brief summary of the typical force field used in simulating liquid surfaces.

The simplest approach is to express the potential energy function as a sum of interactions between all pairs of particles belonging to different molecules (nonbonded interactions), plus a sum of the bonding interactions in each molecule:

$$U(r_1, r_2, \ldots, r_N) = \sum_{i<j} u(r_{ij}) + \sum_k U_{\text{intra}}^{(k)} \qquad [2]$$

where u is called the pair potential and $U_{\text{intra}}^{(k)}$ is the function describing the intramolecular bonding interactions in molecule k. The pair potential u depends on the identity of the two particles i and j, which belong to two different molecules, and on the distance r_{ij} between these particles. A typical form of this pair potential represents the interaction between two particles as a sum of coulomb electrostatic energy and a Lennard-Jones 6–12 term:

$$u(r_{ij}) = 4\varepsilon_{ij}[(\sigma_{ij}/r_{ij})^{12} - (\sigma_{ij}/r_{ij})^6] + q_i q_j / r_{ij} \qquad [3]$$

In this simple approach, the two particles are assigned fixed charges q_i and q_j, which are determined by solving the Schrodinger equation for the individual molecule (or a small cluster), by fitting to experimental dipole moment values, or by fitting to other experimentally measured properties. The Lennard-Jones term includes r^{-12} repulsion and a $-r^{-6}$ attraction terms, which represent approximately the polarizabilities of the two particles and their sizes. These terms are expressed by the parameters ε_{ij} and σ_{ij}, respectively. In a system containing n different particle types, there are in principle $n(n+1)$ independent ε_{ij} and σ_{ij} parameters (e.g., $n = 5$ is needed to simulate the water/chloroform interface). However, the general practice is to assign to each particle type the "self" terms ε_{ii} and σ_{ii} and to use the "mixing rules"[71] to obtain all the other parameters:

$$\varepsilon_{ij} = \sqrt{\varepsilon_{ii}\varepsilon_{jj}}, \quad \sigma_{ij} = (\sigma_{ii} + \sigma_{jj})/2 \qquad [4]$$

This is convenient for reducing the number of needed parameters. It is also useful for establishing the so-called transferable force field in which a set of parameters are assigned to atom types that can then be used for similar molecules.

Because the single particle self-parameters ε and σ are typically optimized to fit the bulk properties of a liquid with the "mixing rule" in place, there is no guarantee that Eq. [4] is optimal when interactions exist between molecules of different liquids, as is the case in simulating the interface between two different liquids (or mixtures of liquids). One approach is to use the mixing rules for the interactions between molecules of the same liquids but to fit the ε_{ij} and σ_{ij} parameters independently for the interactions between molecules of different liquids to reproduce interfacial properties such as surface tension.[72] As an extreme example, a simple model of two immiscible Lennard-Jones atomic liquids might be one in which one uses a Lennard-Jones potential where the attractive term is altogether missing or reduced in size instead of Eq. [4] to describe the interactions between the two different atoms.[73,74] A more general issue that has not received much consideration to date is whether potential energy functions fitted to properties of bulk liquids can even be used to simulate interfaces. We discuss this below in the context of many-body force fields.

While the pair approximation of Eq. [2] is efficient for computer simulations, a better agreement with experimental data can sometimes be achieved by utilizing more general force fields. *n*-Body potentials, which depend on the simultaneous positions of n particles with $n > 2$, provide a more refined description of condensed phase systems,[75] but only in a few cases have they been used for liquid surfaces.[76,77] An obvious case where three-(or higher)-body potentials are necessary is when classical MD is used to model a chemical reaction. The simple A + BC atom exchange reaction, for example, has been modeled with the three-body LEPS potential.[78] The topic of potentials used to model chemical reactions will be further discussed in the section on reactivity at liquid interfaces.

An important subset of many-body potentials shown to be important for simulating interfacial systems are those referred to as polarizable force fields.[79–96] Various aspects of polarizable force fields, especially for use in biomolecular modeling, is explained by Ren et al. in Chapter 3 of this volume. If one treats the fixed charges in Eq. [3] as parameters to be fitted to obtain the best agreement of the condensed phase simulations with experiments, in many cases one finds that the optimal values are considerably different from those obtained from a fit to a molecular (gas phase) dipole moment or from quantum calculations on isolated molecules. This is because in a condensed medium, the local electric field $\mathbf{E}_i$ (at the location of a particle i) is determined by all the fixed charges q_i and by all the induced dipoles $\mathbf{m}_i$ in the system:

$$\mathbf{E}_i = \sum_{j \neq i} \left[\frac{q_i \mathbf{r}_{ij}}{r_{ij}^3} + \mathbf{T}_{ij} \cdot \mathbf{m}_j \right], \quad \mathbf{T}_{ij} = \frac{1}{r_{ij}^3} \left(\frac{3\mathbf{r}_{ij} \circ \mathbf{r}_{ij}}{r_{ij}^2} - 1 \right) \quad [5]$$

where $\mathbf{m}_i = \alpha_i \mathbf{E}_i$, $\mathbf{r}_{ij} = \mathbf{r}_i - \mathbf{r}_j$, and $\mathbf{T}_{ij}$ is the dipole–dipole matrix. Because the value of $\mathbf{E}_i$ needed to determine $\mathbf{m}_i$ depends on the value of $\mathbf{m}_i$ itself, Eq. [5] must be solved iteratively. Once the values of $\mathbf{m}_i$ converge, the additional energy due to this many-body effect is given by

$$U_{\text{pol}} = -\frac{1}{2} \sum_i \mathbf{m}_i \cdot \mathbf{E}_i \qquad [6]$$

A simpler approach that is directly related to the observation that the effective charges in the two-body approximation are different from the fixed gas phase charges is to treat the charges in Eq. [2] as fluctuating dynamic variables. These approaches and others are discussed at length in another chapter in this series.[97]

It is important to keep in mind that the simple pair approximation in Eq. [2] is in fact an "effective" two-body description that takes into account the real many-body nature of the system implicitly through the fitting procedure. If this expression is modified by adding any of the above types of many-body interactions, the parameters that go into the two-body term must be refitted.

The second term in Eq. [2] represents the intramolecular potential energy. This term accounts for vibrations and internal rotations in the molecule relative to the equilibrium configuration. When internal motion is of no interest or consequence, it is possible to keep the molecule rigid using several well-known algorithms.[63,64,98] If the vibrational motion is important (e.g., when calculating vibrational spectra or when accurate modeling of large energy transfer is required), vibrational terms must to be included. The standard choice is to model the bond stretching and bending using harmonic (quadratic) terms, the internal rotations around bonds (torsions) using a series of cosine terms, and to include Lennard-Jones interactions between nonbonded atoms. References to the force field literature mentioned above contain detailed information about these types of potential energy functions and extensive tables of parameters.[67–70] In closing this section, we note that special effort has been made over the past decades to develop accurate force fields for simulating bulk water, and, in recent years, these force fields have been carefully evaluated for their ability to describe interfacial water. For a detailed survey of water models up to the turn of the twenty-first century, see Ref. 99. More recent water models developed specifically to address properties of interfacial water will be discussed below.

Boundary Conditions and the Treatment of Long-Range Forces

Molecular dynamics or Monte Carlo simulations of bulk liquids employ 3-dimensional (3D) periodic boundary conditions (PBC) (using typically a cubic or a truncated octahedron box), which are designed specifically

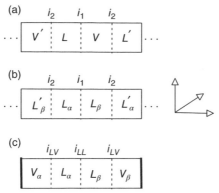

Figure 1 A schematic representation of three typical boundary conditions used in the simulation of the liquid/vapor and liquid/liquid interfaces. See text for details.

to *eliminate* surface effects. We must ensure that the boundary conditions maintain a well-defined and stable interface region. In a laboratory setup, the planar interface between a liquid and its vapors or between two liquids is maintained by gravity. Most simulations of liquid/vapor and liquid/liquid interfaces establish a planar surface by using period boundary conditions in two or three dimensions as depicted in Figure 1. In all of the simulation geometries described below and for the rest of this chapter, we will take the planar interface to be perpendicular to the z axis.

In Figure 1a, a 3D PBC with rectangular symmetry is used to ensure the planar liquid/vapor interface. L and V denote the regions in the central simulation box containing the liquid and vapor phases, respectively. L' and V' are the periodic replicas. In this geometry, there are two independent liquid/vapor interfaces: i_1 and i_2. (The two planes denoted by i_2 are the opposite faces of the central simulation box, so they are the same plane.) This geometry is typically prepared by starting with a bulk liquid cubic box and extending it along the z direction. Depending on the density of the starting box and the temperature of the system, a stable liquid and vapor phase will coexist.

In Figure 1b, the interface between two immiscible liquids, L_α and L_β, is prepared using a 3D PBC similar to that in Figure 1a. i_1 and i_2 are two independent liquid/liquid interfaces. The central simulation box contains the molecules labeled L_α and L_β, and their replicas are along the z directions. This simulation geometry is created by bringing together two bulk cubic boxes containing the two liquids and equilibrating the system. It has the advantage of using both liquid/liquid surfaces for increased statistics. However, because this simulation geometry typically equilibrates to $p \neq 1$ atm, it must be done at constant pressure (see below). There are additional complications when one studies solute molecules that are constrained to the interface, which will be discussed in the section on Solutes at Interfaces.

To avoid possible interaction between the two liquid/liquid interfaces and other complications, a 2D PBC geometry to simulate a single liquid/liquid interface is depicted in Figure 1c. Here, after bringing the two liquid boxes into contact, the simulation box is extended so that each liquid forms an interface with its own vapor. Only one liquid/liquid interface (denoted in the figure by i_{LL}) exists, as the system is not replicated along the z direction. To prevent mixing of the two vapor phases, a reflecting wall is set at the opposite faces of the box (thick lines).

Related to the simulation geometry is the treatment of long-range forces in molecular simulations. It has long been recognized that due to the finite and relatively small size of the simulation boxes, treating intermolecular interactions at large distances must be approximated, and correction for these approximations must be included when calculating thermodynamic properties. The problem is especially acute for the slowly varying electrostatic interactions. While this topic is discussed extensively in the general references given above,[61–64] we note below some specific points.

The short-range Lennard-Jones interactions may be truncated at a distance R_c (typically done with a continuous switching function with a continuous first derivative[61]), which can be as small as 2.5σ. One expects an error of only about 1% in the total internal energy for this value of R_c. However, the error in liquid-surface-related properties is much higher: 5% and 20% errors in the liquid and vapor densities, respectively, and 50% error in the surface tension.[100] Thus, longer cutoff distances and corrections must be included to obtain reliable and consistent results.[101,102]

In contrast, truncation at a computationally feasible (i.e., small) value of R_c produces artifacts when the system is strongly ionic because the potential energy is dominated by the slowly varying $1/r$ terms.[103,104] To address this, a popular approach known as Ewald or lattice sum (similar to the one used to calculate the lattice energy of ionic crystals) is used to sum the electrostatic interactions in the simulation box and all of its replicas. This is done by rewriting the sum of the $1/r$ terms as a sum of a rapidly converged series in real space (so a small cutoff can be used for these terms) and a much more slowly varying smooth function that can be approximated by a few cosine and sine terms in reciprocal (k) space.[61,105,106] These are expensive calculations that scale like $N^{3/2}$, where N is the number of particles but can be made more efficient (scales like $N \log N$) by approximating the k-space calculations with a discrete convolution on an interpolating grid, using the discrete Fast-Fourier transforms (FFT). Several implementations have been discussed.[107–111]

To sum the long-range coulomb interactions for an interfacial system, one can use the 3-dimensional Ewald (3DE) method and its variants mentioned above, provided that the 3D-periodic boundary condition geometry is used (see Figure 1a and b). If the system is only periodic in two dimensions (a "slab" geometry, see Figure 1c), one must use a 2-dimensional version (2DE), which is computationally expensive. There are ways to approximate the Ewald sum in 2D using the 3DE equations by adding a large empty space in the direction normal to the interface,[112] or by using other correctional terms.[113,114] Promising new method for fast Ewald summation in planar/slab-like geometry, which uses spectral representation in terms of both Fourier series and integrals, has been suggested recently.[115]

It should also be mentioned that approaches for treating long-range coulomb interactions with a pairwise compensation scheme have been developed.[116–118] They are based on shifting and damping the pair potential energy such that this function and its first and second derivatives decay continuously to zero at the cutoff distance. Physically, this method is equivalent to placing countercharges at the cutoff sphere. Variations of this method were described and tested for different systems[118] and for liquid interfaces[119] by comparison to infinite lattice sums and to the Ewald method.

Statistical Ensembles for Simulating Liquid Interfaces

In classical molecular dynamics, phase space is sampled by following the particles' trajectories by solving the deterministic equations of motion using the forces on the particles. Constant energy trajectories provide a microcanonical sampling of phase space. Several algorithms were developed for keeping the temperature and pressure fixed by a modification to these dynamics, which allows for the sampling of the isothermal (NVT, also called canonical) and the isothermal-isobaric (NPT) ensembles.[61–64,120,121] In the thermodynamic limit ($N \rightarrow \infty$), ensemble averages calculated by the different ensembles should be equal, but, in simulations with relatively small N, the proper choice of an ensemble can accelerate the convergence to the thermodynamic equilibrium value.

When simulating a liquid/vapor interface, an NVT ensemble is a straightforward choice, because in this two-phase, single component system, fixing the temperature determines the vapor pressure. The choice of volume determines the number of molecules in the vapor phase and thus its density. The situation at a liquid/liquid interface is more complicated and depends on the type of boundary conditions used. Zhang et al. discussed five different statistical ensembles that can be used to simulate the liquid/liquid interface with the boundary conditions of Figure 1b.[122] The available choice is in part because of the fact that the normal pressure, tangential pressure, and surface tension can be fixed independently. Different ensembles can be useful for different applications. For example, to study the water/oil interface a convenient ensemble is constant normal pressure and constant surface area ensemble. But for computing pressure/area isotherms of adsorbed monolayers, the ensemble in which the tangential pressure and the length normal to the surface are held constant is more appropriate. Computing the surface tension is discussed in the section "The Neat Interface."

Comments About Monte Carlo Simulations

If one is interested in equilibrium canonical (fixed temperature) properties of liquid interfaces, an approach to sample phase space is the Monte Carlo (MC) method. Here, only the potential energy function $U(r_1, r_2, \ldots, r_N)$ is required to calculate the probability of accepting random particle displacement moves (and additional moves depending on the ensemble type[61,123]). All of the discussion above regarding the boundary conditions, treatment of long-range interactions, and ensembles applies to MC simulations as well. Because the MC method does not require derivatives of the potential energy function, it is simpler to implement and faster to run, so early simulations of liquid interfaces used it.[124,125] However, dynamical information is not available with this method. We also point out here that the MC method can be used to simulate phase equilibria without creating physical contact between the phases (no interface present). This is done in the Gibbs Ensemble Monte Carlo (GEMC) method proposed by Panagiotopoulos[126] and applied to many systems (e.g., water phase equilibria[127]). The GEMC method involves two simulation boxes, one of which includes a bulk liquid phase and the other a vapor phase at the same temperature. In addition to the usual particle displacement moves in each box, the volume is allowed to change, and particles can be exchanged between the boxes.

THE NEAT INTERFACE

The structure and dynamics of the neat (no solute present) interface between a liquid and its vapor or between two immiscible liquids have been the subjects of numerous theoretical, computational, and experimental studies. Molecular-level insight was first provided by MD and MC simulations in the late 1970s. The first MC simulation of a liquid/vapor interface was used to calculate the density profile and surface tension of a system containing 129 Lennard-Jones particles by applying an external potential to keep the system stable.[124] The first MD simulation followed a few years later.[128] The system studied is the liquid–vapor interface of molten potassium chloride containing either 288 or 504 ions using the two-dimensional periodic boundary condition scheme. The computed surface tension at a temperature just above the melting point was within 30% of the experimental value. Self-diffusion at the interface is enhanced by 50% or more in comparison with the bulk, particularly for the direction perpendicular to the interface. The first study of a neat liquid/liquid interface was an MC investigation by Linse of the water/benzene interface.[125] That was followed by molecular dynamics studies by Smit[129] and by Meyer, Mareschal, and Hayoun[130] of a model liquid/liquid interface made from Lennard-Jones particles that were kept immiscible by tuning their intermolecular interactions. Density profiles and structural properties were calculated in all of these studies.

This activity has accelerated significantly in the past few decades and has been documented in several dozen articles, whose main results and insights are discussed below. A noteworthy point is that much of the computationally derived insight preceded experimental verification. This has been forthcoming in recent years with the development of nonlinear spectroscopy and other highly specific and sensitive experimental techniques. Today the experimental and computational approaches are an integral part of research in this field. We now summarize the insight gained from simulations and, whenever available, briefly provide consistent experimental evidence for it. Our main interest in the neat interface is to provide the background needed for understanding the behavior and reactivity of solute molecules adsorbed at the interface that will be covered later.

Density, Fluctuations, and Intrinsic Structure

In a bulk homogeneous liquid, the longtime (or ensemble) average of the number of particles $\langle n \rangle$ occupying any region of volume v depends on the temperature and pressure, not on the spatial location. The density of the liquid $\rho_L = \langle n \rangle / v$ is constant. In contrast, at a planar liquid/vapor or liquid/liquid interface the average density depends on the location z along the interface normal. The density profile is formally given by

$$\rho(z) = \left\langle \frac{1}{A} \sum_{i=1}^{N} \delta \left(z - z_i \right) \right\rangle \qquad [7]$$

where A is the surface area, z_i is the coordinate of particle i along the interface normal, N is the total number of particles, and δ is the Dirac delta function. In practice, the density profile is calculated by a coarse-graining procedure: The simulation system is divided into slabs of thickness d parallel to the surface. If $n(z)$ is the average number of particles in a slab whose center is at z, then $\rho(z) = n(z)/(Ad)$. d needs to be large enough to get a smoothly varying ρ but small enough to be on the scale where ρ changes substantially. Typical values of d used are $1-3\,\text{Å}$.

Figure 2 depicts typical shapes of the density profile of liquids. Panel A shows the density profile (reduced units) of a Lennard-Jones liquid at equilibrium with its vapor at two different (reduced) temperatures. These profiles have been calculated and reported extensively in the literature using MD, MC, and several approximate statistical mechanical theories.[55,57] The profiles shown in panel A are reproduced from the most accurate molecular dynamics simulations to date.[131] Similar profiles are obtained for many other molecular liquids such as water[132,133] and alcohols.[134] The density changes monotonically from the bulk liquid density ρ_L to the bulk vapor density ρ_V over a distance of a few molecular diameters, that is, $4-10\,\text{Å}$. It can be fitted with reasonably good

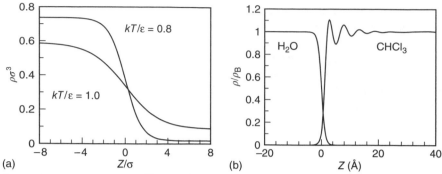

Figure 2 (a) Density profiles (reduced units) of the liquid/vapor interface of a Lennard-Jones fluid at two temperatures (reduced units). Data adapted from Ref. 131. (b) Density profile of the water/chloroform liquid/liquid interface at $T = 298K$. (Adapted with permission from Ref. 635. Copyright 2010 American Chemical Society.)

accuracy to the following expression (obtained from the van der Waals mean field approximation[55]):

$$\rho(z) = 1/2(\rho_L + \rho_V) - 1/2(\rho_L - \rho_V)\tanh(z/2\delta) \qquad [8]$$

where the liquid bulk phase occupies the region $z < 0$, $z = 0$ is the position where the density is equal to the average density of the two bulk phases, and δ is a measure of the interface width. A common way to express this width is to note the distance over which the density changes from 90% to 10% of the bulk density. This is equal to 4.4δ if the mean field expression for ρ (Eq. [8]) is used. As the temperature increases, the width of the liquid/vapor interface increases (further discussed below).

The density profile is used to define the Gibbs Dividing Surface (GDS), used extensively in thermodynamic analysis of the system as well as (a somewhat crude) expression for the average location of the interface. The GDS is defined as the plane $z = z_G$ parallel to the surface, such that the molecular "deficit" on the bulk side ($z < z_G$) is balanced exactly by the "surplus" on the vapor side ($z > z_G$):

$$\int_{z_L}^{z_G} (\rho(z) - \rho_B)dz + \int_{z_G}^{z_V} (\rho(z) - \rho_V)dz = 0 \qquad [9]$$

where z_L and z_V are locations in the bulk liquid and vapor phases, respectively. It is easy to confirm that for the density profile of Eq. [8], $z_G = 0$. The GDS is approximately where the water density is half the bulk liquid value.

In Figure 2 panel B we show a typical liquid/liquid density profile. While the water density looks similar to a liquid/vapor density profile, the organic liquid (chloroform, in this case) density exhibits dampened oscillations. While the source of these oscillations (which is well understood to represent molecular packing when a liquid is in contact with a solid) is not quite clear here; these oscillations have been observed in many liquid/liquid simulations[93,125,135–143] and have also been predicted by classical statistical mechanics density functional calculations.[144] For example, molecular dynamics simulations of two immiscible Lennard-Jones liquids revealed stable equilibrium oscillatory structures in the density profiles, but those oscillations were reduced significantly when the surface area was increased.[137] A more recent MD study and density functional calculation were carried out on a similar model liquid/liquid interface where Lennard-Jones particles are kept immiscible by reducing the attractive interaction between unlike particles. The density profiles of the liquids display oscillations that vanish when approaching the liquid–vapor coexistence, while the total density of the system shows a significant depletion at the interface.[145] However, X-ray reflectivity studies of two different water/oil interfaces do not support with the existence of this depletion.[146]

At the liquid/liquid interface, no unique GDS exists because an equation similar to Eq. [9] can be used to define a dividing surface with respect to either liquid by setting $\rho_V = 0$. If z_A and z_B are locations in bulk liquid A and liquid B, respectively, the GDS with respect to liquid A is given by

$$z_G(A) = z_A + \frac{1}{\rho_A^{bulk}} \int_{z_A}^{z_B} \rho_A(z)dz \qquad [10]$$

where $\rho_A(z)$ is the density profile of liquid A assumed to vary from $\rho_A = \rho_A^{bulk}$ at $z = z_A$ to $\rho_A = 0$ at $z = z_B$. An analogous expression for the GDS defined with respect to liquid B can be written. In discussing the water/liquid interface, we will define the GDS with respect to the water density. As in the case of the liquid/vapor interface, the GDS is close to the plane where the water density is about half the bulk value.

The average density profile gives a useful indication of the thickness of the interface region, but any fluctuations in density along the normal and in the xy plane parallel to the interface are averaged out. While these fluctuations exist even in a homogeneous liquid, their size and the range of their correlations are much more evident at an interface. We will also see later that they are important for understanding solute behavior at the interface. A simple framework for characterizing density fluctuations at the interface is the continuum-level Capillary Wave (CW) theory. This theory has been discussed extensively.[55,56] Here, we note briefly some useful relations and focus on molecular dynamics simulations that have been used to characterize the density fluctuations and to test the validity of CW theory down to a molecular-length scale.

Capillary wave theory considers the density variation at the interface to be the result of the superposition of thermally excited density fluctuations on a bare intrinsic profile. Mathematically, the instantaneous local density at a location $\mathbf{r} \equiv (x, y, z) \equiv (\mathbf{s}, z)$ is given by

$$\rho(\mathbf{r}) = \rho_A + (\rho_A - \rho_B)H[\zeta(\mathbf{s}) - z)] \qquad [11]$$

where $H(u)$ is the step function: $H(u < 0) = 0$, $H(u > 0) = 1$ and $z = \zeta(\mathbf{s})$ is a two-dimensional curved surface separating the two phases whose bulk densities are ρ_A and ρ_B. Snapshots of molecular dynamics simulations of liquid/vapor and liquid/liquid interfaces are qualitatively consistent with this ideal view, and this will be quantified below. The curved surface fluctuates due to thermal motion, and the average over all these fluctuations gives rise to the density profile. CW theory approximates the free energy associated with these fluctuations[55,56] and shows that ζ is a random Gaussian variable with probability distribution $P(\zeta) = (2\pi\langle\zeta^2\rangle)^{-1/2} \exp[-\zeta^2/2\langle\zeta^2\rangle]$ (without loss of generality $\langle\zeta\rangle$ is taken to be zero), in which the average square fluctuations are given by

$$\langle\zeta^2\rangle = \frac{k_B T}{4\pi\gamma} \ln \frac{1 + 2\pi^2 l_c^2/\xi_b^2}{1 + 2\pi^2 l_c^2/A} \qquad [12]$$

where k_B is the Boltzmann constant, T the temperature, γ the surface tension, A the system surface area, ξ_b a molecular length scale taken to be the bulk liquid correlation length (defined as the distance where the bulk liquid radial distribution function's asymptotic value is equal to 1), and $l_c = [2\gamma/(g|m_A\rho_A - m_B\rho_B|)]^{1/2}$ is called the capillary length and is on the order of a few millimeters (g is the gravity constant and m_A and m_B are molecular masses). Performing the average of Eq. [11] over ζ gives a density profile that is mathematically similar in shape to the mean field expression (Eq. [8]):

$$\rho(z) = 1/2(\rho_A + \rho_B) - 1/2(\rho_A - \rho_B)\mathrm{erf}(z/\sqrt{2\langle\zeta^2\rangle}) \qquad [13]$$

where erf is the error function. In the case of the liquid/liquid interface, this gives the total density at the location z. $\sqrt{\langle\zeta^2\rangle}$ is a measure of the interface width (more precisely, the distance over which the density changes from 90% to 10% of the bulk density, which is $1.51\sqrt{\langle\zeta^2\rangle}$). Equation [12] shows that in the thermodynamic limit ($A \to \infty$), the surface width diverges as $(-\ln g)^{1/2}$, while at zero gravity one obtains

$$\langle\zeta^2\rangle = \frac{k_B T}{4\pi\gamma} \ln(A/\xi_b^2) \qquad [14]$$

and the width diverges in the thermodynamic limit. While the issue of the existence of the interface at zero gravity is not completely clear,[147] using Eq. [14] for systems with sizes typically used in computer simulations is acceptable because for these systems $\xi_b \ll l_c \gg \sqrt{A}$, and Eq. [12] reduces to Eq. [14].

Molecular dynamics simulations have been used to test the validity of the CW theory down to distances comparable to ξ_b. Equation [14] predicts a specific dependence of the interface width on the temperature. Simulations at different temperatures can be used to determine $\langle \zeta^2 \rangle$ (by fitting the density profile to Eq. [13]). This, combined with surface tension calculations (see below), can be used to verify that $\sqrt{\langle \zeta^2 \rangle}$ is proportional to $\sqrt{T/\gamma}$. Figure 3 shows this plot generated using the data published in the very recent million particles simulation of the Lennard-Jones liquid/vapor interface.[131] As can be seen, the relation in Eq. [14] holds quite well at low T. Another simple approach is to obtain ξ_b from the bulk radial distribution function ($g(\xi_b) \approx 1$) and confirm the validity of Eq. [14] using the independently calculated surface tension and $\langle \zeta^2 \rangle$, as has been done for several liquid/liquid interfaces.[135,138] Alternatively, if several simulations with different surface areas are performed, Eq. [14] suggests that a plot of $\langle \zeta^2 \rangle$ versus ln A should be a straight line with a slope of $k_B T/4\pi\gamma$.[142,143,148]

A more direct test of the underlying assumptions of CW theory can be done by computing the spectral representation of the fluctuating surface $z = \zeta(\mathbf{s})$. This surface is represented by the superposition of capillary waves of different wave numbers $\mathbf{q}$:

$$\zeta(\mathbf{s}) = \sum_{\mathbf{q}} \alpha(\mathbf{q})e^{i\mathbf{q}\cdot\mathbf{s}}, \quad \alpha(0) = 0 \qquad [15]$$

where $\mathbf{q} = (2\pi/L)(n_x, n_y)$, L is the box length and ($n_x, n_y = \pm 1, \pm 2, \ldots L / \xi_b$), so each mode $\mathbf{q}$ contributes $\langle \alpha(\mathbf{q})\alpha(-\mathbf{q}) \rangle$ to the average of the square width. The

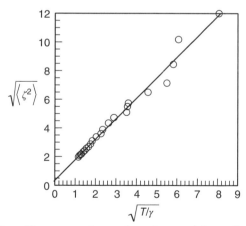

Figure 3   A test of capillary wave theory for a Lennard-Jones liquid/vapor interface. The circles represent independent simulations of the interface width at different temperatures. γ is the surface tension. (Data adapted from Ref. 131. Copyright 2012 American Institute of Physics.)

detailed calculations of CW theory give:[55]

$$\langle \alpha(\mathbf{q})\alpha(-\mathbf{q})\rangle = \frac{kT}{\gamma L^2(\mathbf{q}^2 + 2/\xi_b^2)} \tag{16}$$

For a typical system size used in simulation, $q = |\mathbf{q}| \gg \xi_b^{-1}$, and a plot of $\log \langle \alpha(\mathbf{q})\alpha(-\mathbf{q})\rangle$ versus $\log q$ should be a straight line with slope 2. The key to this approach is determining the instantaneous surface $z = \zeta(\mathbf{s})$. This can be accomplished by finding the contact points (on a grid) between a probe sphere of a fixed radius and the interfacial molecules[149] once the set of interfacial molecules is determined.[133,150,151] Another approach is to construct a coarse-grain density by a convolution of the 3D density field $\sum_{i=1}^{N} \delta(\mathbf{r} - \mathbf{r}_i)$ with a Gaussian function.[152] Jorge et al. recently reviewed these methods.[153]

Experimental measurement of the density variation at liquid/vapor and liquid/liquid interfaces down to the nanometer length scale is possible with X-ray and neutron reflectivity measurements.[43] In a test of capillary wave theory, X-ray reflectivity was used to study the interface between water and n-alkanes, C_nH_{2n+2} with $n = 6$–10, 12, 16, and 22. For all interfaces except the water–hexane ($n = 6$) interface, the interfacial width disagrees with the prediction of capillary wave theory. However, the variation of the observed interfacial width $\langle \zeta^2 \rangle_{obs}$ with the number of carbon atoms can be described by combining the capillary wave prediction for the width $\langle \zeta^2 \rangle_{cap}$ (Eq. [12]) with a contribution that takes into account the finite molecular size:

$$\langle \zeta^2 \rangle_{obs} = \langle \zeta^2 \rangle_{cap} + \sigma_0^2 \tag{17}$$

where σ_0 was found to match the radius of gyration of the shorter alkane molecules or the bulk correlation length for the longer alkanes.[154] A number of simulations have later demonstrated the applicability of Eq. [17].[142,143,148]

An additional contribution to the interface width due to the finite molecular size can also be demonstrated by the calculation of the so-called intrinsic density profile. This profile can be thought of as a generalization of the simple zero thickness assumption, which is implicit in the step function representation in Eq. [11]. The intrinsic density profile can be obtained by computing the density relative to the instantaneous location of the dividing surface $\zeta(\mathbf{s})$:[56,140,142,143,155,156]

$$\rho_{int}(z, q) = \left\langle \frac{1}{A} \sum_{i=1}^{N} \delta(z - z_i + \zeta(\mathbf{s}_i; q)) \right\rangle \tag{18}$$

where $q = |\mathbf{q}|$ is the wavevector value (see Eq. [15]) determining the resolution at which the instantaneous surface ζ is computed. The intrinsic density profile essentially removes from the average smooth density profile the blurring

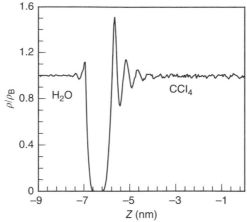

Figure 4 Intrinsic normalized density profiles for water and CCl$_4$ obtained by selecting the optimal grid resolution (see text). (Adapted with permission from Ref. 153. Copyright 2010 American Chemical Society.)

effect of the capillary fluctuations. An example of the intrinsic density profile of water and of carbon tetrachloride at the liquid/liquid interface is presented in Figure 4. In this example, the intrinsic surface $\zeta(\mathbf{s})$ is determined (for each liquid independently) using the grid method mentioned above, with a grid size given by the largest atomic site in each liquid. Other methods are reviewed in the paper by Jorge et al., from which this figure was reproduced.[153] Note the oscillatory density profile of the water, reflecting an underlying layered structure similar to that from the oxygen–oxygen radial distribution function $g(r)$ (see the discussion of $g(r)$ below). Demonstrated below is that other properties of the neat interface and of solute molecules adsorbed at the interface can also be described relative to the intrinsic surface.

Several other approaches for demonstrating the fact that a liquid surface is a rough but relatively sharp region of a two-phase system have been discussed. For example, by showing that the probability distribution of the surface height (defined as the maximum value of $\zeta(\mathbf{s})$ when $\mathbf{s}$ is sampled over a grid of varying sizes[125,135,142,143]) is a Gaussian. As will be discussed later, the local density and its fluctuations have a profound effect on the spectroscopy, dynamics, and reactivity of adsorbed solute molecules.

Surface Tension

The surface tension of a liquid/vapor or liquid/liquid interface can be calculated readily from MD or MC simulations. The goal of these calculations has typically been to test the validity of the force fields utilized, since experimental data on surface tension are readily available. As discussed earlier, the force fields used in simulations of interfacial systems are often optimized to reproduce bulk liquid properties, so there is no guarantee that they will independently reproduce this surface-specific property. It turns out that most simulations that use the simple mixing rules to determine the interactions between the molecules of two different liquids seem to give reasonable results, however.[72]

Several approaches for calculating the surface tension have been developed and are briefly summarized here. The fundamental definition of the surface tension γ depends on the statistical mechanical ensemble used. For example, at constant N, V, T:

$$\gamma = \left(\frac{\partial F}{\partial A} \right)_{N,V,T} \qquad [19]$$

where F is the Helmholtz free energy. Other expressions may be written, as discussed in Chapter 2 by Rowlinson and Widom.[55] Beginning with this definition, several approaches for calculating the surface tension between two phases from molecular simulations have been developed, and they are generally called the "thermodynamic route." These methods are based on directly calculating the change in the system's Helmholtz free energy associated with a change in the interfacial area at constant particle number, volume, and temperature. The performance of these thermodynamic techniques was recently presented by Errington and Kofke.[157]

Other approaches for computing the surface tension start from the statistical mechanical expression for the Helmholtz free energy or for the pressure. The Kirkwood–Buff formula for the surface tension of a liquid/vapor interface of an atomic liquid described by the pair potential approximation is:

$$\gamma = \frac{1}{2A} \left\langle \sum_{i>j} \frac{r_{ij}^2 - 3z_{ij}^2}{2r_{ij}} \frac{\partial u}{\partial r_{ij}} \right\rangle \qquad [20]$$

where the sum is over all pairs of atoms and the average is in the canonical ensemble. The factor of $1/2$ accounts for the existence of two surfaces. Generalized to molecular fluids, the formula is[125]

$$\gamma = \frac{1}{2A} \left\langle \sum_{i>j} \frac{r_{ij}^2 - 3z_{ij}^2}{2r_{ij}} \sum_{m,n} \frac{\mathbf{r}_{ij} \cdot \mathbf{r}_{mn}}{r_{ij} r_{mn}} \frac{\partial u}{\partial r_{mn}} \right\rangle \qquad [21]$$

where r_{mn} is the distance between atom m that belongs to molecule i and atom n that belongs to molecule j, and r_{ij} is the distance between the centers of mass of molecules i and j. This formula can also be used at the liquid/liquid interface (without the factor of 1/2 if we have a single interface). Long-range corrections to the surface tension calculated with this expression assume that for distances $r > r_c$ (the simulation cutoff distance) one may set the radial distribution function to 1 when calculating the ensemble average in the above equations.[158]

These long-range corrections take the form of integrals involving the pair inter-action energy, and they can contribute significantly (10–30%), especially at high temperatures. Early experience using the Kirkwood–Buff formula resulted in slow convergence due to large fluctuations,[125] but with the brute force com-putational power available today, this no longer seems to be an issue.

In the Irving and Kirkwood method[159] the surface tension is expressed as the integral over the difference between the local components of the pressure tensor

$$\gamma = \int (p_N(z) - p_T(z)) \, dz \qquad [22]$$

where $p_N(z)$ and $p_T(z)$ are the normal and tangential components of the pres-sure tensor along the normal to the surface, respectively, and the integral is over the region that includes the interface (in a bulk homogeneous fluid, $p_N(z) = p_T(z)$). In practice, the components of the pressure tensor are calculated in slabs parallel to the interface, and the integral is replaced by a sum over these slabs. The expression for the pressure tensor is similar to that for the pres-sure calculated in a simulation of homogeneous systems, except that there is ambiguity in the interfacial term because no unique way exists to determine which intermolecular forces contribute to the force across a given slab. These choices are discussed briefly in the book by Rowlinson and Widom[55] and in more detail in Refs. 160–162. Expressions for the long-range corrections in the Irving–Kirkwood method were also developed.[163] We should mention that it is also possible to rewrite the Kirkwood–Buff formula as a sum of local contributions from different slabs.[164]

It is important to point out that being an equilibrium property, the surface tension can be calculated in a molecular dynamics or Monte Carlo simulation. The force calculations are of course already done in the MD code, but the calculation of the potential energy derivatives needs to be added to the MC code if the Kirkwood–Buff or Irving–Kirkwood methods are used.

The methods discussed above have been used extensively in the literature, recent example of which is a calculation of the surface tension of six common water models at different temperatures.[165] Particular noteworthy is the recent detailed calculation of the surface tension at the liquid/vapor interface for a series of alcohols by MC simulations, using many of the mechanical and ther-modynamics methods described above. The surface tension, saturated liquid densities, and the critical points compare well with experiments.[134]

Molecular Structure

The interface region is characterized at the molecular level by a strong asym-metry in the molecular interactions due to the gradient in the number density along the interface normal. The pair-interactions themselves may effectively be different in the bulk versus the interface. For example, if a polarizable force field is used, the different local electric fields in these two regions will induce a different electric dipole moment on the interfacial molecules, resulting in different effective charges on these molecules. Consequently, molecular structure and dynamics at the interface may differ significantly from the bulk. Molecular simulations have been particularly useful for elucidating this effect especially because experimental probes of interfacial molecular structure is challenging. In this section, we briefly discuss the most important molecular level structural properties derived from MD and MC simulations over the past 2–3 decades. This is followed by a discussion of dynamical properties of the neat interface.

Molecular Orientation

The orientation of molecules at interfaces impacts other structural properties (such as hydrogen bonding) and surface reactivity and, accordingly, has been the subject of statistical mechanical theories,[166] simulations, and experiments.[19,167–173] Any axis that is fixed in the molecular frame will be randomly oriented in bulk homogeneous liquids, but will typically have a nonuniform distribution at the interface due to the asymmetry in forces experienced by the molecule there. Because of the cylindrical symmetry of the planar interface, only the angle between a molecular axis and the interface normal is typically of interest. However, because three linearly independent vectors are generally needed to fully determine the orientation of a rigid body, a complete description of surface molecular orientation involves determining the joint probability that a molecule (whose center of mass is at z) will have its three vectors at angles α, β, and γ with respect to the interface normal. For example, in the case of water molecular orientation, only two vectors are needed for a complete specification, that is, the water molecular dipole (the vector in the molecule plane, bisecting the HOH angle) and one of the OH bonds. Most reports of water orientation at interfaces provide the orientational profile $P(\alpha, z)$, which gives the probability that the water dipole or the water OH bond (or the HH vector) forms an angle α with respect to the interface normal when the molecule is in a slab located at z. Besides this two-dimensional probability distribution, it is also useful to show the orientational order parameter profile, defined as:

$$S(z) = (3\langle \cos^2\alpha \rangle - 1)/2 \qquad [23]$$

where the angular brackets denote the equilibrium average over all the molecules in the slab located at z. $S = 0$ corresponds to an isotropic orientation, while values of $S = +1$ and -0.5 correspond to orientations that are perpendicular ($\alpha = 90°$) and parallel ($\alpha = 0$) to the interface normal, respectively.

The orientational profiles of water and other liquids at interfaces have been reported in many simulation studies.[93,125,132,135,138,140,142,143,151,174–179]

It is interesting to note that the orientational profiles of the water dipole at the water liquid/vapor interface and at the interface between water and a non-polar liquid all exhibit similar behavior: The water dipole tends to lie parallel to the interface (possibly with a slight tilt toward the bulk) but with a broad distribution. At this orientation, the water molecule is able to maximize its ability to hydrogen bond with other water molecules. The orientation becomes fully isotropic in the bulk at distances that are 1 nm or less from the Gibbs surface. More refined orientational profiles without the capillary wave broadening can be obtained with respect to the intrinsic surface. These intrinsic orientational profiles can better identify the orientational ordering at the interface.[140,142,143,151,176,177,179] Complementary (and sometimes consistent) experimental information has been provided in recent years mainly by nonlinear spectroscopic methods (second harmonic and sum frequency generation), discussed below.[19,167−173]

While the profiles of a single orientational variable can be useful, Jedlovsky et al.,[175] point out that care must be exercised in using them to draw conclusions about the complete orientational structure; a bivariate distribution (represented, e.g., by two-dimensional maps) calculated in successive slabs may be necessary.[142,143,151,175−177,179] In particular, the independent profiles of a single variable cannot be used to reconstruct the full joint distribution.[175]

Pair Distributions

Unlike density and orientational profiles that describe the distribution of single particle properties in space (albeit influenced by interactions with other particles), pair distributions or pair densities describe the mutual correlation between two particles. Specifically, the pair distribution function $\rho^{(2)}(\mathbf{r}_1,\mathbf{r}_2)$ is proportional to the probability of simultaneously finding two particles at locations $\mathbf{r}_1$ and $\mathbf{r}_2$. In a bulk homogeneous liquid (of density ρ_L), $\rho^{(2)}(\mathbf{r}_1,\mathbf{r}_2)$ is a function of the distance $r = |\mathbf{r}_2 - \mathbf{r}_1|$ alone, because of translational and rotational symmetry. The mutual correlation of two particles is then described more conveniently in terms of the *radial distribution function* $g(r)$, such that $4\pi r^2 g(r)\rho_L dr$ is the number of particles in a spherical shell of thickness dr and radius r centered on a given particle. At a flat liquid interface, the spherical symmetry is replaced by a cylindrical symmetry, and the pair distribution becomes a function of the distance r between the two particles as well as their locations z_1 and z_2 along the interface normal. Equivalently, one can use (in addition to r) the location z of the given particle and the angle θ between the interface normal and the vector $\mathbf{r}_2 - \mathbf{r}_1$.

Statistical mechanical approaches to the structure of the neat interface generally employ some approximations of the pair distribution function to develop a theory of density profile or surface tension. An example of such an approximation[55] is using the pair distribution of the homogeneous liquid at the density appropriate to the location z. For a direct calculation of $\rho^{(2)}(r, z, \theta)$, one

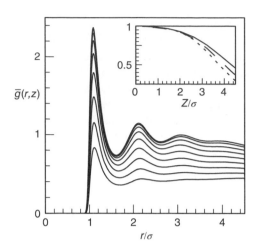

Figure 5   The orientationally averaged $\bar{g}(r,z)$ radial distribution functions for the liquid/vapor interface of a Lennard-Jones fluid at $T^* = 1.0$. The different lines correspond to 0.58σ-thick slabs parallel to the interface. The insert shows the height of the first peak of $\bar{g}(r,z)$ relative to the height in bulk liquid. (Adapted with permission from Ref. 384.)

typically must resort to simulations. Instead of attempting to compute the full 3-dimensional dependency of this distribution function, however, a more practical approach for gaining insights into the particles' mutual correlations at the interface has been to compute the orientationally averaged $g(r)$ within a slab of some thickness parallel to the interface. Thus, while the bulk $g(r)$ is computed by binning all the pair distances, we can obtain an average $\bar{g}(r,z) = \overline{g(r,z,\theta)}$ at the interface by binning the pair distances when one particle is located inside a narrow slab centered at z. Figure 5 shows the $\bar{g}(r,z)$ for a Lennard-Jones liquid/vapor interface at $T^* = 1.0$. The different lines correspond to 0.58σ-thick slabs parallel to the interface. $z = 0$ is the middle of the bulk region, and $z = 4\sigma$ is the location of the Gibbs surface. As the surface is approached, a monotonic decrease in the value of $\bar{g}(r,z)$ is found at each value of r but with no appreciable change in the *location* of the maxima (around 1.1σ). Because approximately half of the configuration space at the Gibbs surface has a near zero density (vapor) as $r \to \infty$, $\bar{g}(r,z_G)$ approaches 0.5 rather than the bulk value of 1. The insert shows how the height of the first peak is scaled down relative to the bulk for three temperatures. The significant reduction in the first peak without a significant change in its position has been demonstrated to some degree for water when a nonpolar solute is near the Gibbs surface.[135,142,143] This reduction reflects the diminished density of the solvent. If one normalizes the local $g(r)$ by the local density[135] (using the density profile), any observed variation in $\bar{g}(r,z)$ as a function of z reflects the intrinsic change in solvation structure. This concept will be useful when examining solute–solvent correlations below.

Hydrogen Bonding

An example of a pair distribution of particular importance is one that deals with hydrogen bonding and specifically hydrogen bonds involving water at interfaces. This topic has been the subject of numerous investigations using MD and MC methods. A discussion of hydrogen bonding requires its definition, and several such definitions have been proposed. The most popular are based on energy[180–182] or geometric[183–185] criteria. For example, two water molecules are considered hydrogen bonded when (a) their mutual interaction energy is more negative than $-10\,kJ/mol$ or (b) when the O–O distance is less than 3.5 Å (the location of the first minimum of the bulk water oxygen–oxygen radial distribution function) with the OHO angle deviating by no more than 30° from linearity. A systematic approach for developing H-bond criteria using a geometric cutoff on the basis of a two-dimensional potential of mean force has been described by Kumar et al.[186] Although these definitions are somewhat arbitrary, the exact choice used does not seem to have much impact on the conclusions discussed below.[187] Regardless which definition is used in the simulation of bulk water, it is found that each water molecule is hydrogen bonded by an average of $N_{HB} \approx 3.5$ hydrogen bonds to other water molecules. Given that the coordination number of water in the bulk is about $N_C = 4$, this shows that the probability for any given hydrogen bond to exist is about $P_{HB} = N_{HB}/N_C \approx 0.87$. At the interface, the number of hydrogen bonds per water molecule decreases. For example, at the water/nitrobenzene interface[138] it is an average of about 2.5 for water molecules near the Gibbs surface. However, at the same time, the coordination number also decreases, in this case to a value of about 2.4. Thus, the probability that any of the water molecules in the first coordination shell is hydrogen bonded is a higher value (0.96) at the interface than in the bulk. The same qualitative picture emerges in several water interfacial systems.[135,136,138,141] It is interesting to note that when the hydrogen bond statistic is computed relative to the intrinsic interface,[140] P_{HB} exhibits a maximum, since there is a significant drop in the coordination number for the interfacial molecules in the outermost layer. The result that on average P_{HB} (interface) > P_{HB} (bulk) is consistent with a less mobile interfacial hydrogen-bonding network at the interface. This can be demonstrated more directly by computing the hydrogen bond lifetime (for more details, see below).

While experimental data about water structure and the hydrogen-bonding environment can be obtained with a variety of techniques (X-ray and neutron diffraction, IR and Raman spectroscopy, and X-ray absorption and emission spectroscopy), experimental data about hydrogen bonding at interfaces is becoming available through the application of SFG spectroscopy. This technique provides information about the vibrational spectra of interfacial molecules, which are influenced strongly by the number, strength, and type of hydrogen bonds, as described in the next section.

The Vibrational Spectrum of Water at Interfaces

A single water molecule has IR- and Raman-active symmetric and antisymmetric OH stretch fundamentals at 3657 and 3756 cm^{-1}, respectively. The bulk liquid IR spectra are broad (full width at half maximum $\approx$ 375 cm^{-1}) and redshifted significantly (peak at 3400 cm^{-1}). The bulk spectrum is typically interpreted by deconvoluting it into overlapping Gaussians representing different classes of water molecules experiencing different local H-bond environments. See Refs. 188 and 189 for two recent examples containing many references to earlier work. However, as can be gleaned from reading these references and considering other issues such as temperature dependence,[190] the interpretation of these spectra has been controversial.

At the liquid/vapor interface, as well as at hydrophobic liquid/water interfaces, the "top" water molecules (farthest away from the bulk) are hydrogen bonded to one or two other water molecules. This is evident from the hydrogen-bonding discussion in the previous section and from the result of water OH bond orientations at interfaces, and it suggests that a significant fraction of interfacial water molecules will have one of their OH bonds "free." This uncoupled stretching mode will appear in the spectra as a sharp peak shifted to frequencies near the gas phase fundamentals. A direct experimental demonstration of the existence of the "free" OH bond at the water liquid/vapor interface was provided first by Shen and coworkers using SFG spectroscopy.[191,192] Since then, it has been demonstrated for several other water surfaces, for example, at the water/CCl$_4$ interface by Richmond and coworkers.[193]

In SFG spectroscopy, a visible, linearly polarized laser beam (with a fixed frequency ω_{VIS} and polarization direction **j**) and a linearly polarized IR laser beam (direction **k** and variable frequency ω_{IR}) are focused at the interface. An output beam with the sum frequency $\omega_{VIS} + \omega_{IR}$ and polarization direction **i** is generated from the interface but not from the bulk, because of symmetry-based selection rules.[13] The output light intensity is proportional to the square of the medium second-order nonlinear susceptibility χ_{ijk}. To a good approximation,

$$\chi_{ijk}(\omega_{VIS}, \omega_{IR}) \approx \chi_{ijk}^{R}(\omega_{IR}) + \chi_{ijk}^{NR}(\omega_{VIS}) \qquad [24]$$

where the two terms on the right-hand side correspond to resonant (R) and nonresonant (NR) parts. As the IR laser frequency is varied, the output signal is enhanced when a symmetry-allowed molecular vibration is in resonance with the IR laser frequency. Since the first application by Shen, the technique has been applied to many molecular systems and reviewed extensively.[16,18,28,194–197] Increased sensitivity has been achieved in recent years by employing broad-band heterodyne-detected sum frequency generation (HD-SFG). Here, the signal intensity decreases linearly (vs quadratically in the older SFG method) with decreasing surface coverage, thus improving the detection limit. An additional improvement is provided by the ability of measuring the phase of the HD-SFG signal. This helps remove the nonresonant background (a problem with the older method).[194] HD-SFG was recently applied[198] to the air/deuterated water interface and revealed that the donor hydrogen bond of the water molecule straddling the interface is only slightly weaker than bulk phase water hydrogen bonds. This suggests an extremely thin interface region, a rare confirmation of two decades of computer simulation results.

Depending on the importance of quantum effects, nonlinear coupling, and dynamical contributions, many methods with varying degrees of sophistication employing fully quantum, semiclassical, and fully classical approaches have been developed for the MD calculation of vibrational spectra of condensed phase molecules. Especially challenging is the problem of calculating the IR and Raman spectrum of bulk water, because this system includes high-frequency coupled anharmonic oscillators, where dynamics and quantum effects are important. Detailed coverage of this important subject is outside the scope of this chapter, and the interested reader can consult a recent review article.[199]

Some of the methods developed for bulk liquids have also been used to calculate the hypothetical IR spectra of water at interfaces. Those studies are consistent with the experimental observation of dangling interfacial OH bonds.[200–202] However, to compare directly with SFG experiments, several groups developed simulation techniques for computing the SFG spectrum. The work in this area is an excellent demonstration of the benefit of combining experiments and simulations to gain insight into the microscopic structure of water at interfaces. In particular, theory has been able to demonstrate the degree to which the spectrum is surface specific and to attribute specific features in the spectra to specific molecular structures.[178,203]

The computational approaches up to 2006 were reviewed by Perry et al.[204] Briefly, these methods are based on representing the SFG spectrum by the Fourier transform of a polarizability–dipole quantum time correlation function (QTCF). A fully classical approach to computing the SFG spectrum is then obtained by replacing the QTCF by a classical expression including a harmonic correction factor:

$$\chi_{ijk}^{R}(\omega) \propto \frac{i}{\hbar} Q_H(\omega) \int_0^{\infty} e^{i\omega t} \langle \alpha_{ij}(t)\mu_k(0)\rangle dt \qquad [25]$$

where $Q_H(\omega) = 1 - e^{-\hbar\omega/kT}$ is a slowly varying function of ω that represents an approximate quantum correction,[204] α_{ij} is the ij component of the molecular polarizability tensor, and μ_k is the component of the electric dipole moment in the k direction. The classical model uses flexible molecules with terms describing the dependence of the molecules' dipole and polarizability on their internal coordinates. This approach was used by Morita and Hynes to calculate the SFG spectrum of the water liquid/vapor interface and is in reasonable agreement with experiments.[205,206] Improvements on the original approach and additional applications are discussed in Ref. 204.

The main problem with the fully classical approach is that the energy of the $0 \to 1$ transition of the OH stretching vibration is significantly larger than kT and corresponds to a region of the potential energy not sampled in classical trajectories. One way to handle this problem is to use a mixed quantum/classical treatment, focusing on a single vibrational mode that is treated quantum mechanically, while the liquid molecules are treated classically. This worked well, for example, in the case of a dilute HOD in liquid D_2O. Using this approach requires information about the dependence of the OH frequency, transition dipoles, and transition polarizability on the instantaneous local environment. Skinner and coworkers used this approach (which was originally developed for computing the bulk IR and Raman spectra) to calculate the real and imaginary parts of the resonant susceptibility.[207] Good agreement with experiments was found. They were able to show that the SFG spectrum is dominated by single-donor molecules with a total of two or three hydrogen bonds. More recently, they also used this approach to interpret phase-sensitive vibrational sum-frequency experiments of the surface consisting of mixtures of HOD and D_2O.[77,208,209] They found that good agreement with experiments (specifically, the imaginary part of the sum-frequency susceptibility) required using a three-body potential for water. Interesting conclusions from this work include the lack of evidence for any special ice-like ordering at the surface of liquid water, and the spectrum can be interpreted as arising from overlapping and canceling positive and negative contributions from molecules in different hydrogen-bonding environments.

We note that in the (inhomogeneous) limit of slow frequency fluctuations, the expression used by Skinner and coworkers is reduced[207] to the one used earlier by a number of researchers to calculate SFG line shapes.[178,210]

Dynamics

The atoms or molecules forming a liquid phase are at constant translational and rotational motion. (We discuss vibrational dynamics later.) These dynamics can be characterized by transport properties and time correlation functions. Significant advances in describing equilibrium dynamics in bulk liquids have been made and applied slowly in recent years to characterizing molecular dynamics at liquid interfaces. We expect the reduced density and the anisotropic forces to have significant effects on the dynamics as in the case of molecular structure. The work to date has been focused mainly on using MD simulations to compute diffusion constants, rotational correlation functions, and hydrogen bond lifetimes. We again limit the scope of the discussion to processes that take place at liquid/vapor and liquid/liquid interfaces and refer the reader to a recent review that focuses on dynamics near solid surfaces and in confined geometries.[211]

Diffusion at the Interface of Neat Liquids

On a timescale longer than a few molecular collisions (tens of femtoseconds), the translational dynamics in most liquids follow a simple diffusion law. This can be characterized in bulk homogeneous liquids by a diffusion coefficient D that can be obtained easily by computing the equilibrium mean square displacement (MSD) time correlation function:

$$\langle R^2(t) \rangle = \langle [\mathbf{r}(t) - \mathbf{r}(0)]^2 \rangle = \frac{1}{N} \frac{1}{T} \sum_{\tau=0}^{T-1} \sum_{i=1}^{N} [\mathbf{r}_i(t + \tau) - \mathbf{r}_i(\tau)]^2 \qquad [26]$$

where $\mathbf{r}_i(t)$ is the vector position of particle i at time t, and the sum is over all N particles and over all T time origins. Statistical mechanical arguments show[212] that for time $t > \tau_{mol}$, $\langle R^2(t) \rangle = 6Dt$, where τ_{mol} is on the order of a few molecular collision times (or more accurately, it is the time it takes for the velocity autocorrelation function to decay to zero). A plot of $\langle R^2(t) \rangle$ versus time shows that at very early times $\langle R^2(t) \rangle \propto t^2$ (inertial motion regime), but at longer times a straight line is obtained with a slope of $6D$.

At an interface, the dynamic parallel and perpendicular to the interface may differ, and both may depend on the distance along the interface normal. The single bulk diffusion coefficient must be replaced by diffusion coefficients in the direction normal (D_{zz}) and parallel ($D_{xx} = D_{yy}$) to the interface: and these (one may naïvely assume) can be calculated from expressions similar to those in Eq. 26:

$$\langle [z(t) - z(0)]^2 \rangle \approx 2D_{zz}t$$
$$\langle [\mathbf{s}(t) - \mathbf{s}(0)]^2 \rangle \approx 4D_{xx}t = 4D_{yy}t \qquad [27]$$

where $\mathbf{s} = (x, y)$ is the component of the vector position in the plane parallel to the interface. D_{zz} and $D_{yy} = D_{xx}$ depend on z, but we expect both to become equal to the bulk diffusion coefficient D when z is far from the Gibbs surface (or from the solid surface in the case of liquid/solid interfaces). The relations in Eq. [27] have been used to estimate the diffusion coefficient at different liquid/vapor and liquid/liquid interfaces with sometimes conflicting results.

The problem with the naïve application of Eq. [27] (especially for computing D_{zz}) is that as an interfacial molecule moves, it changes its z location. For example, the typical time required for a Lennard-Jones atom to traverse a molecular diameter is similar to its surface residence time.[213] Thus, the validity of using the asymptotic behavior of the parallel and perpendicular MSDs breaks down and, in particular, $\langle [z(t) - z(0)]^2 \rangle$ reaches a plateau.[214-216] Berne and coworkers proposed a method to overcome this problem.[217] The diffusion coefficient parallel to the interface is determined from the MSD and from the (time-dependent) probability that a molecule remains in the slab centered at z. The diffusion coefficient perpendicular to the interface (D_{zz}) is determined by finding the value of the friction coefficient ζ that gives the best match of the survival probabilities in each slab calculated by MD and by solving the Langevin equation. The diffusion coefficient is then determined from the Stokes Einstein relation $D_{zz} = k_B T/\zeta$ as a function of distance of the layers from the interface. Calculating the diffusion coefficient of water as a function of distance from the liquid vapor interface shows[217] that, as expected, the diffusion is found to be isotropic far from the interface, and the diffusion coefficient has the value $D = 0.22$ Å^2/ps, in agreement with what is found from independent bulk simulations. The diffusion at the interface is anisotropic and faster, with values of $D_{xx} = D_{yy} = 0.8$ Å^2/ps and $D_{zz} = 0.5$ Å^2/ps. This reflects the fact that with fewer hydrogen bonds per molecule at the interface, the barrier to diffusion is smaller in either parallel or perpendicular directions.

An approach was reported by Duque et al.[213] for calculating the diffusion coefficient perpendicular to the interface. It takes into account the finite residence time (by employing a solution to a stochastic differential equation), but avoids using slabs and considers the molecule positions relative to the intrinsic surface. The main conclusion of their study is that, even if a diffusion coefficient can still be computed, the turnover processes by which molecules enter and leave the intrinsic surface are as important as diffusion itself.

A detailed study of water translational dynamics at the water/hydrocarbon interface was carried out by Chowdhary and Ladanyi.[218] The dynamics were probed in the usual laboratory frame as well as the intrinsic frame to provide insight about the effect of capillary waves on the dynamics. The distribution of residence times was fitted by stretched exponentials. The diffusion constant parallel to the interface was determined as above. In agreement with other studies, they found diffusion to be faster at the interface than in the bulk and also faster when viewed in the intrinsic frame.

Reorientation Dynamics

A useful and common way of describing the reorientation dynamics of molecules in the condensed phase is to use single molecule reorientation correlation functions.[219,220] These will be described later when we discuss solute molecular reorientational dynamics. Indirect experimental probes of the reorientation dynamics of molecules in neat bulk liquids include techniques such as IR, Raman, and NMR spectroscopy.[220] More direct probes involve a variety of time-resolved methods such as dielectric relaxation, time-resolved absorption and emission spectroscopy, and the optical Kerr effect.[221] The basic idea of time-resolved spectroscopic techniques is that a short polarized laser pulse removes a subset of molecular orientations from the equilibrium orientational distribution. The relaxation of the perturbed distribution is monitored by the absorption of a second time-delayed pulse or by the time-dependent change in the fluorescence depolarization.

All these methods are not surface specific and generally cannot be used to study the reorientation dynamics at liquid interfaces. Although the effect of orientational dynamics on the SFG line shape was discussed sometime ago,[222,223] using direct experimental probes of reorientation dynamics of interfacial molecules has gained significant momentum only recently by utilizing time-resolved SFG and SHG spectroscopy.[224−227] The basic idea is similar to the pump-probe studies in bulk liquids, except that one probes the SFG signal emitted specifically by the surface molecules. For example, Bonn and coworkers have shown that recovery of the SFG signal following vibrational excitation with linearly polarized light can be used to extract the timescale of surface molecular reorientational diffusive motion.[225] They found that the reorientation of interfacial water occurs on the subpicosecond timescale, which is several times faster than in the bulk. This behavior is consistent with fewer hydrogen bonds at the interface. Molecular dynamics simulations of interfacial water dynamics are in quantitative agreement with experimental observations and show that, unlike in bulk where the dynamics can be mostly described as random jumps, the interfacial reorientation is largely diffusive in character.[227]

The first molecular dynamics computation of single molecule orientational correlation functions at liquid interfaces was reported by Benjamin.[228] In bulk water, the water dipole correlation time (4 ± 0.2 ps) and the water HH vector correlation time (1.5 ± 0.1 ps, which can be approximately deduced from the NMR line shape) are in reasonable agreement with experiments. The reorientation was found to be faster at the water liquid/vapor interface. The reorientation dynamics of water molecules at the water/1,2-dichloroethane interface is, in contrast, slightly slower (to 6 ± 0.3 and 2.3 ± 0.2 ps for the dipole and the HH vectors, respectively).[135] Similar results were found in a recent study by Chowdhary and Ladanyi of water reorientation near hydrocarbon liquids having different structure (different branching).[218] The slower reorientation was limited to water molecules immediately next to the organic phase. Slower dynamics were observed when the reorientation was calculated in the intrinsic frame (thus eliminating the effect of capillary fluctuations).

It is interesting to note that recent simulations[229] show the reorientation dynamics of water next to surfaces of varying polarity to exhibit a nonmonotonic behavior, with slower dynamics observed next to both highly polar and highly nonpolar surfaces and faster dynamics next to a surface of intermediate polarity. Insight into this behavior as well as into the mechanism governing the reorientation was provided recently by MD simulations and analytic modeling.[230,231] Two competing factors – hydration structure, which controls the number of "free" OH bonds and surface binding energy – explain the nonmonotonic reorientation dynamics. The dynamics can be explained using the extended jump model of Lagge and Hynes, where a relatively large jump in the orientation is followed by a slow diffusion of the O−O vector of two hydrogen-bonded molecules.[232]

Hydrogen Bond Dynamics

The reorientation and translational dynamics of water molecules is closely related to the dynamics of making and breaking hydrogen bonds, because translations and rotations are responsible for bringing two water molecules to the distance and orientation necessary for forming the bond. This topic has been discussed extensively experimentally[233–239] and theoretically[185–187,240–247] in bulk water and in aqueous solutions. It is important for understanding spectroscopy, reactivity, and the behavior of hydrated solute molecules. With the advent of time-resolved SFG spectroscopy, there has been some interest in understanding the hydrogen-bonding dynamics at aqueous interfaces.[230,248–257]

Hydrogen bond dynamics can be studied theoretically using the time correlation function approach originally developed for calculating reaction rate.[258] Using any of the definitions of a hydrogen bond described earlier, a dynamical random variable h is then defined this way: Consider two *tagged* water molecules that are hydrogen bonded at $t = 0$. $h(t)$ is equal to 1 if these two molecules are still bonded at a later time t; otherwise it is equal to zero. Other definitions are possible. For example, one can ignore the past history of $h(t)$ (so $h(t)$ may become 1 again later if *these* two water molecules reform their bond), or one may ignore a short time interval where the bond is broken and immediately reforms. The reader should consult Refs. 187, 240, and 241 for more details.

A normalized equilibrium autocorrelation function $c(t)$ is defined using the random variable h:

$$c(t) = \frac{\langle h(0)h(t)\rangle}{\langle h(0)h(0)\rangle} \qquad [28]$$

where the ensemble average is over all pairs and all time origins. (Note that $\langle h(0)h(0)\rangle = \langle h\rangle$, as $h^2 = h$.) The function $c(t)$ gives the probability that a given hydrogen bond exists at time t, provided it existed at time zero. Clearly, $c(t)$ starts at 1 and decays to zero with a time constant τ_{rx}, where $c(\tau_{rx}) = 1/e$. This decay involves the breakup of a hydrogen bond due to the mutual diffusion and reorientation of a pair of initially hydrogen-bonded water molecules. A detailed analysis can provide information about the intrinsic forward (k_f) and backward (k_b) rate constants for breaking a hydrogen bond and thus the lifetime $\tau = 1/k_f$ of a hydrogen bond.[187,240,241] The function $c(t)$ is also useful for a qualitative account of the lifetime.

To examine hydrogen bond dynamics as a function of its location normal to the interface, the simulation box can be divided into slabs. $c_n(t)$ is computed for the nth slab by including the contributions of all the hydrogen bonds between pairs of water molecules, where at least one of the molecules belongs to the given slab. If a water molecule changes its initial slab during the simulation, its time history then contributes to the average in the new slab.

A hydrogen bond's lifetime is found to be longer at the water/metal[251] and water/protein[248–250,255] interfaces than in bulk water. Berne and coworkers examined the hydrogen bond dynamics at the water liquid/vapor interface.[252] They found faster hydrogen bond dynamics at the interface than in bulk water for the polarizable water models, but slower dynamics if nonpolarizable models are used. Even with the polarizable models, however, the shorter lifetime at the interface was attributed to more rapid translational diffusion at the interface rather than to faster kinetics of hydrogen bond formation. This can be demonstrated by removing the contribution of the relative diffusion of water molecules by defining a new correlation function:

$$c_{nd}(t) = \frac{\langle h(0)h(t)H(t)\rangle}{\langle h(0)H(t)\rangle} \qquad [29]$$

where $H(t) = 1$ if the O−O distance is less than the hydrogen bond defining distance of 3.5 Å (but for any orientation), and using the relation $h(0)h(0) = h(0)$. $c_{nd}(t)$ is the conditional probability that a hydrogen bond exists at time t given that it existed at $t = 0$ *and* that the two water molecules have not diffused away from one another. The relaxation of this correlation function is slower at the interface than in the bulk. This is explained by the fact that the number of water molecules next to a hydrogen-bonded pair that are available to accept or donate a hydrogen bond is smaller at the interface than in the bulk.

The hydrogen bond dynamics at the interface between water and a series of organic liquids of varying polarity and surface structure were examined by Benjamin.[253] The hydrogen bond population relaxation time was found to be slower than in bulk water and strongly dependent on the nature of the organic phase. Profiling the lifetime shows dependence of the lifetime on the location of the water molecules along the interface normal, with the bulk relaxation rate reached at approximately two layers "below" the Gibbs surface. The slowest relaxation was found at the interface that is characterized by the tendency to form the most water protrusions. This is due to the lack of adjacent but non-hydrogen-bonded molecules, which increases the timescale for a water molecule to reorient itself toward a new hydrogen-bonding "partner."

Chowdhary and Ladanyi investigated how the chain length and branching of liquid alkanes influence the dynamics of hydrogen bonds in water next to these liquids.[256] They found that the hydrogen bond dynamics of interfacial water are weakly influenced by the identity of the hydrocarbon phase and by capillary waves. However, they found that hydrogen bond dynamics are sensitive to the initial orientation of the molecules participating in the hydrogen bond.

SOLUTES AT INTERFACES: STRUCTURE AND THERMODYNAMICS

As with neat liquids at interfaces, the molecular structure of solvated molecules at an interface can be discussed using probability distribution functions.[55] We focus here on the single particle distribution (or solute density profile) and the pair distribution, because they have received much attention. We also remark briefly on the subject of solute orientational profiles.

Solute Density

Basic Definitions and the Gibbs Adsorption Equation

The single particle distribution function $\rho^{(1)}(\mathbf{r}_s)$ gives the probability density of finding the center of mass of a solute molecule at the location $\mathbf{r}_s$. In bulk homogeneous solutions, $\rho^{(1)}(\mathbf{r}_s) = \rho_b = $ constant, but at a planar liquid/vapor or liquid/liquid interface $\rho^{(1)}(\mathbf{r}_s)$ depends on the solute location along the interface normal z_s. For simplicity of notation we will denote it by $\rho_s(z)$. Computationally, one counts the number of solute molecules per unit volume in a thin slab centered at z to obtain the density distribution $\rho_s(z)$. The integral of $\rho_s(z)$ along z gives the total number of solute molecules per unit area, and in general this is not equal to the integral of the constant density ρ_b over the same interval (which includes the interface region). The difference is called the surface excess density Γ, which is related to the surface tension of the solution γ and the solute chemical potential μ using the Gibbs adsorption equation:[55]

$$d\gamma = -\Gamma d\mu \quad \text{(fixed temperature)} \qquad [30]$$

If there are c components, the left-hand side is generalized to $d\gamma = -\sum_{i=1}^{c} \Gamma_i d\mu_i$.

The surface excess density Γ can be measured easily. It is typically done as a function of the bulk solute concentration ρ_b to obtain the so-called *adsorption isotherm*.[11] If $\Gamma > 0$, the solute is considered surface-active, and according to Eq. [30] this is associated with a decrease in the surface tension relative to pure solvent(s).

A direct simulation of a solute's density profile is feasible for relatively concentrated solutions. For example, a 1 M aqueous solution containing 100 solute molecules (so that reasonable statistical accuracy can be achieved) must include approximately 5500 water molecules. Such simulations were reported for neutral[259,260] as well as ionic solutions, with the goal of elucidating the structure of the interface as well as establishing the ability of the simulation methodologies to reproduce the relationship between surface tension and surface excess. Of particular recent interest has been the study of ionic distributions at the water liquid/vapor interface because of its relevance to the heterogeneous photochemistry of sea salt aerosols, a system implicated in ozone depletion. These types of simulations were the first to indicate the presence of ions at the top-most layer of aqueous solutions,[261–269] followed by a number of possible experimental confirmations.[264,265,270,271] A molecular dynamics study of the SFG spectra of ionic solutions shows, in agreement with experiments, that adding sodium iodide to water leads to an increase in

SFG intensity in the spectral region that is associated with an increase in the ordering of hydrogen-bonded water molecules.[272] The increase in the SFG intensity was attributed to an increase in the ordering of water molecules caused by the formation of an ionic double layer at the interface.

The Potential of Mean Force – Methodology

The solute density profile is related closely to the local free energy or the free energy profile:[55]

$$A(z) = -RT \ln \rho_s(z) \qquad [31]$$

where R is the gas constant (8.314 J/mol/K), and T is the temperature. $A(z)$ is also sometimes called the Potential of Mean Force (PMF), because it represents the canonical ensemble average of the effective potential experienced by the solute. $A(z) - A(z_{bulk})$ is the free energy change associated with moving the solute from the bulk to the location z. If the solute is surface active, $A(z)$ will typically have a minimum at some interface location z_{int}, and $\Delta A_{ads} = A(z_{int}) - A(z_{bulk})$ is the solute adsorption free energy (negative). It can be obtained from the experimentally determined adsorption isotherm.[11]

Unlike the adsorption isotherm, the experimental determination of $\rho_s(z)$ is much more difficult, and only recently have X-ray reflectivity measurements been used as the first experimental technique to measure ions' density profiles directly at the water–organic liquid interface.[273,274] Thus, most of our knowledge of $\rho_s(z)$ and $A(z)$ has been obtained from computer simulations.

If solute–solute interactions can be neglected, $\rho_s(z)$ can be calculated by a canonical ensemble average of a system that includes a single solute molecule:

$$\rho_s(z) = \langle \delta(z - z_{cm}) \rangle = \frac{\int e^{-\beta H} \delta(z - z_{cm}) d\mathbf{r}}{\int e^{-\beta H} d\mathbf{r}} \qquad [32]$$

where z_{cm} is the center of mass of the solute along the interface normal calculated relative to the system's center of mass. If, in a given interval, $A(z)$ varies significantly (more than several RTs), calculating the ensemble average in Eq. [32] by a direct sampling of z_{cm} is not feasible, and well-known "tricks" for simulating rare events must be employed. In the "windowing" (also called "umbrella sampling") method,[61,212,275] the interval of interest is divided into a series of n overlapping "windows," such that within each window a statistically meaningful sampling of z_{cm} can be obtained. A series of $A_i(z)$ segments $(i = 1, 2, \ldots n)$ can then be combined by using their overlapping regions[276,277] to arrive at the final free energy profile for the entire region of interest. The solute is typically restricted to a given window by a window potential (which is zero when the solute is inside that window but rises rapidly when the solute's center of mass attempts to escape the window).

If $A(z)$ varies rapidly in a given region, the requirement of many overlapping windows becomes computationally inefficient. It is possible to increase the sampling statistics by utilizing a "biasing potential." This is based on the fact that if one adds to the Hamiltonian H in Eq. [32] a term $U(z_{cm})$, which is a function of z_{cm} alone, the resulting biased distribution is related to the original distribution by:[61]

$$\rho_s^{bias}(z) = e^{-\beta U(z)} \rho_s(z) \qquad [33]$$

If U is selected to be a reasonable guess of $-A(z)$, the biased distribution will be nearly flat, and the original distribution can be recovered from this relation. The umbrella sampling method has been applied mainly to chemical reactions (see below) and in some cases to solute adsorption.[278]

Another approach used to compute the PMF for solute crossing an interface is based directly on the concept of mean potential:[266,273,279−282] We denote by $f_N(z)$ the normal component of the average total force on the solute's center of mass, when the solute center of mass is fixed at the location z. The series of $f_N(z)$ values computed along a fine grid allows for calculating the integral, which is the negative of the PMF:

$$A(z_2) - A(z_1) = -\int_{z_1}^{z_2} f_N(z)dz \qquad [34]$$

Expanding this to the case of a more general coordinate and to the case when this coordinate is unconstrained has been discussed by Darve and Pohorille.[279] We point out that calculating the PMF is just one example of free energy calculation methods that have been developed and discussed extensively in the literature. The reader can consult the many excellent sources on general simulation techniques mentioned earlier in this chapter and others,[283] as well as a chapter in this book series dedicated to this topic.[284]

The Potential of Mean Force – Applications

The methods discussed above have been applied mainly to calculating the potential of mean force for the adsorption of ions at the liquid/vapor interface of water. This was motivated by fundamental questions in atmospheric chemistry and by the problem of ions and other solutes' transfer across the water/organic liquid interface, found in electrochemistry, solvent extraction, and phase transfer catalysis.

The Gibbs adsorption equation and the fact that the surface tension of most inorganic aqueous salt solutions increases with the salt's bulk concentration suggest a negative surface excess for these ions and thus a positive free energy of ion adsorption at the water surface. Continuum electrostatic model calculations give rise to a monotonic increase in the free energy

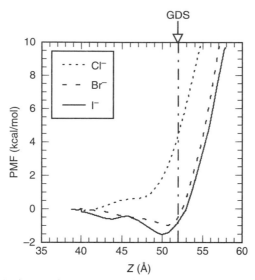

Figure 6 Potential of mean force of several halide ions across the water liquid/vapor interface at $T = 300\,K$. GDS is the GDS. (Reprinted with permission from Ref. 280. Copyright 2002 American Chemical Society.)

profile as an ion approaches the interface from the higher dielectric medium region.[228,285-289] This effect is known as the image charge repulsion. Early molecular dynamics calculations gave qualitatively similar results for small nonpolarizable ions[228,285] such as Li^+, Na^+, and F^-. Those results have been confirmed in more recent calculations using polarizable ions and water models.[95,280] Contrary to the simple electrostatic picture, however, molecular dynamics calculations[261-268,290] and experiments[264,265,270,271,291,292] suggest a local surface enhancement of larger polarizable anions such as I^-, thiocyanide anion,[264,265] and dicarboxylate dianions.[262] An example is shown in Figure 6. The small local minimum in the iodine ion PMF near the Gibbs surface corresponds to an increase in local concentration. Note that the total surface excess, which includes the depleted sublayer, still gives rise to an overall negative surface excess in agreement with the Gibbs adsorption equation.[268,290] In addition to the mutual induced polarizabilities of the water and the ions, other entropic and energetic[293] considerations and surface water fluctuations[294] have been implicated in the local surface enhancement of ions.

Of particular fundamental and practical interest has been the recent and sometimes conflicting work on the surface behavior of the hydronium and hydroxide ions.[264,295-301] Experimental and theoretical work has established that the concentration of hydronium cations is enhanced at the water liquid/vapor interface compared with the bulk. Some surface tension measurements as a function of bulk water pH along with electrophoretic and titration measurements of air bubbles or oil droplets in water suggest strong affinity of hydroxide anions for the water/vapor interface. Surface-selective nonlinear spectroscopies, photoelectron spectroscopy, and molecular simulations, in contrast, show at most a weak surface affinity of hydroxide ions.[297,298]

PMF calculations of ion and solute transfer across the water/immiscible liquid interface generally show a monotonic change between the two phases.[273,274,281,282,302−310] The net change in the PMF − A(bulk liquid)− A(bulk water) $\approx \Delta G_{transfer}$ − gives the Gibbs free energy of transfer from the water to the immiscible liquid phase. Results are in qualitative agreement with experimental data and with continuum electrostatic models,[288,289,305] although in the latter case adjustable parameters (the size of the spherical cavity used to model the ion) were needed to obtain a good fit.[305]

While most comparisons of the free energy profile between experiments and simulations have been limited to just the net free energy of transfer, X-ray reflectivity measurements can probe the total ionic density across the interface between two immiscible electrolyte solutions and can be used to make a comparison with the full PMF. The PMF of Br^- and tetra butyl ammonium cation (TBA^+) across the water/nitrobenzene interface, calculated by MD, have been used to derive the ionic density distribution. The total distribution was found to agree well with the measured one over a wide range of bulk electrolyte concentration.[273,274] This contrasts with the predictions of the continuum electrostatic Gouy–Chapman theory, where ionic density distributions vary substantially from the X-ray reflectivity measurements, and this underscores the importance of molecular-scale structure at the interface.

Of course, if one is interested only in the net free energy of transfer, then calculating the full PMF is not necessary. Instead, one may calculate the absolute free energy of solute solvation in the bulk of each solvent and simply take the difference. This, however, ignores the possibility that ions (especially small hydrophilic ones) may retain a partial hydration shell while being transferred into the organic phase. This approach was demonstrated for the free energy of transfer of alkali and halide ions from water to 1,2-dichloroethane (DCE). Free energy calculations of different-size ion-water clusters (different numbers of water molecules) in bulk DCE[311] reproduce the free energy of transfer (ΔG_t) of ions from water to this solvent, in reasonable agreement with the experimental data. We will see below that accounting for the possibility of ions retaining part of their hydration shell is critical for a correct interpretation of several dynamical phenomena as well.

The topic of ion transfer free energy across the liquid/liquid interface is closely related to the problem of ion channels in biological membranes,[312,313] but this is outside the scope of this chapter.

Solute–Solvent Correlations

The solute–solvent pair correlation function $\rho^{(2)}(\mathbf{r}_s, \mathbf{r}_l)$ gives the probability of finding the solute molecule center of mass at location $\mathbf{r}_s$, given a solvent molecule at r_l. In bulk liquid, $\rho^{(2)}(r_s, r_l)$ is a function of only the solvent–solute distance $r = |r_s - r_l|$ and is proportional to the solvent solute radial distribution function $g_{sl}(r)$. However, at a planar interface $\rho^{(2)}(r_s, r_l)$ is a function of r, z (the solute center of mass location) and the angle θ that the solute–solvent vector $r_s - r_l$ forms with the interface normal. Direct information about this quantity has been provided mainly by computer simulations. As was the case for the pair correlations in neat liquid surfaces, one typically computes $g_{sl}(r;z)$ by averaging $g_{sl}(r, z, \theta)$ over θ with the solute center of mass allowed to move in a slab of some small width (a few Å) centered at z.

In Figure 7 we demonstrate an important principle governing ion behavior at aqueous interfaces, namely, that ions are able to keep their hydration shell intact to some degree and that depends on their charge and size.[228,280,285,290,314,315] This figure shows the orientationally averaged $g_{X\text{-}O}(r;z)$ at three locations: in bulk water (solid line), near the Gibbs surface of the water liquid/vapor interface (dashed line), and at a slab whose center is located 3 Å "above" the Gibbs surface (dotted line). The top panels show the results for X = Li$^+$ (small ion) and X = I$^-$ (large ion). The small ion is able to keep its hydration shell intact when it is moved to the liquid/vapor interface. The interfacial orientationally averaged $g(r)$ is almost identical to the bulk $g(r)$ for $r < r_{min}$, where r_{min} is the location of the first minimum of

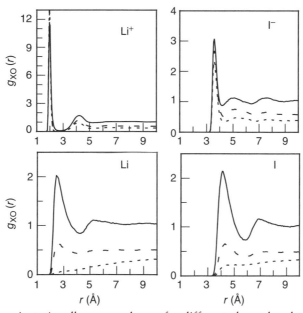

Figure 7 The orientationally averaged $g_{X\text{-}O}$ for different charged and uncharged ions at three locations: in bulk water (solid line), near the Gibbs surface of the water liquid/vapor interface (dashed line), and 3 Å "above" the Gibbs surface (dotted line).

$g(r)$, which is the region of the first hydration shell. The ability of the larger ion to do this is significant but somewhat diminished. The bottom panels depict calculations done for fictitious particles identical to Li^+ and I^- in their size, but with no charge. Here, in contrast with the case of charged particles, there is a significant depletion in the number of water molecules in the first hydration shell. Note that in all cases, the second hydration shell is diminished significantly. Note also that the asymptotic value of $g(r)$ reflects the average density of the medium. Another way to demonstrate this effect is to compute the total solvation energy profile of the ion $E(z)$ and the energy of the first hydration shell $E_{1shell}(z)$, calculated when the ion is in different z locations. While $E(z)$ drops significantly as the ion is moved to the liquid/vapor interface, $E_{1shell}(z)$ remains essentially constant.

This semi-invariance of the ion's hydration structure manifests itself during ion transfer across the liquid/liquid interface. Depending on its size and charge, the ion is able to drag all or part of its hydration shell as it is transferred from the aqueous to the organic phase.[281,282,302,306−308,314,316−318] Closely related to this concept is the role played by the fluctuations in the solvent−ion interactions in driving small ions toward and away from the interface.[294]

The ability of ions to interact with liquid surfaces in this manner has important implications for understanding several aspects of spectroscopy and dynamics involving ionic solutes, to be discussed below.

Solute Molecular Orientation

As in the case of molecular orientation at the interface of neat liquids, solute molecular orientation can provide insight into the local intermolecular interactions at the interface, which, in turn, is useful for interpreting dynamics, spectroscopy, and reactivity. The simple picture that the hydrophilic part of an asymmetric solute molecule tends to point toward the bulk aqueous phase, while the hydrophobic part points toward the opposite direction, has been confirmed in both simulations and experiments. Polarization-dependent SHG and SFG nonlinear spectroscopy can be used to determine relative as well as absolute orientations of solute molecules with significant nonlinear hyperpolarizability.[16,17,172,173,319,320] The technique is based on the fact that the SFG and the SHG signals coming from an interface depend on the polarization of the two input and one output lasers. Because an interface with a cylindrical symmetry has only four elements of the 27-element second-order susceptibility tensor being nonzero, these elements (which depend on the molecular orientation) can be measured. This enables the determination of different moments of the orientational distribution:

$$\overline{\cos^n \phi} = \int_0^\pi \cos^n \phi P(\phi) \sin \phi d\phi \qquad [35]$$

from which one can model $P(\phi)$. These orientational distributions are typically found to be broad, in part due to the capillary broadening of the interface region.

Much more detailed information can be obtained from molecular dynamics and Monte Carlo simulations. This includes the solute orientational profile, which can be expressed using the orientational probability distribution $P(\phi; z)$. If **d** is a vector fixed in the molecular frame of a solute molecule, the probability distribution of the angle ϕ between this vector and the normal to the interface is calculated easily using computer simulations as a function of the solute location z. For relatively large dye molecules with a slow reorientation time, convergence can be slow, so it is important to verify that the computed $P(\phi; z)$ is uniform for z values in the bulk region. Only a few molecular dynamics studies have been reported, with results that generally show an orientational preference with broad distributions.[321-323]

SOLUTES AT INTERFACES: ELECTRONIC SPECTROSCOPY

Experimental and theoretical studies of visible and UV spectra of solute molecules in the condensed phase have a very long history. These techniques provide detailed and rich information about the molecular structure, dynamics, and interactions governing solvation.[324-334] The development of nonlinear optical spectroscopic techniques[13] has enabled the study of electronic transitions in solute molecules adsorbed at liquid interfaces, and in this section we focus on some computational aspects of solute electronic spectroscopy at liquid/vapor and liquid/liquid interfaces. We consider here static spectroscopy and postpone the discussion of related dynamic phenomena to the section on solute dynamics at interfaces.

A Brief General Background on Electronic Spectroscopy in the Condensed Phase

For simplicity we focus on transitions between two electronic states of a solute molecule $|a> \rightarrow |b>$, in a dilute solution (so solute–solute interactions and energy transfer are not considered). We can think of the states $|a>$ and $|b>$ as the ground and excited states, respectively, for an absorption spectrum and having the reverse role for discussing the emission spectrum. The solvent–solute interactions will modify the energy of the two states, so the transition energy will change compared with that in the gas phase, $\Delta E_0 = E_0^b - E_0^a$. The relatively sharp transition in the gas phase (with linewidth generated by lifetime and vibrational broadening) acquires an additional width due to fluctuations in the positions of the solvent molecules relative to the solute. Thus, solvent effects on electronic spectra contain valuable information about solvation structure (and dynamics) and solvent–solute interactions.

Because our focus is on the solvent effects at interfaces, we consider as a simple model a two-level quantum system coupled to a classical bath. In a more realistic system one must include the contributions of intramolecular solute vibration and their coupling to a semiclassical or quantum bath.[333,334] The classical treatment allows us to obtain the transition energy for a particular solvent configuration by $\Delta E = H_b(\mathbf{r}) - H_a(\mathbf{r})$, where H_a and H_b are the total (solvent + solute) Hamiltonian of the system when the solute is in the $| a >$ and $| b >$ electronic states, respectively. For a weak electromagnetic field, neglecting solvent dynamics and assuming an infinite excited state lifetime, the (area normalized) static line shape is given in the Franck–Condon approximation by the distribution of transition energies in the initial state:

$$I(\omega) = \langle \delta[\omega - \Delta E(\mathbf{r})/\hbar]\rangle_a \qquad [36]$$

where δ is the Dirac delta function and $\langle \dots \rangle_a = \dfrac{\displaystyle\int e^{-\beta H_a} \dots \, d\mathbf{r}}{\displaystyle\int e^{-\beta H_a} d\mathbf{r}}$ represents the

canonical ensemble average when the system's Hamiltonian is H_a. Eq. [36] is the basis for a classical computer simulation calculation of the line shape: One runs MD or MC simulations with the Hamiltonian H_a and bins the instantaneous energy gap to obtain the spectrum. By replacing the delta function with its Fourier representation,

$$\langle \delta(\omega - x)\rangle = (2\pi)^{-1} \int_{-\infty}^{\infty} e^{-i\omega t}\langle e^{ixt}\rangle dt \qquad [37]$$

and using the second-order cumulant expansion of $\langle e^{ixt}\rangle$, it is straightforward to show that these line shapes are Gaussians, with peak positions given by $\omega = \langle \Delta E/\hbar \rangle_a$:[329,335,336]

$$I(\omega) = \frac{1}{\sqrt{2\pi\sigma^2}} e^{-(\omega - \langle \Delta E/\hbar\rangle_a)^2/2\sigma^2}, \quad \hbar^2\sigma^2 = \langle \Delta E^2\rangle_a - \langle \Delta E\rangle_a^2 \qquad [38]$$

These expressions have been used as the starting point for the development of a number of approximate statistical mechanical and continuum model formulas for the solvent-induced spectral shift. Because a detailed discussion of this topic is outside our scope, we refer the reader to the literature for a few examples of these results.[81,92,324,329,330,335–341]

However, it is useful for future reference to highlight some results. If the solute is modeled as a point dipole inside a spherical cavity of radius R

and the solvent as a dielectric continuum with a dielectric constant ε, the solvent-induced shift in the peak spectrum is then given by

$$\Delta\omega = \frac{4\mu_g(\mu_g - \mu_e)}{R^3}\frac{\varepsilon - 1}{\varepsilon + 2} \tag{39}$$

where μ_g and μ_e are the solute's electric dipole moments in the ground and excited states, respectively. This formula shows that if the excited state dipole is larger than that of the ground state, the spectrum is redshifted (*positive solvatochromism*), while if $\mu_e < \mu_g$ the spectrum is blueshifted (*negative solvatochromism*). This follows simply from the fact that the state with the larger dipole is stabilized to a greater degree by a polar solvent.

It is well known experimentally that the ability of a solvent to interact differently with the ground and excited states typically involves much more than just its dielectric constant; it may also depend on the details of the solvent–solute interaction and the solvent structure. The solvent *polarity scale* is an empirically based approach to express quantitatively the differential solvation of the ground and excited states of a solute.[327,342–344] It uses the electronic spectral shift as a convenient one-parameter characterization of the ability of the solvent to interact with the solute. Several solvent polarity scales have been developed on the basis of the spectral shift of several different dye molecules. For example, one of the most widely used scales, called the $E_T(30)$ scale, is equal to the spectral shift in kcal/mol for the $\pi \rightarrow \pi^*$ transition in pyridinium N-phenolate betaine dye. The $E_T(30)$ values, as well as other polarity scales, have been tabulated for many solvents.[332]

It is important to point out that using the polarity scale to quantify the liquid's solvation power works well if the solvent–solute interactions are nonspecific. It will fail if specific interactions, such as hydrogen bonding, play important roles.[332,344,345] If that happens one can introduce additional parameters to describe these specific interactions. For example, in the Kamlet–Taft empirical approach, one introduces a parameter to describe hydrogen-bond donor acidity (alpha) and another parameter to describe hydrogen bond acceptor basicity (beta).[342,346,347] The ability of such an approach to describe the hydrogen bonding (or other specific interactions) of the solvent generally depends on finding an indicator molecule that interacts with the solvent primarily through hydrogen bonding (or other specific interactions of interest). For a recent review with extensive references, see Ref. 332.

Experimental Electronic Spectroscopy at Liquid Interfaces

To measure electronic spectra of solute molecules at interfaces, one must overcome the problem that most of the UV and visible signals absorbed or emitted from solute molecules are generated in the bulk. Unless most of the solute molecules are at the interface, which is the case for systems such as adsorbed monolayers, micelles, and microemulsions,[348–354] the bulk signal masks the much weaker interface signal. Resonant SHG and Electronic Sum Frequency Generation (ESFG) are ideal techniques for studying surface electronic spectra because, as a second-order nonlinear optical processes, they are dipole forbidden in bulk media. In the SHG technique, a laser beam of frequency ω and intensity $I(\omega)$ (and a specific polarization direction) is focused on the interface. The intensity $I(2\omega)$ of the detected output beam with a frequency 2ω (and a specific polarization direction) is proportional to $I^2(\omega)$ and to the solute surface density squared, and it depends on the squares of the nonzero elements of the second-order susceptibility tensor of the medium $|\chi^{(2)}|^2$ (as well as on the orientation of the incident beam with respect to the surface normal). Each of these elements depends on the solute's molecular hyperpolarizability, and it is resonantly enhanced when $2\omega = \Delta E/\hbar$, giving rise to an effective electronic excitation spectrum. Because a nonresonant term also contributes to the signal, the peak of the *observed* SHG spectrum does not necessarily match the electronic transition energy, so it must be removed by a fitting procedure.[355]

SHG has been used mainly to measure solute surface concentration, molecular orientations, and relaxation, as mentioned earlier. The application of SHG to electronic spectra at liquid interfaces has been mainly limited to measuring the peak spectrum for insight into the polarity of the interface region.[23,355–361] By comparing the peak position of the SHG spectrum at the interface with the peak UV absorption spectra of the same dye molecule in the bulk of different solvents, one can place the interface polarity on an established polarity scale. For example, the gas phase intramolecular $\pi \rightarrow \pi^*$ charge transfer band of *N,N'*-diethyl-*p*-nitroaniline (DEPNA) is centered at 329 nm.[362] It is redshifted to 429 nm in bulk water[355] but to only 359 nm in the nonpolar bulk hexane.[346] The positive solvatochromic shift of DEPNA is mainly due to the large increase in the electric dipole moment on excitation (from 5.1 D to 12.9 D). At the air/water interface, the SHG peak is at the intermediate value of 373 nm, which is similar to the peak position in bulk carbon tetrachloride (375 nm) and butyl ether (372 nm).[346] This indicates that the water/air interface has a polarity similar to that of these two liquids (at least as it concerns the interaction with DEPNA). A similar polarity of the air/water interface was obtained[356] using the $E_T(30)$ polarity indicator (which exhibits negative solvatochromism).

By applying this approach to adsorbed dyes at liquid/liquid interfaces, Eisenthal and coworkers noticed that the polarity of several liquid/liquid interfaces (between water and 1,2-dichloroethane, chlorobenzene, or heptane) is close to the arithmetic average of the polarities of the two bulk phases.[356] Because it is known that most of the contribution to solvation energy comes from the first solvent–solute coordination shell, the simple arithmetic rule indicates that DEPNA and the $E_T(30)$ indicator molecules have a mixed solvation environment at the water/liquid interface.

The idea of assigning a specific polarity value to a liquid interface region on the basis of a solvatochromic shift of an adsorbed dye molecule involves several assumptions. Although these assumptions have been found to be violated sometimes, they provide valuable new insights into the structure of and molecular interactions at interfaces. The assumptions typically made include the following:

1. *Dependence of the perceived polarity on the solute probe.* Different probe molecules may interact differently with the interfacial solvent molecules, thus "reporting" different solvation environments. For example, the DEPNA polarity indicator suggests that the air/water interface is less polar than bulk 1,2-dichloroethane (DCE),[356] while the SHG spectrum of the doubly charged eosin B dye at the air/water interface and in bulk DCE[358] gives the opposite result. Here, despite the loss of some hydration at the interface, the strong field arising from the two charges in eosin B enables it to form a local hydration shell that is more stable than in bulk DCE.

   For another example at the liquid/liquid interface, Steel and Walker[359] used two different solvatochromic probe molecules, *para*-nitrophenol (PNP) and 2,6-dimethyl-*para*-nitrophenol (dmPNP), to study the polarity of the water–cyclohexane interface. These probes give spectral shifts as a function of bulk solvent polarity that are very similar because both solutes are mainly sensitive to the nonspecific solvent dipolar interactions. However, when these two dye molecules are adsorbed at the water/cyclohexane interface, they experience quite different polarities. The more polar solute (PNP) has a maximum SHG peak that is close to that of bulk water, and thus it reports a high-polarity environment. In contrast, the less polar solute (dmPNP) reports a much lower interface polarity, having a maximum SHG peak close to that of bulk cyclohexane. Clearly, the more polar solute is adsorbed on the water side of the interface, keeping most of its hydration shell, and thus reports a higher polarity than does the nonpolar solute. Other examples of the surface polarity dependence on probe molecules are discussed in Ref. 363.

2. *Dependence on probe location and orientation.* From our discussion of the neat interface we now know that this region is very narrow, and the SHG peak spectrum will likely depend strongly on the solute location and orientation. This was proved by Steel and Walker who designed a series of surfactant solvatochromic probes they call "molecular rulers."[364] Each of these surfactant molecules consists of an anionic hydrophilic sulphate group (which is restricted to the aqueous phase), attached to a hydrophobic solvatochromic probe moiety by a variable length alkyl spacer. The probe is based on *para*-nitroanisole (PNA), whose bulk solution excitation maximum shifts monotonically with solvent polarity.[343] When these surfactant molecules adsorb at the interface, the anionic end is in the aqueous phase, and the probe moiety resides at variable positions relative to the interface,

depending on the length of the alkyl spacer. The SHG spectra of these sur-
factants were measured at different water/liquid alkane interfaces.[23,360,361]
As the length of the alkyl spacer changes, the polarity "reported" by the
PNA probe varies from the polarity of bulk water to that of bulk alkane.
This is direct evidence that the interface is sharp – the width is no greater
than the size of a C_6 "ruler", that is, about 1 nm.

3. *Solute association and aggregation may affect the spectra significantly.* An
   example of this is the work of Teramae and coworkers who found that the
   SHG of rhodamine dyes adsorbed at the heptane/water interface exhibit
   spectral shifts that vary with bulk concentration[365] due to association and
   structural rearrangements at the interface.

4. *Electronic structure is affected by the transfer of the solute probe to the interface.* A change in the solvent environment may affect the solute electronic
   structure[366,367] and change the reported polarity. If the ground state solute
   polarizability is especially large, for instance, a positive solvatochromism
   may become negative with an increase in solvent polarity.[332,368] As an
   example, Nagatani et al. have measured the SHG spectra of the free base and
   zinc(II) complex of cationic *meso*-tetrakis(N-methyl-4-pyridyl)porphyrins,
   H_2TMPyP^{4+}, and $ZnTMPyP^{4+}$ at the water/DCE interface, and compared
   them to the UV–vis spectra of these two species in the bulk of the two
   liquids.[369] Although the molecular hyperpolarizability of the symmetrically
   substituted porphyrins is very small,[370] the adsorbed species give rise to
   intense SH signals, indicating that its electronic structure at the interface is
   modified relative to the bulk structure.

   While the resonant SHG is a powerful and sensitive technique for studying the structure of a neat interface using a solute probe, the signal-to-noise
ratio of published data is of relatively poor quality. More importantly, the
$|\chi^{(2)}|^2$ spectra cannot be compared directly with UV absorption spectra, and
thus line shape information is not readily available, although qualitative correlations between SHG spectral width and the heterogeneity of the system can
be made.[23,360,361] Recently, Tahara and coworkers used heterodyne-detected
electronic sum frequency generation (HD-ESFG) spectroscopy to measure the
imaginary $\chi^{(2)}$ spectra, which, for the first time, can be compared with the
bulk absorption line shape.[371] The effective polarity of the air/water interface, which was determined by the peak position, depended on the type of
dye molecule used. Interestingly, the $Im[\chi^{(2)}]$ spectra at the air/water interface
showed substantially broader bandwidths than do the UV absorption spectra
taken in equally polar bulk solvents or even in bulk water. This suggests that
the solvation environment at the air/water interface is more heterogeneous than
that in bulk solvents, a point that will be further discussed below.

Computer Simulations of Electronic Transitions at Interfaces

A limitation of all the experimental methods used to learn about the interface structure from the behavior of adsorbed solute molecules is that the solute perturbs the interface to some degree. Computer simulations that reproduce the experimental data and provide molecular insight help provide a link between the neat and the perturbed structures. For example, simulations that reproduce the spectral shift of an adsorbed solute at a liquid/liquid interface can reveal the relative contribution of each solvent and provide an explanation of the average polarity rule mentioned earlier.[372,373]

The methodology for computing electronic spectra discussed earlier has been used extensively to calculate absorption and emission spectra in bulk liquids. Exactly the same procedures can be used at liquid interfaces. We report here a few results that have provided complementary insight to the experiments. A more detailed review of this topic is provided in Ref. 363.

The key to the simulations involves developing accurate potential energy functions for describing the interaction of the two solute electronic states with the solvent(s). One simple approach is to use *ab initio* quantum calculations to determine the gas phase solute charge distribution in both the ground and excited states. These distributions are then adjusted, if necessary, to reproduce the ground state solute dipole moment and the peak position of the absorption spectra in the bulk. The Lennard-Jones parameters in the ground and excited states are typically set to be equal, but they can be adjusted if a significant change in atomic polarizability between the two electronic states exists. Alternatively, dynamic atomic polarizabilities in a many-body description of the potentials can be used. This is warranted if the experimental value of the solute's dipole moment in the bulk phase differs significantly from that of the gas phase.

The approaches described in the previous paragraph were used to compute the electronic absorption spectra of DEPNA (*N,N'*-diethyl-*p*-nitroaniline) at the water liquid/vapor and water/DCE interfaces, as well as in the bulk of these solvents.[373] The same solute's potential energy functions were used in the calculation of the electronic spectra in all the systems. The excited state dipole moment of DEPNA was chosen so that the peak position of its spectrum in bulk water agrees with the experimental value of 429 nm. The half-width at half-height of the bulk spectrum was 34 nm, in good agreement with the experimental value of 37 nm,[355] suggesting that the pure inhomogeneous broadening assumption made and the potential energy functions used are acceptable. The calculated peak position of the spectrum at the water liquid/vapor interface is 382 nm. Compared with the experimental value of 373 ± 4 nm,[355] this is good agreement considering the crude model used for the excited state charge distribution. The calculated peak position at the water/DCE interface using a nonpolarizable model is 405 nm, compared with the experimental value of 415 ± 15 nm. While the calculations correctly suggest that this interface is more polar than the water liquid/vapor interface, its polarity is underestimated. The problem is that the nonpolarizable DCE model used underestimates the ability of the highly polarizable DCE molecules to stabilize the large excited state dipole (17D), and indeed, the calculated peak spectrum in bulk DCE (370 nm) is significantly different from the experimental value of 398 nm. Switching to a polarizable DCE potential (using the method discussed in Eqs. [5] and [6]) improves the results: the peak position in bulk DCE is at 385 nm and 417 nm at the water/DCE interface. These simulations help explain why the peak of the interfacial spectrum is near the average of the two bulk phases: Due to the depletion of the hydration shell at the interface, the water contribution, which accounts for most of the spectral shift, decreases by about 50% relative to the bulk. The roughness of the water/DCE interface is also an important factor, because forcing the interface to remain flat by removing capillary fluctuations makes the interface much less polar.[373]

The role of many-body polarizable potentials in MD-based predictions of electronic spectra at the water liquid/vapor interface was examined by Benjamin[374] for a model dipolar solute. The peak absorption spectrum was found to be more sensitive to the solvent polarizability than to the solute polarizability. The contribution of many-body polarizabilities was found to be more pronounced as the excited state dipole was increased, and more in the bulk than at the interface.

In addition to the intrinsic polarity of the bulk liquid(s), the solute location, and the solute structure, electronic spectral shifts may depend on some liquid surface structural features. In a study to demonstrate this,[372] a diatomic solute with charges and Lennard-Jones parameters designed to mimic approximately DEPNA was restricted to a narrow window at two different interface locations between water and four organic liquids representing various polarities and molecular structures: 1-octanol, DCE, n-nonane, and carbon tetrachloride. The electronic transition involves a change in the solute charge distribution that corresponds approximately to the experimental dipole moment jump in DEPNA. The spectral shifts relative to the gas phase are given in Figure 8. They clearly demonstrate that the effective surface polarity depends on the molecular structure of the interface (in addition to the polarity of the bulk liquids and the solute's location). For example, the water/CCl_4 interface is more polar than is the water/n-nonane interface, despite (in these model calculations) the two liquids being purely nonpolar and nonpolarizable (as reflected by the values $\Delta\omega = 0$ in the bulk of these two liquids). An increased roughness of the interface between water and the nearly spherical CCl_4 molecules exist relative to the case where water is next to the long-chain hydrocarbon molecules. This roughening allows for better access of the water molecules to the solute molecule. Note also that despite the bulk octanol giving rise to a larger shift than does bulk DCE, the opposite is true when the probe is at the Gibbs surface, reflecting the greater roughness of the water/DCE interface.

The 1-octanol/water interface polarity was studied experimentally by Walker and coworkers using the molecular ruler idea discussed earlier.[23]

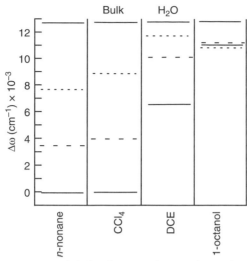

Figure 8   The electronic spectral shift relative to the gas phase of a model chromophore located in the bulk of several liquids (solid lines), at the Gibbs surface (dotted lines), and on the organic side of the Gibbs surface (dashed line). (Data taken from Ref. 372.)

Results suggest that at the water/octanol interface, a region exists whose polarity is *less* than that of bulk 1-octanol (and of bulk water). This finding was interpreted (and supported by simulations[375]) by considering the orientational preference of the long-chain 1-octanol molecules at the interface: The OH groups point toward the aqueous phase to form a monolayer, thus creating a purely hydrocarbon region, consisting of alcohol tails, extending a few Å away from the OH layer.

The above computational techniques can be used to study electronic transitions and the effective polarity of other interfaces, such as that between water and self-assembled monolayers, in micelles, and at liquid/solid interfaces. The interested reader can consult Ref. 363.

In closing this discussion of using electronic spectra to assess the polarity of the surface region, it should be mentioned that a cruder but sometimes useful approach one could take is to assign a dielectric constant to the surface region. While the static dielectric constant ε (and more generally, the frequency-dependent complex dielectric function $\varepsilon(\omega)$) of a bulk liquid is a well-defined quantity that is experimentally and computationally accessible,[79,376,377] it is not clear how to generalize this for application to the inhomogeneous region. This problem reflects the difficulties associated with using a direct experimental probe to examine the buried interface, and of applying the statistical mechanical theory of the dielectric response to an inhomogeneous region. The simplest, albeit inadequate, solution that has been used extensively[288,363,378−381] is to assume the dielectric constant to be equal to the bulk value of one phase up to a mathematically defined sharp interface, at which point it jumps to the constant bulk value of the second phase. A solution of the electrostatic boundary problem can be done in terms of image charges, and the genesis of the interface effect is the result of interactions between the solute molecules and these image charges. Most continuum electrostatic treatments of interfaces use this model, as mentioned earlier in our discussion of ion solvation. Its application to electronic spectra has been reviewed.[363] A continuum description becomes increasingly inaccurate as distances from the interface become comparable to the size of a solute molecule. Another crude concept used in simple continuum models of interfaces for calculating adsorption free energy and electronic spectra involves the use of an "effective" interfacial dielectric constant. For example, the reduced orientational freedom of interfacial water molecules and their reduced density result in a smaller effective dielectric constant than in the bulk. This is consistent with assigning the water liquid/vapor a polarity value similar to that of CCl_4. Finally, we mention that a molecular theory of a local dielectric constant, which reproduces interfacial electric fields, can be developed with the aid of molecular dynamics simulation as described by Shiratori and Morita.[382]

Our discussion thus far has been limited to the spectral peak shift. However, much information about solvent–solute interactions and the structural heterogeneity of the interface region can be derived from the electronic adsorption bandwidth.[92,324,325,328−330,335,383] This information is also available from simulations, but as mentioned earlier, this topic has been only briefly investigated due to the lack of experimental data. One problem with the experimental data is that the electronic sum frequency spectra bandwidth cannot be compared directly with bulk absorption spectra. Another issue is that because spectral widths depend on the polarity of the medium,[324] one must correct for this effect when using these linewidths to compare the degree of solvent inhomogeneity in *different* media.

The imaginary part of the second-order nonlinear susceptibility $(Im[\chi^{(2)}])$ of several coumarin dyes at the water–air interface has been measured using heterodyne-detected electronic sum frequency generation (HD-ESFG) spectroscopy.[371] The width of these spectra was found to be broader than the width of the $Im[\chi^{(1)}]$ derived from the UV spectra of the same chromophore in bulk solvents of similar polarity. This was attributed to the solute molecule sampling a wider distribution of solvent configurations at the interface than in the bulk. However, while the heterogeneity of the local solvation environment at the interface (e.g., due to orientational anisotropy) is an important factor, the contribution of molecules from across the whole interface region is as, or even more, important, as recent molecular dynamics simulations have demonstrated.[384,385] To better address questions about electronic transition spectral widths, it will be important to develop accurate computational approaches to calculate $\chi^{(2)}$ directly.[386]

We note in closing that spectral line shape can also be determined within a continuum model description by taking into account the relationship between the Gaussian spectral width and the free energy of the relevant charge distribution.[324,381]

SOLUTES AT INTERFACES: DYNAMICS

Understanding the dynamical behavior of solute molecules at interfaces, together with the structure, thermodynamics, and spectroscopy discussed above, is a prerequisite for understanding their reactivity. Here we are concerned with the pathways by which solute molecules exchange energy with the solvent environment. Specifically, we describe three related phenomena: (i) solute vibrational energy and phase relaxation; (ii) rotational energy and reorientational relaxation, and (iii) solvation dynamics. These three phenomena are critical for understanding chemical reactivity in bulk liquids and are likely also to be important for understanding chemical reactivity at liquid interfaces. Unlike the materials covered in previous sections, little has been done in this area experimentally, and most of our current knowledge has been obtained from computational work. We now focus on computational methodology, the insight gained from it, and make comparisons with the much more established body of work in bulk liquids.

Solute Vibrational Relaxation at Liquid Interfaces

The rate and mechanism of vibrational energy and phase relaxation in condensed media are both determined by the structure and the nature of the solute–solvent interactions. Vibrational relaxation is of fundamental importance for understanding chemical reaction dynamics. Although many experimental and theoretical studies have focused on the relaxation of solute molecules in bulk liquids,[209,387−413] far less has been done at liquid surfaces.[403,414−421]

No direct measurements of the vibrational lifetime of solute molecules at liquid/vapor and liquid/liquid interfaces have been reported to date. However, vibrational energy relaxation times of N_3^-, NCO^-, and NCS^- inside the water pools of reverse micelles were measured and found to be about three times longer than in bulk water.[417] There are also several studies using time-resolved SFG measurements at liquid interfaces,[422,423] which can follow vibrational spectral evolution, examining neat water at liquid/vapor and liquid/solid interfaces.[195,196,423]

Given our current knowledge about the neat liquid interface structure and dynamics, the basic question regarding vibrational relaxation at liquid interfaces is: how do these structures and dynamics affect the relaxation rate? Are the same factors responsible for relaxation in bulk liquids applicable to relaxation at liquid interfaces, or, are there unique surface effects that also need to be taken into account?

The computational methods that have been developed for studying vibrational energy relaxation rate in bulk liquids can often be used with minor adjustments for studying vibrational relaxation at the interface. A typical system under study usually includes one solute molecule adsorbed at a liquid/vapor interface or at the interface between two immiscible liquids. To improve statistical accuracy and to evaluate the relaxation rate as a function of distance along the interface normal, one may use the windowing method discussed above to constrain the solute center of mass location to a slab of some narrow width. It is important that whatever method is used to study this process at the interface, the same should be used when calculating the relaxation in the bulk of the relevant liquid(s). This provides an accurate benchmark for assessing surface effects when one uses the same system potential energy functions the same way.

To focus on the question of surface effects on vibrational energy relaxation rate, without the complications of intramolecular energy flow, Benjamin and coworkers studied the vibrational relaxation of a diatomic solute molecule (single vibrational mode) at various liquid/vapor and liquid/liquid interfaces.[424] The solute is modeled using the Morse potential:

$$V(R) = D_e[e^{-\alpha(R-R_{eq})} - 1]^2 \tag{40}$$

which describes an anharmonic oscillator. The parameters D_e and R_{eq} are typically taken to match the experimental dissociation energy and equilibrium bond length, respectively, and the parameter α is selected to reproduce the gas phase vibrational frequency ω_0.

The computational approach employs *nonequilibrium* classical trajectories to determine the rate. The solute is prepared with an initial vibrational energy $E_0 = \frac{p_R^2(0)}{2\mu} + V(R_0)$, where $p_R(0)$ is the initial momentum along the diatomic bond, R_0 is the initial bond length, and μ is the diatomic reduced mass. E_0 is significantly larger than the classical equilibrium value of kT per vibrational mode. All other solute and solvent degrees of freedom are selected from a thermal distribution at the temperature T. The trajectory is run in the NVE ensemble for a period of time long enough for the energy to relax to the equilibrium value of $E_\infty = kT$. A few hundred trajectories with a distribution of initial conditions are run to determine the normalized nonequilibrium correlation function:[212]

$$S(t) = \frac{\overline{E}_t - \overline{E}_\infty}{\overline{E}_0 - \overline{E}_\infty} \tag{41}$$

where $\overline{E}_t$ is the average vibrational energy of the solute at time t. In some cases, $S(t)$ is found to be a simple exponential $S(t) = e^{-t/\tau}$, describing a relaxation with a lifetime τ. Otherwise, the average lifetime may be calculated from $S(t)$ using:

$$\tau_{NE} = \int_0^\infty S(t)dt \tag{42}$$

Independently, the vibrational lifetime can be estimated by using classical equilibrium MD to approximate the quantum first-order perturbation theory expression for the vibrational relaxation rate (inverse of the lifetime). The quantum relaxation rate of a vibrational mode coupled to a bath is proportional to the Fourier transform of a force along the vibrational coordinate correlation function.[387] The idea is to approximate this correlation function using classical equilibrium MD trajectories.[395,397,425-428]

In the case of a diatomic solute, if F_A and F_B are the total forces on the two atoms (calculated while the bond length is kept rigid), the force along the vibrational coordinate is given by[429] $F = \mu(F_A/m_A - F_B/m_B) \cdot n_{AB}$, where n_{AB} is a unit vector in the direction of the diatomic bond and the Fourier transform is

$$\zeta(\omega) = \int_{-\infty}^{\infty} \langle \delta F(t)\delta F(0)\rangle \cos(\omega t)dt \qquad [43]$$

where $\delta F = F - \langle F \rangle$ and $\langle \dots \rangle$ denotes an equilibrium ensemble average. The lifetime is then given by the Landau–Teller formula:[387,430,431]

$$\tau_{LT} = \mu kT/\zeta(\omega_{eq}) \qquad [44]$$

where ω_{eq} is the equilibrium oscillator frequency (typically different from the gas phase value ω_0). The Landau–Teller result is a reasonable approximation for low-frequency vibrations at high temperatures. An approximate quantum correction is obtained by multiplying Eq. [43] by the correction factor $\beta\hbar\omega/(1 - e^{-\beta\hbar\omega})$. This correction gives the exact quantum results for a harmonic oscillator coupled to a harmonic bath.[432,433]

The above nonequilibrium and equilibrium approaches have been used to study the vibrational relaxation of ionic and nonionic solutes at the liquid/vapor interface of water,[434,435] as well as at the liquid/vapor interface of the weakly polar solvent acetonitrile (CH_3CN),[436] and at different locations at the water/CCl_4 liquid/liquid interface.[437] The bulk relaxation rates were found generally to be in good agreement with experimental data, giving some support for the methodology selected and the potential energy surfaces used. Fair agreement was found between the nonequilibrium trajectory calculations of the lifetime and the Landau–Teller results (especially for the low-frequency oscillators). These latter results were used to gain insight into the molecular factors influencing the rate, because in the Landau–Teller model the force correlation function can be split into electrostatic and a nonelectrostatic interaction contributions, as well as the contributions from the different solvents (in the case of the liquid/liquid interface).[437]

Lifetimes at the liquid/vapor interface for several diatomic solutes compared with the same calculations carried out in the bulk are summarized in Table 1. We note that the vibrational lifetime is always greater at the liquid/vapor interface than in the bulk. However, while for neutral solutes (polar or nonpolar) the surface effect is large, $\tau_{surf}/\tau_{bulk} \approx 3.1–3.5$, for the

Table 1 Calculated Vibrational Lifetime of Different Solute Molecules at the Liquid/Vapor Interface of Several Solvents at 298 K[434,436]

Solute	Solvent	τ_{bulk} (ps)	τ_{surf} (ps)	τ_{surf} (ps)/τ_{bulk} (ps)	$g_{max}^{bulk}/g_{max}^{surf}$
I_2^-	H_2O	0.60	0.90	1.5	1.6
I_2	H_2O	5.4	16.6	3.1	3.0
ClO	H_2O	7.9	28.2	3.5	3.2
ClO^-	H_2O	0.64	0.69	1.1	1.2
ClO	CH_3CN	179	555	3.1	2.8
ClO^-	CH_3CN	13.9	14.3	1.0	1.0

ionic solutes the surface effect is much smaller. For I_2^- in water, for example, $\tau_{surf}/\tau_{bulk} \approx 1.5$, and for ClO^- in water and in acetonitrile, $\tau_{surf}/\tau_{bulk} \approx 1.1$. Table 1 also shows remarkable correlations between the surface effect on the lifetime and on the peak value of the solvent–solute radial distribution function: $\tau_{surface}/\tau_{bulk} \approx \max(g_{bulk})/\max(g_{surface})$. This latter correlation and our discussion of molecular structure of interfacial solvation suggests a simple explanation for the surface effect on vibrational lifetime.

A key concept to consider in this regard is that vibrational relaxation is dominated by short-range repulsive forces, despite the fact that most of the contribution to the solvation energy is electrostatic for ionic solutes. The vibrational relaxation is determined by the magnitude of the *fluctuations* in the instantaneous force along the oscillator coordinate. Fluctuations in the rapidly varying repulsive interactions are much larger than those produced by the slowly varying electrostatic interactions.[394,395] The strong attractive electrostatic forces do have an important role: they place the solvent–solute equilibrium intermolecular distance within the rapidly varying repulsive region of the Lennard-Jones potential. We noted earlier that ionic solutes are better able to preserve their solvation shell than neutral solutes as they are moved from the bulk to the interface, and this is reflected in the ratio of the peak values of the radial distribution functions shown in Table 1. Retaining a solvation shell helps preserve the contribution of the short-range repulsive forces as an ionic solute is transferred to the surface, but less so when a nonionic solute is transferred. The degree to which the short-range repulsive forces are preserved is directly proportional to the peak value of $g(r)$, thus explaining the correlation indicated in the table. To say it differently, the reduced local density at the liquid/vapor interface decreases the vibrational friction and increases the lifetime, but an ionic solute is able to preserve somewhat the bulk local solvent density and thus mute the surface effect on the lifetime. Recall that the ability of the ionic solute to preserve its hydration shell increases as its size is reduced. This explains the difference in behavior between I_2^- and ClO^-. The increase in vibrational lifetime when nonionic solutes are transferred from the bulk to the liquid/vapor interface was also observed in several triatomic molecules.[435,438]

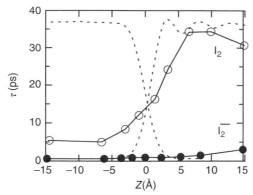

Figure 9 The vibrational lifetime of I_2^- (solid circles) and I_2 (open circles) at different locations across the water/CCl$_4$ interface. The density profiles of water (on the left) and CCl$_4$ (on the right) are shown in dotted lines. (Figures adapted with permission from Ref. 437. Copyright 2004 American Institute of Physics.)

We next consider vibrational relaxation at the interface between water and an immiscible liquid. Though total density remains relatively constant on transferring the solute from bulk water across the interface, the change in the hydration shell structure produces a change in the vibrational lifetime that follows the same principle discussed above, namely, that an ionic solute is able to keep its hydration shell. This principle manifests itself in the increase in surface roughness on ion transfer across the liquid/liquid interface, as noted earlier.

Consider the example in Figure 9 of vibrational lifetime *profile* of I_2^- and I_2 across the water/CCl$_4$ interface, calculated when the solutes are in parallel slabs of 3 Å width.[437] The vibrational lifetimes of these solute molecules in the two bulk liquids are in reasonable agreement with experiments and are also in good agreement with the simple Landau Teller model. The LT model attributes the faster relaxation in bulk water than in bulk CCl$_4$ to its greater density of phonon states in the region of the solute vibrational frequency.[437] In addition, the electrostatic forces pull the water molecules tighter around the ionic solute, increasing the Lennard-Jones component of the friction.

As I_2 is moved from bulk water to bulk CCl$_4$, the relaxation time increases monotonically and uniformly across the interface from a value of 5 ps in bulk water to around 30 ps in bulk CCl$_4$. The profile approximately tracks the change in the density profile of water. Contrarily, as the I_2^- crosses the interface, its relaxation time remains close to the value in bulk water, reaching the value expected in bulk CCl$_4$ only when it is deep into the organic side of the interface.

Here is the explanation for this behavior: as the nonionic I_2 is transferred from bulk water to bulk CCl$_4$, the number of water molecules in the solute's hydration shell (representing the main contribution to the fluctuating force on the I–I bond) decreases rapidly and monotonically. This induces an increase in the I_2 vibrational lifetime. In contrast, as the ionic solute crosses the interface, the partly retained hydration shell sustains a significant portion of the fluctuating force, and only a weak variation of the vibrational lifetime across the interface is observed. This latter effect can be shown by calculating directly the contributions of the two solvents to the total friction. As was the case at the liquid/vapor interface of water, a high degree of correlation exists between the height of the first peak of the water-solute radial distribution function and the vibrational relaxation rate.[437]

The normal dynamic capillary roughness of a neat liquid/liquid interface is magnified when an ionic solute (especially a small one) crosses the interface. The results in Table 1 suggest a correlation between the solute's vibrational lifetime and the interface roughness. This can be demonstrated more directly by forcing the liquid/liquid interface to remain flat, a technique often used to investigate the role of surface roughness. This is accomplished by adding to the system Hamiltonian a small external potential that is coupled to surface fluctuations and keeps the interface molecularly sharp and flat. The results, described in detail in the original paper,[437] show that removing capillary fluctuations has no effect on the neutral solute vibrational relaxation while the relaxation time of the ionic solute rises significantly as it crosses the flat interface.

Solute Rotational Relaxation at Liquid Interfaces

Solute rotational energy and orientation relaxation have been heavily studied in bulk liquids experimentally using a variety of techniques and also theoretically mainly with continuum models and MD.[219−221,439−468] In contrast, time-resolved studies at liquid interfaces are much more difficult to do, so they have been limited to several time-resolved SHG and SFG measurements addressing the reorientation dynamics of dye molecules at the water liquid/vapor interface.[223,469−475] A few theoretical studies using molecular dynamics simulations have appeared.[228,476−479] They have been motivated by some conflicting reports about the ability of the interface region to enhance[472] or to slow down[469−471,473] the reorientation dynamics of adsorbed solute molecules.

The many studies in bulk liquids have clarified contributing factors that influence rotational dynamics in condensed media. A natural question is: how do these factors play out at liquid interfaces? In particular, liquid density and polarity have been identified as two of the most important factors affecting rotational energy and reorientation relaxations in the bulk, and, as discussed earlier in this chapter, both are modified significantly at liquid interfaces.

The reduced density at a liquid/vapor interface is expected to lower the collision frequency and to reduce the rate of rotational energy relaxation. At the same time, fewer collisions enable faster scrambling of molecular orientations and thus are expected to increase the rate of orientational relaxation (as long as the density is not too low[219]). These contributing influences can be quantified using the concept of mechanical (or hydrodynamic) friction.[221]

In polar liquids, a polar solute experiences an additional friction, called the dielectric friction, produced by a lag in the electrostatic forces as the solute dipole rotates away from its equilibrium orientation.[221,440,442,443,480] The reduced polarity at the liquid–vapor and water–organic liquid interfaces[355,363] is thus expected to slow energy relaxation and speed up reorientation. However, surface roughness, capillary fluctuations, and the ability of an ionic solute to keep its hydration shell can complicate this picture.

The reorientation dynamics can be studied theoretically by computing the equilibrium orientational correlation function,[220] defined as:

$$C_l(t) = \langle P_l[\mathbf{d}(t) \cdot \mathbf{d}(0)] \rangle \qquad [45]$$

where $\mathbf{d}(t)$ is a unit vector fixed in the molecular frame of reference (or simply along the bond in the case of the diatomic solute), and P_l is the lth order Legendre polynomial. For example, $P_1(x) = x$, so $C_1(t) = \langle \cos \theta(t) \rangle$. The average orientational relaxation time is taken as the integral:

$$\tau_l = \int_0^\infty C_l(t)dt \qquad [46]$$

In bulk isotropic media, experiments such as IR and NMR spectroscopy and fluorescence anisotropy decay can give information about these correlation functions or their moments.[220,221] At an interface with a cylindrical symmetry, SHG and SFG spectroscopies give information about out-of-plane and in-plane reorientation, and they involve more complicated time correlation functions.[196,223,226,481] Nonetheless, the simple $C_l(t)$ are still useful for a direct comparison between bulk and surface reorientation.

At an early time, prior to the "first" collision between the solute and the solvent molecules, the orientational correlation function has a typical Gaussian time-dependency, reflecting a period of free inertial rotation:[219]

$$C_l(t) \approx e^{-(kT/2I)l(l+1)t^2}, \ t < \tau_0 \qquad [47]$$

where τ_0 is the time it takes the solute to complete one rotation. At longer times, if the rotation is highly hindered, it can be viewed as a succession of very small angle jumps around a randomly oriented axis, and the dynamics follow a simple diffusion model giving:[219]

$$C_l(t) = \exp[-l(l+1)D_r t] \qquad [48]$$

Whether the actual dynamics follow these two expressions can be checked by MD simulations to gain insight into the solute reorientational mechanism.

To investigate rotational energy relaxation, one can compute the rotational energy equilibrium correlation function and the corresponding energy relaxation time τ_E:

$$C_E(t) = \frac{\langle \delta E(t + \tau) \delta E(\tau) \rangle}{\langle \delta E(\tau) \delta E(\tau) \rangle}, \quad \delta E = E - \langle E \rangle \quad \text{[49]}$$

$$\tau_E = \int_0^\infty C_E(t) dt \quad \text{[50]}$$

For a diatomic solute, the rotational energy is $E = J^2/2I$, where I is the solute's moment of inertia and J its angular momentum. $\langle E \rangle = RT$ is the classical equilibrium average of the diatomic solute rotational energy at the temperature T.

Rotational relaxation can also be investigated by nonequilibrium simulations. The solute molecule is subjected to an instantaneous jump in its angular momentum, and the energy and orientations dynamics are followed. This is repeated for an ensemble of initial solute positions and velocities, from which time-dependent averages are computed.

To discern what controls rotational dynamics at liquid interfaces, molecular dynamics simulations were performed on a model diatomic solute at different locations in both the water liquid/vapor interface[478] and the water/CCl$_4$ interface.[479] The solute was modeled as two identical atoms separated by a rigid bond of length 4 Å (using the SHAKE algorithm[482]). The assigned mass of each atom is 35 amu, giving the value $\tau_0 = (2\pi/9)(I/kT)^{1/2} = 1.05$ ps for the time it takes the molecule to complete one rotation in the gas phase. Various solute electric dipole moments were considered by placing partial charges q_s, $-q_s$ onto the two atoms, with $q_s = 0$, 0.1, 0.2, 0.3, 0.4, 0.6, 0.8 (atomic units). The resulting dipole moment, $\mu(D) = 19.212 q_s$, spans approximately the range of dipoles encountered in experimental studies of rotational dynamics in bulk liquids and at interfaces. The solute molecule center of mass was typically constrained to 3–4 Å-wide slabs centered at different surface locations.

The top panel of Figure 10 shows the reorientational relaxation time τ_2 as a function of the solute's electric dipole moment, when the solute's center of mass is constrained to a slab of width 4 Å centered at the Gibbs surface (labeled G), 3.5 Å "above" the Gibbs surface (S), and in bulk water (B). As expected, in every location the relaxation time increases with the increase of the solute dipole moment, reflecting the increase in the dielectric friction. For relatively small solute dipole moments ($\mu < 6D$), the friction is dominated by the "mechanical" density-dependent contribution, and the relaxation in the higher density bulk region is much slower than the relaxation at the interface. However, as the dipole moment is increased, the bulk and the surface reorientation relaxation times become similar. This behavior, which mirrors that of the

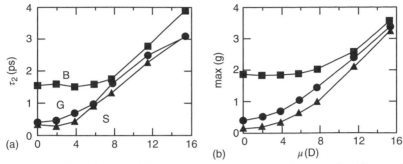

Figure 10 Dipolar solute equilibrium reorientation time at the water liquid/vapor interface. (a) The reorientation time τ_2 as a function of the solute's dipole moment. The solute is located in the bulk (B, squares), at the Gibbs surface region (G, circles) and 3.5Å "above" the Gibbs surface (S, triangles). (b) the peak value of $g(r)$ versus the dipole moment for all solute molecules studied. Squares, circles, and triangles correspond to regions B, G, and S, respectively. (Reprinted with permission from Ref. 478. Copyright 2007 American Institute of Physics.)

vibrational relaxation discussed in the previous section, is due to the solute's ability to keep the hydration shell to a degree that increases with the increase in its dipole moment. This keeps the surface rotational friction similar to that in the bulk. The correlation between the relaxation time and the peak value of the solute radial distribution function, shown in the bottom panel, is further evidence for the role played by the first hydration shell, much like what is observed for the vibrational relaxation.

The energy relaxation time, Eq. [50], as a function of the magnitude of the solute's dipole and its location exhibits an opposite trend.[478] The increased collision rate (which slows down reorientation) enhances the rotational energy transfer to the solvent molecules. Thus, for a solute with a small dipole, the energy relaxation at the interface is much slower than in the bulk. However, the difference between the bulk and surface energy relaxation rates decreases as the dipole is increased because preserving the solute hydration shell makes the interfacial friction similar to that of the bulk, despite the fact that the average solvent density just outside the solute hydration shell is smaller than in the bulk.

Nonequilibrium energy relaxation rates, calculated using nonequilibrium MD, are somewhat slower than the equilibrium rates in the case of a nonpolar solute but are almost the same for polar solutes.[478] However, the trends in relaxation times as a function of solute dipole and location are essentially the same as the equilibrium trends discussed above. The difference between equilibrium and nonequilibrium rotational relaxation reflects, in part, the difference between the equilibrium structure of the solute–solvent complex.[468,483] The insight gained from studying a simple diatomic solute has been useful for understanding the rotational behavior of large dye molecules.[484]

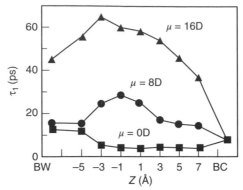

Figure 11   Dipolar solute equilibrium reorientation time τ_1 as a function of the solute's center of mass location along the water/CCl$_4$ interface normal. The solute dipole in Debye is indicated. BW and BC refer to bulk water and bulk CCl$_4$, respectively. (Reprinted with permission from Ref. 479. Copyright 2008 American Chemical Society.)

Let us now consider the reorientation dynamics at the liquid/liquid interface. The most significant contrast with the liquid/vapor interface is that there is no longer a low-density region (although the bulk viscosity of the two solvents may be quite different). Instead, a large variation in solvent polarity may introduce a marked difference in the dielectric friction for a highly polar solute.

A systematic study of the rotational relaxation, of the diatomic solute described above, at the water/CCl$_4$ interface has been carried out.[479] The solute molecules' equilibrium reorientation correlation functions, Eq. [45], were evaluated at different interface locations and in the bulk of the two solvents. Some of the results for the reorientation time τ_1 are reproduced in Figure 11.

The reorientation of the nonpolar solute ($\mu = 0$, squares) in bulk water is slightly slower than in bulk CCl$_4$, reflecting the somewhat greater viscosity of water. Because CCl$_4$ is nonpolar, increasing the solute dipole has no effect on the reorientation time in that medium. In contrast, the reorientation time in bulk water increases as the solute dipole is increased, due to the greater dielectric friction.[221,440,442,443,445,480,485−487]

As the solute is moved from the bulk across the interface, the reorientational behavior depends on the solute dipole in a surprising way. Contrary to the expectation that the surface reorientation time would lie between the two bulk phase values, Figure 11 shows that the reorientation at the interface is slightly faster than in either of the two bulk phases for a nonpolar solute but is significantly slower than either of the bulk phases in the case of a polar solute.

Lower effective viscosity at a liquid/liquid interface has been correlated experimentally with spherical molecular shapes[46] and theoretically with high surface tension.[74] A local dip in density is also a feature of liquids that partially wet hydrophobic surfaces.[488] These observations may explain the reduced effective viscosity experienced by the nonpolar solute at the water/CCl$_4$ interface, but, this remains an open issue, as the X-ray reflectivity measurements of the water/liquid hydrocarbon interface, discussed earlier, failed to detect a lower density region.[146]

The longer reorientation time of a polar solute at some surface locations compared to being in bulk water suggests the existence of a higher local dielectric friction despite the lower polarity at the interface. A similar effect was observed in the MD simulation of an actual dye molecule. It was found that *N,N'*-diethyl-p-nitroaniline rotates *slower* at the water/DCE interface than in either of the two bulk liquids.[373] A possible explanation for this invokes a coupling between dielectric and hydrodynamic friction and a concept we are already familiar with: the strong electrostatic interactions enable the solute to retain a significant fraction of its hydration shell, depending on the magnitude of its dipole. This gives rise to a solute-water hydration complex that is quite stable at the interface. Its larger volume results in a significantly larger hydrodynamic friction and slower rotation.[455] When the solute is moved deeper into the organic side of the interface or when the magnitude of the dipole is reduced, the hydration complex breaks down, and its rotation speeds up.

Finally, we note that the behavior of different correlation functions $C_l(t)$ (Eq. [48]) with $l = 1-5$ shows diffusional behavior for the largest dipole solute (16D) at both the water liquid/vapor interface and at the water/CCl_4 interface. Breakdown of this relationship is observed when the solute hydration shell is diminished, for example, in bulk CCl_4 and for the nonpolar solute at both interfaces. In these cases, the dynamics can no longer be viewed as being small successive angular jumps around a randomly oriented axis. Instead, individual trajectories show sudden and large free rotation segments around a fixed axis. This has also been observed for the rotation of large dye molecules at the water liquid/vapor interface.[477]

Solvation Dynamics

Given a solute in equilibrium with solvent molecules, a sudden change in the solute's electronic structure due to an absorption of electromagnetic radiation or an electron transfer will generally create a nonequilibrium state. The solvent electronic and nuclear degrees of freedom will respond to reestablish equilibrium. These solvent dynamics can be monitored experimentally. Assuming an instantaneous response of the solvent electronic degrees of freedom, the slower solvent response involves translation, rotation, and vibration of the solvent molecules, which can be followed by classical molecular dynamics. Because experimental and theoretical studies of solvation dynamics can reveal important phenomena needed to understand solvent dynamics and solute–solvent interactions, they have been reviewed extensively.[377,390,489−492] Solvation dynamics studies at liquid surfaces have also been reviewed.[27,363] Here we give only a brief summary focusing on the unique surface effects, which, we will see, are similar to the effects discussed above regarding vibrational and rotational relaxation.

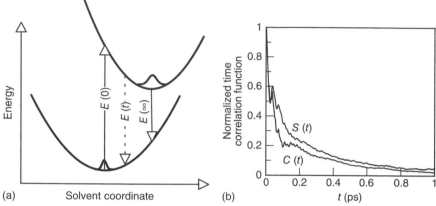

Figure 12 (a) A schematic representation of the energy changes involved in a time-resolved fluorescence experiment. (b) The equilibrium $C(t)$ and nonequilibrium $S(t)$ solvation dynamics correlation functions corresponding to the creation of a 12D dipole in bulk water. (Adapted with permission from Ref. 537.)

Consider two solute electronic states, $|a>$ and $|b>$. At time $t = 0$, the solvent is at equilibrium with state $|a>$. Let $\Delta E(0)$ denote the average energy difference between these two states, including the interaction energy with the solvent, immediately following the transition $|a> \rightarrow |b>$. As the solvent equilibrates to the state $|b>$, the energy difference evolves in time, reaching the value $\Delta E(\infty)$ when equilibrium is reached. The dynamical response of the solvent(s) can be followed using the nonequilibrium time correlation function:

$$S(t) = \frac{\Delta E(t) - \Delta E(\infty)}{\Delta E(0) - \Delta E(\infty)} \qquad [51]$$

In Time Resolved Fluorescence (TRF) experiments, depicted schematically in Figure 12, the emission spectrum line shape changes from a peak value of $\Delta E(0)$ to the final $\Delta E(\infty)$ peak of the equilibrium emission spectrum as the solvent responds to the new solute electronic state.[493,494] Experimental results in bulk liquids show that the nonequilibrium correlation functions initially exhibit a very fast (less than 50 fs[495]) inertial component, which may account for 60–80% of the total relaxation in water. This is followed by a multiexponential relaxation on the subpicosecond to picosecond timescale,[496] corresponding to reorientation and translation of solvent molecules, or, to particular intramolecular solvent modes[493,497] around the solute. Slower dynamics are found in more viscous liquids.[498]

While TRF and other techniques have been used extensively to probe solvent dynamics in bulk liquids[377,390,489,490,499] and in micelles and reverse micelles and other complex environments,[377,498,500–509] it is not surface specific and cannot generally be used to study solvent dynamics at liquid interfaces, unless the solute is adsorbed strongly at the interface with little bulk concentration. (Exceptions include using TRF in total reflection geometry.[510–512]) Solvation dynamics at liquid interfaces has been mainly studied using Time-Resolved Resonance Second Harmonic Generation (TRSHG)[513–517] and recently using Time-Resolved Sum Frequency Generation (TRSFG) spectroscopy.[518] In TRSHG, an optical pump pulse excites the solute molecules to an electronic excited state. A time-delayed probe pulse with a frequency ω generates a second harmonic pulse with frequency 2ω from only the interfacial molecules. As the solvent molecules reorganize around the excited molecule, the resonant SH signal changes with the delay time, allowing for the construction of the $S(t)$ correlation function.[513,514] The solvation dynamics of coumarin 314 (C314) adsorbed at the air–water interface was measured using TRSHG[513] and found to be similar to that in bulk water (0.8 ps). Experiments with polarized pump pulses in the direction parallel and perpendicular to the interface showed that the solvation dynamics depend on the solute orientation, being faster when the pump pulse is parallel to the interface.[514] The solvation dynamics of C314 adsorbed at the air–water interface in the presence of neutral, anionic, and cationic surfactants show that electric field of the dye and interactions with specific hydrophilic groups can slow down interfacial water dynamics.[515,516,519,520]

Solvation dynamics has been studied computationally by nonequilibrium MD as well as equilibrium classical MD techniques. Let H_a and H_b denote the total Hamiltonian of the system when the solute is in the $|a>$ and $|b>$ electronic states, respectively. Following a sudden transition, the initial equilibrium distribution with respect to H_a evolves under the Hamiltonian H_b. Clearly, the time-dependent energy gap needed to compute $S(t)$ is $\Delta E(t) = H_b(\mathbf{r}) - H_a(\mathbf{r})$, where $\mathbf{r}$ is the instantaneous system atomic positions. A set of initial conditions selected from the $|a>$ state equilibrium distribution is propagated under the Hamiltonian H_b, and the energy gap at each time step is averaged over all the independent trajectories. If H_b contains polarizable terms, they must be equilibrated before the energy gap is calculated. Because the energy gap is calculated as a difference between two different potential energy functions at the same nuclear configuration, in general one must compute both at each step. In practice, however, because the sudden change in the electronic state involves a jump in the value of a few parameters like the solute dipole moment, $H_b(\mathbf{r}) - H_a(\mathbf{r})$ is a simple expression that in many cases is already evaluated during the normal simulations with the Hamiltonian H_b.

The equilibrium calculation of solvation dynamics involves computing the equilibrium time correlation function:

$$C_v(t) = \frac{\langle \delta E(t)\delta E(0)\rangle_v}{\langle \delta E(0)\delta E(0)\rangle_v}, \quad \delta E(t) = \Delta E(t) - \langle \Delta E(t)\rangle_v \qquad [52]$$

where $\langle\rangle_v$ means the equilibrium ensemble averages with the dynamics governed by the Hamiltonian H_v ($v = |a>$ or $|b>$) and $\Delta E(t) = H_b(\mathbf{r}) - H_a(\mathbf{r})$.

If one invokes the linear response approximation, then[336,521]

$$S(t) = C_a(t) = C_b(t) \qquad [53]$$

While in many cases solvent dynamics follow the linear response approximation, even for large perturbations away from equilibrium, it is expected to fail when the equilibrium fluctuations do not sample important regions of phase space in which the nonequilibrium dynamics takes place.[468,483,522–526] Thus, calculations of both $S(t)$ and $C_v(t)$ are useful for elucidating the mechanism of solvent dynamics. Figure 12 shows as an example the computed $S(t)$ and $C(t)$ for a sudden creation of a 12D dipole in bulk water (see below for more details about these calculations).

Approximate analytical theories of solvation dynamics are typically based on the linear response approximation and additional statistical mechanics or continuum electrostatic approximations to $C_v(t)$. The continuum electrostatic approximation requires the frequency-dependent solvent dielectric response $\varepsilon(\omega)$.[390] For example, the Debye model, for which $\varepsilon(\omega) = \varepsilon_\infty + (\varepsilon_0 - \varepsilon_\infty)/(1 + i\tau_D\omega)$, predicts that the solvation dynamics will follow a single exponential relaxation:

$$S(t) = e^{-t/\tau_L}, \tau_L = \varepsilon_\infty \tau_D/\varepsilon_0 \qquad [54]$$

where τ_L is called the longitudinal relaxation time, τ_D is the Debye relaxation time, and ε_0 and ε_∞ are the static and infinite-frequency dielectric constants of the liquid. This simple Debye model can be improved by selecting a more complicated expression for $\varepsilon(\omega)$ (e.g., as a sum of Debye-like terms for different molecular motions[493]). Other approximations take into account the finite size of the liquid molecules.[527–530]

Extensive theory and computer simulation work has been able to clarify the molecular mechanisms of solvation dynamics in bulk liquids over the past three decades.[390,491] One of the most important conclusions from this body of work is that most of the contribution to polar solvation dynamics comes from the solute's first solvation shell.[531] This conclusion and the earlier discussion about the prominent role the solute hydration shell plays in understanding vibrational and rotational dynamics at liquid interfaces suggest that surface effects on solvation dynamics will be muted as the solute's polarity is increased. An experimental validation of this are the similar solvation dynamics of C314 at the water liquid/vapor interface and in bulk water, mentioned above, where the highly polar excited state ($\mu = 12D$) implicates an interfacial hydration structure similar to the bulk.

The first simulations of solvation dynamics at liquid interfaces involved the hypothetical sudden "charging" of an ion.[228,532] The dynamical response at the water liquid/vapor interface and in the bulk are almost identical. Each $S(t)$ shows a very rapid initial relaxation, corresponding to inertial solvent motion,[495,521,533] followed by a nearly exponential decay. The major factor explaining the similar bulk and surface response is the ability of the ion to keep its hydration shell almost intact.[228] Interestingly, this suggests that a dielectric continuum model parameterized to fit the bulk relaxation will fail to account for the surface dynamics. Twelve years passed until the next simulation of solvation dynamics at the water liquid/vapor interface was done by Pantano and Laria.[477] They found the solvation dynamics of C314 to be only slightly slower at the interface (0.79 ps) than in the bulk (0.56 ps), agreeing with experiment. These results are consistent with having an interfacial hydration structure similar to the one in the bulk. The initial fast inertial drop accounted for 50% of the overall relaxation in the bulk compared with 35% at the interface. The linear response approximation was found to be valid, as the equilibrium time correlation function was in excellent agreement with the nonequilibrium one.

In addition to the important role played by the structure of the polar (and ionic) solute's hydration shell, a new factor comes into play at the liquid/liquid interface: the relatively slow solvent translational motion associated with capillary fluctuations and the related larger structural deformation of the interface. These slow dynamics were demonstrated in the first study of solvation dynamics at a model liquid/liquid interface, where a diatomic nonpolar solvent is in contact with a diatomic polar solvent.[534] The electronic transitions studied were a charge separation ($A–D \rightarrow A^+–D^-$) and the reverse, a charge recombination process, taking place between donor–acceptor pairs. The dynamics at the interface were significantly slower than those in the bulk, especially for the charge separation process when the A–D vector is perpendicular to the interface. This retardation arises from the sudden creation of a large dipole that requires large structural reorganization to reach equilibrium, while a significant portion of configuration space is occupied by nonpolar solvent molecules. A significant deviation from the linear response results, but only for the charge separation process, because the equilibrium fluctuations in the final state sample very different structures from those that are required to reach equilibrium beginning from a neutral AD pair.

When two liquids with very different bulk Debye relaxation times are in contact, a very sharp liquid/liquid interface leads to a large dependence of the solvation dynamics on the solute's location. This is similar to the solvatochromic shift dependence of an adsorbed solute on its location as discussed earlier. This position dependence was demonstrated by following charge creation at different locations in the water/octanol interface.[535] The relaxation time changed by two orders of magnitude when its position was varied by a few Å relative to the interface. Because the dynamics depends on different solvents, the interface behavior is typically characterized by a multiexponential relaxation (following an inertial component). This solvent dependence and dependence on surface location were illustrated by a systematic study of dynamics behavior following a change in the permanent dipole of a dipolar solute at four liquid/liquid interfaces consisting of solvents with different polarities and molecular structures.[536] As above, a slow component was found not to

Table 2 Calculated Solvation Dynamics from Equilibrium and Nonequilibrium Simulations Following Two Types of Electronic Transitions[537]

Electronic Transition $\mu_a \rightarrow \mu_b$	Bulk H_2O	Liquid/Vapor H_2O, Gibbs Surface	Vapor Side of the Gibbs Surface	Gibbs Surface H_2O/CCl_4	Organic Side of the Gibbs Surface
12D → 0D (nonequilib-rium)	0.05	0.07	0.04	0.09	0.04
12D → 0D (equilibrium)	0.06	0.13	0.16	0.1	0.16
0D → 12D (nonequilib-rium)	0.2	0.6	1.4	0.6	5.9
0D → 12D (equilibrium)	0.07	0.11	0.15	0.10	0.30

be present in the bulk of either liquid, corresponding to the relatively slow diffusion of finger-like water structures at the interface. In this case, marked deviations from a linear response were observed.

The model used in the previous two sections shows how charged groups maintain their hydration shell at aqueous interfaces and how that was used to understand rotational and vibrational relaxation. The same model can be used for solvation dynamics.[537] We compute the solvation dynamics following the transition $(\mu = 0) \rightarrow (\mu = 12D)$ and the reverse transition $(\mu = 12D) \rightarrow (\mu = 0)$ of a diatomic solute held at different locations in the water liquid/vapor interface and in the water/CCl$_4$ interface. The equilibrium and nonequilibrium correlation functions both approximate a biexponential relaxation. Table 2 summarizes the average relaxation times in picoseconds.

The nonequilibrium solvation dynamics following the $(\mu = 12D) \rightarrow (\mu = 0)$ transition are essentially independent of the solute's location. The very rapid subpicosecond relaxation is complete within a few hundred femtoseconds. The linear response approximation is satisfied in all locations except on the vapor side of the water liquid/vapor interface and on the organic side of the water/CCl$_4$ interface. The main reason for the deviation from linear response is the different contribution of the inertial component to the total relaxation, as the equilibrium relaxation is determined from the solvent fluctuations around the neutral solute.

Much more interesting and revealing is the reverse process: $(\mu = 0) \rightarrow (\mu = 12D)$. The nonequilibrium solvation dynamics becomes much slower and sensitive to the location. The nonequilibrium relaxation in bulk water and at the water liquid/vapor interface gives $\tau(\text{bulk}) < \tau(\text{Gibbs}) < \tau(\text{above Gibbs})$. As previously discussed, the structure of the hydration shell around the large dipole at the interface is similar to that in the bulk. This structure is created from a very different solvent configuration when the solute suddenly acquires a 12D dipole; this takes longer to form at the interface than in the bulk. In contrast, the equilibrium correlation function is computed from the energy fluctuations in the final $\mu = 12D$ state, and those fluctuations are similar in the three locations.

The nonequilibrium dynamics on the organic side of the water/CCl_4 Gibbs surface are much slower than in all other locations. This is because the final equilibrium state requires significant perturbation to the water structure. This also explains why in this case a breakdown of linear response is observed.

Finally, it is interesting to note that the results at the Gibbs surface of the water/CCl_4 interface are almost identical to those at the Gibbs surface of the water liquid/vapor interface for both transitions. This is consistent with previous simulations, showing that the polarity of these systems is very similar.[536]

SUMMARY

Because of the close connection between the three phenomena discussed in this section, it is useful to summarize here the main results. When comparing the bulk and the surface vibrational, rotational, and solvation dynamics of adsorbed solute molecules, two controlling structural motifs emerge: (i) the interface is a narrow, rough region broadened by density fluctuations (nanoscale capillaries) that may couple to solute modes; (ii) a charged solute at the interface tends to have a hydration shell that is similar in structure to the one in bulk water, the similarity of which depends on the solute size and charge. These two factors also provide a foundational understanding of surface effects on solute thermodynamics and spectroscopy. However, while they have been the subject of several largely consistent experimental and computational studies, many of the dynamical results presented in the last three sections await experimental confirmation. In particular, it would be interesting to confirm that the surface effect on vibrational lifetime is sensitive to the polarity of the solute and that the solvation dynamics following the creation of a large dipole on the organic side of the interface between water and a nonpolar liquid can be used to probe slow surface density fluctuations. Such experiments can add significantly to our understanding of the structure and dynamics of the interface. Extending the computations to larger dye molecules with anticipated large nonlinear responses would be useful in the search for systems that could be studied experimentally.

The tendency of solute molecules to keep their hydration shell, the roughness of the liquid surface and its dynamic fluctuations are not only important for understanding solute spectroscopy, thermodynamic, and relaxation phenomena, but also, as we show below, they play a major role in elucidating the surface effect on chemical reactivity.

REACTIVITY AT LIQUID INTERFACES

Introduction

Theoretical[538–541] and experimental[221] studies of chemical reaction dynamics and thermodynamics in bulk liquids have demonstrated in recent years that one must take into account the molecular structure of the liquid to fully understand solvation and reactivity. The solvent is not to be viewed as simply a static medium but as playing an active role at the microscopic level. Our discussion thus far underscores the unique molecular character of the interface region: asymmetry in the intermolecular interactions, nonrandom molecular orientation, modifications in the hydrogen-bonding network, and other such structural features. We expect these unique molecular structure and dynamics to influence the rate and equilibrium of interfacial chemical reactions. One can also approach solvent effects on interfacial reactions at a continuum macroscopic level where the interface region is characterized by gradually changing properties such as density, viscosity, dielectric response, and other properties that are known to influence reactivity.

Computational studies of neat liquid surfaces are becoming a mature area of study, but investigation of chemical reaction thermodynamics and dynamics is much more limited. This is due, in part, to the scarcity of molecular-level experimental data. While some computational work focused on reactions that were also studied experimentally, most of the published computational work relied on simple model reactions to address these two important general questions:

(a) How does the interface region affect the rate and the equilibrium of different types of reactions?
(b) Can solvent effects on reactions at interfaces be understood by simply scaling bulk effects, or should it be treated in a unique manner?

Methods used to study reactivity in bulk liquids are relatively well developed and generally can be used without major modification to study reactions at interfaces. The computational approach typically involves these steps:[538–541]

1. *Define a reaction coordinate X(r), which in general is a function of some (or even many) of the atomic positions in the systems.* Examples include a torsional angle in a molecule for a conformational transition, a bond distance for a simple dissociation reaction, or, a function of many solvent atomic positions in the case of an electron transfer reaction. Keep in mind that in many cases the simple classical force fields described earlier in this chapter are inadequate for describing the proper dynamics of the system along the coordinate X. A quantum description at some level is likely needed

to account for the change in the electronic structure that is typically involved in a reaction.

2. Perform *equilibrium free energy calculations to determine the free energy profile A(x) along the coordinate X*. This follows the methodology discussed earlier. The direct sampling of X, from which $A(x) = -\beta^{-1} \ln\langle \delta[X(\mathbf{r}) - x] \rangle$ can be determined, is rarely successful because a large free energy barrier can prevent efficient (or any!) sampling of phase space. Umbrella sampling with a biasing potential is therefore a standard "trick" to use. Knowledge of the free energy profile provides the activation free energy and thus an estimate of the rate constant from transition state theory (TST). The free energy difference between the reactants and products with proper accounts for the thermodynamic standard state allows for a computation of the equilibrium constant.

3. Perform *nonequilibrium trajectory calculations to explore possible dynamical contributions to the rate, energy flow, and mechanism*. The actual chemical reaction rate constant differs from the TST value because not every trajectory reaching the transition state will end up as products; the transmission coefficient gives the fraction of successful trajectories. Methods to calculate it[538−541] can be used with no modification at liquid interfaces, an example of which will be discussed below. Besides determining the transmission coefficient, reactive trajectories, connecting reactants, and products provide information about the reaction mechanism and energy flow in the system.[541−543]

Computational studies of chemical reactions dynamics at liquid/vapor and liquid/liquid interfaces to date include the following types of reactions: isomerization, photodissociation, acid dissociation, electron transfer, proton transfer, ion transfer, and nucleophilic substitution. These studies have been motivated by experimental observations and fundamental scientific interest in understanding how the unique surface properties affect chemical reactivity. Some of these studies have been reviewed.[544,545] Here we present two examples selected to demonstrate the computational steps described above and their relation to the concepts developed in earlier sections. The focus is on contrasting the surface reactivity with that in the bulk and on examining surface effects in light of the knowledge about the structure and dynamics of neat interface and interfacial solvation, discussed earlier in the chapter.

Electron Transfer Reactions at Liquid/Liquid Interfaces

Electron transfer (ET) at the interface between two immiscible electrolyte solutions (IES) is important in electrochemistry,[12,48,546,547] solar energy conversion, "artificial photosynthesis,"[548,549] phase transfer catalysis, and is relevant to biological processes at membrane interfaces and in DNA environments so it is especially important to understand.[550] The experimental study of ET reactions at IES interfaces has a long history.[12] Until recently, measurements of those electron transfer rates involved mainly conventional electrochemical methods, where the interface is under external potential control and a steady state current versus voltage is measured.[12,551] These measurements suffer from drawbacks, including an inability to distinguish clearly between electron and ion transfer, a narrow potential window, distortion due to the charging current, and the large resistivity of the organic phase. These drawbacks limit the number of experimental systems that can be studied, and thus very few reliable rate constants have been reported at interfaces.

There has been a recent surge in experimental activity due to the availability of new methods in which those drawbacks can be minimized or controlled. These methods include scanning electrochemical microscopy (SECM),[552–560] thin-layer cyclic voltammetry,[561,562] and spectroelectrochemical methods (some taking advantage of recent advances in nonlinear optics).[563–578] In the SECM technique, one has the ability to widen the potential window and that increases significantly the range of the driving force (ΔG_{rxn}) for the ET reactions studied at the liquid/liquid interface. Because of that it has enabled the observation of the Marcus-inverted region (see below) and provided a reliable determination of the reorganization free energy.[560,579–581]

Spectroelectrochemical methods have been used in recent years to study fast-photoinduced electron transfer at the liquid/liquid interface.[566–570,582] Of particular importance is extending the idea of employing solvent (typically N,N-dimethylaniline or DMA) as an electron donor to the liquid/liquid interface.[571,583–585] The advantage of this approach is that complications due to ion transfer across the interface and to diffusion are obviated. Several studies of ET between coumarin dyes and electron-donating solvents in micelles, reverse micelles, at the surface of proteins, and in nanocavities have demonstrated ultrafast electron transfer that is faster than solvation due to the close proximity of the redox pair. These experiments provided additional evidence for the existence of the Marcus-inverted region at liquid interfacial systems.[572–578,586]

The basic theory of ET in bulk liquid and at liquid/metal interfaces is well-developed[587–590] but applications of that theory to ET at IES were slow to be adopted due to insufficient knowledge about the molecular structure of the liquid/liquid interface and due to experimental difficulties. ET rate constants under steady state conditions are typically obtained from current/voltage measurements. A potential difference V is established across the interface, and the current I is measured. If the basic theory of electron transfer at the solution/metal interface[591] is applicable to the liquid/liquid interface, one then expects the following Butler–Volmer relation between the voltage and the current:

$$I = I_0 \left(e^{(1-\alpha)nF(V-V_{eq})/RT} - e^{\alpha nF(V-V_{eq})/RT} \right) \qquad [55]$$

where T is the temperature, R is the gas constant, F is the Faraday constant, n is the number of electrons in the balanced half-reaction at the anode or the cathode, V_{eq} is determined from the Nernst equation ($V_{eq} = V_0 + \frac{RT}{nF} \ln \frac{a_O^s}{a_R^s}$, where V_0 is the standard potential and a_O^s, a_R^s are the activities of the oxidized and reduced agents at the interface), α is a constant called the transfer coefficient (see below), and I_0 is the so-called exchange current. The exchange current is the current that flows at equilibrium ($V = V_{eq}$) at the cathode (or the anode, they must be equal in magnitude and opposite in sign), and it is directly related to the heterogeneous rate constant k: $I_0 = nF[O]^{1-\alpha}[R]^\alpha k$, where [O] and [R] are the equilibrium concentrations. The key point is that the Butler–Volmer relation can be derived from the Marcus theory of electron transfer. This derivation[587] shows that the transfer coefficient is given by $\alpha = 1/2 + \Delta G/2\lambda$, where ΔG is the reaction free energy, and λ is the reorganization free energy (see below).

Current voltage measurements of ET at IES sometimes conform to the Butler–Volmer equation and sometimes not.[561,562] This is not surprising because some of the assumptions on which this equation is based may fail at the liquid/liquid interface. These assumptions include: (1) The potential drop across the interface is close to that imposed on the electrodes, or if not, a correction is included to properly account for the potential carried by the diffuse layers of ions at the interface; (2) The current is due to ET alone, and if not, a correction due to ion transfer must be included; (3) Marcus theory is valid.

The basic assumption of Marcus theory is that the solvent free energy functions controlling ET are paraboli with equal curvatures (linear response). On the basis of this, making an assumption about the structure of the interface, and using a continuum electrostatic expression for the activation free energy,[546,592,593] Marcus derived expressions for the bimolecular rate constant of an ET at IES.[594,595] To date, the computational work in this area has focused on testing the assumptions underlying the above theory. We briefly summarize Marcus theory to present the computational work.

Marcus Theory of ET at IES Interfaces
We consider the ET reaction DA → D$^+$A$^-$ between an electron donor (D) and an electron acceptor (A) that are adsorbed at the interface between two immiscible liquids. For simplicity, we ignore the contribution of the vibrational modes of D and A because the contribution of these modes to the activation free energy is not expected to be modified significantly by the interface. At a given distance R between the reactants, the probability of an electron transfer depends on the overlap of the reactants' wave functions (electronic coupling) and on the probability that a solvent fluctuation will equalize the energy of the two diabatic states $|DA\rangle$ and $|D^+A^-\rangle$. For weak coupling, the rate constant for the electron transfer reaction is given by:[593–595]

$$k_r = \kappa \nu V_r e^{-\beta \Delta G^\#}$$ [56]

where κ is the Landau–Zener factor between the two diabatic electronic states,[596] and ν is a "collision" frequency, which is determined from the equilibrium solvent fluctuations in the reactant state. V_r is the reaction volume, which accounts for all of the possible configurations of the reactant pair per unit area of the interface. For example, if one assumes that the two reactants are spherical (radii a_1 and a_2) and are restricted to being on opposite sides of a flat interface, then: $V_r = 2\pi(a_1 + a_2)(\delta R)^3$ and δR is a length scale for the distance-dependent electronic coupling between the two diabatic states.[593]

The activation energy $\Delta G^{\#}$ is given by[593–595]

$$\Delta G^{\#} = W_r + \frac{(\lambda + \Delta G + W_p - W_r)^2}{4\lambda} \qquad [57]$$

where ΔG is the reaction free energy, W_r is the reversible work required to bring the reactants from the bulk of each phase to the interface, $-W_p$ is the reversible work required to separate the products, and λ is the reorganization free energy. ΔG is controlled externally by varying the voltage across the interface. W_r and W_p may be estimated from experimental adsorption isotherms, or they can be calculated using continuum electrostatic models or by the MD or MC methods described earlier.

The key quantity, the reorganization free energy λ, is defined as the reversible work needed to change the equilibrium solvent configuration around the reactants to be that around the products at a fixed electronic state. It was first calculated analytically by Kharkats[597] for the case of two reactants located at various positions along the line normal to the interface (modeled as a mathematically ideal plane separating the two bulk phases). Marcus generalized the calculations to include any orientation of the reactants relative to the interface normal.[593] Kharkats and Benjamin[598] investigated the case where the reactants may cross the interface, accounting for a mixed solvation shell at the interface. They show that the reorganization free energy is affected markedly by the possibility of the ions crossing the interface, and that this will have a significant effect on the rate.

To test this theory, Marcus estimated λ by using the rate constant of the half reaction at the solution/metal interface. This estimate and other assumptions gave reasonable agreement with the experimental rate constant for the reaction between the $Fe(CN)_6^{4-/3-}$ couple in water and for the $Lu(PC)_2^{+/2+}$ (hexacyanoferrate-lutetium biphthalocyanine) couple in DCE.[593,594]

Microscopic Models

Marcus's model assumes the validity of a linear response approximation and that a continuum electrostatic description of the interface is suitable for the purpose of calculating the activation free energy. Furthermore, to obtain expressions for the rate constant, the interface is assumed to be either a mathematically sharp plane or a broad homogeneous phase. Unfortunately, an insufficient number of experimental data exist to test these assumptions. Thus, the main focus of using atomistic approaches to interfacial ET has been to gain insight into how important the molecular structure of the interface actually is for influencing the ET rate.

Key to the microscopic description of ET used in several MD simulations is a definition of the reaction coordinate. Because solvent fluctuations control the probability of electron transfer, this coordinate must be a function of solvent configuration. A useful one-dimensional definition, used extensively in simulations, is the energy gap when the system is in either of the two electronic states:[599–602]

$$X(\mathbf{r}) = U_P(\mathbf{r}) - U_R(\mathbf{r}) \qquad [58]$$

where U_R and U_P are the potential energies of the reactant state $\psi_R = |DA\rangle$ and the product state $\psi_P = |D^+A^-\rangle$, respectively, and $\mathbf{r}$ represents the positions of all the atoms. It can be shown that this coordinate is equivalent (in some sense) to the solvent polarization coordinate used by Marcus to derive the continuum electrostatic expressions for the reorganization free energy.[603,604] The probability $P(x)$ that $X(\mathbf{r})$ is equal to some value x is:

$$P_v(x) = \langle \delta[X(\mathbf{r}) - x] \rangle_v \qquad [59]$$

where the ensemble average is over the reactant $(v = R)$ or the product $(v = P)$ state. The solvent free energy associated with solvent fluctuations in the state v is given by

$$G_v(x) = -\beta^{-1} \ln P_v(x) \qquad [60]$$

When the system is in the state v, most fluctuations in the solvent coordinate are near the vicinity of its equilibrium value $x_v = \langle X(\mathbf{r}) \rangle_v$, so direct sampling of Eq. [59] will only provide the free energy for small fluctuations around the equilibrium. However, the reorganization free energy is the difference:

$$\lambda_R = G_R(x_P) - G_R(x_R) = -\beta^{-1} \ln \frac{P_R(x_P)}{P_R(x_R)} \qquad [61]$$

with an equivalent expression for the product state. Because x_R and x_P are typically very different, an umbrella sampling procedure is required. One approach developed by King and Warshel[600] is to simply consider a set of intermediate virtual electronic states, $v_m, m = 1, 2, \ldots$, and calculate the free energy function for each individual state. These functions are then "stitched" together to reconstruct the desired full functions G_R and G_P. If the different virtual electronic states correspond to a transfer of a fixed fraction of an electron, this procedure is equivalent to adding a biasing potential that is linear in the coordinate x.[600,605]

At the transition state, $x = x^{\#}$, the energy of the two states is the same. This nonequilibrium state can be obtained from the intersection $G_R(x^{\#}) = G_P(x^{\#})$.

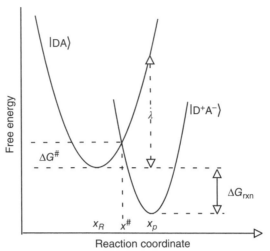

Figure 13   A schematic representation of the free energy functions involved in the thermally activated electron transfer reaction $DA \rightarrow D^+A^-$. See text for details.

Then, the activation free energy for the forward ET reaction is $\Delta G^{\#} = G_R\,(x^{\#}) - G_R\,(x_R)$. Figure 13 depicts the quantities discussed above.

Equation [57] for the activation free energy is based on the assumption that the solvent free energies $G_R(x)$ and $G_P(x)$ are paraboli with identical curvatures k:

$$G_R(x) = \frac{1}{2}k(x - x_R)^2; \; G_P(x) = \frac{1}{2}k(x - x_P)^2 + \Delta G \qquad [62]$$

where ΔG is the reaction free energy. The parabolic assumption can be checked by an MD umbrella calculation of the full solvent free energy. This was done for ET reactions in bulk water,[600,601] at the interface between two simple liquids,[606] and at the water/self-assembled monolayer interface.[607] Note that the solvent force constants k_v can be calculated directly from the fluctuations in the solvent coordinate in the two equilibrium states,[602] $k_v = 1/\beta \langle (x - x_v)^2 \rangle_v$. If $k_R = k_P$, the assumption of a parabolic free energy can then be used to compute the activation free energy on the basis of simulations in the initial and final states alone. If the force constants are different, it is still possible to find the activation free energy approximately by an extrapolation procedure developed by Voth and coworkers.[96,608]

The solvent free energies for an ET reaction between two charge transfer centers adsorbed at the water/1,2-dichloroethane interface were investigated by MD simulations.[609] The charge centers were modeled as Lennard-Jones spheres with the parameters $\sigma = 5\,\text{Å}$ and $\epsilon = 0.1\,\text{kcal/mol}$. In bulk water, the free energy curves calculated from the molecular dynamics simulations are approximately well described by paraboli. While the curvature of the free energy function for the ion pair state D^+A^- (the constant k in Eq. [62]) is smaller than that for the neutral pair (DA state), the activation free energy predicted by the molecular dynamics results is almost identical to the one predicted by the quadratic approximation for the case of zero reaction free energy ($\Delta G = 0$), a region where most ET rate measurements are carried out. At the water/1,2-dichloroethane interface, the deviation between the molecular dynamics calculations and the linear response theory is even smaller. The reorganization free energy calculated from the continuum dielectric model[593] using the MD-derived static dielectric constants of water ($\varepsilon_0 = 82.5$) and DCE ($\varepsilon_0 = 10$) and other geometrical parameters is $\lambda = 74$ kcal/mol compared to the MD value of 80 kcal/mol.[609] This agreement, however, is due to an error cancellation: the water contribution to the electrostatic potential at the location of the charge transfer centers is underestimated by the continuum model (due to the neglect of the specific hydration structure), while the DCE potential is overestimated.[532,610]

Other assumptions made in deriving the ET rate constant, such as the interface being flat and the inability of the ions to penetrate the interface, have also been investigated by molecular dynamics simulations. As discussed earlier in this chapter, the roughness of the interface and the ability of ions to cross the interface and be partially solvated by both liquids are features of the real system that simple continuum models cannot account for.

Finally, we note that fast electron transfer reactions in a polar environment may be strongly controlled by the rate at which solvent dipoles are able to reorient in response to the electron transfer. Thus solvent dynamics at the interface may be tightly coupled with the electron transfer. This is the case for photoinduced electron transfer reactions in which one of the reactants is photoexcited prior to the ET act itself. For example, Eisenthal and coworkers used SHG[571] (and more recently SFG[611]) to study the ultrafast excited state electron transfer at a water/organic liquid interface. In the SHG experiment, a 424 nm photon excites C314 adsorbed at the water/dimethylaniline (DMA) interface. The electron transfer from an interfacial DMA molecule to the excited state coumarin (C314*) was explored by measuring the resonant SHG signal produced from the C314 by a probe pulse. The fast signal change was attributed to the solvation dynamics of C314* on the subpicosecond timescale, followed by ET on a 14–16 ps timescale.

Nucleophilic Substitution Reactions and Phase Transfer Catalysis (PTC)

Many chemical reactions that take place at the interface between two immiscible liquids are coupled to reactant/product transfer processes, both to and from the bulk phases and across the interface. Sometimes, depending on the reaction, these transfer processes complicate the theoretical and the experimental analysis, as is the case for coupled electron and ion transfer mentioned in the previous section. However, in other cases this coupling is the basis of the process being studied and cannot be avoided, as in Liquid–Liquid Phase Transfer Catalysis[5,6,612–614] (LLPTC). This process is used widely in organic synthesis,[615] in pharmaceutical and agrochemical industries,[5] in materials science,[616] and in "green" chemistry applications.[614] In LLPTC, a water-soluble reactant is transferred, with the aid of a phase transfer catalyst, from an aqueous phase into an organic phase, where it reacts with a water-insoluble reactant. Once complete, the catalyst, which is typically a quaternary ammonium cation, transfers the product to the aqueous phase, and the catalytic cycle repeats.

One of the most common reactions carried out under PTC conditions is nucleophilic substitution (S_N2). It is well known that the strong reactivity of anionic nucleophiles in gas phase S_N2 reactions is reduced markedly when the reaction is carried out in a polar protic solvent such as water.[327,617–627] The condensed phase reactivity can be enhanced if the reaction is done in a low-polarity aprotic solvent, like chloroform, which can lower the barrier compared with that in bulk water significantly. A PT catalyst can transfer small nucleophiles such as F^- and Cl^- into the low-polarity solvents and enable the reaction.

Our earlier discussion on solute behavior at the liquid/liquid interface raises several fundamental questions about LLPTC. For example, how important is it to consider the hydration state of the nucleophilic ions for reactions carried out under PTC conditions? We already demonstrated that the transfer of small hydrophilic anions from water to an organic solvent is accompanied by a few water molecules.[6] Experimental studies in bulk nonpolar solvents suggest that the hydration state of the anion strongly influences its nucleophilicity.[628–630] Much theoretical support for this has also been provided from studies of S_N2 reactions in gas phase clusters[631–633] and in bulk liquids.[626,634] Another question revolves around the finding about the strong dependence of solvent polarity on the surface location and orientation. How would this affect the reactivity of an S_N2 reaction given its strong dependence on polarity?

To address these questions, molecular dynamics simulations were carried out for a simple benchmark symmetric S_N2 reaction, $Cl^- + CH_3Cl \rightarrow CH_3Cl + Cl^-$, at different locations of the water/chloroform interface.[635] The reaction was modeled using the Empirical Valence Bond (EVB) approach.[618,619,622,636] Briefly, EVB assumes the electronic state of the reactive system can be described using two orthonormal valence states, $\psi_1 = Cl:^- CH_3 - Cl$ and $\psi_2 = Cl - CH_3 Cl:^-$

$$\Psi = c_1\psi_1 + c_2\psi_2, \quad \langle \psi_i | \psi_j \rangle = \delta_{ij} \qquad [63]$$

The total Hamiltonian is written as:

$$\hat{H} = \begin{pmatrix} H_{11}\left(\mathbf{r}_i, \mathbf{r}_d, \mathbf{r}_s\right) & H_{12}(r_1, r_2, \theta) \\ H_{12}(r_1, r_2, \theta) & H_{22}(\mathbf{r}_i, \mathbf{r}_d, \mathbf{r}_s) \end{pmatrix} \qquad [64]$$

$$H_{11} = E_k + H_{11}^0(r_1, r_2, \theta) + U_{ss}(\mathbf{r}_s) + U_{si}(\mathbf{r}_s, \mathbf{r}_i) + U_{sd}(\mathbf{r}_s, \mathbf{r}_d) \qquad [65]$$

where H_{11} is the diabatic Hamiltonian describing the system in the state ψ_1. It includes E_k – the kinetic energy of all atoms; $H_{11}^0(r_1, r_2, \theta)$ – the gas phase interaction between the Cl^- ion and the CH_3Cl molecule; $U_{ss}(\mathbf{r}_s)$ – the individual solvents and the solvent–solvent potential energies; $U_{si}(\mathbf{r}_s, \mathbf{r}_i)$ – the solvent–ion potential energy; and $U_{sd}(\mathbf{r}_s, \mathbf{r}_d)$ – the solvent–CH_3Cl potential energy. In Eqs. [64] and [65], $\mathbf{r}_i$ is the vector position of the Cl^- ion, $\mathbf{r}_d$ is the vector position of the CH_3Cl atoms, and $\mathbf{r}_s$ stands for the positions of all the solvent atoms. r_1 is the distance between the Cl^- ion and the carbon atom, r_2 is the C–Cl bond distance in CH_3Cl, and θ is the Cl^- – C–Cl angle. H_{22} is the diabatic Hamiltonian describing the system in the state ψ_2, and due to the symmetry of the reaction, H_{22} has the same functional form as H_{11} but with the two chlorine atom labels interchanged.

The detailed functional forms and parameter values of all the potential energy terms in Eq. [65] can be found elsewhere.[636] Here we note that the gas phase potential energy, $H_{11}^0(r_1, r_2, \theta)$, is a generalization to noncollinear geometries of the form suggested by Mathis et al.[622] It includes a Morse potential for the CH_3Cl bond, an exponential repulsive term for the interaction between the Cl^- ion and the CH_3 radical, and an ion-dipole term for combined short-range repulsion and long-range attractive interactions between the Cl^- ion and the CH_3Cl bond. These terms are obtained from a fit to the *ab initio* calculations of Tucker and Truhlar[637] and to experimental data.[622] The generalization to nonlinear geometry is accomplished by making some of the parameters θ-dependent and adding a bending energy term with parameters determined by a best fit to the gas phase *ab initio* values of the energy, to the location of the transition state, and to the ion-dipole depth as a function of θ. $U_{ss}(\mathbf{r}_s)$, $U_{si}(\mathbf{r}_s, \mathbf{r}_i)$, and $U_{sd}(\mathbf{r}_s, \mathbf{r}_d)$ are all given by the sum of Lennard-Jones plus Coulomb interactions between every two sites on different molecules. The solvents' Lennard-Jones parameters, the intramolecular potential terms, and the corresponding intramolecular parameters can be found elsewhere.[638] The water model was selected to be the model used previously to study the bulk and interfacial properties of water. The water and chloroform potentials used gives rise to a stable liquid/liquid interface with a surface tension of 25±3 dynes/cm, in reasonable agreement with the experimental value of 26.6 dynes/cm. The off-diagonal electronic coupling term H_{12} in Eq. [64] is the one suggested by Hynes and coworkers:[622,639]

$$H_{12} = -QS(r_1)S(r_2) \qquad [66]$$

where $S(r)$ is the overlap integral for the sigma orbital formed from the carbon 2p and chlorine 3p atomic orbitals. $S(r)$ is determined using Slater-type orbitals and the approximation of Mulliken et al.,[636,640] and $Q = 678.0$ kcal/mol is a parameter that is fitted to obtain the correct gas phase activation energy.

The diagonalization of Eq. [64] yields the electronic ground state adiabatic Hamiltonian as a function of all nuclear coordinates:

$$H_{ad} = \frac{1}{2}(H_{11} + H_{22}) - \frac{1}{2}\left[(H_{11} - H_{22})^2 + 4H_{12}^2\right]^{1/2} \qquad [67]$$

The classical trajectory calculations are carried out using this Hamiltonian. The reaction coordinate is defined by

$$\xi = r_1 - r_2 \qquad [68]$$

so the reactants and products' states correspond to $\xi \ll 0$ and $\xi \gg 0$, respectively. The minimum energy path along ξ for the collinear geometry is shown in Figure 14. The total wavefunction (the values for c_1 and c_2 in Eq. [63]) shows that at the transition state in a vacuum (and on average in solution) each Cl atom in the reaction system carries a partial charge of $\delta \approx 0.5$: [$Cl^{-\delta} - CH_3 - Cl^{-\delta}$]. As ξ varies from $-\infty$ to $+\infty$, the charge on the nucleophile varies from -1 to the charge on the Cl atom in the isolated CH_3Cl molecule.

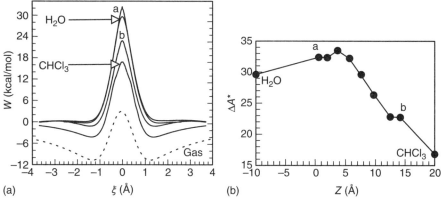

Figure 14 (a) The reaction free energy profile for the $Cl^- + CH_3Cl$ reaction at different locations of the water/chloroform interface. The lines labeled H_2O and $CHCl_3$ depict the profiles in the corresponding bulk solvents. The dotted line is the minimum energy path along the collinear geometry in the gas phase. (b) The activation free energy of this reaction versus all the locations studied along the interface normal. (Reprinted with permission from Ref. 479. Copyright 2010, American Chemical Society.)

The reaction free energy profile as a function of the reaction coordinate was calculated using umbrella sampling with overlapping windows and a biasing potential,[212] as discussed earlier. The specific expressions used are:

$$W(\xi) = -\beta^{-1} \ln P(\xi) - U_b(\xi) \qquad [69]$$

$$P(\xi) = \frac{\int \delta(r_1 - r_2 - \xi) \exp[-\beta(H_{ad} + U_b(\xi)]d\Gamma}{\int \exp[-\beta(H_{ad} + U_b(\xi)]d\Gamma} \qquad [70]$$

A good choice for the biasing potential $U_b(\xi)$ (an analytic function of ξ chosen to approximate $- W(\xi)$ to accelerate the convergence of the ensemble average in Eq. [70]) was a Gaussian fit of the transition-state region of the free energy profile in bulk chloroform.[636] The $W(\xi)$ were calculated at seven different interface locations by restricting the reactants' center of mass to slabs parallel to the liquid/liquid interface. Other details about the calculations are given in Ref. 635.

The results are shown in Figure 14. The top panel includes the free energy profiles at two interface locations, the gas phase potential energy along the minimum energy path for the collinear reaction geometry (dotted line) and the free energy profiles in bulk chloroform and in bulk water. The bottom panel shows the activation free energy barrier ΔA^* for all the locations. As the polarity of the bulk medium increases when going from the gas phase to bulk chloroform and bulk water, there is a significant increase in the activation free energy because the reactants and products experience a much greater lowering of their free energy than does the transition state. This, in turn, is because the separate reactants (or products) having the localized charge distribution on the chloride ion are much more favorably solvated than the delocalized charges of the transition state. Because the polarity of the interface region is expected to be somewhere between that of the two bulk phases,[356,381] one would expect the activation free energy of the reaction at the different interface locations to fall in between the values in bulk water and in bulk chloroform. The bottom panel of Figure 14 shows the unexpected result that as the reactants' center of mass first moves past the Gibbs surface ($Z = 0$) toward the organic phase, the activation free energy becomes *greater* than in bulk water. As the reactants are moved deeper into the organic phase ($Z > 10\,\text{Å}$), the activation free energy begins to decrease. While the barrier is lower than that in bulk water, at the largest Z studied, it is still higher than in bulk chloroform. This suggests that *for the reaction rate to be significantly larger than in bulk water, the reaction must take place deep in the organic phase.*

The fundamental reason for the relatively high barrier at or near the interface is the ability of the nucleophile to retain some number of water molecules when it is in the vicinity of the interface and to keep the reactants hydrated. The resulting deformation of the interface is coupled strongly to the solute charge distribution and thus to the reaction coordinate. The behavior of the reaction far from the Gibbs surface is similar to that of the reaction $Cl^-(H_2O)_n + CH_3Cl$ in bulk chloroform as a function of n. For example, at $Z = 15$ Å, $\Delta A^* = 22$ kcal/mol, which is similar to $\Delta A^* = 21$ kcal/mol calculated for the reaction in bulk chloroform with $n = 1$. As n increases, the activation free energy increases monotonically to the value in bulk water.[641] Thus, the higher barrier of the reaction proximal to the Gibbs surface arises from a unique surface effect that can be traced to the orientational dependence of the reaction barrier.[635] When the system is at or near the transition state ($\xi = 0$), the $Cl^{-0.5} - CH_3 - Cl^{-0.5}$ vector tends to lie parallel to the interface, but when the charge on the nucleophile is more or less fully developed (at $\xi \geq 0.3$ Å), the vector $Cl^- - CH_3 - Cl$ tends to lie perpendicular to the interface, with the Cl^- pointing toward the water phase. Thus, the transition state experiences an environment that is significantly less polar than the reactants, explaining the high barrier in the $Z < 5$ Å region.

The activation free energy ΔA^* can be used to compute the TST approximation of the rate constant $k_{TST} = Ce^{-\beta \Delta A^*}$, where C is the preexponential factor. Because not every trajectory that reaches the transition state ends up as products, the actual rate is reduced by a factor κ (the transmission coefficient) as described earlier. The transmission coefficient can be calculated using the reactive flux correlation function method.[540,541,642] Starting from an equilibrated ensemble of the solute molecules constrained to the transition state ($\xi = 0$), random velocities in the direction of the reaction coordinate are assigned from a flux-weighted Maxwell–Boltzmann distribution, and the constraint is released. The value of the reaction coordinate is followed dynamically until the solvent-induced recrossings of the transition state cease (in less than 0.1 ps). The normalized flux correlation function can be calculated using[642]

$$\kappa(t) = N_+^{-1} \sum_{i=1}^{N_+} \theta[\xi_i^+(t)] - N_-^{-1} \sum_{i=1}^{N_-} \theta[\xi_i^-(t)] \qquad [71]$$

where $\xi_i^\pm$ is the value of the reaction coordinate for the ith trajectory at time t, given that at $t = 0$, $d\xi/dt$ is positive (negative), N_+ (N_-) is the corresponding number of trajectories, and θ is the unit step function. $\kappa(t)$ converges to a fixed value when the recrossing of the transition state is complete. The total number of trajectories, $N = N_+ + N_-$, should be about 1000 for this procedure to converge.

The TST assumes no recrossings ($\kappa = 1$); it may fail due to interaction of the solute with the solvent molecules, causing the trajectory to recross the top of the barrier.[539,540] Specifically, the solvent molecules that are equilibrated to the transition state charge distribution are not in the proper orientation necessary to solvate the product charge distribution. This might give rise to an additional temporary solvent barrier preventing the system from proceeding directly to the product side and induce recrossings.[643] The results of the calculations[635] are consistent with the above picture. In general, however, deviation from TST values are not large, and the κ values fall in between the values in bulk water ($\kappa = 0.57$) and bulk chloroform ($\kappa = 0.76$) when the reaction takes place at the interface.

Given that the main role played by the phase transfer catalyst is to bring the hydrophilic reactant into contact with the hydrophobic reactant confined to the organic phase, an important question is: To what extent does the catalyst influence the reaction itself? Calculations were performed of the free energy profile of the benchmark $Cl^- + CH_3Cl$ S_N2 reaction at the water/chloroform interface in the presence of the phase transfer catalyst tetramethylammonium cation (TMA$^+$).[644] TMA$^+$ moderately *increases* the barrier height of this reaction when it is associated with the Cl^- nucleophile. This is especially evident when the nucleophile is hydrated by a few water molecules. This suggests that the most effective role of the phase transfer catalyst is to bring the nucleophile deep into the organic phase with a minimal number of associated water molecules, followed by dissociation of the ion pair in the bulk organic phase before reaction.

CONCLUSIONS

Computational and experimental studies reveal that the interface between two fluid phases is a few nm thick highly anisotropic region, characterized by rapidly varying density, dielectric response, and molecular structure. It is also dynamic. Molecular motions are anisotropic, and density fluctuations create instantaneous structures that are significant on the narrow scale of the interface. These characteristics result in a marked influence on the solvation of solute molecules and on the equilibrium and rate of chemical reactions that take place in this region. In some cases, the effect of the interface region can be understood by the direct application of theories developed for understanding solvation and reactivity in bulk liquids. For example, by introducing interfacial solvent friction (in analogy with the bulk concept), one can understand the solvent effect on the rate of molecular reorientation and isomerization reactions. In other cases, one must take into account the unique interface characteristics to explain surface effects. For example, surface roughness at the interface between two immiscible liquids can have a significant effect on solvation dynamics and activation free energy for S_N2 reactions.

Despite the significant computational and experimental progress made in recent years, much more research is needed at both fronts to gain a molecular-level understanding of structure, thermodynamics, and dynamics at liquid interfaces. In particular, molecular-level, time-resolved experimental studies of solvation, relaxation, and reactions at liquid interfaces are needed.

At the theoretical level, simulations are needed of systems with more realistic potentials and of quantities that can be directly compared with experimental observations (such as from electronic sum frequency), while pushing the limit of system size and timescale. The development of multiscale theoretical models combining simple continuum models with some microscopic structure of the interface would also be desirable. Progress in these areas will advance our fundamental understanding of liquid interfacial phenomena while providing support for technological development in the areas of energy storage, catalysis, and the environment.

ACKNOWLEDGMENTS

This work was supported by the National Science Foundation (most recently by grant CHE-0809164). I would like to thank my collaborators: Daniel Rose, Karl Schweighofer, David Michael, Ilya Chorny, John Vieceli Nicholas Winter, and Katherine Nelson.

REFERENCES

1. J. O. M. Bockris and A. Gonzalez-Martin, in *Spectroscopic and Diffraction Techniques in Interfacial Electrochemistry*, C. Gutierrez and C. Melendres (Eds.), Kluwer Academic Publishers, Dordrecht, 1990, p. 1, The Advancing Frontier in the Knowledge of the Structure of the Interphases.

2. A. J. Bard and L. R. Faulkner, *Electrochemical Methods: Fundamentals and Applications*, New York, Wiley, 1980.

3. M. Cox, M. Hidalgo, and M. Valiente, *Solvent Extraction for the 21st Century*, London, SCI, 2001.

4. H. Watarai, N. Teramae, and T. Sawada, *Interfacial Nanochemistry: Molecular Science and Engineering at Liquid-Liquid Interfaces*, Kluwer Academic/Plenum, New York, 2005.

5. C. M. Starks, C. L. Liotta, and M. Halpern, *Phase Transfer Catalysis*, New York, Chapman & Hall, 1994.

6. A. G. Volkov, *Interfacial Catalysis*, New York, Marcel Dekker, 2003.

7. K. Arai, M. Ohsawa, F. Kusu, and K. Takamura, *Bioelectrochemistry and Bioenergetics*, **31**, 65 (1993). Drug Ion Transfer Across an Oil–Water Interface and Pharmacological Activity.

8. R. B. Gennis, *Biomembranes*, Springer, New York, 1989.

9. R. W. Johnson and G. E. Gordon, *The Chemistry of Acid Rain: Sources and Atmospheric Processes*, ACS Symposium Series, Washington, 1987.

10. B. J. Finlayson-Pitts, *Chem. Rev.*, **103**, 4801 (2003). The Tropospheric Chemistry of Sea Salt: A Molecular-Level View of the Chemistry of NaCl and NaBr.

11. A. W. Adamson, *Physical Chemistry of Surfaces*, Fifth edition, Wiley, New York, 1990.

12. H. H. Girault and D. J. Schiffrin, in *Electroanalytical Chemistry*, A. J. Bard (Ed.), Dekker, New York, 1989, p. 1, Electrochemistry of Liquid-Liquid Interfaces.

13. Y. R. Shen, *The Principles of Nonlinear Optics*, New York, Wiley, 1984.

14. Y. R. Shen, *Annu. Rev. Phys. Chem.*, **40**, 327 (1989). Optical 2nd Harmonic-Generation at Interfaces.

15. Y. R. Shen, *Nature*, **337**, 519 (1989). Surface Properties Probed by 2nd Harmonic and Sum Frequency Generation.

16. K. B. Eisenthal, *Annu. Rev. Phys. Chem.*, **43**, 627 (1992). Equilibrium and Dynamic Processes at Interfaces by Second Harmonic and Sum Frequency Generation.

17. K. B. Eisenthal, *Chem. Rev.*, **96**, 1343 (1996). Liquid Interfaces by Second Harmonic and Sum-Frequency Spectroscopy.

18. G. L. Richmond, *Chem. Rev.*, **102**, 2693 (2002). Molecular Bonding and Interactions at Aqueous Surfaces as Probed by Vibrational Sum Frequency Spectroscopy.

19. S. G. Grubb, M. W. Kim, T. Raising, and Y. R. Shen, *Langmuir*, **4**, 452 (1988). Orientation of Molecular Monolayers at the Liquid-Liquid Interface as Studied by Optical Second Harmonic Generation.

20. R. M. Corn and D. A. Higgins, *Chem. Rev.*, **94**, 107 (1994). Optical Second Harmonic Generation as a Probe of Surface Chemistry.

21. P. F. Brevet and H. H. Girault, in *Liquid-Liquid Interfaces*, A. G. Volkov and D. W. Deamer (Eds.), CRC Press, Boca Raton, 1996, p. 103, Second Harmonic Generation at Liquid/Liquid Interfaces.

22. G. L. Richmond, *Annu. Rev. Phys. Chem.*, **52**, 257 (2001). Structure and Bonding of Molecules at Aqueous Surfaces.

23. W. H. Steel and R. A. Walker, *Nature*, **424**, 296 (2003). Measuring Dipolar Width across Liquid-Liquid Interfaces with 'Molecular Rulers'.

24. M. A. Mendez, R. Partovi-Nia, I. Hatay, B. Su, P. Y. Ge, A. Olaya, N. Younan, M. Hojeij, and H. H. Girault, *Phys. Chem. Chem. Phys.*, **12**, 15163 (2010). Molecular Electrocatalysis at Soft Interfaces.

25. S. Ong, X. Zhao, and K. B. Eisenthal, *Chem. Phys. Lett.*, **191**, 327 (1992). Polarization of Water Molecules at a Charged Interface: Second Harmonic Studies of the Silica/Water Interface.

26. P. R. Fischer, J. L. Daschbach, and G. L. Richmond, *Chem. Phys. Lett.*, **218**, 200 (1994). Surface Second Harmonic Studies of Si(111)/Electrolyte and Si(111)/SiO$_2$/Electrolyte Interfaces.

27. K. B. Eisenthal, *Chem. Rev.*, **106**, 1462 (2006). Second Harmonic Spectroscopy of Aqueous Nano- and Microparticle Interfaces.

28. F. M. Geiger, *Annu. Rev. Phys. Chem.*, **60**, 61 (2009). Second Harmonic Generation, Sum Frequency Generation, and $\chi^{(3)}$: Dissecting Environmental Interfaces with a Nonlinear Optical Swiss Army Knife.

29. M. B. Pomfret, J. C. Owrutsky, and R. A. Walker, *Ann. Rev. Anal. Chem.*, **3**, 151 (2010). In Situ Optical Studies of Solid-Oxide Fuel Cells.

30. A. Lewis, A. Khatchatouriants, M. Treinin, Z. P. Chen, G. Peleg, N. Friedman, O. Bouevitch, Z. Rothman, L. Loew, and M. Sheres, *Elegans. Chem. Phys.*, **245**, 133 (1999). Second-Harmonic Generation of Biological Interfaces: Probing the Membrane Protein Bacteriorhodopsin and Imaging Membrane Potential around GFP Molecules at Specific Sites in Neuronal Cells of C.

31. J. S. Salafsky, *Phys. Chem. Chem. Phys.*, **9**, 5704 (2007). Second-Harmonic Generation for Studying Structural Motion of Biological Molecules in Real Time and Space.

32. D. Beaglehole, in *Fluid Interfacial Phenomena*, C. A. Croxton (Ed.), Wiley, New York, 1986, p. 523, Experimental Studies of Liquid Interfaces.

33. P. S. Pershan, *Faraday Discuss. Chem. Soc.*, **89**, 231 (1990). Structure of Surfaces and Interfaces as Studied Using Synchrotron Radiation: Liquid Surfaces.

34. D. K. Schwartz, M. L. Schlossman, E. H. Kawamoto, J. G. Kellogg, P. S. Pershan, and B. M. Ocko, *Phys. Rev. A*, **41**, 5687 (1990). Thermal Diffuse X-Ray-Scattering of the Water-Vapor Interface.

35. K. Yasumoto, N. Hirota, and M. Terazima, *Phys. Rev. B*, **60**, 9100 (1999). Surface and Molecular Dynamics at Gas–liquid Interfaces Probed by Interface-Sensitive Forced Light Scattering in the Time Domain.

36. M. G. Munoz, M. Encinar, L. J. Bonales, F. Ortega, F. Monroy, and R. G. Rubio, *J. Phys. Chem. B*, **109**, 4694 (2005). Surface Light-Scattering at the Air-Liquid Interface: From Newtonian to Viscoelastic Polymer Solutions.

37. J. Als-Nielsen, *Physica*, **140A**, 376 (1986). Synchrotron X-Ray Studies of Liquid–Vapor Interfaces.

38. M. K. Sanyal, S. K. Sinha, K. G. Huang, and B. M. Ocko, *Phys. Rev. Lett.*, **66**, 628 (1991). X-Ray Scattering Study of Capillary-Wave Fluctuations at a Liquid Surface.

39. L. T. Lee, D. Langevin, and B. Farnoux, *Phys. Rev. Lett.*, **67**, 2678 (1991). Neutron Reflectivity of an Oil–Water Interface.

40. R. A. Cowley, in *Equilibrium Structure and Properties of Surfaces and Interfaces*, A. Gonis and G. M. Stocks (Eds.), Plenum, New York, 1992, p. 1, X-Ray Scattering from Surfaces and Interfaces.

41. P. S. Pershan, *Physica A*, **231**, 111 (1996). Liquid Surface Order: X-Ray Reflectivity.

42. D. M. Mitrinovic, Z. Zhang, S. M. Williams, Z. Huang, and M. L. Schlossman, *J. Phys. Chem. B*, **103**, 1779 (1999). X-Ray Reflectivity Study of the Water-Hexane Interface.

43. M. L. Schlossman, *Curr. Opin. Coll. Interf. Sci.*, **7**, 235 (2002). Liquid-Liquid Interfaces: Studied by X-Ray and Neutron Scattering.

44. G. Luo, S. Malkova, S. V. Pingali, D. G. Schultz, B. Lin, M. Meron, T. J. Graber, J. Gebhardt, P. Vanysek, and M. L. Schlossman, *Faraday Discuss*, **129**, 23 (2005). X-Ray Studies of the Interface between Two Polar Liquids: Neat and with Electrolytes.

45. M. E. Saecker, S. T. Govoni, D. V. Kowalski, M. E. King, and G. M. Nathanson, *Science*, **252**, 1421 (1991). Molecular Beam Scattering from Liquid Surfaces.

46. M. J. Wirth and J. D. Burbage, *J. Phys. Chem.*, **96**, 9022 (1992). Reorientation of Acridine Orange at Liquid Alkane Water Interfaces.

47. J. M. Kovaleski and M. J. Wirth, *J. Phys. Chem.*, **99**, 4091 (1995). Lateral Diffusion of Acridine Orange at Liquid Hydrocarbon/Water Interfaces.

48. A. J. Bard and M. V. Mirkin, *Scanning Electrochemical Microscope*, New York, Marcel Dekker, 2001.

49. F. O. Laforge, P. Sun, and M. V. Mirkin, *Adv. Chem. Phys.*, **139**, 177 (2008). Physicochemical Applications of Scanning Electrochemical Microscopy.

50. R. P. Sperline and H. Freiser, *Langmuir*, **6**, 344 (1990). Adsorption at the Liquid-Liquid Interface Analyzed by in Situ Infrared Attenuated Total Reflection Spectroscopy.

51. A. B. Horn and J. Sully, *J. Chem. Soc. Faraday Trans.*, **93**, 2741 (1997). Reaction and Diffusion in Heterogeneous Atmospheric Chemistry Studied by Attenuated Total Internal Reflection IR Spectroscopy.

52. K. R. Wilson, B. S. Rude, T. Catalano, R. D. Schaller, J. G. Tobin, D. T. Co, and R. J. Saykally, *J. Phys. Chem. B.*, **105**, 3346 (2001). X-Ray Spectroscopy of Liquid Water Microjets.

53. K. R. Wilson, R. D. Schaller, D. T. Co, R. J. Saykally, B. S. Rude, T. Catalano, and J. D. Bozek, *J. Chem. Phys.*, **117**, 7738 (2002). Surface Relaxation in Liquid Water and Methanol Studied by X-Ray Absorption Spectroscopy.

54. C. A. Croxton, *Statistical Mechanics of the Liquid Surface*, New York, Wiley, 1980.

55. J. S. Rowlinson and B. Widom, *Molecular Theory of Capillarity*, Oxford, Clarendon, 1982.

56. J. K. Percus and G. O. Williams, in *Fluid Interfacial Phenomena*, C. A. Croxton (Ed.), Wiley, New York, 1986, p. 1, The Intrinsic Interface.

57. D. Henderson, *Fundamentals of Inhomogeneous Fluids*, New York, Marcel Dekker, 1992.

58. A. Kovalenko and F. Hirata, in *Interfacial Nanochemistry: Molecular Science and Engineering at Liquid-Liquid Interfaces*, H. Watarai, N. Teramae, and T. Sawada (Eds.), Kluwer Academic/Plenum, New York, 2005, p. 97, A Molecular Theory of Solutions at Liquid Interfaces.

59. J. C. Shelley and D. R. Berard, in *Reviews in Computational Chemistry*, K. B. Lipkowitz and D. B. Boyd (Eds.), VCH, New York, 1998, Vol. **12**, p. 137, Computer Simulation of Water Physisorption at Metal-Water Interfaces.

60. M. Schoen and S. H. L. Klapp, in *Reviews in Computational Chemistry*, K. B. Lipkowitz and D. B. Boyd (Eds.), 2007, Vol. **24**, p. 1, Nanoconfined Fluids: Soft Matter between Two and Three Dimensions.

61. M. P. Allen and D. J. Tildesley, *Computer Simulation of Liquids*, Clarendon, Oxford, 1987.

62. A. R. Leach, *Molecular Modeling: Principles and Applications*, 2nd edition, Pearson, New York, 2001.

63. D. Frenkel and B. Smit, *Understanding Molecular Simulation: From Algorithms to Applications*, 2nd edition, Academic Press, San Diego, 2002.

64. D. C. Rapaport, *The Art of Molecular Dynamics Simulations*, 2nd edition, Cambridge University Press, Cambridge, 2004.

65. D. B. Boyd, in *Reviews in Computational Chemistry*, K. B. Lipkowitz and D. B. Boyd (Eds.), Wiley-VCH, New York, 1997, Vol. **11**, p. 373, Appendix: Compendium of Software and Internet Tools for Computational Chemistry.

66. S. Nosé, *J. Phys. Condens. Matter*, **2**, SA115 (1991). Constant-Temperature Molecular Dynamics.

67. U. Burkert and N. L. Allinger, *Molecular Mechanics*, Washington, ACS, 1982.

68. U. Dinur and A. T. Hagler, in *Reviews in Computational Chemistry*, K. B. Lipkowitz and D. B. Boyd (Eds.), Wiley-VCH, New York, 1991, Vol. **2**, p. 99, New Approaches to Empirical Force Fields.

69. M. Jalaie and K. B. Lipkowitz, in *Reviews in Computational Chemistry*, K. B. Lipkowitz and D. B. Boyd (Eds.), Wiley-VCH, New York, 2000, Vol. **14**, p. 441, Appendix: Published Force Field Parameters for Molecular Mechanics, Molecular Dynamics, and Monte Carlo Simulations.

70. W. D. Cornell, P. Cieplak, C. I. Bayly, I. R. Gould, K. M. Merz, D. M. Ferguson, D. C. Spellmeyer, T. Fox, J. W. Caldwell, and P. A. Kollman, *J. Am. Chem. Soc.*, **117**, 5179 (1996). A Second Generation Force Field for the Simulation of Proteins, Nucleic Acids, and Organic Molecules.

71. J.-P. Hansen and I. R. McDonald, *Theory of Simple Liquids*, 2nd edition, Academic, London, 1986, p. 179.

72. A. R. Vanbuuren, S. J. Marrink, and H. J. C. Berendsen, *J. Phys. Chem.*, **97**, 9206 (1993). A Molecular Dynamics Study of the Decane Water Interface.

73. M. Hayoun, M. Meyer, and M. Mareschal, in *Chemical Reactivity in Liquids*, G. Ciccotti and P. Turq (Eds.), Plenum, New York, 1987, p. 279, Molecular Dynamics Simulation of a Liquid-Liquid Interface.

74. A. W. Hill and I. Benjamin, *J. Phys. Chem. B*, **108**, 15443 (2004). Influence of Surface Tension on Adsorbate Molecular Rotation at Liquid/Liquid Interfaces.

75. R. J. Sadus, *Molecular Simulation of Fluids: Theory, Algorithms and Object-Orientation*, Elsevier, Amsterdam, 1999.

76. J. A. Barker, *Mol. Phys.*, **80**, 815 (1993). Surface Tension and Atomic Interactions in Simple Liquids.

77. P. A. Pieniazek, C. J. Tainter, and J. L. Skinner, *J. Am. Chem. Soc.*, **133**, 10360 (2011). Surface of Liquid Water: Three-Body Interactions and Vibrational Sum-Frequency Spectroscopy.

78. R. D. Levine, *Molecular Reaction Dynamics*, Cambridge University Press, Cambridge, UK, 2005.

79. B. J. Alder and E. L. Pollock, *Ann. Rev. Phys. Chem.*, **32**, 311 (1981). Simulation of Polar and Polarizable Fluids.

80. M. Sprik and M. L. Klein, *J. Chem. Phys.*, **89**, 7556 (1988). A Polarizable Model for Water Using Distributed Charge Sites.

81. Y. C. Chen, J. Lebowitz, and P. Nielaba, *J. Chem. Phys.*, **91**, 340 (1989). Line Shifts and Broadenings in Polarizable Liquids.

82. P. Ahlstrom, A. Wallqvist, S. Engstrom, and B. Jonsson, *Mol. Phys.*, **68**, 563 (1989). A Molecular Dynamics Study of Polarizable Water.

83. J. Caldwell, L. X. Dang, and P. A. Kollman, *J. Am. Chem. Soc.*, **112**, 9144 (1990). Implementation of Nonadditive Intermolecular Potentials by Use of Molecular-Dynamics - Development of a Water–Water Potential and Water Ion Cluster Interactions.

84. A. Wallqvist, *Chem. Phys. Lett.*, **165**, 437 (1990). Polarizable Water at a Hydrophobic Wall.

85. A. Wallqvist, *Chem. Phys.*, **148**, 439 (1990). Incorporating Intramolecular Degrees of Freedom in Simulations of Polarizable Liquid Water.

86. K. Motakabbir and M. Berkowitz, *Chem. Phys. Lett.*, **176**, 61 (1991). Liquid Vapor Interface of TIP4P Water: Comparison between a Polarizable and a Nonpolarizable Model.

87. L. X. Dang, J. E. Rice, J. Caldwell, and P. A. Kollman, *J. Am. Chem. Soc.*, **113**, 2481 (1991). Ion Solvation in Polarizable Water- Molecular Dynamics Simulations.

88. L. X. Dang, *J. Chem. Phys.*, **96**, 6970 (1992). Development of Nonadditive Intermolecular Potentials Using Molecular-Dynamics - Solvation of Li^+ and F^- Ions in Polarizable Water.

89. O. A. Karim, *J. Chem. Phys.*, **96**, 9237 (1992). Potential of Mean Force for an Aqueous Chloride Ion Pair - Simulation with a Polarizable Model.

90. G. Corongiu and E. Clementi, *J. Chem. Phys.*, **98**, 4984 (1993). Molecular-Dynamics Simulations with a Flexible and Polarizable Potential - Density of States for Liquid Water at Different Temperatures.

91. D. E. Smith and L. X. Dang, *J. Chem. Phys.*, **100**, 3757 (1994). Computer Simulations of NaCl Association in Polarizable Water.

92. J. S. Bader and B. J. Berne, *J. Chem. Phys.*, **104**, 1293 (1996). Solvation Energies and Electronic Spectra in Polar, Polarizable Media: Simulation Tests of Dielectric Continuum Theory.

93. T. M. Chang and L. X. Dang, *J. Chem. Phys.*, **104**, 6772 (1996). Molecular Dynamics Simulations of CCl_4-H_2O Liquid-Liquid Interface with Polarizable Potential Models.

94. C. Tsun-Mei, L. X. Dang, and K. A. Peterson, *J. Phys. Chem. B*, **101**, 3413 (1997). Computer Simulation of Chloroform with a Polarizable Potential Model.

95. P. Jungwirth and D. J. Tobias, *J. Phys. Chem. A*, **106**, 379 (2002). Chloride Anion on Aqueous Clusters, at the Air-Water Interface, and in Liquid Water: Solvent Effects on Cl^- Polarizability.

96. D. W. Small, D. V. Matyushov, and G. A. Voth, *J. Am. Chem. Soc.*, **125**, 7470 (2003). The Theory of Electron Transfer Reactions: What May Be Missing?

97. S. W. Rick and S. J. Stuart, in *Reviews in Computational Chemistry*, K. B. Lipkowitz and D. B. Boyd (Eds.), Wiley-VCH, Hoboken, NJ, 2002, Vol. **18**, p. 89, Potentials and Algorithms for Incorporating Polarizability in Computer Simulations.

98. G. Ciccotti and J. P. Ryckaert, *Computer Physics Reports*, **4**, 345 (1986). Molecular Dynamics Simulation of Rigid Molecules.

99. A. Wallqvist and R. D. Mountain, in *Reviews in Computational Chemistry*, K. B. Lipkowitz and D. B. Boyd (Eds.), Wiley-VCH, New York, 1999, Vol. **13**, p. 183, Molecular Models of Water: Derivation and Description.

100. A. Trokhymchuk and J. Alejandre, *J. Chem. Phys.*, **111**, 8510 (1999). Computer Simulations of Liquid/Vapor Interface in Lennard-Jones Fluids: Some Questions and Answers.

101. J. Janecek, *J. Phys. Chem. B*, **110**, 6264 (2006). Long Range Corrections in Inhomogeneous Simulations.

102. P. Grosfils and J. F. Lutsko, *J. Chem. Phys.*, **130** (2009). Dependence of the Liquid–Vapor Surface Tension on the Range of Interaction: A Test of the Law of Corresponding States.

103. P. Auffinger and D. L. Beveridge, *Chem. Phys. Lett.*, **234**, 413 (1995). A Simple Test for Evaluating the Truncation Effects in Simulations of Systems Involving Charged Groups.

104. C. Sagui and T. A. Darden, *Annu. Rev. Biophys. Biomol. Struct.*, **28**, 155 (1999). Molecular Dynamics Simulations of Biomolecules: Long-Range Electrostatic Effects.

105. P. Ewald, *Ann. Phys.*, **64**, 253 (1921). Die Berechnung Optischer Und Elektrostatischer Gitterpotentiale.

106. S. W. d. Leeuw, J. W. Perram, and E. R. Smith, *Proc. Roy. Soc. London. Series A, Mathematical and Physical Sciences*, **373**, 27 (1980). Simulation of Electrostatic Systems in Periodic Boundary Conditions.

107. T. Darden, D. York, and L. Pedersen, *J. Chem. Phys.*, **98**, 10089 (1993). Particle Mesh Ewald – an N Log(N) Method for Ewald Sums in Large Systems.

108. U. Essmann, L. Perera, M. L. Berkowitz, T. Darden, H. Lee, and L. G. Pedersen, *J. Chem. Phys.*, **103**, 8577 (1995). A Smooth Particle Mesh Ewald Method.

109. A. Y. Toukmaji and J. A. Board, *Comput. Phys. Commun.*, **95**, 73 (1996). Ewald Summation Techniques in Perspective: A Survey.

110. M. Deserno and C. Holm, *J. Chem. Phys.*, **109**, 7678 (1998). How to Mesh up Ewald Sums. I. A Theoretical and Numerical Comparison of Various Particle Mesh Routines.

111. M. Deserno and C. Holm, *J. Chem. Phys.*, **109**, 7694 (1998). How to Mesh up Ewald Sums. II. An Accurate Error Estimate for the Particle-Particle-Particle-Mesh Algorithm.

112. A. Brodka and A. Grzybowski, *J. Chem. Phys.*, **117**, 8208 (2002). Electrostatic Interactions in Computer Simulations of a Three-Dimensional System Periodic in Two Directions: Ewald-Type Summation.

113. I. C. Yeh and M. L. Berkowitz, *J. Chem. Phys.*, **111**, 3155 (1999). Ewald Summation for Systems with Slab Geometry.

114. M. Kawata and U. Nagashima, *Chem. Phys. Lett.*, **340**, 165 (2001). Particle Mesh Ewald Method for Three-Dimensional Systems with Two-Dimensional Periodicity.

115. D. Lindbo and A. K. Tornberg, *J. Chem. Phys.*, **136** (2012). Fast and Spectrally Accurate Ewald Summation for 2-Periodic Electrostatic Systems.

116. D. Zahn, B. Schilling, and S. M. Kast, *J. Phys. Chem. B*, **106**, 10725 (2002). Enhancement of the Wolf Damped Coulomb Potential: Static, Dynamic, and Dielectric Properties of Liquid Water from Molecular Simulation.

117. D. Wolf, P. Keblinski, S. R. Phillpot, and J. Eggebrecht, *J. Chem. Phys.*, **110**, 8254 (1999). Exact Method for the Simulation of Coulombic Systems by Spherically Truncated, Pairwise R^{-1} Summation.

118. S. Kale and J. Herzfeld, *J. Chem. Theory Comput.*, **7**, 3620 (2011). Pairwise Long-Range Compensation for Strongly Ionic Systems.

119. K. Z. Takahashi, T. Narumi, and K. Yasuoka, *J. Chem. Phys.*, **134** (2011). Cutoff Radius Effect of the Isotropic Periodic Sum and Wolf Method in Liquid–Vapor Interfaces of Water.

120. H. C. Andersen, *J. Chem. Phys.*, **72**, 2384 (1980). Molecular Dynamics Simulation at Constant Pressure and/or Temperature.

121. G. J. Martyna, D. J. Tobias, and M. L. Klein, *J. Chem. Phys.*, **101**, 4177 (1994). Constant-Pressure Molecular-Dynamics Algorithms.

122. Y. Zhang, S. E. Feller, B. R. Brooks, and R. W. Pastor, *J. Chem. Phys.*, **103**, 10252 (1995). Computer Simulation of Liquid/Liquid Interfaces. 1. Theory and Application to Octane/Water.

123. J. P. Valleau and G. M. Torrie, in *Statistical Mechanics, Part A: Equilibrium Techniques*, B. J. Berne (Ed.), Plenum, New York, 1977, A Guide to Monte Carlo for Statistical Mechanics. 2. Byways.

124. K. S. Liu, *J. Chem. Phys.*, **60**, 4226 (1974). Phase Separation of Lennard-Jones Systems - Film in Equilibrium with Vapor.

125. P. Linse, *J. Chem. Phys.*, **86**, 4177 (1987). Monte Carlo Simulation of Liquid-Liquid Benzene-Water Interface.

126. A. Z. Panagiotopoulos, *Mol. Phys.*, **61**, 813 (1987). Direct Determination of Phase Coexistence Properties of Fluids by Monte-Carlo Simulation in a New Ensemble.

127. L. X. Dang, T. M. Chang, and A. Z. Panagiotopoulos, *J. Chem. Phys.*, **117**, 3522 (2002). Gibbs Ensemble Monte Carlo Simulations of Coexistence Properties of a Polarizable Potential Model of Water.

128. D. M. Heyes and J. H. R. Clarke, *J. Chem. Soc.-Faraday Trans. II*, **75**, 1240 (1979). Molecular-Dynamics Model of the Vapor–Liquid Interface of Molten Potassium-Chloride.

129. B. Smit, *Phys. Rev. A.*, **37**, 3431 (1988). Molecular Dynamics Simulation of Amphiphilic Molecules at a Liquid-Liquid Interface.

130. M. Meyer, M. Mareschal, and M. Hayoun, *J. Chem. Phys.*, **89**, 1067 (1988). Computer Modeling of a Liquid-Liquid Interface.

131. H. Watanabe, N. Ito, and C.-K. Hu, *J. Chem. Phys.*, **136**, 204102 (2012). Phase Diagram and Universality of the Lennard-Jones Gas–Liquid System.

132. R. S. Taylor, L. X. Dang, and B. C. Garrett, *J. Phys. Chem. B.*, **100**, 11720 (1996). Molecular Dynamics Simulations of the Liquid/Vapor Interface of SPC/E Water.

133. L. B. Pártay, G. Hantal, P. Jedlovszky, A. Vincze, and G. Horvai, *J. Comput. Chem.*, **29**, 945 (2008). A New Method for Determining the Interfacial Molecules and Characterizing the Surface Roughness in Computer Simulations. Application to the Liquid–Vapor Interface of Water.

134. F. Biscay, A. Ghoufi, V. Lachet, and P. Malfreyt, *J. Phys. Chem. C*, **115**, 8670 (2011). Prediction of the Surface Tension of the Liquid–Vapor Interface of Alcohols from Monte Carlo Simulations.

135. I. Benjamin, *J. Chem. Phys.*, **97**, 1432 (1992). Theoretical Study of the Water/1,2-Dichloroethane Interface: Structure, Dynamics and Conformational Equilibria at the Liquid-Liquid Interface.

136. D. Michael and I. Benjamin, *J. Phys. Chem.*, **99**, 1530 (1995). Solute Orientational Dynamics and Surface Roughness of Water/Hydrocarbon Interfaces.

137. S. Toxvaerd and J. Stecki, *J. Chem. Phys.*, **102**, 7163 (1995). Density Profiles at a Planar Liquid-Liquid Interface.

138. D. Michael and I. Benjamin, *J. Electroanal. Chem.*, **450**, 335 (1998). Molecular Dynamics Simulation of the Water/Nitrobenzene Interface.

139. P. A. Fernandes, M. N. D. S. Cordeiro, and J. A. N. F. Gomes, *J. Phys. Chem. B.*, **103**, 6290 (1999). Molecular Dynamics Simulation of the Water/2-Heptanone Liquid-Liquid Interface.

140. J. Chowdhary and B. M. Ladanyi, *J. Phys. Chem. B*, **110**, 15442 (2006). Water-Hydrocarbon Interfaces: Effect of Hydrocarbon Branching on Interfacial Structure.

141. M. N. D. S. Cordeiro, *Molecular Simulation*, **29**, 817 (2003). Interfacial Tension Behaviour of Water/Hydrocarbon Liquid-Liquid Interfaces: A Molecular Dynamics Simulation.

142. M. Jorge and M. N. D. S. Cordeiro, *J. Phys. Chem. C*, **111**, 17612 (2007). Intrinsic Structure and Dynamics of the Water/Nitrobenzene Interface.

143. M. Jorge and M. N. D. S. Cordeiro, *J. Phys. Chem. B*, **112**, 2415 (2008). Molecular Dynamics Study of the Interface between Water and 2-Nitrophenyl Octyl Ether.

144. I. Napari, A. Laaksonen, V. Talanquer, and D. W. Oxtoby, *J. Chem. Phys.*, **110**, 5906 (1999). A Density Functional Study of Liquid-Liquid Interfaces in Partially Miscible Systems.

145. P. Geysermans, N. Elyeznasni, and V. Russier, *J. Chem. Phys.*, **123** (2005). Layered Interfaces Between Immiscible Liquids Studied by Density-Functional Theory and Molecular-Dynamics Simulations.

146. K. Kashimoto, J. Yoon, B. Y. Hou, C. H. Chen, B. H. Lin, M. Aratono, T. Takiue, and M. L. Schlossman, *Phys. Rev. Lett.*, **101** (2008). Structure and Depletion at Fluorocarbon and Hydrocarbon/Water Liquid/Liquid Interfaces.

147. M. Requardt, *J. Stat. Phys.*, **64**, 807 (1991). Does the Three-Dimensional Capillary Wave Model Lead to a Universally Valid and Pathology-Free Description of the Liquid–vapor Interface near $g = 0$? A Controversial Point of View.

148. S. Senapati and M. L. Berkowitz, *Phys. Rev. Lett.*, **87**, 176101 (2001). Computer Simulation Study of the Interface Width of the Liquid/Liquid Interface.

149. B. Lee and F. M. Richards, *J. Mol. Bio.*, **55**, 379 (1971). Interpretation of Protein Structures – Estimation of Static Accessibility.

150. E. Chacón, P. Tarazona, and J. Alejandre, *J. Chem. Phys.*, 125, 014709 (2006). The Intrinsic Structure of the Water Surface.

151. G. Hantal, P. Terleczky, G. Horvai, L. Nyulaszi, and P. Jedlovszky, *J. Phys. Chem. C*, 113, 19263 (2009). Molecular Level Properties of the Water-Dichloromethane Liquid/Liquid Interface, as Seen from Molecular Dynamics Simulation and Identification of Truly Interfacial Molecules Analysis.

152. A. P. Willard and D. Chandler, *J. Phys. Chem. B*, 114, 1954 (2010). Instantaneous Liquid Interfaces.

153. M. Jorge, G. Hantal, P. Jedlovszky, and M. N. D. S. Cordeiro, *J. Phys. Chem. C*, 114, 18656 (2010). Critical Assessment of Methods for the Intrinsic Analysis of Liquid Interfaces. 2. Density Profiles.

154. D. M. Mitrinovic, A. M. Tikhonov, M. Li, Z. Huang, and M. L. Schlossman, *Phys. Rev. Lett.*, 85, 582 (2000). Noncapillary-Wave Structure at the Water-Alkane Interface.

155. E. Chacón and P. Tarazona, *Phys. Rev. Lett.*, 91, 166103 (2003). Intrinsic Profiles Beyond the Capillary Wave Theory: A Monte Carlo Study.

156. F. Bresme, E. Chacón, P. Tarazona, and K. Tay, *Phys. Rev. Lett.*, 101 (2008). Intrinsic Structure of Hydrophobic Surfaces: The Oil–Water Interface.

157. J. R. Errington and D. A. Kofke, *J. Chem. Phys.*, 127, 174709 (2007). Calculation of Surface Tension Via Area Sampling.

158. E. M. Blokhuis, D. Bedeaux, and C. D. Holcomb, *Mol. Phys.*, 85, 665 (1995). Tail Corrections to the Surface Tension of a Lennard-Jones Liquid-Vapour Interface.

159. J. H. Irving and J. G. Kirkwood, *J. Chem. Phys.*, 18, 817 (1950). The Statistical Mechanical Theory of Transport Processes. 4. The Equations of Hydrodynamics.

160. J. P. R. B. Walton, D. J. Tildesley, J. S. Rowlinson, and J. R. Henderson, *Mol. Phys.*, 48, 1357 (1983). The Pressure Tensor at the Planar Surface of a Liquid.

161. A. Ghoufi, F. Goujon, V. Lachet, and P. Malfreyt, *J. Chem. Phys.*, 128, 154716 (2008). Multiple Histogram Reweighting Method for the Surface Tension Calculation.

162. F. Biscay, A. Ghoufi, V. Lachet, and P. Malfreyt, *Phys. Chem. Chem. Phys.*, 11, 6132 (2009). Calculation of the Surface Tension of Cyclic and Aromatic Hydrocarbons from Monte Carlo Simulations Using an Anisotropic United Atom Model.

163. M. X. Guo and B. C. Y. Lu, *J. Chem. Phys.*, 106, 3688 (1997). Long Range Corrections to Thermodynamic Properties of Inhomogeneous Systems with Planar Interfaces.

164. A. Ghoufi, F. Goujon, V. Lachet, and P. Malfreyt, *Phys. Rev. E*, 77 (2008). Expressions for Local Contributions to the Surface Tension from the Virial Route.

165. F. Chen and P. E. Smith, *J. Chem. Phys.*, 126 (2007). Simulated Surface Tensions of Common Water Models.

166. K. E. Gubbins, in *Fluid Interfacial Phenomena*, C. A. Croxton (Ed.), Wiley, New York, 1986, p. 477, Molecular Orientation at the Free Liquid Surface.

167. T. Raising, Y. R. Shen, M. W. Kim, P. Valint, and J. Bock, *Phys. Rev. A.*, 31, 537 (1985). Orientation of Surfactant Molecules at a Liquid-Air Interface Measured by Optical Second-Harmonic Generation.

168. M. C. Goh, J. M. Hicks, K. Kemnitz, G. R. Pinto, K. Bhattacharyya, K. B. Eisenthal, and T. F. Heinz, *J. Phys. Chem.*, 92, 5074 (1988). Absolute Orientation of Water Molecules at the Neat Water Surface.

169. D. A. Higgins, R. R. Naujok, and R. M. Corn, *Chem. Phys. Lett.*, 213, 485 (1993). Second Harmonic Generation Measurements of Molecular Orientation and Coadsorption at the Interface between Two Immiscible Electrolyte Solutions.

170. R. R. Naujok, D. A. Higgins, D. G. Hanken, and R. M. Corn, *J. Chem. Soc. Faraday Trans.*, 91, 1411 (1995). Optical Second-Harmonic Generation Measurements of Molecular Adsorption and Orientation at the Liquid/Liquid Electrochemical Interface.

171. F. Eisert, O. Dannenberger, and M. Buck, *Phys. Rev. B.*, 58, 10860 (1998). Molecular Orientation Determined by Second-Harmonic Generation: Self-Assembled Monolayers.

172. Y. R. Shen and V. Ostroverkhov, *Chem. Rev.*, **106**, 1140 (2006). Sum-Frequency Vibrational Spectroscopy on Water Interfaces: Polar Orientation of Water Molecules at Interfaces.

173. Y. Rao, M. Comstock, and K. B. Eisenthal, *J. Phys. Chem. B*, **110**, 1727 (2006). Absolute Orientation of Molecules at Interfaces.

174. M. Matsumoto and Y. Kataoka, *J. Chem. Phys.*, **88**, 3233 (1988). Study on Liquid–Vapor Interface of Water. I. Simulational Results of Thermodynamic Properties and Orientational Structure.

175. P. Jedlovszky, Á. Vincze, and G. Horvai, *J. Chem. Phys.*, **117**, 2271 (2002). New Insight into the Orientational Order of Water Molecules at the Water/1,2-Dichloroethane Interface: A Monte Carlo Simulation Study.

176. P. Jedlovszky, A. Vincze, and G. Horvai, *J. Mol. Liq.*, **109**, 99 (2004). Properties of Water/Apolar Interfaces as Seen from Monte Carlo Simulations.

177. P. Jedlovszky, A. Vincze, and G. Horvai, *Phys. Chem. Chem. Phys.*, **6**, 1874 (2004). Full Description of the Orientational Statistics of Molecules near to Interfaces. Water at the Interface with CCl_4.

178. D. S. Walker, D. K. Hore, and G. L. Richmond, *J. Phys. Chem. B*, **110**, 20451 (2006). Understanding the Population, Coordination, and Orientation of Water Species Contributing to the Nonlinear Optical Spectroscopy of the Vapor-Water Interface through Molecular Dynamics Simulations.

179. G. Hantal, M. Darvas, L. B. Partay, G. Horvai, and P. Jedlovszky, *J. Phys.-Condensed Matter*, **22**, 284112 (2010). Molecular Level Properties of the Free Water Surface and Different Organic Liquid/Water Interfaces, as Seen from ITIM Analysis of Computer Simulation Results.

180. W. L. Jorgensen, *Chem. Phys. Lett.*, **70**, 326 (1980). Monte-Carlo Results for Hydrogen-Bond Distributions in Liquid Water.

181. M. Matsumoto and I. Ohmine, *J. Chem. Phys.*, **104**, 2705 (1996). A New Approach to the Dynamics of Hydrogen Bond Network in Liquid Water.

182. F. W. Starr, J. K. Nielsen, and H. E. Stanley, *Phys. Rev. Lett.*, **82**, 2294 (1999). Fast and Slow Dynamics of Hydrogen Bonds in Liquid Water.

183. M. G. Sceats and S. A. Rice, *J. Chem. Phys.*, **72**, 3236 (1980). The Water-Water Pair Potential near the Hydrogen-Bonded Equilibrium Configuration.

184. M. Mezei and D. L. Beveridge, *J. Chem. Phys.*, **74**, 622 (1981). Theoretical-Studies of Hydrogen-Bonding in Liquid Water and Dilute Aqueous-Solutions.

185. A. Luzar and D. Chandler, *Nature*, **379**, 55 (1996). Hydrogen-Bond Kinetics in Liquid Water.

186. R. Kumar, J. R. Schmidt, and J. L. Skinner, *J. Chem. Phys.*, **126**, 204107 (2007). Hydrogen Bonding Definitions and Dynamics in Liquid Water.

187. F. W. Starr, J. K. Nielsen, and H. E. Stanley, *Phys. Rev. E*, **62**, 579 (2000). Hydrogen-Bond Dynamics for the Extended Simple Point-Charge Model of Water.

188. J. B. Brubach, A. Mermet, A. Filabozzi, A. Gerschel, and P. Roy, *J. Chem. Phys.*, **122**, 184509 (2005). Signatures of the Hydrogen Bonding in the Infrared Bands of Water.

189. D. A. Schmidt and K. Miki, *J. Phys. Chem. A*, **111**, 10119 (2007). Structural Correlations in Liquid Water: A New Interpretation of IR Spectroscopy.

190. P. L. Geissler, *J. Am. Chem. Soc.*, **127**, 14930 (2005). Temperature Dependence of Inhomogeneous Broadening: On the Meaning of Isosbestic Points.

191. Q. Du, R. Superfine, E. Freysz, and Y. R. Shen, *Phys. Rev. Lett.*, **70**, 2313 (1993). Vibrational Spectroscopy of Water at the Vapor/Water Interface.

192. Q. Du, E. Freysz, and Y. R. Shen, *Science*, **264**, 826 (1994). Surface Vibrational Spectroscopic Studies of Hydrogen Bonding and Hydropohobicity.

193. L. F. Scatena, M. G. Brown, and G. L. Richmond, *Science*, **292**, 908 (2001). Water at Hydrophobic Surfaces: Weak Hydrogen Bonding and Strong Orientation Effects.

194. I. V. Stiopkin, H. D. Jayathilake, A. N. Bordenyuk, and A. V. Benderskii, *J. Am. Chem. Soc.*, **130**, 2271 (2008). Heterodyne-Detected Vibrational Sum Frequency Generation Spectroscopy.

195. M. Sovago, R. K. Campen, G. W. H. Wurpel, M. Muller, H. J. Bakker, and M. Bonn, *Phys. Rev. Lett.*, **100**, 173901 (2008). Vibrational Response of Hydrogen-Bonded Interfacial Water Is Dominated by Intramolecular Coupling.

196. C. S. Tian and Y. R. Shen, *Chem. Phys. Lett.*, **470**, 1 (2009). Sum-Frequency Vibrational Spectroscopic Studies of Water/Vapor Interfaces.

197. A. M. Jubb, W. Hua, and H. C. Allen, *Annu. Rev. Phy. Chem.*, **63**, 107 (2012). Environmental Chemistry at Vapor/Water Interfaces: Insights from Vibrational Sum Frequency Generation Spectroscopy.

198. I. V. Stiopkin, C. Weeraman, P. A. Pieniazek, F. Y. Shalhout, J. L. Skinner, and A. V. Benderskii, *Nature*, **474**, 192 (2011). Hydrogen Bonding at the Water Surface Revealed by Isotopic Dilution Spectroscopy.

199. J. L. Skinner, B. M. Auer, and Y. S. Lin, *Adv. Chem. Phys.*, **142**, 59 (2009). Vibrational Line Shapes, Spectral Diffusion, and Hydrogen Bonding in Liquid Water.

200. I. Benjamin, *Phys. Rev. Lett.*, **73**, 2083 (1994). Vibrational Spectrum of Water at the Liquid/Vapor Interface.

201. I. F. W. Kuo and C. J. Mundy, *Science*, **303**, 658 (2004). An Ab Initio Molecular Dynamics Study of the Aqueous Liquid–Vapor Interface.

202. V. Buch, *J. Phys. Chem. B*, **109**, 17771 (2005). Molecular Structure and OH-Stretch Spectra of Liquid Water Surface.

203. F. G. Moore and G. L. Richmond, *Acc. Chem. Res.*, **41**, 739 (2008). Integration or Segregation: How Do Molecules Behave at Oil/Water Interfaces?

204. A. Perry, C. Neipert, B. Space, and P. B. Moore, *Chem. Rev.*, **106**, 1234 (2006). Theoretical Modeling of Interface Specific Vibrational Spectroscopy: Methods and Applications to Aqueous Interfaces.

205. A. Morita and J. T. Hynes, *J. Phys. Chem. B.*, **106**, 673 (2002). A Theoretical Analysis of the Sum Frequency Generation Spectrum of the Water Surface. II. Time-Dependent Approach.

206. A. Morita, *J. Phys. Chem. B*, **110**, 3158 (2006). Improved Computation of Sum Frequency Generation Spectrum of the Surface of Water.

207. B. M. Auer and J. L. Skinner, *J. Chem. Phys.*, **129**, 214705 (2008). Vibrational Sum-Frequency Spectroscopy of the Liquid/Vapor Interface for Dilute HOD in D_2O.

208. C. J. Tainter, P. A. Pieniazek, Y. S. Lin, and J. L. Skinner, *J. Chem. Phys.*, **134**, 184501 (2011). Robust Three-Body Water Simulation Model.

209. J. L. Skinner, P. A. Pieniazek, and S. M. Gruenbaum, *Acc. Chem. Res.*, **45**, 93 (2012). Vibrational Spectroscopy of Water at Interfaces.

210. D. S. Walker and G. L. Richmond, *J. Phys. Chem. C.*, **111**, 8321 (2007). Understanding the Effects of Hydrogen Bonding at the Vapor-Water Interface: Vibrational Sum Frequency Spectroscopy of H_2O/HOD/D_2O Mixtures Studied Using Molecular Dynamics Simulations.

211. M. D. Fayer, *Acc. Chem. Res.*, **45**, 3 (2012). Dynamics of Water Interacting with Interfaces, Molecules, and Ions.

212. D. Chandler, *Introduction to Modern Statistical Mechanics*, Oxford University Press, Oxford, 1987.

213. D. Duque, P. Tarazona, and E. Chacón, *J. Chem. Phys.*, **128**, 134704 (2008). Diffusion at the Liquid–Vapor Interface.

214. R. G. Winkler, R. H. Schmid, A. Gerstmair, and P. Reineker, *J. Chem. Phys.*, **104**, 8103 (1996). Molecular Dynamics Simulation Study of the Dynamics of Fluids in Thin Films.

215. E. A. J. F. Peters and T. M. A. O. M. Barenbrug, *Phys. Rev. E*, **66** (056701) (2002). Efficient Brownian Dynamics Simulation of Particles near Walls. I. Reflecting and Absorbing Walls.

216. M. Sega, R. Vallauri, and S. Melchionna, *Phys. Rev. E*, **72**, 041201 (2005). Diffusion of Water in Confined Geometry: The Case of a Multilamellar Bilayer.

217. P. Liu, E. Harder, and B. J. Berne, *J. Phys. Chem. B*, **108**, 6595 (2004). On the Calculation of Diffusion Coefficients in Confined Fluids and Interfaces with an Application to the Liquid–Vapor Interface of Water.

218. J. Chowdhary and B. M. Ladanyi, *J. Phys. Chem. B*, **112**, 6259 (2008). Water/Hydrocarbon Interfaces: Effect of Hydrocarbon Branching on Single-Molecule Relaxation.

219. W. A. Steele, *Adv. Chem. Phys.*, **34**, 1 (1976). The Rotation of Molecules in Dense Phases.

220. A. I. Burshtein and S. I. Temkin, *Spectroscopy of Molecular Rotation in Gases and Liquids*, Cambridge University Press, Cambridge, 1994.

221. G. R. Fleming, *Chemical Applications of Ultrafast Spectroscopy*, New York, Oxford University, 1986.

222. X. Wei and Y. R. Shen, *Phys. Rev. Lett.*, **86**, 4799 (2001). Motional Effect in Surface Sum-Frequency Vibrational Spectroscopy.

223. J. T. Fourkas, R. A. Walker, S. Z. Can, and E. Gershgoren, *J. Phys. Chem. C*, **111**, 8902 (2007). Effects of Reorientation in Vibrational Sum-Frequency Spectroscopy.

224. M. Hayashi, Y. J. Shiu, K. K. Liang, S. H. Lin, and Y. R. Shen, *J. Phys. Chem. A*, **111**, 9062 (2007). Theory of Time-Resolved Sum-Frequency Generation and Its Applications to Vibrational Dynamics of Water.

225. H. K. Nienhuys and M. Bonn, *J. Phys. Chem. B*, **113**, 7564 (2009). Measuring Molecular Reorientation at Liquid Surfaces with Time-Resolved Sum-Frequency Spectroscopy: A Theoretical Framework.

226. Z. Gengeliczki, D. E. Rosenfeld, and M. D. Fayer, *J. Chem. Phys.*, **132**, 244703 (2010). Theory of Interfacial Orientational Relaxation Spectroscopic Observables.

227. C. S. Hsieh, R. K. Campen, A. C. V. Verde, P. Bolhuis, H. K. Nienhuys, and M. Bonn, *Phys. Rev. Lett.*, **107**, 116102 (2011). Ultrafast Reorientation of Dangling OH Groups at the Air-Water Interface Using Femtosecond Vibrational Spectroscopy.

228. I. Benjamin, *J. Chem. Phys.*, **95**, 3698 (1991). Theoretical Study of Ion Solvation at the Water Liquid–Vapor Interface.

229. S. R. V. Castrillon, N. Giovambattista, I. A. Aksay, and P. G. Debenedetti, *J. Phys. Chem. B*, **113**, 1438 (2009). Effect of Surface Polarity on the Structure and Dynamics of Water in Nanoscale Confinement.

230. G. Stirnemann, P. J. Rossky, J. T. Hynes, and D. Laage, *Faraday Discuss.*, **146**, 263 (2010). Water Reorientation, Hydrogen-Bond Dynamics and 2D-IR Spectroscopy Next to an Extended Hydrophobic Surface.

231. G. Stirnemann, S. R. V. Castrillon, J. T. Hynes, P. J. Rossky, P. G. Debenedetti, and D. Laage, *Phys. Chem. Chem. Phys.*, **13**, 19911 (2011). Non-Monotonic Dependence of Water Reorientation Dynamics on Surface Hydrophilicity: Competing Effects of the Hydration Structure and Hydrogen-Bond Strength.

232. D. Laage and J. T. Hynes, *Science*, **311**, 832 (2006). A Molecular Jump Mechanism of Water Reorientation.

233. G. M. Gale, G. Gallot, F. Hache, N. Lascoux, S. Bratos, and J.-C. Leicknam, *Phys. Rev. Lett.*, **82**, 1068 (1999). Femtosecond Dynamics of Hydrogen Bonds in Liquid Water: A Real Time Study.

234. J. B. Asbury, T. Steinel, C. Stromberg, K. J. Gaffney, I. R. Piletic, A. Goun, and M. D. Fayer, *Phys. Rev. Lett.*, **9123**, 7402 (2003). Hydrogen Bond Dynamics Probed with Ultrafast Infrared Heterodyne-Detected Multidimensional Vibrational Stimulated Echoes.

235. C. J. Fecko, J. D. Eaves, J. J. Loparo, A. Tokmakoff, and P. L. Geissler, *Science*, **301**, 1698 (2003). Ultrafast Hydrogen-Bond Dynamics in the Infrared Spectroscopy of Water.

236. E. T. J. Nibbering and T. Elsaesser, *Chem. Rev.*, **104**, 1887 (2004). Ultrafast Vibrational Dynamics of Hydrogen Bonds in the Condensed Phase.

237. S. Park and M. D. Fayer, *Proc. Nat. Acad. Sci. U. S. A.*, **104**, 16731 (2007). Hydrogen Bond Dynamics in Aqueous NaBr Solutions.

238. F. D'amico, F. Bencivenga, A. Gessini, and C. Masciovecchio, *J. Phys. Chem. B*, 114, 10628 (2010). Temperature Dependence of Hydrogen-Bond Dynamics in Acetic Acid-Water Solutions.

239. A. A. Bakulin, M. S. Pshenichnikov, H. J. Bakker, and C. Petersen, *J. Phys. Chem. A*, 115, 1821 (2011). Hydrophobic Molecules Slow Down the Hydrogen-Bond Dynamics of Water.

240. A. Luzar and D. Chandler, *Phys. Rev. Lett.*, 76, 928 (1996). Effect of Environment on Hydrogen Bond Dynamics in Liquid Water.

241. A. Luzar, *J. Chem. Phys.*, 113, 10663 (2000). Resolving the Hydrogen Bond Dynamics Conundrum.

242. A. Chandra, *Phys. Rev. Lett.*, 85, 768 (2000). Effects of Ion Atmosphere on Hydrogen-Bond Dynamics in Aqueous Electrolyte Solutions.

243. H. Xu, H. A. Stern, and B. J. Berne, *J. Phys. Chem. B.*, 106, 2054 (2002). Can Water Polarizability Be Ignored in Hydrogen Bond Kinetics?

244. K. B. Mller, R. Rey, and J. T. Hynes, *J. Phys. Chem. A*, 108, 1275 (2004). Hydrogen Bond Dynamics in Water and Ultrafast Infrared Spectroscopy: A Theoretical Study.

245. B. Nigro, S. Re, D. Laage, R. Rey, and J. T. Hynes, *J. Phys. Chem. A.*, 110, 11237 (2006). On the Ultrafast Infrared Spectroscopy of Anion Hydration Shell Hydrogen Bond Dynamics.

246. F. Sterpone, G. Stirnemann, J. T. Hynes, and D. Laage, *J. Phys. Chem. B*, 114, 2083 (2010). Water Hydrogen-Bond Dynamics around Amino Acids: The Key Role of Hydrophilic Hydrogen-Bond Acceptor Groups.

247. B. S. Mallik and A. Chandra, *J. Chem. Sci.*, 124, 215 (2012). Hydrogen Bond Dynamics and Vibrational Spectral Diffusion in Aqueous Solution of Acetone: A First Principles Molecular Dynamics Study.

248. S. Balasubramanian, S. Pal, and B. Bagchi, *Phys. Rev. Lett.*, 89, 115505/1 (2002). Hydrogen-Bond Dynamics near a Micellar Surface: Origin of the Universal Slow Relaxation at Complex Aqueous Interfaces.

249. M. Tarek and D. J. Tobias, *Phys. Rev. Lett.*, 88, 138101 (2002). Role of Protein-Water Hydrogen Bond Dynamics in the Protein Dynamical Transition.

250. C. F. Lopez, S. O. Nielsen, M. L. Klein, and P. B. Moore, *J. Phys. Chem, B.*, 108(6603) (2004). Hydrogen Bonding Structure and Dynamics of Water at the Dimyristoylphosphatidylcholine Lipid Bilayer Surface from a Molecular Dynamics Simulation.

251. S. Paul and A. Chandra, *Chem. Phys. Lett.*, 386, 218 (2004). Hydrogen Bond Dynamics at Vapour-Water and Metal-Water Interfaces.

252. P. Liu, E. Harder, and B. J. Berne, *J. Phys. Chem. B.*, 109, 2949 (2005). Hydrogen-Bond Dynamics in the Air-Water Interface.

253. I. Benjamin, *J. Phys. Chem. B*, 109, 13711 (2005). Hydrogen Bond Dynamics at Water/Organic Liquid Interfaces.

254. N. Winter, J. Vieceli, and I. Benjamin, *J. Phys. Chem. B*, 112, 227 (2008). Hydrogen-Bond Structure and Dynamics at the Interface between Water and Carboxylic Acid-Functionalized Self-Assembled Monolayers.

255. B. Jana, S. Pal, and B. Bagchi, *J. Phys. Chem. B*, 112, 9112 (2008). Hydrogen Bond Breaking Mechanism and Water Reorientational Dynamics in the Hydration Layer of Lysozyme.

256. J. Chowdhary and B. M. Ladanyi, *J. Phys. Chem. B*, 113, 4045 (2009). Hydrogen Bond Dynamics at the Water/Hydrocarbon Interface.

257. D. E. Rosenfeld and C. A. Schmuttenmaer, *J. Phys. Chem. B*, 115, 1021 (2011). Dynamics of the Water Hydrogen Bond Network at Ionic, Nonionic, and Hydrophobic Interfaces in Nanopores and Reverse Micelles.

258. T. Yamamoto, *J. Chem. Phys.*, 33, 281 (1960). Quantum Statistical Mechanical Theory of the Rate of Exchange Chemical Reactions in the Gas Phase.

259. I. Benjamin, *J. Chem. Phys.*, 110, 8070 (1999). Structure Thermodynamics and Dynamics of the Liquid/Vapor Interface of Water/DMSO Mixtures.

260. T.-M. Chang and L. X. Dang, *J. Phys. Chem. B.*, **109**, 5759 (2005). Liquid–Vapor Interface of Methanol–Water Mixtures: A Molecular Dynamics Study.

261. P. Salvador, J. E. Curtis, D. J. Tobias, and P. Jungwirth, *Phys. Chem. Chem. Phys.*, **5**, 3752 (2003). Polarizability of the Nitrate Anion and Its Solvation at the Air/Water Interface.

262. B. Minofar, M. Mucha, P. Jungwirth, X. Yang, Y. J. Fu, X. B. Wang, and L. S. Wang, *J. Am. Chem. Soc.*, **126**, 11691 (2004). Bulk Versus Interfacial Aqueous Solvation of Dicarboxylate Dianions.

263. L. Vrbka, M. Mucha, B. Minofar, P. Jungwirth, E. C. Brown, and D. J. Tobias, *Curr. Opin. Coll. Interf. Sci.*, **9**, 67 (2004). Propensity of Soft Ions for the Air/Water Interface.

264. P. B. Petersen and R. J. Saykally, *J. Phys. Chem. B*, **109**, 7976 (2005). Evidence for an Enhanced Hydronium Concentration at the Liquid Water Surface.

265. P. B. Petersen, R. J. Saykally, M. Mucha, and P. Jungwirth, *J. Phys. Chem. B*, **109**, 10915 (2005). Enhanced Concentration of Polarizable Anions at the Liquid Water Surface: SHG Spectroscopy and MD Simulations of Sodium Thiocyanide.

266. T. M. Chang and L. X. Dang, *Chem. Rev.*, **106**, 1305 (2006). Recent Advances in Molecular Simulations of Ion Solvation at Liquid Interfaces.

267. P. Jungwirth and D. J. Tobias, *Chem. Rev.*, **106**, 1259 (2006). Specific Ion Effects at the Air/Water Interface.

268. P. Jungwirth and B. Winter, *Ann. Rev. Phys. Chem.*, **59**, 343 (2008). Ions at Aqueous Interfaces: From Water Surface to Hydrated Proteins.

269. L. Sun, X. Li, T. Hede, Y. Q. Tu, C. Leck, and H. Agren, *J. Phys. Chem. B*, **116**, 3198 (2012). Molecular Dynamics Simulations of the Surface Tension and Structure of Salt Solutions and Clusters.

270. S. Gopalakrishnan, P. Jungwirth, D. J. Tobias, and H. C. Allen, *J. Phys. Chem. B*, **109**, 8861 (2005). Air-Liquid Interfaces of Aqueous Solutions Containing Ammonium and Sulfate: Spectroscopic and Molecular Dynamics Studies.

271. P. B. Petersen and R. J. Saykally, *Annu. Rev. Phys. Chem.*, **57**, 333 (2006). On the Nature of Ions at the Liquid Water Surface.

272. E. C. Brown, M. Mucha, P. Jungwirth, and D. J. Tobias, *J. Phys. Chem. B*, **109**, 7934 (2005). Structure and Vibrational Spectroscopy of Salt Water/Air Interfaces: Predictions from Classical Molecular Dynamics Simulations.

273. G. Luo, S. Malkova, J. Yoon, D. G. Schultz, B. Lin, M. Meron, I. Benjamin, P. Vanysek, and M. L. Schlossman, *Science*, **311**, 216 (2006). Ion Distributions near a Liquid-Liquid Interface.

274. G. Luo, S. Malkova, J. Yoon, D. G. Schultz, B. Lin, M. Meron, I. Benjamin, P. Vanysek, and M. L. Schlossman, *J. Electroanal. Chem.*, **593**, 142 (2006). Ion Distributions at the Nitrobenzene-Water Interface Electrified by a Common Ion.

275. A. Pohorille and M. A. Wilson, *J. Mol. Struct.*, **103**, 271 (1993). Molecular Structure of Aqueous Interfaces.

276. S. Kumar, J. M. Rosenberg, D. Bouzida, R. H. Swendsen, and P. A. Kollman, *J. Comput. Chem.*, **16**, 1339 (1995). Multidimensional Free-Energy Calculations Using the Weighted Histogram Analysis Method.

277. J. Kastner, *Wiley Interdisciplinary Reviews-Computational Molecular Science*, **1**, 932 (2011). Umbrella Sampling.

278. K. J. Schweighofer and I. Benjamin, *Chem. Phys. Lett.*, **202**, 379 (1993). Dynamics of Ion Desorption from the Liquid–Vapor Interface of Water.

279. E. Darve and A. Pohorille, *J. Chem. Phys.*, **115**, 9169 (2001). Calculating Free Energies Using Average Force.

280. L. X. Dang, *J. Phys. Chem. B*, **106**, 10388 (2002). Computational Study of Ion Binding to the Liquid Interface of Water.

281. C. D. Wick and L. X. Dang, *J. Phys. Chem. B*, **110**, 6824 (2006). Distribution, Structure, and Dynamics of Cesium and Iodide Ions at the H_2O-CCl_4 and H_2O-Vapor Interfaces.

282. C. D. Wick and L. X. Dang, *J. Phys. Chem. C*, **112**, 647 (2008). Molecular Dynamics Study of Ion Transfer and Distribution at the Interface of Water and 1,2-Dichlorethane.

283. A. Pohorille, C. Jarzynski, and C. Chipot, *J. Phys. Chem. B*, **114**, 10235 (2010). Good Practices in Free-Energy Calculations.

284. T. P. Straatsma, in *Reviews in Computational Chemistry*, K. B. Lipkowitz and D. B. Boyd (Eds.), VCH Publishers, New York, 1996, Vol. 9, p. 81, Free Energy by Molecular Simulation.

285. M. A. Wilson and A. Pohorille, *J. Chem. Phys.*, **95**, 6005 (1991). Interaction of Monovalent Ions with the Water Liquid–Vapor Interface: A Molecular Dynamics Study.

286. M. N. Tamashiro and M. A. Constantino, *J. Phys. Chem. B*, **114**, 3583 (2010). Ions at the Water-Vapor Interface.

287. Y. Levin, *J. Chem. Phys.*, **129** (2008). "Phantom Ion Effect" And the Contact Potential of the Water-Vapor Interface.

288. Y. Levin, A. P. dos Santos, and A. Diehl, *Phys. Rev. Lett.*, **103** (2009). Ions at the Air-Water Interface: An End to a Hundred-Year-Old Mystery?

289. A. P. dos Santos and Y. Levin, *Langmuir*, **28**, 1304 (2012). Ions at the Water–Oil Interface: Interfacial Tension of Electrolyte Solutions.

290. G. L. Warren and S. Patel, *J. Phys. Chem. C*, **112**, 7455 (2008). Comparison of the Solvation Structure of Polarizable and Nonpolarizable Ions in Bulk Water and near the Aqueous Liquid–vapor Interface.

291. E. A. Raymond and G. L. Richmond, *J. Phys. Chem. B*, **108**, 5051 (2004). Probing the Molecular Structure and Bonding of the Surface of Aqueous Salt Solutions.

292. D. F. Liu, G. Ma, L. M. Levering, and H. C. Allen, *J. Phys. Chem. B*, **108**, 2252 (2004). Vibrational Spectroscopy of Aqueous Sodium Halide Solutions and Air-Liquid Interfaces: Observation of Increased Interfacial Depth.

293. C. Caleman, J. S. Hub, P. J. van Maaren, and D. van der Spoel, *Proc. Natl. Acad. Sci. U.S.A.*, **108**, 6838 (2011). Atomistic Simulation of Ion Solvation in Water Explains Surface Preference of Halides.

294. J. Noah-Vanhoucke and P. L. Geissler, *Proc. Nat. Acad. Sci. U. S. A.*, **106**, 15125 (2009). On the Fluctuations That Drive Small Ions toward, and Away from, Interfaces between Polar Liquids and Their Vapors.

295. T. L. Tarbuck, S. T. Ota, and G. L. Richmond, *J. Am. Chem. Soc.*, **128**, 14519 (2006). Spectroscopic Studies of Solvated Hydrogen and Hydroxide Ions at Aqueous Surfaces.

296. R. Vacha, D. Horinek, M. L. Berkowitz, and P. Jungwirth, *Phys. Chem. Chem. Phys.*, **10**, 4975 (2008). Hydronium and Hydroxide at the Interface between Water and Hydrophobic Media.

297. B. Winter, M. Faubel, R. Vacha, and P. Jungwirth, *Chem. Phys. Lett.*, **474**, 241 (2009). Behavior of Hydroxide at the Water/Vapor Interface.

298. B. Winter, M. Faubel, R. Vacha, and P. Jungwirth, *Chem. Phys. Lett.*, **481**, 19 (2009). Reply to Comments on Frontiers Article 'Behavior of Hydroxide at the Water/Vapor Interface'.

299. J. K. Beattie, *Chem. Phys. Lett.*, **481**, 17 (2009). Comment on 'Behaviour of Hydroxide at the Water/Vapor Interface' [Chem. Phys. Lett. 474 (2009) 241].

300. A. Gray-Weale, *Chem. Phys. Lett.*, **481**, 22 (2009). Comment on 'Behaviour of Hydroxide at the Water/Vapor Interface' [Chem. Phys. Lett. 474 (2009) 241].

301. C. J. Mundy, I. F. W. Kuo, M. E. Tuckerman, H. S. Lee, and D. J. Tobias, *Chem. Phys. Lett.*, **481**, 2 (2009). Hydroxide Anion at the Air-Water Interface.

302. K. J. Schweighofer and I. Benjamin, *J. Phys. Chem.*, **99**, 9974 (1995). Transfer of Small Ions across the Water/1,2-Dichloroethane Interface.

303. A. Pohorille, P. Cieplak, and M. A. Wilson, *Chem. Phys.*, **204**, 337 (1996). Interactions of Anesthetics with the Membrane-Water Interface.

304. C. Chipot, M. A. Wilson, and A. Pohorille, *J. Phys. Chem. B*, **101**, 782 (1997). Interactions of Anesthetics with the Water-Hexane Interface. A Molecular Dynamics Study.

305. I. Benjamin, *Annu. Rev. Phys. Chem.*, **48**, 401 (1997). Molecular Structure and Dynamics at Liquid-Liquid Interfaces.

306. K. J. Schweighofer and I. Benjamin, *J. Phys. Chem. A*, **103**, 10274 (1999). Transfer of a Tetra Methyl Ammonium Ion across the Water-Nitrobenzene Interface: Potential of Mean Force and Non-Equilibrium Dynamics.

307. L. X. Dang, *J. Phys. Chem. B*, **103**, 8195 (1999). Computer Simulation Studies of Ion Transport across a Liquid/Liquid Interface.

308. C. D. Wick and L. X. Dang, *Chem. Phys. Lett.*, **458**, 1 (2008). Recent Advances in Understanding Transfer Ions across Aqueous Interfaces.

309. C. D. Wick and L. X. Dang, *J. Phys. Chem. C*, **112**, 647 (2008). Molecular Dynamics Study of Ion Transfer and Distribution at the Interface of Water and 1,2-Dichlorethane.

310. N. Kikkawa, T. Ishiyama, and A. Morita, *Chem. Phys. Lett.*, **534**, 19 (2012). Molecular Dynamics Study of Phase Transfer Catalyst for Ion Transfer through Water-Chloroform Interface.

311. D. Rose and I. Benjamin, *J. Phys. Chem. B*, **113**, 9296 (2009). Free Energy of Transfer of Hydrated Ion Clusters from Water to an Immiscible Organic Solvent.

312. M. Saraniti, S. Aboud, and R. Eisenberg, in *Reviews in Computational Chemistry*, K. B. Lipkowitz, T. R. Cundari, and V. J. Gillet (Eds.), Wiley, New York, 2006, Vol. **22**, p. 229, The Simulation of Ionic Charge Transport in Biological Ion Channels: An Introduction to Numerical Methods.

313. M. A. Wilson, C. Y. Wei, P. Bjelkmar, B. A. Wallace, and A. Pohorille, *Biophysical Journal*, **100**, 2394 (2011). Molecular Dynamics Simulation of the Antiamoebin Ion Channel: Linking Structure and Conductance.

314. I. Chorny and I. Benjamin, *J. Phys. Chem. B*, **109**, 16455 (2005). Hydration Shell Exchange Dynamics During Ion Transfer across the Liquid/Liquid Interface.

315. C. A. Wick and S. S. Xantheas, *J. Phys. Chem. B*, **113**, 4141 (2009). Computational Investigation of the First Solvation Shell Structure of Interfacial and Bulk Aqueous Chloride and Iodide Ions.

316. I. Benjamin, *Science*, **261**, 1558 (1993). Mechanism and Dynamics of Ion Transfer across a Liquid-Liquid Interface.

317. I. Benjamin, *J. Phys. Chem. B*, **112**, 15801 (2008). Structure and Dynamics of Hydrated Ions in a Water-Immiscible Organic Solvent.

318. C. D. Wick and O. T. Cummings, *Chem. Phys. Lett.*, **513**, 161 (2011). Understanding the Factors That Contribute to Ion Interfacial Behavior.

319. P. B. Miranda and Y. R. Shen, *J. Phys. Chem. B.*, **103**, 3292 (1999). Liquid Interfaces: A Study by Sum-Frequency Vibrational Spectroscopy.

320. Y. Rao, S. Y. Hong, N. J. Turro, and K. B. Eisenthal, *J. Phys. Chem. C*, **115**, 11678 (2011). Molecular Orientational Distribution at Interfaces Using Second Harmonic Generation.

321. A. Pohorille and I. Benjamin, *J. Chem. Phys.*, **94**, 5599 (1991). Molecular Dynamics of Phenol at the Liquid–Vapor Interface of Water.

322. P. Jedlovszky, I. Varga, and T. Gilanyi, *J. Chem. Phys.*, **119**, 1731 (2003). Adsorption of Apolar Molecules at the Water Liquid–Vapor Interface: A Monte Carlo Simulations Study of the Water-*n*-Octane System.

323. B. C. Garrett, G. K. Schenter, and A. Morita, *Chem. Rev.*, **106**, 1355 (2006). Molecular Simulations of the Transport of Molecules across the Liquid/Vapor Interface of Water.

324. R. A. Marcus, *J. Chem. Phys.*, **43**, 1261 (1965). On the Theory of Shifts and Broadening of Electronic Spectra of Polar Solutes in Polar Media.

325. N. Mataga and T. Kubota, *Molecular Interactions and Electronic Spectra*, New York, Dekker, 1970.

326. B. Martire and R. Gilbert, *Chem. Phys.*, **56**, 241 (1981). Langevin Simulation of Picosecond-Resolved Electronic Spectra in Solution.

327. C. Reichardt, *Solvents and Solvent Effects in Organic Chemistry*, 2nd edition, Springer-Verlag, Weinheim, 1988.

328. R. F. Loring, *J. Phys. Chem.*, **94**, 513 (1990). Statistical Mechanical Calculation of Inhomogeneously Broadened Absorption-Line Shapes in Solution.

329. N. E. Shemetulskis and R. F. Loring, *J. Chem. Phys.*, **95**, 4756 (1991). Electronic Spectra in a Polar Solvent: Theory and Simulation.

330. H. Ågren and K. V. Mikkelsen, *J. Mol. Struct.*, **234**, 425 (1991). Theory of Solvent Effects on Electronic Spectra.

331. J. G. Saven and J. L. Skinner, *J. Chem. Phys.*, **99**, 4391 (1993). A Molecular Theory of the Line-Shape - Inhomogeneous and Homogeneous Electronic-Spectra of Dilute Chromophores in Nonpolar Fluids.

332. C. Reichardt, *Chem. Rev.*, **94**, 2319 (1994). Solvatochromic Dyes as Solvent Polarity Indicators.

333. S. A. Egorov, E. Gallicchio, and B. J. Berne, *J. Chem. Phys.*, **107**, 9312 (1997). The Simulation of Electronic Absorption Spectrum of a Chromophore Coupled to a Condensed Phase Environment: Maximum Entropy Versus Singular Value Decomposition Approaches.

334. S. A. Egorov, E. Rabani, and B. J. Berne, *J. Chem. Phys.*, **108**, 1407 (1998). Vibronic Spectra in Condensed Matter: A Comparison of Exact Quantum Mechanical and Various Semiclassical Treatments for Harmonic Baths.

335. H. M. Sevian and J. L. Skinner, *J. Chem. Phys.*, **97**, 8 (1992). Molecular Theory of Transition Energy Correlations for Pairs of Chromophores in Liquids or Glasses.

336. M. D. Stephens, J. G. Saven, and J. L. Skinner, *J. Chem. Phys.*, **106**, 2129 (1997). Molecular Theory of Electronic Spectroscopy in Nonpolar Fluids: Ultrafast Solvation Dynamics and Absorption and Emission Line Shapes.

337. E. G. McRae, *J. Phys. Chem.*, **61**, 562 (1957). Theory of Solvent Effects on Molecular Electronic Spectra - Frequency Shifts.

338. N. G. Bakhshiev, *Opt. Spektrosk.*, **16**, 821 (1964). Universal Intermolecular Interactions and Their Effect on the Position of the Electronic Spectra of Molecules in Two Component Solutions

339. D. Chandler, K. S. Schweizer, and P. G. Wolynes, *Phys. Rev. Lett.*, **49**, 1100 (1982). Electronic States of a Topologically Disordered System - Exact Solution of the Mean Spherical Model for Liquids.

340. L. S. P. Mirashi and S. S. Kunte, *Spectrochimica Acta, Part A*, **45**(1147) (1989). Solvent Effects on Electronic Absorption Spectra of Nitrochlorobenzenes, Nitrophenols and Nitroanilines. 3. Excited State Dipole Moments and Specific Solute Solvent Interaction Energies Employing Bakhshiev Approach.

341. J. J. Aaron, M. Maafi, C. Párkányi, and C. Boniface, *Spectrochimica Acta, Part A*, **51**(603) (1995). Quantitative Treatment of the Solvent Effects on the Electronic Absorption and Fluorescence-Spectra of Acridines and Phenazines - The Ground and First Excited Singlet-State Dipole-Moments.

342. M. J. Kamlet, J. L. Abboud, and R. W. Taft, in *Progress in Physical Organic Chemistry*, S. G. Cohen, A. Streitwieser, and R. W. Taft (Eds.), Wiley, New York, 1981, Vol. 13, p. 485, An Examination of Linear Solvation Energy Relationships.

343. C. Laurence, P. Nicolet, M. T. Dalati, J. L. M. Abboud, and R. Notario, *J. Phys. Chem.*, **98**, 5807 (1994). The Empirical-Treatment of Solvent Solute Interactions - 15 Years of π^*.

344. D. V. Matyushov, R. Schmid, and B. M. Ladanyi, *J. Phys. Chem. B*, **101**, 1035 (1997). A Thermodynamic Analysis of the π^* and ET(30) Polarity Scale.

345. Y. Marcus, *Chem. Soc. Rev.*, **22**, 409 (1993). The Properties of Organic Liquids That Are Relevant to Their Use as Solvating Solvents.

346. M. J. Kamlet and R. W. Taft, *J. Am. Chem. Soc.*, **98**, 377 (1976). Solvatochromic Comparison Method. 1. Beta-Scale of Solvent Hydrogen-Bond Acceptor (HBA) Basicities.

347. M. J. Kamlet, T. N. Hall, J. Boykin, and R. W. Taft, *J. Org. Chem.*, **44**, 2599 (1979). Linear Solvation Energy Relationships. 6. Additions to and Correlations with the π^* Scale of Solvent Polarities.

348. G. Saroja and A. Samanta, *Chem. Phys. Lett.*, **246**, 506 (1995). Polarity of the Micelle-Water Interface as Seen by 4-Aminophthalimide, a Solvent Sensitive Fluorescence Probe.

349. K. Kalyanasundaram and J. K. Thomas, *J. Am. Chem. Soc.*, **99**, 2039 (1977). Environmental Effects on Vibronic Band Intensities in Pyrene Monomer Fluorescence and Their Application in Studies of Micellar Systems.

350. K. Kalyanasundaram and J. K. Thomas, *J. Phys. Chem.*, **81**, 2176 (1977). Solvent-Dependent Fluorescence of Pyrene-3-Carboxaldehyde and Its Applications in Estimation of Polarity at Micelle-Water Interfaces.

351. C. A. T. Laia and S. M. B. Costa, *Phys. Chem. Chem. Phys.*, **1**, 4409 (1999). Probing the Interface Polarity of AOT Reversed Micelles Using Centro-Symmetrical Squaraine Molecules.

352. M. F. Vitha, J. D. Weckwerth, K. Odland, V. Dema, and P. W. Carr, *J. Phys. Chem.*, **100**, 18823 (1996). Study of the Polarity and Hydrogen Bond Ability of Sodium Dodecyl Sulfate Micelles by the Kamlet-Taft Solvatochromic Comparison Method.

353. M. F. Vitha and P. W. Carr, *J. Phys. Chem. B*, **102**, 1888 (1998). Study of the Polarity and Hydrogen-Bond Ability of Dodecyltrimethylammonium Bromide Micelles by the Kamlet-Taft Solvatochromic Comparison Method.

354. R. E. Riter, J. R. Kimmel, E. P. Undiks, and N. E. Levinger, *J. Phys. Chem. B*, **101**, 8292 (1997). Novel Reverse Micelles Partitioning Nonaqueous Polar Solvents in a Hydrocarbon Continuous Phase.

355. H. Wang, E. Borguet, and K. B. Eisenthal, *J. Phys. Chem.*, **101**, 713 (1997). Polarity of Liquid Interfaces by Second Harmonic Generation Spectroscopy.

356. H. F. Wang, E. Borguet, and K. B. Eisenthal, *J. Phys. Chem. B*, **102**, 4927 (1998). Generalized Interface Polarity Scale Based on Second Harmonic Spectroscopy.

357. A. A. Tamburello-Luca, P. Hébert, P. F. Brevet, and H. H. Girault, *J. Chem. Soc., Faraday Trans.*, **92**, 3079 (1996). Resonant-Surface Second-Harmonic Generation Studies of Phenol Derivatives at Air/Water and Hexane/Water Interfaces.

358. A. A. Tamburello-Luca, P. Hébert, R. Antoine, P. F. Brevet, and H. H. Girault, *Langmuir*, **13**, 4428 (1997). Optical Surface Second Harmonic Generation Study of the Two Acid/Base Equilibria of Eosin B at the Air/Water Interface.

359. W. H. Steel and R. A. Walker, *J. Am. Chem. Soc.*, **125**, 1132 (2003). Solvent Polarity at an Aqueous/Alkane Interface: The Effect of Solute Identity.

360. W. H. Steel, Y. Y. Lau, C. L. Beildeck, and R. A. Walker, *J. Phys. Chem. B*, **108**, 13370 (2004). Solvent Polarity across Weakly Associating Interfaces.

361. W. H. Steel, C. L. Beildeck, and R. A. Walker, *J. Phys. Chem. B*, **108**, 16107 (2004). Solvent Polarity Across Strongly Associating Interfaces.

362. M. Essfar, G. Guiheneuf, and J. M. Abboud, *J. Am. Chem. Soc.*, **104**, 6786 (1982). Electronic Absorption-Spectra of Polarity Polarizability Indicators in the Gas-Phase.

363. I. Benjamin, *Chem. Rev.*, **106**, 1212 (2006). Static and Dynamic Electronic Spectroscopy at Liquid Interfaces.

364. W. H. Steel, R. Nolan, F. Damkaci, and R. A. Walker, *J. Am. Chem. Soc.*, **124**, 4824 (2002). Molecular Rulers: A New Family of Surfactants for Measuring Interfacial Widths.

365. T. Uchida, A. Yamaguchi, T. Ina, and N. Teramae, *J. Phys. Chem. B*, **104**, 12091 (2000). Observation of Molecular Association at Liquid/Liquid and Solid/Liquid Interfaces by Second Harmonic Generation Spectroscopy.

366. H. J. Kim and J. T. Hynes, *J. Chem. Phys.*, **93**, 5194 (1990). Equilibrium and Nonequilibrium Solvation and Solute Electronic Structure. I. Formulation.

367. H. J. Kim and J. T. Hynes, *J. Chem. Phys.*, **93**, 5211 (1990). Equilibrium and Nonequilibrium Solvation and Solute Electronic Structure. II. Strong Coupling Limit.

368. P. Jacques, *J. Phys. Chem.*, **90**, 5535 (1986). On the Relative Contributions of Nonspecific and Specific Interactions to the Unusual Solvatochromism of a Typical Merocyanine Dye.

369. H. Nagatani, A. Piron, P. F. Brevat, D. J. Fermin, and H. H. Girault, *Langmuir*, **18**, 6647 (2002). Surface Second Harmonic Generation of Cationic Water-Soluble Porphyrins at the Polarized Water/1,2-Dichloroethane Interface.

370. K. S. Suslick, C. T. Chen, G. R. Meredith, and L. T. Cheng, *J. Am. Chem. Soc.*, **114**, 6928 (1992). Push-Pull Porphyrins as Nonlinear Optical-Materials.

371. S. K. Mondal, S. Yamaguchi, and T. Tahara, *J. Phys. Chem. C*, **115**, 3083 (2011). Molecules at the Air/Water Interface Experience a More Inhomogeneous Solvation Environment Than in Bulk Solvents: A Quantitative Band Shape Analysis of Interfacial Electronic Spectra Obtained by HD-ESFG.

372. D. Michael and I. Benjamin, *J. Chem. Phys.*, **107**, 5684 (1997). Electronic Spectra of Dipolar Solutes at Liquid-Liquid Interfaces. Effect of Interface Structure and Polarity.

373. D. Michael and I. Benjamin, *J. Phys. Chem.*, **102**, 5154 (1998). Structure, Dynamics and Electronic Spectrum of N,N'-Diethyl-p-Nitroaniline at Water Interfaces. A Molecular Dynamics Study.

374. I. Benjamin, *Chem. Phys. Lett.*, **287**, 480 (1998). Electronic Spectra in Bulk Water and at the Water Liquid/Vapor Interface. Effect of Solvent and Solute Polarizabilities.

375. I. Benjamin, *Chem. Phys. Lett.*, **393**, 453 (2004). Polarity of the Water/Octanol Interface.

376. F. Franks, *Water: A Comprehensive Treatise*, Plenum, New York, 1982, Vol. 4.

377. N. Nandi, K. Bhattacharyya, and B. Bagchi, *Chem. Rev.*, **100**, 2013 (2000). Dielectric Relaxation and Solvation Dynamics of Water in Complex Chemical and Biological Systems.

378. B. E. Conway, in *The Liquid State and Its Electrical Properties*, E. E. Kunhardt, L. G. Christophorou, and L. H. Luessen (Eds.), NATO ASI series B, Plenum, New York, 1988, Vol. **193**, p. 323, Electrical Aspects of Liquid/Vapor, Liquid/Liquid, and Liquid/Metal Interfaces.

379. R. R. Dogonadze, E. Kalman, A. A. Kornyshev, and J. Ulstrup, *The Chemical Physics of Solvation; Part C*, Elsevier, Amsterdam, 1988.

380. A. M. Kuznetsov, *Charge Transfer in Physics, Chemistry and Biology*, Gordon and Breach, Amsterdam, 1995.

381. I. Benjamin, *J. Phys. Chem. A*, **102**, 9500 (1998). Solvent Effects on Electronic Spectra at Liquid Interfaces. A Continuum Electrostatic Model.

382. K. Shiratori and A. Morita, *J. Chem. Phys.*, **134**, 234705 (2011). Molecular Theory on Dielectric Constant at Interfaces: A Molecular Dynamics Study of the Water/Vapor Interface.

383. R. Kubo, *J. Phys. Soc. Jpn.*, **17**, 1100 (1962). Generalized Cumulant Expansion Method.

384. I. Benjamin, *Chem. Phys. Lett.*, **515**, 56 (2011). Inhomogeneous Broadening of Electronic Spectra at Liquid Interfaces.

385. K. V. Nelson and I. Benjamin, *J. Phys. Chem. B*, **116**, 4286 (2012). Electronic Absorption Line Shapes at the Water Liquid/Vapor Interface.

386. H. Watanabe, S. Yamaguchi, S. Sen, A. Morita, and T. Tahara, *J. Chem. Phys.*, **132**, 144701 (2010). "Half-Hydration" At the Air/Water Interface Revealed by Heterodyne-Detected Electronic Sum Frequency Generation Spectroscopy, Polarization Second Harmonic Generation, and Molecular Dynamics Simulation.

387. D. W. Oxtoby, *Adv. Chem. Phys.*, **47**, 487 (1981). Vibrational Population Relaxation in Liquids.

388. J. Chesnoy and G. M. Gale, *Adv. Chem. Phys.*, **70**(part 2), 297 (1988). Vibrational Relaxation in Condensed Phases.

389. J. C. Owrutsky, D. Raftery, and R. M. Hochstrasser, *Annu. Rev. Phys. Chem.*, **45**, 519 (1994). Vibrational Relaxation Dynamics in Liquids.

390. R. M. Stratt and M. Maroncelli, *J. Phys. Chem.*, **100**, 12981 (1996). Nonreactive Dynamics in Solution: The Emerging Molecular View of Solvation Dynamics and Vibrational Relaxation.

391. M. Stratt and M. Maroncelli, *J. Phys. Chem.*, **100**, 12981 (1996). Nonreactive Dynamics in Solution – The Emerging Molecular View of Solvation Dynamics and Vibrational Relaxation.

392. E. J. Heilweil, M. P. Casassa, R. R. Cavanagh, and J. C. Stephenson, *Annu. Rev. Phys. Chem.*, **40**, 143 (1989). Picosecond Vibrational Energy Transfer Studies of Surface Adsorbates.

393. J. C. Tully, *Annu. Rev. Phys. Chem.*, **51**, 153 (2000). Chemical Dynamics at Metal Surfaces.

394. B. M. Ladanyi and R. M. Stratt, *J. Chem. Phys.*, **111**, 2008 (1999). On the Role of Dielectric Friction in Vibrational Energy Relaxation.

395. R. Rey and J. T. Hynes, *J. Chem. Phys.*, **108**, 142 (1998). Vibrational Phase and Energy Relaxation of CN⁻ in Water.

396. B. M. Ladanyi and R. M. Stratt, *J. Phys. Chem. A.*, **102**, 1068 (1998). Short-Time Dynamics of Vibrational Relaxation in Molecular Fluids.

397. S. A. Egorov and J. L. Skinner, *J. Chem. Phys.*, **112**, 275 (2000). Vibrational Energy Relaxation of Polyatomic Solutes in Simple Liquids and Supercritical Fluids.

398. C. P. Lawrence and J. L. Skinner, *J. Chem. Phys.*, **119**, 1623 (2003). Vibrational Spectroscopy of HOD in Liquid D_2O. VI. Intramolecular and Intermolecular Vibrational Energy Flow.

399. R. D. Coalson and D. G. Evans, *Chem. Phys.*, **296**, 117 (2004). Condensed Phase Vibrational Relaxation: Calibration of Approximate Relaxation Theories with Analytical and Numerically Exact Results.

400. H. J. Bakker, *J. Chem. Phys*, **121**, 10088 (2004). Vibrational Relaxation in the Condensed Phase.

401. A. Ma and R. M. Stratt, *J. Chem. Phys*, **121**, 11217 (2004). Multiphonon Vibrational Relaxation in Liquids: Should It Lead to an Exponential-Gap Law?

402. J. C. Bolinger, T. J. Bixby, and P. J. Reid, *J. Chem. Phys.*, **123**, 084503 (2005). Time-Resolved Infrared Absorption Studies of the Solvent-Dependent Vibrational Relaxation Dynamics of Chlorine Dioxide.

403. G. M. Sando, K. Dahl, Q. Zhong, and J. C. Owrutsky, *J. Phys. Chem. A*, **109**, 5788 (2005). Vibrational Relaxation of Azide in Formamide Reverse Micelles.

404. A. D. Koutselos, *J. Chem. Phys.*, **125**, 244304 (2006). Mixed Quantum-Classical Molecular Dynamics Simulation of Vibrational Relaxation of Ions in an Electrostatic Field.

405. S. Z. Li, J. R. Schmidt, and J. L. Skinner, *J. Chem. Phys.*, **125**, 244507 (2006). Vibrational Energy Relaxation of Azide in Water.

406. I. Navrotskaya and E. Geva, *J. Chem. Phys.*, **127**, 054504 (2007). Comparison between the Landau-Teller and Flux-Flux Methods for Computing Vibrational Energy Relaxation Rate Constants in the Condensed Phase.

407. S. Shigeto, Y. Pang, Y. Fang, and D. D. Dlott, *J. Phys. Chem. B*, **112**, 232 (2008). Vibrational Relaxation of Normal and Deuterated Liquid Nitromethane.

408. Y. S. Lin, S. G. Ramesh, J. M. Shorb, E. L. Sibert, and J. L. Skinner, *J. Phys. Chem. B*, **112**, 390 (2008). Vibrational Energy Relaxation of the Bend Fundamental of Dilute Water in Liquid Chloroform and D-Chloroform.

409. D. J. Shaw, M. R. Panman, and S. Woutersen, *Phys. Rev. Lett.*, **103**, 227401 (2009). Evidence for Cooperative Vibrational Relaxation of the NH⁻, OH⁻, and OD⁻ Stretching Modes in Hydrogen-Bonded Liquids Using Infrared Pump-Probe Spectroscopy.

410. A. Kandratsenka, J. Schroeder, D. Schwarzer, and V. S. Vikhrenko, *J. Chem. Phys.*, **130**, 174507 (2009). Nonequilibrium Molecular Dynamics Simulations of Vibrational Energy Relaxation of HOD in D_2O.

411. B. C. Pein, N. H. Seong, and D. D. Dlott, *J. Phys. Chem. A*, **114**, 10500 (2010). Vibrational Energy Relaxation of Liquid Aryl-Halides X-C_6H_5 (X = F, Cl, Br, I).

412. A. Eftekhari-Bafrooei and E. Borguet, *J. Am. Chem. Soc.*, **132**, 3756 (2010). Effect of Hydrogen-Bond Strength on the Vibrational Relaxation of Interfacial Water.

413. F. X. Vazquez, S. Talapatra, and E. Geva, *J. Phys. Chem. A*, **115**, 9775 (2011). Vibrational

Energy Relaxation in Liquid HCl and DCl Via the Linearized Semiclassical Method: Electrostriction Versus Quantum Delocalization.

414. J. Yi and J. Jonas, *J. Phys. Chem.*, **100**, 16789 (1996). Raman Study of Vibrational and Rotational Relaxation Liquid Benzene-D(6) Confined to Nanoporous Silica Glasses.

415. C. Matranga and P. Guyot-Sionnest, *J. Chem. Phys.*, **112**, 7615 (2000). Vibrational Relaxation of Cyanide at the Metal/Electrolyte Interface.

416. Q. Zhong, A. P. Baronavski, and J. C. Owrutsky, *J. Chem. Phys.*, **118**, 7074 (2003). Vibrational Energy Relaxation of Aqueous Azide Ion Confined in Reverse Micelles.

417. Q. Zhong, A. P. Baronavski, and J. C. Owrutsky, *J. Chem. Phys.*, **119**, 9171 (2003). Reorientation and Vibrational Energy Relaxation of Pseudohalide Ions Confined in Reverse Micelle Water Pools.

418. A. G. Kalampounias, S. N. Yannopoulos, W. Steffen, L. I. Kirillova, and S. A. Kirillov, *J. Chem. Phys.*, **118**, 8340 (2003). Short-Time Dynamics of Glass-Forming Liquids: Phenyl Salicylate (Salol) in Bulk Liquid, Dilute Solution, and Confining Geometries.

419. S. M. Li, T. D. Shepherd, and W. H. Thompson, *J. Phys. Chem. A*, **108**, 7347 (2004). Simulations of the Vibrational Relaxation of a Model Diatomic Molecule in a Nanoconfined Polar Solvent.

420. A. M. Dokter, S. Woutersen, and H. J. Bakker, *Phys. Rev. Lett.*, **94**, 178301 (2005). Anomalous Slowing Down of the Vibrational Relaxation of Liquid Water Upon Nanoscale Confinement.

421. A. Ghosh, M. Smits, M. Sovago, J. Bredenbeck, M. Muller, and M. Bonn, *Chem. Phys.*, **350**, 23 (2008). Ultrafast Vibrational Dynamics of Interfacial Water.

422. E. L. Hommel, G. Ma, and H. C. Allen, *Analytical Sciences*, **17**, 1325 (2001). Broadband Vibrational Sum Frequency Generation Spectroscopy of a Liquid Surface.

423. A. N. Bordenyuk and A. V. Benderskii, *J. Chem. Phys*, **122**, 134713 (2005). Spectrally and Time-Resolved Vibrational Surface Spectroscopy: Ultrafast Hydrogen-Bonding Dynamics at D_2O/CaF_2 Interface.

424. I. Benjamin, *J. Phys. Chem. B.*, **106**, 9375 (2006). Theoretical Studies of Solute Vibrational Energy Relaxation at Liquid Interfaces.

425. H. Gai and G. A. Voth, *J. Chem. Phys.*, **99**, 740 (1993). Vibrational-Energy Relaxation of Si-H Stretching Modes on the H/Si(111)1x1 Surface.

426. S. A. Egorov and J. L. Skinner, *J. Chem. Phys.*, **105**, 7047 (1996). A Theory of Vibrational Energy Relaxation in Liquids.

427. R. Rey and J. T. Hynes, *J. Chem. Phys.*, **104**, 2356 (1996). Vibrational Energy Relaxation of HOD in Liquid D_2O.

428. J. L. Skinner and K. Park, *J. Phys. Chem. B*, **105**, 6716 (2001). Calculating Vibrational Energy Relaxation Rates from Classical Molecular Dynamics Simulations: Quantum Correction Factors for Processes Involving Vibration-Vibration Energy Transfer.

429. R. M. Whitnell, K. R. Wilson, and J. T. Hynes, *J. Chem. Phys.*, **96**, 5354 (1992). Vibrational Relaxation of a Dipolar Molecule in Water.

430. L. Landau and E. Teller, *Phys. Z. Sovietunion*, **10**, 34 (1936). Theory of Sound Dispersion.

431. R. Zwanzig, *Ann. Rev. Phys. Chem.*, **16**, 67 (1965). Time-Correlation Functions and Transport Coefficients in Statistical Mechanics.

432. J. S. Bader and B. J. Berne, *J. Chem. Phys.*, **100**, 8359 (1994). Quantum and Classical Relaxation Rates from Classical Simulations.

433. J. L. Skinner, *J. Chem. Phys.*, **107**, 8717 (1997). Semiclassical Approximations to Golden Rule Rate Constants.

434. J. Vieceli, I. Chorny, and I. Benjamin, *J. Chem. Phys.*, **117**, 4532 (2002). Vibrational Relaxation at Water Surfaces.

435. I. Chorny, J. Vieceli, and I. Benjamin, *J. Phys. Chem. B*, **107**, 229 (2003). Photodissociation and Vibrational Relaxation of OClO at Liquid Surfaces.

436. I. Chorny and I. Benjamin, *J. Mol. Liq.*, **110**, 133 (2004). Molecular Dynamics Study of the Vibrational Relaxation of OCl and OCl⁻ in the Bulk and the Surface of Water and Acetonitrile.

437. I. Benjamin, *J. Chem. Phys.*, **121**, 10223 (2004). Vibrational Relaxation at the Liquid/Liquid Interface.

438. J. Vieceli, I. Chorny, and I. Benjamin, *Chem. Phys. Lett.*, **364**, 446 (2002). Vibrational Relaxation of ICN in Bulk and Surface Chloroform.

439. R. G. Gordon, *J. Chem. Phys.*, **42**, 3658 (1965). Relations between Raman Spectroscopy and Nuclear Spin Relaxation.

440. C. Hu and R. Zwanzig, *J. Chem. Phys.*, **60**, 4354 (1974). Rotational Friction Coefficients for Spheroids with the Slipping Boundary Condition.

441. V. E. Bondybey and L. E. Brus, *J. Chem. Phys.*, **62**, 620 (1975). Rigid Cage Effect on ICl Photodissociation and *B* O⁺ Fluorescence in Rare-Gas Matrices.

442. J. T. Hynes, R. Kapral, and M. Weinberg, *J. Chem. Phys.*, **67**, 3256 (1977). Microscopic Boundary Layer Effects and Rough Sphere Rotation.

443. J. L. Dote, D. Kivelson, and R. N. Schwartz, *J. Phys. Chem.*, **85**, 2169 (1981). A Molecular Quasi-Hydrodynamic Free-Space Model for Molecular Rotational Relaxation in Liquids.

444. A. J. Barnes, W. J. Orville-Thomas, and J. Yarwood, *Molecular Liquids Dynamics and Interactions*, Dordrecht, Reidel, 1984.

445. G. van der Zwan and J. T. Hynes, *J. Phys. Chem.*, **89**, 4181 (1985). Time-Dependent Fluorescence Solvent Shifts, Dielectric Friction, and Nonequilibrium Solvation in Polar Solvents.

446. C. H. Wang, *Spectroscopy of Condensed Media*, New York, Academic Press, 1985.

447. B. J. Berne and R. Pecora, *Dynamic Light Scattering*, Robert E. Krieger, Malabar, 1990.

448. Y. Hu and G. R. Fleming, *J. Chem. Phys.*, **94**, 3857 (1991). Molecular Dynamics Study of Rotational Reorientation of Tryptophan and Several Indoles in Water.

449. J. M. Polson, J. D. D. Fyfe, and K. R. Jeffrey, *J. Chem. Phys.*, **94**, 3381 (1991). The Reorientation of t-Butyl Groups in Butylated Hydroxytoluene: A Deuterium Nuclear Magnetic Resonance Spectral and Relaxation Time Study.

450. M. Cho, S. J. Rosenthal, N. F. Scherer, L. D. Ziegler, and G. R. Fleming, *J. Chem. Phys.*, **96**, 5033 (1992). Ultrafast Solvent Dynamics: Connection between Time Resolved Fluorescence and Optical Kerr Measurements.

451. Y. J. Chang and E. W. Castner, *J. Chem. Phys.*, **99**, 113 (1993). Femtosecond Dynamics of Hydrogen-Bonding Solvents. Formamide and N-Methylformamide in Acetonitrile, DMF, and Water.

452. A. I. Krylov and R. B. Gerber, *J. Chem. Phys.*, **100**, 4242 (1994). Photodissociation of ICN in Solid and Liquid Ar⁻ Dynamics of the Cage Effect and of Excited-State Isomerization.

453. M. P. Heitz and F. V. Bright, *J. Phys. Chem.*, **100**, 6889 (1996). Probing the Scale of Local Density Augmentation in Supercritical Fluids: A Picosecond Rotational Reorientation Study.

454. G. S. Jas, Y. Wang, S. W. Pauls, C. K. Johnson, and K. Kuczera, *J. Chem. Phys.*, **107**, 8800 (1997). Influence of Temperature and Viscosity on Anthracene Rotational Diffusion in Organic Solvents: Molecular Dynamics Simulations and Fluorescence Anisotropy Study.

455. M.-L. Horng, J. A. Gardecki, and M. Maroncelli, *J. Phys, Chem, A.*, **101**, 1030 (1997). Rotational Dynamics of Coumarin 153: Time-Dependent Friction, Dielectric Friction, and Other Nonhydrodynamic Effects.

456. C. J. Bardeen, J. Che, K. R. Wilson, V. V. Yakovlev, V. A. Apkarian, C. C. Martens, R. Zadoyan, B. Kohler, and M. Messina, *J. Chem. Phys*, **106**, 8486 (1997). Quantum Control of I_2 in the Gas Phase and in Condensed Phase Solid Kr Matrix.

457. A. Blokhin and M. F. Gelin, *J. Phys. Chem. B*, **101**, 236 (1997). Rotation of Nonspherical Molecules in Dense Fluids: A Simple Model Description.

458. J. S. Baskin, M. Chachisvilis, M. Gupta, and A. H. Zewail, *J. Phys. Chem. A*, **102**, 4158 (1998). Femtosecond Dynamics of Solvation: Microscopic Friction and Coherent Motion in Dense Fluids.

459. M. G. Kurnikova, N. Balabai, D. H. Waldeck, and R. D. Coalson, *J. Am. Chem. Soc.*, **120**, 6121 (1998). Rotational Relaxation in Polar Solvents. Molecular Dynamics Study of Solute-Solvent Interaction.

460. A. Idrissi, M. Ricci, P. Bartolini, and R. Righini, *J. Chem. Phys.*, **111**, 4148 (1999). Optical Kerr-Effect Investigation of the Reorientational Dynamics of CS_2 in CCl_4 Solutions.

461. D. J. Cook, J. X. Chen, E. A. Morlino, and R. M. Hochstrasser, *Chem. Phys. Lett.*, **309**, 221 (1999). Terahertz-Field-Induced Second-Harmonic Generation Measurements of Liquid Dynamics.

462. O. Kajimoto, *Chem. Rev.*, **99**, 355 (1999). Solvation in Supercritical Fluids: Its Effects on Energy Transfer and Chemical Reactions.

463. J. Jang and R. M. Stratt, *J. Chem. Phys.*, **112**, 7524 (2000). The Short-Time Dynamics of Molecular Reorientation in Liquids. 1. The Instantaneous Generalized Langevin Equation.

464. G. B. Dutt, T. K. Ghanty, and K. K. Singh, *J. Chem. Phys.*, **115**, 10845 (2001). Rotational Dynamics of Neutral Red in Dimethylsulfoxide: How Important Is the Solute's Charge in Causing Additional Friction?

465. J. E. Adams and A. Siavosh-Haghighi, *J. Phys. Chem. B*, **106**, 7973 (2002). Rotational Relaxation in Supercritical CO_2.

466. Y. Zhang, J. Jiang, and M. A. Berg, *J. Chem. Phys.*, **118**, 7534 (2003). Ultrafast Dichroism Spectroscopy of Anthracene in Solution. IV. Merging of Inertial and Diffusive Motions in Toluene (and References Therein for Earlier Work by the Same Authors).

467. T.-M. Chang and L. X. Dang, *J. Chem. Phys.*, **118**, 8813 (2003). On Rotational Dynamics of an NH_4^+ Ion in Water.

468. A. C. Moskun, A. E. Jailaubekov, S. E. Bradforth, G. Tao, and R. M. Stratt, *Science*, **311**, 1907 (2006). Rotational Coherence and a Sudden Breakdown in Linear Response Seen in Room-Temperature Liquids.

469. A. Castro, E. V. Sitzmann, D. Zhang, and K. B. Eisenthal, *J. Phys. Chem.*, **95**, 6752 (1991). Rotational Relaxation at the Air/Water Interface by Time-Resolved Second Harmonic Generation.

470. K. B. Eisenthal, *J. Phys. Chem.*, **100**, 12997 (1996). Photochemistry and Photophysics of Liquid Interfaces by Second Harmonic Spectroscopy.

471. D. Zimdars, J. I. Dadap, K. B. Eisenthal, and T. F. Heinz, *J. Phys. Chem. B.*, **103**, 3425 (1999). Anisotropic Orientational Motion of Molecular Adsorbates at the Air-Water Interface.

472. R. Antoine, A. A. Tamburello-Luca, P. Hebert, P. F. Brevet, and H. H. Girault, *Chem. Phys. Lett.*, **288**, 138 (1998). Picosecond Dynamics of Eosin B at the Air/Water Interface by Time-Resolved Second Harmonic Generation: Orientational Randomization and Rotational Relaxation.

473. K. T. Nguyen, X. Shang, and K. B. Eisenthal, *J. Phys. Chem. B.*, **110**, 19788 (2006). Molecular Rotation at Negatively Charged Surfactant/Aqueous Interfaces.

474. Y. Rao, D. H. Song, N. J. Turro, and K. B. Eisenthal, *J. Phys. Chem. B*, **112**, 13572 (2008). Orientational Motions of Vibrational Chromophores in Molecules at the Air/Water Interface with Time-Resolved Sum Frequency Generation.

475. X. M. Shang, K. Nguyen, Y. Rao, and K. B. Eisenthal, *J. Phys. Chem. C*, **112**, 20375 (2008). In-Plane Molecular Rotational Dynamics at a Negatively Charged Surfactant/Aqueous Interface.

476. K. Raghavan, K. Foster, K. Motakabbir, and M. Berkowitz, *J. Chem. Phys.*, **94**, 2110 (1991). Structure and Dynamics of Water at the Pt(111) Interface: Molecular Dynamics Study.

477. D. A. Pantano and D. Laria, *J. Phys. Chem. B*, **107**, 2971 (2003). Molecular Dynamics Study of Solvation of Coumarin-314 at the Water/Air Interface.

478. I. Benjamin, *J. Chem. Phys.*, **127**, 204712 (2007). Solute Rotational Dynamics at the Water Liquid/Vapor Interface.

479. I. Benjamin, *J. Phys. Chem. C.*, **112**, 8969 (2008). Solute Orientational Dynamics at the Water/Carbon Tetrachloride Interface.

480. P. Debye, *Polar Molecules*, New York, Dover, 1945.

481. S. J. Byrnes, P. L. Geissler, and Y. R. Shen, *Chem. Phys. Lett.*, **516**, 115 (2011). Ambiguities in Surface Nonlinear Spectroscopy Calculations.

482. J. P. Ryckaert, G. Ciccotti, and H. J. C. Berendsen, *J. Comput. Phys.*, **23**, 327 (1977). Numerical Integration of the Cartesian Equations of Motion of a System with Constraints: Molecular Dynamics of n-Alkanes.

483. G. Tao and R. M. Stratt, *J. Chem. Phys.*, **125**, 114501 (2006). The Molecular Origins of Nonlinear Response in Solute Energy Relaxation: The Example of High-Energy Rotational Relaxation.

484. M. L. Johnson, C. Rodriguez, and I. Benjamin, *J. Phys. Chem. A*, **113**, 2086 (2009). Rotational Dynamics of Strongly Adsorbed Solute at the Water Surface.

485. T. W. Nee and R. Zwanzig, *J. Chem. Phys.*, **52**, 6353 (1970). Theory of Dielectric Relaxation in Polar Liquids.

486. D. S. Alavi and D. H. Waldeck, *J. Chem. Phys.*, **94**, 6196 (1991). Rotational Dielectric Friction on a Generalized Charge Distribution.

487. D. S. Alavi and D. H. Waldeck, *J. Chem. Phys.*, **98**, 3580 (1993). Erratum: Rotational Dielectric Friction on a Generalized Charge Distribution [J. Chem. Phys. 94, 6196 (1991).].

488. L. R. Pratt and A. Pohorille, *Chem. Rev.*, **102**, 2671 (2002). Hydrophobic Effects and Modeling of Biophysical Aqueous Solution Interfaces.

489. M. Maroncelli, *J. Mol. Liq*, **57**, 1 (1993). The Dynamics of Solvation in Polar Liquids.

490. P. J. Rossky and J. D. Simon, *Nature*, **370**, 263 (1994). Dynamics of Chemical Processes in Polar Solvents.

491. B. Bagchi and B. Jana, *Chem. Soc. Rev.*, **39**, 1936 (2010). Solvation Dynamics in Dipolar Liquids.

492. W. H. Thompson, *Ann. Rev. Phys. Chem.*, **62**, 599 (2011). Solvation Dynamics and Proton Transfer in Nanoconfined Liquids.

493. J. D. Simon, *Acc. Chem. Res.*, **21**, 128 (1988). Time-Resolved Studies of Solvation in Polar Media.

494. P. F. Barbara and W. Jarzeba, *Adv. Photochem.*, **15**, 1 (1990). Ultrafast Photochemical Intramolecular Charge and Excited State Solvation.

495. R. Jimenez, G. R. Fleming, P. V. Kumar, and M. Maroncelli, *Nature*, **369**, 471 (1994). Femtosecond Solvation Dynamics of Water.

496. G. C. Walker, W. Jarzeba, T. J. Kang, A. E. Johnson, and P. F. Barbara, *J. Opt. Soc. Am. B*, **7**, 1521 (1990). Ultraviolet Femtosecond Fluorescence Spectroscopy - Techniques and Applications.

497. B. M. Ladanyi and M. Maroncelli, *J. Chem. Phys.*, **109**, 3204 (1998). Mechanisms of Solvation Dynamics of Polyatomic Solutes in Polar and Nondipolar Solvents: A Simulation Study.

498. M. N. Kobrak, *J. Chem. Phys.*, **127** (2007). A Comparative Study of Solvation Dynamics in Room-Temperature Ionic Liquids.

499. D. F. Underwood and D. A. Blank, *J. Phys. Chem. A*, **107**, 9736 (2003). Ultrafast Solvation Dynamics: A View from the Solvent's Prospective Using a Novel Resonant-Pump, Nonresonant-Probe Technique.

500. N. E. Levinger, *Curr. Opin. Coll. Interf. Sci.*, **5**, 118 (2000). Ultrafast Dynamics in Reverse Micelles, Microemulsions, and Vesicles.

501. N. E. Levinger, *Science*, **298**, 1722 (2002). Water in Confinement.

502. E. M. Corbeil and N. E. Levinger, *Langmuir*, **19**, 7264 (2003). Dynamics of Polar Solvation in Quaternary Microemulsions.

503. J. A. Ingram, R. S. Moog, N. Ito, R. Biswas, and M. Maroncelli, *J. Phys. Chem. B*, **107**, 5926 (2003). Solute Rotation and Solvation Dynamics in a Room-Temperature Ionic Liquid.

504. H. Shirota and H. Segawa, *Langmuir*, **20**, 329 (2004). Solvation Dynamics of Formamide and N,N-Dimethylformamide in Aerosol OT Reverse Micelles.

505. M. Sakurai and A. Yoshimori, *J. Chem. Phys.*, **122**, 104509 (2005). Bandwidth Analysis of Solvation Dynamics in a Simple Liquid Mixture.

506. S. Dey, D. K. Sasmal, D. K. Das, and K. Bhattacharyya, *Chemphyschem*, **9**, 2848 (2008). A Femtosecond Study of Solvation Dynamics and Anisotropy Decay in a Catanionic Vesicle: Excitation-Wavelength Dependence.

507. T. N. Burai and A. Datta, *J. Phys. Chem. B*, **113**, 15901 (2009). Slow Solvation Dynamics in the Microheterogeneous Water Channels of Nafion Membranes.

508. H. K. Kashyap and R. Biswas, *J. Phys. Chem. B*, **114**, 254 (2010). Solvation Dynamics of Dipolar Probes in Dipolar Room Temperature Ionic Liquids: Separation of Ion-Dipole and Dipole-Dipole Interaction Contributions.

509. P. Setua, C. Ghatak, V. G. Rao, S. K. Das, and N. Sarkar, *J. Phys. Chem. B*, **116**, 3704 (2012). Dynamics of Solvation and Rotational Relaxation of Coumarin 480 in Pure Aqueous-AOT Reverse Micelle and Reverse Micelle Containing Different-Sized Silver Nanoparticles inside Its Core: A Comparative Study.

510. M. Yanagimachi, N. Tamai, and H. Masuhara, *Chem. Phys. Lett.*, **200**, 469 (1992). Solvation Dynamics of a Coumarin Dye at Liquid Solid Interface Layer - Picosecond Total Internal-Reflection Fluorescence Spectroscopic Study.

511. T. Yamashita, T. Uchida, T. Fukushima, and N. Teramae, *J. Phys. Chem. B*, **107**, 4786 (2003). Solvation Dynamics of Fluorophores with an Anthroyloxy Group at the Heptane/Water Interface as Studied by Time-Resolved Total Internal Reflection Fluorescence Spectroscopy.

512. T. Yamashita, Y. Amino, A. Yamaguchi, and N. Teramae, *Chem. Lett.*, **34**, 988 (2005). Solvation Dynamics at the Water/Mica Interface as Studied by Time-Resolved Fluorescence Spectroscopy.

513. D. Zimdars, J. I. Dadap, K. B. Eisenthal, and T. F. Heinz, *Chem. Phys. Lett.*, **301**, 112 (1999). Femtosecond Dynamics of Solvation at the Air/Water Interface.

514. D. Zimdars and K. B. Eisenthal, *J. Phys. Chem. A*, **103**, 10567 (1999). Effect of Solute Orientation on Solvation Dynamics at the Air/Water Interface.

515. A. V. Benderskii and K. B. Eisenthal, *J. Phys. Chem. B.*, **104**, 11723 (2000). Effect of Organic Surfactant on Femtosecond Solvation Dynamics at the Air-Water Interface.

516. A. V. Benderskii and K. B. Eisenthal, *J. Phys. Chem. B.*, **105**, 6698 (2001). Aqueous Solvation Dynamics at the Anionic Surfactant Air/Water Interface.

517. A. V. Benderskii and K. B. Eisenthal, *J. Phys. Chem. B*, **108**, 3376 (2004). Effect of Organic Surfactant on Femtosecond Solvation Dynamics at the Air-Water Interface (Vol 104, p. 11723, 2000.).

518. Y. Rao, N. J. Turro, and K. B. Eisenthal, *J. Phys. Chem. C.*, **114**, 17703 (2010). Solvation Dynamics at the Air/Water Interface with Time-Resolved Sum-Frequency Generation.

519. A. V. Benderskii and K. B. Eisenthal, *J. Phys. Chem. A.*, **106**, 7482 (2002). Dynamical Time Scales of Aqueous Solvation at Negatively Charged Lipid/Water Interface.

520. A. V. Benderskii, J. Henzie, S. Basu, X. M. Shang, and K. B. Eisenthal, *J. Phys. Chem. B*, **108**, 14017 (2004). Femtosecond Aqueous Solvation at a Positively Charged Surfactant/Water Interface.

521. E. A. Carter and J. T. Hynes, *J. Chem. Phys.*, **94**, 5961 (1991). Solvation Dynamics for an Ion Pair in a Polar Solvent: Time Dependent Fluorescence and Photochemical Charge Transfer.

522. T. Fonseca and B. M. Ladanyi, *J. Phys. Chem.*, **95**, 2116 (1991). Breakdown of Linear Response for Solvation Dynamics in Methanol.

523. T. Fonseca and B. M. Ladanyi, *J. Mol. Liq.*, **60**, 1 (1994). Solvation Dynamics in Methanol – Solute and Perturbation Dependence.

524. M. S. Skaf and B. M. Ladanyi, *J. Phys. Chem.*, **100**, 18258 (1996). Molecular Dynamics Simulation of Solvation Dynamics in Methanol–Water Mixtures.

525. P. L. Geissler and D. Chandler, *J. Chem. Phys.*, **113**, 9759 (2000). Importance Sampling and the Theory of Nonequilibrium Solvation Dynamics in Water.

526. A. E. Bragg, M. C. Cavanagh, and B. J. Schwartz, *Science*, **321**, 1817 (2008). Linear Response Breakdown in Solvation Dynamics Induced by Atomic Electron-Transfer Reactions.

527. P. G. Wolynes, *J. Chem. Phys.*, **86**, 5133 (1987). Linearized Microscopic Theories of Nonequilibrium Solvation.

528. I. Rips, J. Klafter, and J. Jortner, *J. Chem. Phys.*, **89**, 4288 (1988). Solvation Dynamics in Polar Liquids.

529. A. L. Nichols and D. F. Calef, *J. Chem. Phys.*, **89**, 3783 (1988). Polar Solvent Relaxation: The Mean Spherical Approximation Approach.

530. B. Bagchi, *Annu. Rev. Phys. Chem.*, **40**, 115 (1989). Dynamics of Solvation and Charge Transfer Reactions in Dipolar Liquids.

531. M. Maroncelli, *J. Chem. Phys.*, **94**, 2084 (1991). Computer Simulations of Solvation Dynamics in Acetonitrile.

532. I. Benjamin, in *Reaction Dynamics in Clusters and Condensed Phases*, J. Jortner, R. D. Levine, and B. Pullman (Eds.), Kluwer, Dordrecht, The Netherlands, 1994, p. 179, Solvation and Charge Transfer at Liquid Interfaces.

533. A. Chandra and B. Bagchi, *J. Chem. Phys.*, **94**, 3177 (1991). Inertial Effects in Solvation Dynamics.

534. I. Benjamin, *Chem. Phys.*, **180**, 287 (1994). Solvent Dynamics Following Charge Transfer at the Liquid-Liquid Interface.

535. D. Michael and I. Benjamin, *J. Phys. Chem.*, **99**, 16810 (1995). Proposed Experimental Probe of the Liquid/Liquid Interface Structure: Molecular Dynamics of Charge Transfer at the Water/Octanol Interface.

536. D. Michael and I. Benjamin, *J. Chem. Phys.*, **114**, 2817 (2001). Molecular Dynamics Computer Simulations of Solvation Dynamics at Liquid/Liquid Interfaces.

537. I. Benjamin, *Chem. Phys. Lett.*, **469**, 229 (2009). Solute Dynamics at Aqueous Interfaces.

538. D. Chandler, *J. Chem. Phys.*, **68**, 2959 (1978). Statistical Mechanics of Isomerization Dynamics in Liquids and the Transition State Approximation.

539. J. T. Hynes, in *The Theory of Chemical Reactions*, M. Baer (Ed.), CRC Press, Boca Raton, FL, 1985, Vol. **4**, p. 171, The Theory of Reactions in Solution.

540. B. J. Berne, M. Borkovec, and J. E. Straub, *J. Phys. Chem.*, **92**, 3711 (1988). Classical and Modern Methods in Reaction Rate Theory.

541. R. M. Whitnell and K. R. Wilson, in *Reviews in Computational Chemistry*, K. B. Lipkowitz and D. B. Boyd (Eds.), VCH, New York, 1993, Vol. **4**, p. 67, Computational Molecular Dynamics of Chemical Reactions in Solution.

542. C. Dellago, P. G. Bolhuis, F. S. Csajka, and D. Chandler, *J. Chem. Phys.*, **108**, 1964 (1998). Transition Path Sampling and the Calculation of Rate Constants.

543. B. Peters, *Molecular Simulation*, **36**, 1265 (2010). Recent Advances in Transition Path Sampling: Accurate Reaction Coordinates, Likelihood Maximisation and Diffusive Barrier-Crossing Dynamics.

544. I. Benjamin, *Prog. React. Kin. Mech.*, **27**, 87 (2002). Chemical Reaction Dynamics at Liquid Interfaces. A Computational Approach.

545. S. Z. Wang, R. Bianco, and J. T. Hynes, *Computational and Theoretical Chemistry*, **965**, 340 (2011). An Atmospherically Relevant Acid: HNO_3.

546. Y. I. Kharkats and A. G. Volkov, *J. Electroanal. Chem.*, **184**, 435 (1985). Interfacial Catalysis: Multielectron Reactions at the Liquid-Liquid Interface.

547. W. Schmickler, *Interfacial Electrochemistry*, Oxford University Press, Oxford, 1996.

548. D. J. Fermin, H. D. Duong, Z. F. Ding, P. F. Brevet, and H. H. Girault, *Electrochem. Commun.*, **1**, 29 (1999). Solar Energy Conversion Using Dye-Sensitised Liquid Vertical Bar Liquid Interfaces.

549. R. Lahtinen, D. J. Fermin, K. Kontturi, and H. H. Girault, *J. Electroanal. Chem.*, **483**, 81 (2000). Artificial Photosynthesis at Liquid Vertical Bar Liquid Interfaces: Photoreduction of Benzoquinone by Water Soluble Porphyrin Species.

550. P. F. Barbara and E. J. J. Olson, *Adv. Chem. Phys.*, **107**, 647 (1999). Experimental Electron Transfer Kinetics in a DNA Environment.

551. A. G. Volkov and D. W. Deamer, *Liquid-Liquid Interfaces*, CRC, Boca Raton, 1996.

552. C. Wei, A. J. Bard, and M. V. Mirkin, *J. Phys. Chem.*, **99**, 16033 (1995). Scanning Electrochemical Microscopy. 31. Application of SECM to the Study of Charge-Transfer Processes at the Liquid-Liquid Interface.

553. M. Tsionsky, A. J. Bard, and M. V. Mirkin, *J. Phys. Chem.*, **100**, 17881 (1996). Scanning Electrochemical Microscopy. 34. Potential Dependence of the Electron-Transfer Rate and Film Formation at the Liquid/Liquid Interface.

554. B. Liu and M. V. Mirkin, *J. Am. Chem. Soc.*, **121**, 8352 (1999). Potential-Independent Electron Transfer Rate at the Liquid/Liquid Interface.

555. B. Liu and M. V. Mirkin, *J. Phys. Chem. B*, **106**, 3933 (2002). Electron Transfer at Liquid/Liquid Interfaces. The Effects of Ionic Adsorption, Electrolyte Concentration, and Spacer Length on the Reaction Rate.

556. J. Zhang and P. R. Unwin, *Phys. Chem. Chem. Phys.*, **4**, 3820 (2002). Microelectrochemical Measurements of Electron Transfer Rates at the Interface between Two Immiscible Electrolyte Solutions: Potential Dependence of the Ferro/Ferricyanide-7,7,8,8-Tetracyanoquinodimethane (TCNQ)/TCNQ System.

557. Y. M. Bai, P. Sun, M. Q. Zhang, Z. Gao, Z. Y. Yang, and Y. H. Shao, *Electrochimica Acta*, **48**, 3447 (2003). Effects of Solution Viscosity on Heterogeneous Electron Transfer across a Liquid/Liquid Interface.

558. D. G. Georganopoulou, M. V. Mirkin, and R. W. Murray, *Nano Letters*, **4**, 1763 (2004). SECM Measurement of the Fast Electron Transfer Dynamics between Au_{38}^{1+} Nanoparticles and Aqueous Redox Species at a Liquid/Liquid Interface.

559. M. Q. Zhang, H. Liu, H. Hu, S. B. Xie, P. Jing, Y. Kou, and Y. H. Shao, *Chemical Journal of Chinese Universities-Chinese*, **27**, 1355 (2006). Studies of Electron Transfer Reactions at the Interface between Room-Temperature Ionic Liquid+1,2-Dichloroethane Solutions and Water by Scanning Electrochemical Microscopy.

560. F. Li, A. L. Whitworth, and P. R. Unwin, *J. Electroanal. Chem.*, **602**, 70 (2007). Measurement of Rapid Electron Transfer across a Liquid/Liquid Interface from 7,7,8,8-Tetracyanoquinodimethane Radical Anion in 1,2-Dichloroethane to Aqueous Tris(2,2-Bipyridyl)-Ruthenium (III).

561. C. Shi and F. C. Anson, *J. Phys. Chem. B.*, **103**, 6283 (1999). Electron Transfer Between Reactants Located on Opposite Sides of Liquid/Liquid Interfaces.

562. C. Shi and F. C. Anson, *J. Phys. Chem. B.*, **105** (2001). Rates of Electron-Transfer Across Liquid/Liquid Interfaces. Effects of Changes in Driving Force and Reaction Reversibility.

563. K. L. Kott, D. A. Higgins, R. J. McMahon, and R. M. Corn, *J. Am. Chem. Soc.*, **115**, 5342 (1993). Observation of Photoinduced Electron Transfer at a Liquid-Liquid Interface by Optical Second Harmonic Generation.

564. R. A. W. Dryfe, Z. Ding, R. G. Wellington, P. F. Brevet, A. M. Kuznetsov, and H. H. Girault, *J. Phys. Chem. A.*, **101**, 2519 (1997). Time-Resolved Laser-Induced Fluorescence Study of Photoinduced Electron Transfer at the Water/1,2-Dichloroethane Interface.

565. K. Weidemaier, H. L. Tavernier, and M. D. Fayer, *J. Phys. Chem. B.*, **101**, 9352 (1997). Photoinduced Electron Transfer on the Surfaces of Micelles.

566. D. J. Fermin, Z. Ding, H. D. Duong, P.-F. Brevet, and H. H. Girault, *J. Phys. Chem. B.*, **102**, 10334 (1998). Photoinduced Electron Transfer at Liquid/Liquid Interfaces. 1. Photocurrent Measurements Associated with Heterogeneous Quenching of Zinc Porphyrins.

567. D. J. Fermin, H. D. Duong, Z. Ding, P.-F. Brevet, and H. H. Girault, *J. Am. Chem. Soc.*, **121**, 10203 (1999). Photoinduced Electron Transfer at Liquid/Liquid Interfaces. Part III. Photoelectrochemical Responses Involving Porphyrin Ion Pairs.

568. H. Jensen, J. J. Kakkassery, H. Nagatani, D. J. Fermin, and H. H. Girault, *J. Am. Chem. Soc.*, **122**, 10943 (2000). Photoinduced Electron Transfer at Liquid/Liquid Interfaces. Part IV. Orientation and Reactivity of Zinc Tetra(4-Carboxyphenyl) Porphyrin Self-Assembled at the Water|1,2-Dichloroethane Junction.

569. N. Eugster, D. J. Fermin, and H. H. Girault, *J. Phys. Chem. B*, **106**, 3428 (2002). Photoinduced Electron Transfer at Liquid/Liquid Interfaces. Part VI. On the Thermodynamic Driving Force Dependence of the Phenomenological Electron-Transfer Rate Constant.

570. N. Eugster, D. J. Fermin, and H. H. Girault, *J. Am. Chem. Soc.*, **125**, 4862 (2003). Photoinduced Electron Transfer at Liquid|Liquid Interfaces: Dynamics of the Heterogeneous Photoreduction of Quinones by Self-Assembled Porphyrin Ion Pairs.

571. E. A. McArthur and K. B. Eisenthal, *J. Am. Chem. Soc.*, **128**, 1068 (2006). Ultrafast Excited-State Electron Transfer at an Organic Liquid/Aqueous Interface.

572. S. Ghosh, K. Sahu, S. K. Mondal, P. Sen, and K. Bhattacharyya, *J. Chem. Phys.*, **125**, 054509 (2006). A Femtosecond Study of Photoinduced Electron Transfer from Dimethylaniline to Coumarin Dyes in a Cetyltrimethylammonium Bromide Micelle.

573. S. Ghosh, S. K. Mondal, K. Sahu, and K. Bhattacharyya, *J. Phys. Chem. A.*, **110**, 13139 (2006). Ultrafast Electron Transfer in a Nanocavity. Dimethylaniline to Coumarin Dyes in Hydroxypropyl Gamma-Cyclodextrin.

574. S. Ghosh, S. K. Mondal, K. Sahu, and K. Bhattacharyya, *J. Chem. Phys.*, **126**, 204708 (2007). Ultrafast Photoinduced Electron Transfer from Dimethylaniline to Coumarin Dyes in Sodium Dodecyl Sulfate and Triton X-100 Micelles.

575. A. Chakraborty, D. Chakrabarty, P. Hazra, D. Seth, and N. Sarkar, *Chem. Phys. Lett.*, **382**, 508 (2003). Photoinduced Intermolecular Electron Transfer between Coumarin Dyes and Electron Donating Solvents in Cetyltrimethylammonium Bromide (CTAB) Micelles: Evidence for Marcus Inverted Region.

576. A. Chakraborty, D. Seth, D. Chakrabarty, P. Hazra, and N. Sarkar, *Chem. Phys. Lett.*, **405**, 18 (2005). Photoinduced Electron Transfer from Dimethyl Aniline to Coumarin Dyes in Reverse Micelles.

577. A. Chakraborty, D. Seth, P. Setua, and N. Sarkar, *J. Phys. Chem. B*, **110**, 16607 (2006). Photoinduced Electron Transfer in a Protein-Surfactant Complex: Probing the Interaction of SDS with BSA.

578. A. Chakraborty, D. Seth, P. Setua, and N. Sarkar, *J. Chem. Phys.*, **124**, 074512 (2006). Photoinduced Electron Transfer from N,N-Dimethylaniline to 7-Amino Coumarins in Protein-Surfactant Complex: Slowing Down of Electron Transfer Dynamics Compared to Micelles.

579. M. Tsionsky, A. J. Bard, and M. V. Mirkin, *J. Am. Chem. Soc.*, **119**, 10785 (1997). Long-Range Electron Transfer through a Lipid Monolayer at the Liquid/Liquid Interface.

580. Y. B. Zu, F. R. F. Fan, and A. J. Bard, *J. Phys. Chem. B*, **103**, 6272 (1999). Inverted Region Electron Transfer Demonstrated by Electrogenerated Chemiluminescence at the Liquid/Liquid Interface.

581. Z. Ding, B. M. Quinn, and A. J. Bard, *J. Phys. Chem. B.*, **105**, 6367 (2001). Kinetics of Heterogeneous Electron Transfer at Liquid/Liquid Interfaces as Studied by SECM.

582. H. Jensen, D. J. Fermin, and H. H. Girault, *Phys. Chem. Chem. Phys.*, **3**, 2503 (2001). Photoinduced Electron Transfer at Liquid/Liquid Interfaces. Part V. Organisation of Water-Soluble Chlorophyll at the Water/1,2-Dichloroethane Interface.

583. K. Yoshihara, *Adv. Chem. Phys.*, **107**, 371 (1999). Ultrafast Intermolecular Electron Transfer in Solution.

584. E. W. Castner, D. Kennedy, and R. J. Cave, *J. Phys. Chem. A*, **104**, 2869 (2000). Solvent as Electron Donor: Donor/Acceptor Electronic Coupling is a Dynamical Variable.

585. V. O. Saik, A. A. Goun, and M. D. Fayer, *J. Chem. Phys.*, **120**, 9601 (2004). Photoinduced Electron Transfer and Geminate Recombination for Photoexcited Acceptors in a Pure Donor Solvent.

586. A. Chakraborty, D. Seth, D. Chakrabarty, and N. Sarkar, *Spectrochimica Acta Part A-Molecular and Biomolecular Spectroscopy*, **64**, 801 (2006). Photoinduced Intermolecular Electron Transfer from Dimethyl Aniline to 7-Amino Coumarin Dyes in the Surface of Beta-Cyclodextrin.

587. R. A. Marcus, *J. Chem. Phys.*, **43**, 679 (1965). On the Theory of Electron-Transfer Reactions. VI. Unified Treatment for Homogeneous and Electrode Reactions.

588. J. Ulstrup, *Charge Transfer Processes in Condensed Media*, Springer, Berlin, 1979.

589. M. D. Newton and N. Sutin, *Ann. Rev. Phys. Chem.*, **35**, 437 (1984). Electron Transfer Reactions in Condensed Phases.

590. M. J. Weaver, *Chem. Rev.*, **92**, 463 (1992). Dynamical Solvent Effects on Activated Electron-Transfer Reactions: Principles, Pitfalls, and Progress.

591. R. A. Marcus, *Annu. Rev. Phys. Chem.*, **15**, 155 (1964). Chemical and Electrochemical Electron-Transfer Theory.

592. A. M. Kuznetsov and Y. I. Kharkats, in *The Interface Structure and Electrochemical Processes at the Boundary between Two Immiscible Liquids*, V. E. Kazarinov (Ed.), Springer, Berlin, 1987, p. 11, Problems of a Quantum Theory of Charge Transfer Reactions.

593. R. A. Marcus, *J. Phys. Chem.*, **94**, 1050 (1990). Reorganization Free Energy for Electron Transfers at Liquid-Liquid and Dielectric Semiconductor-Liquid Interfaces.

594. R. A. Marcus, *J. Phys. Chem.*, **94**, 4152 (1990). Theory of Electron-Transfer Rates across Liquid-Liquid Interfaces.

595. R. A. Marcus, *J. Phys. Chem.*, **95**, 2010 (1991). Theory of Electron-Transfer Rates across Liquid-Liquid Interfaces. 2. Relationships and Application.

596. C. Zener, *Proc. R. Soc. London, Ser. A*, **137**, 696 (1932). Non-Adiabatic Crossing of Energy Levels.

597. Y. I. Kharkats, *Elektrokhimiya*, **12**, 1370 (1976). On the Calculation of Probability of Electron Transfer through the Interface between Two Dielectric Media.

598. I. Benjamin and Y. I. Kharkats, *Electrochimica Acta*, **44**, 133 (1998). Reorganization Free Energy for Electron Transfer Reactions at Liquid/Liquid Interfaces.

599. J. K. Hwang and A. Warshel, *J. Am. Chem. Soc.*, **109**, 715 (1987). Microscopic Examination of Free-Energy Relationships for Electron Transfer in Polar Solvents.

600. G. King and A. Warshel, *J. Chem. Phys.*, **93**, 8682 (1990). Investigation of the Free Energy Functions for Electron Transfer Reactions.

601. R. A. Kuharski, J. S. Bader, D. Chandler, M. Sprik, M. L. Klein, and R. W. Impey, *J. Chem. Phys.*, **89**, 3248 (1988). Molecular Model for Aqueous Ferrous-Ferric Electron Transfer.

602. E. A. Carter and J. T. Hynes, *J. Phys. Chem.*, **93**, 2184 (1989). Solute-Dependent Solvent Force Constants for Ion Pairs and Neutral Pairs in Polar Solvent.

603. R. A. Marcus, *J. Chem. Phys.*, **24**, 979 (1956). Electrostatic Free Energy and Other Properties of States Having Nonequilibrium Polarization.

604. J. C. Rasaiah and J. J. Zhu, *J. Chem. Phys.*, **129**, 214503 (2008). Reaction Coordinates for Electron Transfer Reactions.

605. M. Tachiya, *J. Phys. Chem.*, **93**, 7050 (1989). Relation between the Electron-Transfer Rate and the Free Energy Change of Reaction.

606. I. Benjamin, *J. Phys. Chem.*, **95**, 6675 (1991). Molecular Dynamics Study of the Free Energy Functions for Electron Transfer Reactions at the Liquid-Liquid Interface.

607. J. Vieceli and I. Benjamin, *Chem. Phys. Lett.*, **385**, 79 (2004). Electron Transfer at the Interface between Water and Self-Assembled Monolayers.

608. D. V. Matyushov and G. A. Voth, *J. Chem. Phys.*, **113**, 5413 (2000). Modeling the Free Energy Surfaces of Electron Transfer in Condensed Phases.

609. I. Benjamin, in *Structure and Reactivity in Aqueous Solution: ACS Symposium Series 568*, C. J. Cramer and D. G. Truhlar (Eds.), American Chemical Society, Washington,

DC, 1994, p. 409, A Molecular Model for an Electron Transfer Reaction at the Water/1,2-Dichloroethane Interface.

610. I. Benjamin, *Acc. Chem. Res.*, **28**, 233 (1995). Theory and Computer Simulations of Solvation and Chemical Reactions at Liquid Interfaces.

611. Y. Rao, M. Xu, S. Jockusch, N. J. Turro, and K. B. Eisenthal, *Chem. Phys. Lett.*, **544**, 1 (2012). Dynamics of Excited State Electron Transfer at a Liquid Interface Using Time-Resolved Sum Frequency Generation.

612. M. Makosza, *Pure Appl. Chem.*, **72**, 1399 (2000). Phase Transfer Catalysis. A General Green Methodology in Organic Synthesis.

613. D. Albanese, *Catalysis Reviews-Science and Engineering*, **45**, 369 (2003). Liquid-Liquid Phase Transfer Catalysis: Basic Principles and Synthetic Applications(Reprinted from Interfacial Catalysis, p. 203–226, 2003).

614. C. A. Eckert, C. L. Liotta, D. Bush, J. S. Brown, and J. P. Hallett, *J. Phys. Chem. B.*, **108**, 18108 (2004). Sustainable Reactions in Tunable Solvents.

615. D. Albanese, *Mini-Reviews in Organic Chemistry*, **3**, 195 (2006). New Applications of Phase Transfer Catalysis in Organic Synthesis.

616. Y. H. Shao, M. V. Mirkin, and J. F. Rusling, *J. Phys. Chem. B*, **101**, 3202 (1997). Liquid/Liquid Interface as a Model System for Studying Electrochemical Catalysis in Microemulsions. Reduction of Trans-1,2-Dibromocyclohexane with Vitamin B-12.

617. C. K. Ingold, *Structure and Mechanism in Organic Chemistry*, 2nd edition, Cornell University, Ithaca, NY, 1969.

618. A. Warshel and R. M. Weiss, *J. Am. Chem. Soc.*, **102**, 6218 (1980). An Empirical Valence Bond Approach for Comparing Reactions in Solutions and in Enzymes.

619. J. K. Hwang, G. King, S. Creighton, and A. Warshel, *J. Am. Chem. Soc.*, **110**, 5297 (1988). Simulation of Free-Energy Relationships and Dynamics of S_N2 Reactions in Aqueous-Solution.

620. J. Chandrasekhar, S. F. Smith, and W. L. Jorgensen, *J. Am. Chem. Soc.*, **106**, 3049 (1984). S_N2 Reaction Profiles in the Gas-Phase and Aqueous-Solution.

621. W. L. Jorgensen and J. K. Buckner, *J. Phys. Chem.*, **90**, 4651 (1986). Effect of Hydration on the Structure of an S_N2 Transition-State.

622. J. R. Mathis, R. Bianco, and J. T. Hynes, *J. Mol. Liq.*, **61**, 81 (1994). On the Activation Free-Energy of the $Cl^- + CH_3Cl$ S_N2 Reaction in Solution.

623. W.-P. Hu and D. G. Truhlar, *J. Am. Chem. Soc.*, **116**, 7797 (1994). Modeling Transition State Solvation at the Single-Molecule Level: Test of Correlated Ab Initio Predictions against Experiment for the Gas-Phase S_N2 Reaction of Microhydrated Fluoride with Methyl Chloride.

624. G. Vayner, K. N. Houk, W. L. Jorgensen, and J. I. Brauman, *J. Am. Chem. Soc.*, **126**, 9054 (2004). Steric Retardation of S_N2 Reactions in the Gas Phase and Solution.

625. G. I. Almerindo and J. R. Pliego, *Chem. Phys. Lett.*, **423**, 459 (2006). Rate Acceleration of S_N2 Reactions through Selective Solvation of the Transition State.

626. J. R. Pliego, *J. Phys. Chem. B*, **113**, 505 (2009). First Solvation Shell Effects on Ionic Chemical Reactions: New Insights for Supramolecular Catalysis.

627. Y. Kim, C. J. Cramer, and D. G. Truhlar, *J. Phys. Chem. A*, **113**, 9109 (2009). Steric Effects and Solvent Effects on S_N2 Reactions.

628. D. Landini, A. Maia, and A. Rampoldi, *J. Org. Chem.*, **54**, 328 (1989). Dramatic Effect of the Specific Solvation on the Reactivity of Quaternary Ammonium Fluorides and Poly(Hydrogen Fluorides), (HF) in Media of Low Polarity.

629. A. Maia, *Pure Appl. Chem.*, **67**, 697 (1995). Anion Activation by Quaternary Onium Salts and Polyether Ligands in Homogeneous and Heterogeneous Systems.

630. D. Albanese, D. Landini, A. Maia, and M. Penso, *Industrial & Engineering Chemistry Research*, **40**, 2396 (2001). Key Role of Water for Nucleophilic Substitutions in Phase-Transfer-Catalyzed Processes: A Mini-Review.

631. D. K. Bohme and A. B. Raksit, *Can. J. Chem.*, **63**, 3007 (1985). Gas-Phase Measurements of the Influence of Stepwise Solvation on the Kinetics of S_N2 Reactions of Solvated F^- with CH_3Cl and CH_3Br and of Solvated Cl^- with CH_3Br.

632. X. G. Zhao, S. C. Tucker, and D. G. Truhlar, *J. Am. Chem. Soc.*, **113**, 826 (1991). Solvent and Secondary Kinetic Isotope Effects for the Microhydrated S_N2 Reaction of $Cl^-(H_2O)_N$ with CH_3Cl.

633. M. Re and D. Laria, *J. Chem. Phys.*, **105**, 4584 (1996). Solvation Effects on a Model S_N2 Reaction in Water Clusters.

634. J. M. Hayes and S. M. Bachrach, *J. Phys. Chem. A*, **107**, 7952 (2003). Effect of Micro and Bulk Solvation on the Mechanism of Nucleophilic Substitution at Sulfur in Disulfides.

635. K. V. Nelson and I. Benjamin, *J. Phys. Chem. C.*, **114**, 1154 (2010). A Molecular Dynamics-Empirical Valence Bond Study of an S_N2 Reaction at the Water/Chloroform Interface.

636. I. Benjamin, *J. Chem. Phys.*, **129**, 074508 (2008). Empirical Valence Bond Model of an S_N2 Reaction in Polar and Non-Polar Solvents.

637. S. C. Tucker and D. G. Truhlar, *J. Am. Chem. Soc.*, **112**, 3338 (1990). A 6-Body Potential-Energy Surface for the S_N2 Reaction $Cl^-(G)+CH_3Cl(G)$ and a Variational Transition-State-Theory Calculation of the Rate-Constant.

638. I. Benjamin, *J. Chem. Phys.*, **103**, 2459 (1995). Photodissociation of ICN in Liquid Chloroform; Molecular Dynamics of Ground and Excited State Recombination, Cage Escape and Hydrogen Abstraction Reaction.

639. J. J. I. Timoneda and J. T. Hynes, *J. Phys. Chem.*, **95**, 10431 (1991). Nonequilibrium Free-Energy Surfaces for Hydrogen-Bonded Proton-Transfer Complexes in Solution.

640. R. S. Mulliken, C. A. Rieke, D. Orloff, and H. Orloff, *J. Chem. Phys.*, **17**, 1248 (1949). Formulas and Numerical Tables for Overlap Integrals.

641. K. V. Nelson and I. Benjamin, *J. Chem. Phys.*, **130**, 194502 (2009). Microhydration Effects on a Model S_N2 Reaction in a Nonpolar Solvent.

642. B. J. Gertner, K. R. Wilson, and J. T. Hynes, *J. Chem. Phys.*, **90**, 3537 (1989). Nonequilibrium Solvation Effects on Reaction-Rates for Model S_N2 Reactions in Water.

643. B. J. Gertner, J. P. Bergsma, K. R. Wilson, S. Lee, and J. T. Hynes, *J. Chem. Phys.*, **86**, 1377 (1987). Nonadiabatic Solvation Model for S_N2 Reactions in Polar Solvents.

644. K. V. Nelson and I. Benjamin, *J. Phys. Chem. C*, **115**, 2290 (2011). Effect of a Phase Transfer Catalyst on the Dynamics of an S_N2 Reaction. A Molecular Dynamics Study.

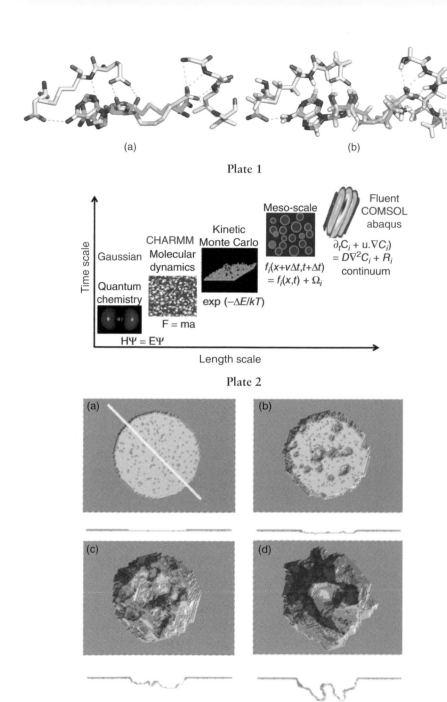

(a)   (b)

Plate 1

Meso-scale

Fluent
COMSOL
abaqus

Kinetic
Monte Carlo

CHARMM
Molecular
dynamics

Gaussian

$\partial_t C_i + u.\nabla C_i)$
$= D\nabla^2 C_i + R_i$
continuum

$f_i(x+v\Delta t, t+\Delta t)$
$= f_i(x,t) + \Omega_i$

Quantum
chemistry

$\exp(-\Delta E/kT)$

$F = ma$

$H\Psi = E\Psi$

Length scale

Time scale

Plate 2

(a)   (b)

(c)   (d)

Plate 3

Reviews in Computational Chemistry, Volume 28, First Edition.
Edited by Abby L. Parrill and Kenny B. Lipkowitz.
© 2015 John Wiley & Sons, Inc. Published 2015 by John Wiley & Sons, Inc.

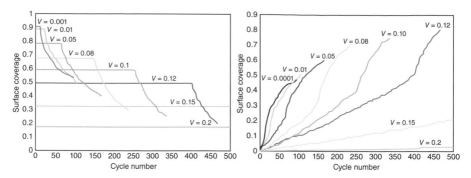

Plate 4

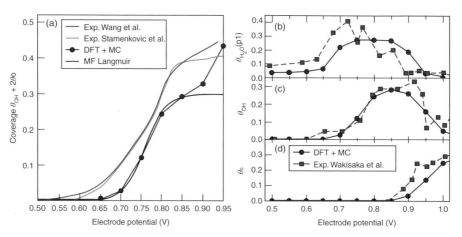

Plate 5

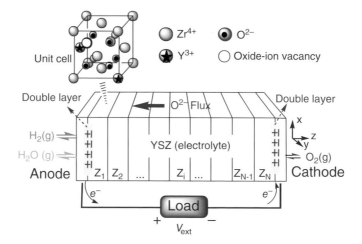

Plate 6

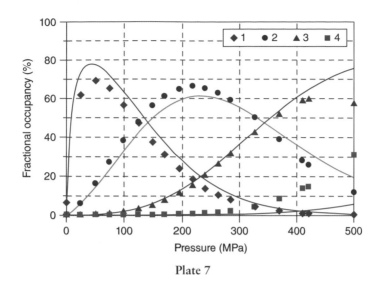

Plate 7

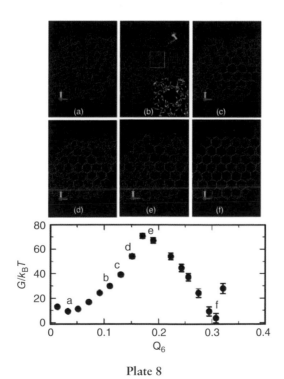

Plate 8

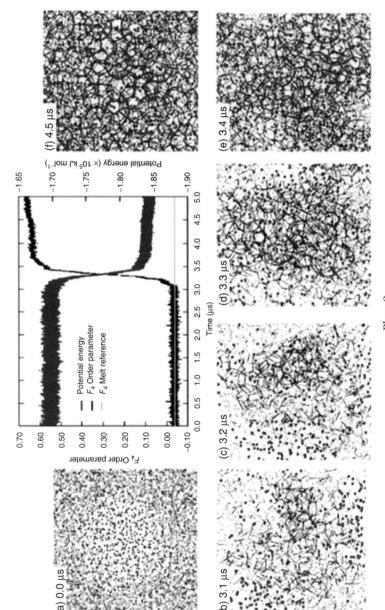

(a) 0.0 μs

(b) 3.1 μs

(c) 3.2 μs

(d) 3.3 μs

(e) 3.4 μs

(f) 4.5 μs

Potential energy
F_4 Order parameter
F_4 Melt reference

Plate 9

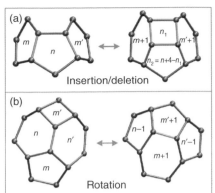
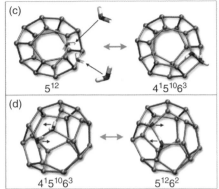

Plate 10

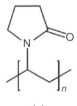

(a)

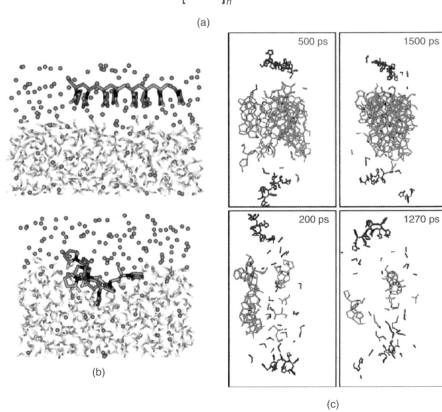

(b)

(c)

Plate 11

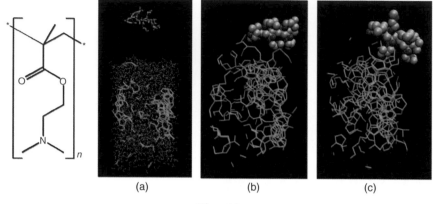

(a)   (b)   (c)

Plate 12

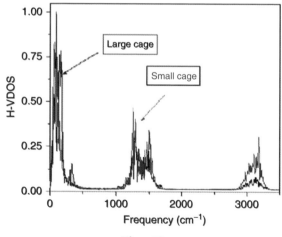

Plate 13

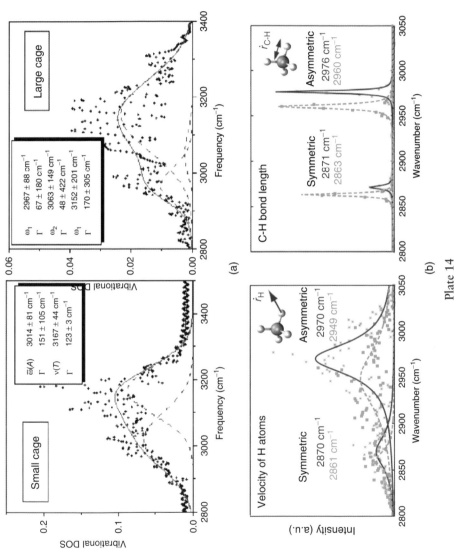

Plate 14

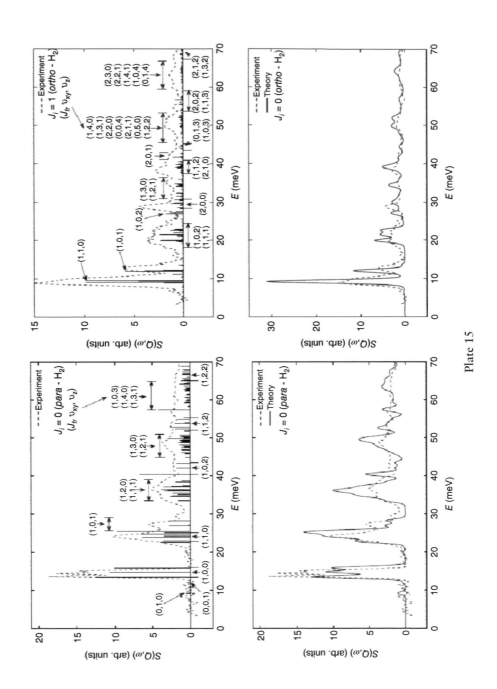

Plate 15

## CHAPTER 7

Computational Techniques in the Study of the Properties of Clathrate Hydrates

John S. Tse

Department of Physics and Engineering Physics, University of Saskatchewan, Saskatoon, Saskatchewan S7N 5B2, Canada

HISTORICAL PERSPECTIVE

Clathrate hydrates are nonstoichiometric framework crystalline solids in which water molecules form the host lattices with cavities that guest atoms or molecules can occupy. Clathrate hydrates are also generally known as "gas hydrates," a term used to describe solids formed by the combination of gas and liquid with a large excess of water. The study of clathrate hydrates has a long history. It started in 1778 when Joseph Priestly synthesized the first gas hydrate by passing sulphur dioxide gas into cold water at atmospheric pressure.[1] In 1810, Sir Humphry Davy found that an aqueous solution of oxymuriatic acid (chlorine) froze more readily than water.[2] In 1982, Faraday[3] estimated the composition of the solid formed to be roughly $Cl_2 \cdot 10H_2O$. Following these discoveries, more than 40 gas hydrates have been prepared and characterized from the late nineteenth century to early twentieth century. The remaining subject of controversy was the crystal structures, which was not resolved until the discovery of X-ray crystallography.[4-6] In the early days, research on gas hydrates was mainly focused on the synthesis and measurement of the gas–liquid–solid phase thermodynamics equilibria. It was unknown that gas hydrates could occur naturally. The practical significance

Reviews in Computational Chemistry, Volume 28, First Edition.
Edited by Abby L. Parrill and Kenny B. Lipkowitz.
© 2015 John Wiley & Sons, Inc. Published 2015 by John Wiley & Sons, Inc.

of gas hydrate was first realized by Hammerschmidt[7] in 1934. He found the formation of crystalline clathrate hydrate from natural gas, and water was the cause for the plugging of pipelines and posed significant safety hazards. Since then, research interest in the determination of other physical properties of gas hydrates have grown.[8] In 2010, the BP Deepwater Horizon wellhead blast leading to the disastrous oil spill in the Gulf of Mexico and the initial failure to cap the oil flow by a metal containment box was attributed to the formation of gas hydrates.[9] The current revival in the intensity of gas hydrate research is motivated by the discovery of a gas hydrate deposit in Siberia, Russia, in 1963.[9] Most significantly, naturally occurred gas hydrate, which is rich in methane, is a valuable energy source. In the recent years, large deposits of gas hydrates have been found in many regions on Earth including the seafloor, ocean sediments and deep lake sediments as well as in permafrost regions. An estimate of the tons of oil equivalent (TOE) reflecting the potential of the terrestrial gas hydrate is over 15×10^{12} TOE.[10] At the current rate of consumption, it was estimated these deposits can supply energy for 200 years.

Under suitable pressure and temperature conditions, most small molecules form clathrate hydrates. This led to the idea that carbon dioxide may be exchanged for methane in the hydrate leading to sequestration of the potential greenhouse gas by injection of carbon dioxide into natural gas hydrate deposits[10,11] and at the same time allowing harvest of the released methane. The potential for simultaneous carbon dioxide sequestration and conversion of natural gas hydrate reservoirs into usable energy source is a major impetus of recent research intensity on this subject. Apart from the importance as energy source, clathrate hydrates also found applications in the desalination of salt water[12,13] or as coolant in air conditioning.[14] It has also been speculated that the sudden climate change in the later Quaternary about 15,000 years ago may be due to methane from disrupted gas hydrates that caused significant global warming and led to mass extinction on the Earth.[15,16] The open lattice of the gas hydrate structure has long been known as a matrix for isolation of molecules. The dynamics of the encaged guests can be studied by nuclear magnetic resonance (NMR) or dielectric spectroscopy.[17]

The history of gas hydrate research, its physical and thermodynamics properties, and the impact on energy and the environment have been presented in several excellent books and reviews.[17−19] This chapter is focused on the molecular aspect, particularly, on the application of different computational methods to investigate structural and thermodynamic stability, guest and lattice dynamics, thermal transport and mechanisms for homogeneous and hetero-geneous crystallization. Even within this restricted choice of topics, there are numerous contributions to the modelling of hydrate properties. This chapter is not intended to provide a comprehensive review on the field of computational gas hydrate research. Even within the topics chosen no attempt was made to exhaustively review available literatures and competing theories. Rather, the focus is to highlight a number of theoretical techniques that have been applied to resolve topical problems in gas hydrates through specific examples.

STRUCTURES

Knowing the compositions and structures of gas hydrates are prerequisites for molecular-level modelling. Although gas hydrates were known in the early 1800s, the structure of these materials remained unknown until the advent of X-ray diffraction methods in the early twentieth century. Stackelberg[4,5] was the first to study the structure of gas hydrate with powder X-ray diffraction. He found that clathrate hydrates exist in two distinct cubic crystalline forms. However, the proposed structures were found to be incorrect. Remarkably, from the building of molecular models, Claussen[20] was the first to arrive at the correct structures, which were later confirmed. Since then many hydrate structures have been studied. Under atmospheric pressure and low temperature, a majority can be classified in one of the two cubic structures termed structure I (S-I) and structure II (S-II).

The unit cell in S-I is made up of 46 water molecules with an average cell size of *ca.* 12 Å. S-II hydrate is composed of 136 water molecules with a large unit cell of *ca.* 17 Å. Each structure has small cages and large cages (Figure 1).[4,5,17−19] The small cages in both structures are pentagonal dodecahedra (5^{12}), the large cages in S-I are tetrakaidecahedra consisting of 12 pentagons and two hexagons ($5^{12}6^2$), and the large cages in S-II are hexakaidecahedra consisting of 12 pentagons and four hexagons ($5^{12}6^4$)

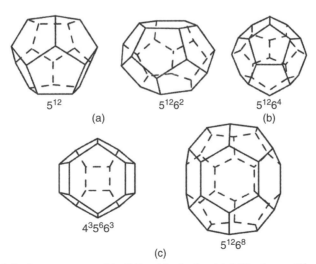

Figure 1 Polyhedra water cages' building blocks for (a) S-I hydrate: 5^{12} and $5^{12}6^2$; (b) S-II hydrate: $5^{12}6^4$; (c) S-H hydrate: 5^{12}, $4^3 5^6 6^3$, and $5^{12}6^8$.

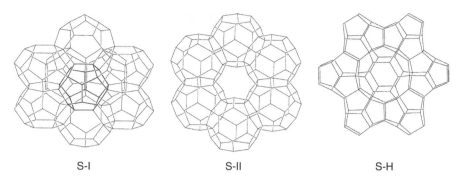

S-I   S-II   S-H

Figure 2   Clathrate hydrate structures and stacking sequences of the cages.

(Figure 2). S-I is usually formed from molecules with diameter in the range 4.1–5.8 Å, whereas S-II is formed by molecules with diameter less than 4.1 Å or greater than about 5.5 Å. In 1987, it was found that a large organic molecule also forms a clathrate hydrate with a hexagonal unit cell (S-H), which consists of 34 water molecules.[21] This hexagonal structure is built from three ubiquitous 5^{12} cages, two medium-size $4^3 5^6 6^3$, and one large $5^{12} 6^8$ cages. Bulky molecules with an average size of 5.71–6.0 Å can be accommodated in the large cavity. It is remarkable that a distinct tetragonal structure (S-T) was found in bromine hydrate[22] with the water framework containing sixteen 14-hedral cavities $5^{12} 6^2$, four 15-hedral cavities $5^{12} 6^3$, and 10 dodecahedral cavities 5^{12} in the unit cell. Gas hydrates with S-I, S-II and S-H have been found in nature.[23]

A comprehensive analysis of a large set of low-temperature neutron and X-ray diffraction data for S-I and S-II hydrates has revealed that the lattice parameters are not sensitive to the shape, size and polarity of the encaged guests.[24] Furthermore, because the structures of gas hydrates were resolved in 1950, it was generally accepted that the large cages in S-I hydrates are smaller than those in S-II, and molecules with small van der Waals (vdW) diameters would prefer this structure. In 1984, it was shown that the two smallest guests that form clathrate hydrates, namely, argon and krypton, do so in the S-II structure.[25] This finding, as will be discussed later, has an important consequence on the vindication of a simple statistical mechanic model describing the thermodynamic stability of gas hydrates.[26–28]

THE VAN DER WAALS–PLATTEEUW SOLID SOLUTION THEORY

Characterization of the phase diagram of gas hydrates dominated the early research in gas hydrates. Although the structures, either S-I or S-II, show little variation with the guests, the structural thermodynamic stabilities, in contrast, can differ significantly. As mentioned, chlorine readily forms a stable hydrate *above* the freezing point of water.[1–3] In comparison, for some guests, for example, argon, the dissociation temperature of the gas hydrate is −124 °C.[29] A general and elegant model to predict the thermodynamic stability of clathrate compounds and, in particular, for gas hydrates, was developed by Van der Waals and Platteeuw.[26–29] This phenomenological theory is based on the observation that all clathrates are guest–host complexes. The empty host itself is thermodynamically unstable, but the clathrate structure is stabilized by inclusion of the guests. Furthermore, for a given class of clathrate hydrate (e.g., S-I, S-II, S-H, etc.), the structure of the host lattice shows little or no dependence on the guests. This suggested that the guest–host intermolecular interactions should be broadly similar. Finally, the guest:water compositions of gas hydrates are variable and dependent on the nature of the guest. Thus, clathrate hydrates are nonstoichiometric compounds. Therefore, it is natural to hypothesize a clathrate as a solid-state solution where the crystal structure of the solvent (water host) remains unchanged by the addition of the solutes (guests) and the "mixture" remains in a single homogeneous phase.

The van der Waals and Platteeuw (vdW-P) statistical thermodynamic model was based on the following assumptions:[28]

(i)   Each cavity of the clathrate can accommodate only a single guest, that is, no multiple occupancy is allowed.
(ii) The encaged guests do not distort the cavity.
(iii) Apart from the interaction between the cavity and the guest, the guest–guest interactions are negligible.

Consider the simplest case of a clathrate (C) with only one type of cavity. The inclusion of a guest (G) into the host (H) can be described by the process *empty cavity* + G → *occupied cavity*. This can be described as a reversible chemical reaction of binding the guest into the host,

$$H + G \Longleftrightarrow C \qquad [1]$$

The equilibrium constant, K, is simply the Langmuir isotherm,

$$K = \frac{[C]}{[H][G]} \qquad [2]$$

where the square brackets denote concentrations. If there is only one type of guest (solute) with concentration $[C] \sim x_G$, then, $[G] \sim 1-x_G$. Substitute into Eq. [2], with $[H] \sim p_H$

$$K = \frac{x_G}{(1 - x_G)p_H} \qquad [3]$$

then

$$p_H = K^{-1} \frac{x_G}{(1 - x_G)} \qquad [4]$$

At thermodynamic equilibrium, for N_Q cavities and N_G guests, the Gibbs–Duhem equation requires

$$N_Q d\mu_Q + N_G d\mu_G = 0 \qquad [5]$$

According to Eq. [4],

$$d\mu_G = kTd \ln p_H = kT[x_G(1 - x_G)]^{-1} dx_G \qquad [6]$$

Because $N_G = v_G N_Q x_G$, and v_G is a fraction of the specific type of cavity, in this case, $v_G = 1$. Substituting into Eq. [5] and integrating reference to the chemical potential of the hypothetical empty cavity (μ_Q^0), one obtains the important relationship relating the chemical potential of the occupied hydrate μ_Q to the occupancy (or concentration) of the guest,

$$\mu_Q - \mu_Q^0 = \Delta\mu = kTv \ln(1 - x_G) \qquad [7]$$

Equation [7] can be generalized to a multicavities system with fractional number of the ith cage v_i and concentration (occupancy) x_i,

$$\mu_Q - \mu_Q^0 = kT \sum_i v_i \ln \left(1 - \sum_K x_{Ki} \right) \qquad [8]$$

This is the fundamental equation of the vdW-P model and has been used to calculate the thermodynamics properties of gas hydrates.

Implicit in the solid solution model is that the guest–host intermolecular interaction is weak compared to the binding energy of the host water lattice. The inclusion reaction[1] can then be regarded as the physical adsorption of the guest onto the water framework. The equilibrium constant K is just the Langmuir constant and from statistical mechanics it is related to the interaction potential, U,

$$K = \frac{1}{kT} \int_{V_{cell}} \exp \left(-\frac{U}{kT} \right) dV \qquad [9]$$

The evaluation of K requires multidimensional integration (the position and orientation of the guest relative to individual water forming the cavity). Employing the Lennard-Jones and Devonshire cell model[30] for liquids by assuming that (i) the guest is spherical (ii) the host potential field is uniformly distributed on a sphere and (iii) the guest can rotate freely about the center of the cage, the interaction potential becomes spherically symmetric and the integral is reduced to a one-dimensional problem that can be solved analytically, provided the precise form of U is known. In spite of the obvious simplicity of the vdW-P/Lennard-Jones cell model, early numerical applications using less than perfect intermolecular potentials, such as the empirical Lennard-Jones potentials, had achieved remarkable successes,[28,31] demonstrating the theory is fundamentally sound and providing a convenient tool for the estimation of clathrate hydrate formation conditions. This theory is still incorporated in some modern commercial engineering process simulators.

A remarkable success of the vdW-P model is the prediction that type II structure should be more stable than type I if the guest is very small.[32] According to Eq. [8], the chemical potentials of water molecules in structures I and II hydrate are

$$\mu_I = \mu_I^0 - \frac{kT}{23}[\ln(1 + K_{S,I}P) + 3\ln(1 + K_{L,I}P)] \qquad [10]$$

$$\mu_I = \mu_{II}^0 - \frac{kT}{23}[2\ln(1 + K_{S,II}P) + \ln(1 + K_{L,II}P)] \qquad [11]$$

where μ^0 refers to chemical potential of water in the hypothetical empty lattice, K_S is the Langmuir constant relating the extent of occupancy by the gas of the small 12-hedral cages to the pressure P (strictly, the fugacity) of the hydrate-forming gas, and $K_{L,I}$, $K_{L,II}$ are corresponding Langmuir constants for the 14- and 16-hedra. Because the 12-hedra are similar in the two structures, $K_{S,I} \approx K_{S,II}$. The prefactors in front of the logarithmic terms are the number of small and large cages in S-I and S-II.

Although accurate values of μ^0 are not known, it has been shown that $\mu_I^0 > \mu_{II}^0$. For relatively large guest S-II is more stable because $K_{L,II} \gg K_{S,I} \approx K_{S,II}$ and $K_{L,I}$, only the 16-hedra are occupied by guest. On the other hand, for very small guests, S-II stability results from the condition $K_S > K_{L,I}, K_{L,II}$, that is, from a preference for the small cages. S-I will generally be preferred when $K_{L,I} > K_S, K_{L,II}$ or $K_{L,I} \approx K_S > K_{L,II}$. This prediction made by Holder and Manganiello[32] was later confirmed by the experimental determination of the structures of argon and krypton hydrates, the two smallest gases that form clathrates, to be S-II. In comparison, the larger xenon hydrate is S-I.[25] The structural preference can be rationalized as follows. For smaller guests, their guest–host interaction is optimal in the small cage. Because there are more small cages in S-II hydrate (16 out of 24 cages) than S-I (2 out of 8 cages), smaller guests, such as argon and krypton, favour type II structure.

COMPUTATIONAL ADVANCEMENTS

Thermodynamic Modelling

In view of the successes of the vdW-P model, many attempts have been made to preserve the simplicity of the theory and to improve the accuracy and reliability through the construction of more reliable interaction potentials and/or a more rigorous evaluation of the configuration integral in Eq. [9].

The first major improvement was to remove the spherical cell approximation. A spherically symmetric intermolecular potential is not appropriate for the oblate geometry of the hydrate cavities. Furthermore, there is no valid reason to approximate all guest molecules as a spherical entity except for the rare gases. Explicit guest–water interactions must be considered and evaluated as sum of binary interactions between guest and host. To evaluate the multidimensional integral, Tester et al.[33] pioneered the first numerical calculation on gas hydrates using a digital computer employing the Metropolis Monte Carlo scheme. The method sampled of the energy surface in which the guest in a cage was moved randomly in uniformly sized steps to different locations inside the clathrate cavity and if the guest is asymmetric, spinning it randomly about its center of mass. At each location, the total potential energy was calculated and the trial move was accepted or rejected on the basis of the Metropolis selection scheme in accordance with the energy probability distribution, $\exp[-U/kT]$. A Markov chain is then generated from a sequence of moves by displacing and spinning the guest. At thermodynamic equilibrium, the fugacity of the different types of cavities must be equivalent. By matching the calculated fugacity, occupancy in different cages can be determined. Consequently, the dissociation pressure can be estimated from Eq. [8]. This method, coded in a machine language and executed on the MANIAC II computer at the Los Alamos Scientific Laboratory, was applied to the study of S-I hydrate formed from the rare gases and several nonspherical polyatomic molecules. Results of the calculations will not be elaborated here. Interested readers are referred to the original article.[33] The calculations, however, revealed one very significant finding. Analysis of the distribution of the random moves showed that the argon is displaced from the respective center of the small 5^{12} and the large $5^{12}6^2$ cages with much greater displacement in the large cage (Figure 3). Thus, the potential surface of argon situated in the large cage is not optimal. The argon has to move closer to the side of the cavity to enhance stability. This is an indication that argon should not exist as S-I, a fact that was later confirmed as argon hydrate indeed adopts the type II structure[25] (*vide supra*).

A logical extension to the Monte Carlo method is to improve on the intermolecular potential. In the past, parameters for the interaction potential functions were often determined by fitting a set of calculated results, for example, dissociation pressures, with experiments. These parameters may not bear any physical meaning nor are they transferable to different

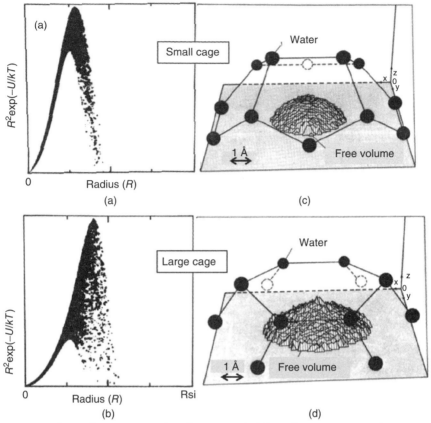

Figure 3 Spherically symmetric distribution probability of argon guest in the (a) small and (b) large cage of S-I hydrate obtained from Monte Carlo calculations. The corresponding free volume occupied by an argon guest in the (c) small and (d) large cage. Adapted from Ref. 33.

systems. It is desirable to determine accurately, which is physically based and modeled, an independent interaction parameter. For this purpose, Tse and Davidson[34] constructed a xenon–water potential with a Born–Mayer repulsive function and improved the dispersive interactions including mutually induced higher moments (dipole–quadrupole, quadrupole–quadrupole). To account for the correct asymptotic behaviour of the attractive term at small separation, a damping factor was added that reduces the contribution of the dispersive energy to give a better potential near the energy minimum. Using the more realistic intermolecular potential constructed from theoretical atomic data, the small/large occupancy ratio in S-I hydrate observed by NMR spectroscopy[35] was reproduced correctly. With the general availability of high-level *ab initio* quantum mechanical methods and software, the potential energy surface (PES) for a binary system can be computed accurately. For example, a methane–water potential surface with the binding energies to within 0.1 kcal/mol of the basis set limit has been achieved using MP2/6-3111G(2d,2p) and then corrected to near the accuracy of MP2/cc-pVQZ.[36] The calculated spherically averaged six-dimensional PES can be fitted to the exponential-6 potential. The numerical radial intermolecular potential profile between methane and water is found to be consistent with experimental vibration–rotation–tunneling spectroscopy and second virial coefficient data. This potential function was used to determine the difference in the chemical potential ($\Delta\mu$, Eq. [8]) and enthalpy (ΔH) of the hypothetical empty S-I framework and water.[37] The predicted values $\Delta\mu^0 = 1236 \pm 4$ J/mol and $\Delta H^0 = 1703 \pm 62$ J/mol can be compared with the currently accepted values for $\Delta\mu^0 = 1287$ J/mol and $\Delta H^0 = 931$ J/mol derived from heat capacity experiments[38] and $\Delta\mu^0 = 1299$ J/mol and $\Delta H^0 = 1861$ J/mol estimated from the analysis of thermodynamic phase equilibria measurements.[39] Subsequently, an argon–water pair potential was determined in a similar manner using MP2/aug-cc-pVQZ calculations corrected for many body and basis set superposition errors.[40] Using this potential, the reference parameters for S-II argon hydrate was determined to be $\Delta\mu^0 = 1077 \pm 5$ J/mol and $\Delta H^0 = 1294 \pm 11$ J/mol, which were comparable to the corresponding accepted values $\Delta\mu^0 = 1069$ J/mol and $\Delta H^0 = 764$ J/mol. It is noteworthy that these reference but fictitious parameters are central to the vdW-P theory and cannot be determined from experiment. High-level *ab initio* calculations help to establish a theoretical bind to these important qualities.

The accurate prediction of thermodynamics properties using the vdW-P statistical model requires accurate interaction potentials. Although *ab initio* calculations are highly accurate for a binary interaction potential, correction to the many body effect has to be made in an *ad hoc* manner when used in the condensed phase. For practical purposes, it is desirable to extract a realistic average intermolecular potential from experimental data. Bazant and Trout[40] recognized that this can be achieved for clathrate hydrates from the standard vdw-P statistical model from the temperature dependence of the Langmuir constants. Rewriting the definition (Eq. [9]) for the Langmuir constant in a spherical cell approximation with a cell radius R ($\beta = 1/kT$)

$$K(\beta) = 4\pi\beta \int_0^R \exp(-\beta U(r)) r^2 dr \qquad [12]$$

It was realized that the experimental Langmuir constant data are described well by a van't Hoff temperature dependence given by

$$K(\beta) = K_0 e^{m\beta} \qquad [13]$$

Equating Eqs [12] and [13] gives an integral equation,

$$K(\beta) = 4\pi\beta \int_0^\infty e^{-\beta U(r)} r^2 \, dr \qquad [14]$$

Note that making the upper limit of integration from $R \to \infty$ introduces negligible error because of the low-temperature accessible in a clathrate experiment.[40] The intermolecular potential $U(r)$ can be obtained by inverting the experimental Langmuir constant data numerically. In principle, Eq. [14] has an infinite number of solutions; however, there exists a unique central-well solution that has physical meaning. The central-well solution has the functional form

$$K(\beta) = \beta F(\beta) e^{-\beta U(r)} \qquad [15]$$

where

$$F(\beta) = \beta \int_0^\infty e^{-\beta y} g(y) \, dy \qquad [16]$$

and $g(y)$ is the Laplace transform of the function

$$g(\beta) = \frac{F(\beta)}{\beta} = \frac{K(\beta) e^{\beta U_0(r)}}{\beta^2} \qquad [17]$$

The general expression for the central-well potential $U(r)$ is

$$U(r) = U_0(r) + g^{-1}\left(\frac{4}{3}\pi r^3\right) \qquad [18]$$

Once $U(r)$ is extracted from experimental data, it can be fitted to a convenient functional form, such as the Lennard-Jones or Kihara potentials, for further computation. The methane Langmuir constants in S-I and S-II hydrates derived from this method are compared with the *ab initio* predictions in Figure 4 over a broad temperature range.[41] The agreement is excellent. This method, in combination with the vdw-P theory, has been applied to a variety of single- and multiple-component gas hydrates, the results of which are generally very reliable. A typical example is the ethane–propane phase diagram shown in Figure 5.

Efforts have been made over time to extend the vdW-P theory to treat multiple occupancy in the hydrate cavities. A simple scheme was to treat the addition of a guest into the cavity as successive chemical reactions with an equilibrium constant associated with each step.[42] For the *m*th cage,

$$C_m + G \rightleftharpoons C_m G \quad (K_{eq,1})$$
$$C_m G + G \rightleftharpoons C_m G_2 \quad (K_{eq,2}) \qquad [19]$$
$$C_m G_{n-1} + G \rightleftharpoons C_m G_n \quad (K_{eq,n})$$

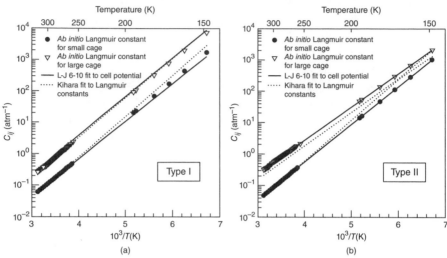

Figure 4 Comparison of Langmuir constants for (a) S-I and (b) S-II methane hydrate calculated by using fitted *ab initio* intermolecular potential and those obtained from other empirical potentials. Reprinted (adapted) with permission from *J. Phys. Chem. B,* **109**, 8153 (2005). Copyright 2005 American Chemical Society.

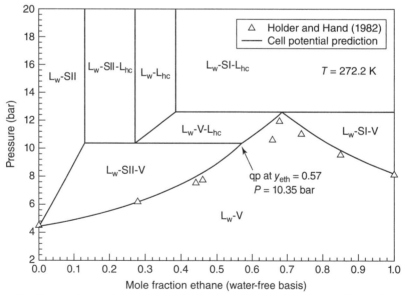

Figure 5 Comparison of experimental (G. D. Holder and J. H. Hand, *AIChE J,* **28**, 440 (1982)) and predicted isothermal phase diagram for ethane and propane hydrate at 277.6 K. Reprinted (adapted) with permission from *J. Phys. Chem. B,* **109**, 8153 (2005). Copyright 2005 American Chemical Society.

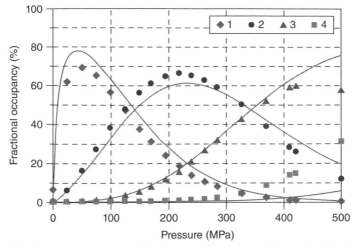

Figure 6 Fractional occupancy of the large cavities of S-II hydrates by hydrogen clusters formed by one to four molecules at 274 K. Symbols represent results of GCMC simulations. N. I. Papadimitriou, I .N. Tsimpanogiannis, A. Th. Paioaannou, and A. K. Stubos, *J. Phys. Chem.*, C **112**, 10294 (2008). Reprinted (adapted) with permission from *J. Phys. Chem. B*, **114**, 9602 (2010). Copyright 2010 American Chemical Society. (For a color version of this figure, please see plate 7 in color plate section.)

The fraction of the cavities occupied by k number of guests (θ_m^k) can be written as

$$\theta_m^k = \frac{F_m^k}{1 + \sum_{i=1}^{k} F_m^k} \qquad [20]$$

where $F_m^j = \prod_{i=1}^{j} K_{eq,j} f^i$, and f is the fugacity (see Eqs [11] and [12] *vide supra*). This method has been applied to the prediction of the fractional occupancy of the large cavities of an S-II hydrate of hydrogen. Figure 6 compares the results of this simple approach using the spherical cell vdW-P model with those obtained from more elaborate grand canonical ensemble Monte Carlo calculations. The overall agreement is reasonable.[43]

Atomistic Simulations

Because the local structure and the three-dimensional water–hydrogen-bond network in gas hydrates is similar to ice Ih, a number of properties of gas hydrates, especially those dominated by the host lattices, are expected to be similar to those of ice, with the caveat that allowance is made for differences between the densities of the hydrates and ice. Thus, the heat capacities, sound velocities, enrichment of heavy isotopes of hydrogen and oxygen in the water framework, and, some spectroscopic, electrical and mechanical properties

Table 1 Experimental Thermal Conductivity of Ice I*h* and Selected Clathrate Hydrates

Compound	T (K)	λ (W m/K)
Ice I*h*	260	2.35
Structure I hydrates		
Xenon	245	0.36
Methane	213	0.45
Ethylene oxide	263	0.49
Structure II hydrates		
Cyclobutanone	260	0.47
Tetrahydrofuran	260	0.51
1,3 Dioxolane	260	0.51

of hydrates are found to be close to that of ice.[44,45] Therefore, skepticism was initially voiced on the report of a much lower thermal conductivity in the methane–water–sand system than in ice.[46] However, subsequent studies indeed confirm this unusual observation.[47] Even more surprisingly, the thermal conductivity was found to be independent of the structure and the guest component of the hydrates (Table 1)! Extension of the thermal conductivity measurement of an S-II tetrahydrofuran (THF) hydrate to low temperature revealed an anomalous temperature dependence that resembles the behaviour of a disordered solid rather than a crystal.[48–50] Moreover, after scaling the thermal conductivity data by the Debye temperature for THF hydrate, it was found that the results coincide with the universal curve for glassy solids.[51] This is a remarkable and important observation (Figure 7).[50] Even today, the mechanism for thermal transport in disordered solids is not clearly understood and the problem still remains an area of active research. The slow progress in this regard is partly because of the absence of accurate structural information on disordered systems, which, in turn, has hindered the development of a quantitative theory. Because the thermal conductivity of a crystalline structure is found to behave like a glass, it may serve as an excellent model to study the mechanism of heat transport in disorder systems. The first atomistic simulation of a gas hydrate using classical molecular dynamics (MD) method was performed for this purpose.[52]

Phonon–phonon interactions are the primary cause for the thermal resistance in a solid.[53] In general, because the thermal energy is relatively low, heat transport is likely through the acoustic modes. A complete knowledge of the phonon density of states (frequency spectra) in both ice and gas hydrates may provide important clues in understanding the thermal conductivity anomaly. MD simulations on S-I methane hydrate were carried out on a single cubic unit cell of length 12.03 Å containing 46 water molecules, with periodic boundary conditions imposed.[52] The initial positions of the oxygen atoms were taken from the structure of ethylene oxide deuterated hydrate at $-25\,°C$ determined

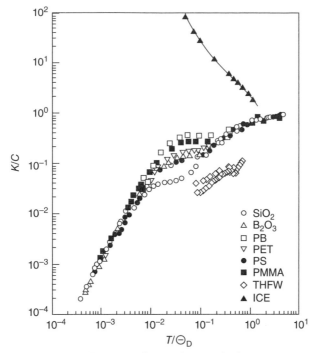

Figure 7 Scaled thermal conductivities for ice Ih, THF hydrate (THFW), and six amorphous solids: PB, polybutadiene; PET, poly(ethylene terephthalate); PS, polystyrene; PMMA, poly(methyl methacrylate). The solid curve is a least squares fit to the ice experimental data. Reprinted (adapted) with permission from *J. Phys. Chem.*, **92**, 5006 (1988). Copyright 1998 American Chemical Society.

by single-crystal neutron diffraction measurement.[54] For the MD calculations on ice Ic, a larger system was used, consisting of 64 water molecules in a comparable cubic box of length 12.70 Å. In this case, the initial oxygen coordinates were taken from an X-ray diffraction study.[55] As the protons of the water molecules in both the clathrate hydrate and ice Ic are positionally disordered, a protocol for proton assignment is required. The initial proton coordinates were selected randomly but conform to the Bernal–Fowler rules[56] and subjected to the condition that there is no residual dipole moment in the simulation model. The simple-point-charge (SPC) intermolecular water potential was used.[57] The methane molecules were treated as spherical particles. The water–methane interaction was described by a Lennard-Jones potential with the methane potential parameters estimated from data on solid krypton.[55] By today's standard, the initial calculation was rather rudimentary, yet, despite the simplicity, the results were very informative.

Relevant thermodynamic properties, like the heat capacities for ice and methane hydrate, are in good agreement with experiments (Table 2). The vibrational density of states (VDOS) can be calculated from the MD

Table 2   Comparison of Theoretical and Experimental Heat Capacity (kcal/mol) for Empty Hydrate, Methane Hydrate, and Ice Ic

System	T (K)	Translation	Libration	Guest	Total	Expt.
Empty hydrate	110	16.4	0.7		17.1	17.2
Empty hydrate	205	21.3	6.9		28.2	28.5
Methane hydrate	145	18.5	2.4	24.2	20.9	21.2
Methane hydrate	214	21.3	7.4	24.5	28.8	29.6
Ice Ic	69	11.5	0.0		11.5	11.4
Ice ic	252	22.0	9.8		31.8	34.0

trajectories by Fourier transform of the atom velocity self-correlation functions. The VDOS of ice obtained from inelastic incoherent neutron scattering (IINS) was reproduced well by the calculations (Figure 8c). An important piece of information revealed from the MD calculations is the existence of low-frequency localized guest vibrations embedded in the long-wavelength $(0-150 \, cm^{-1})$ acoustic lattice modes of the host lattice. These localized vibrations can be attributed to the rattling of methane in the cages. The methane vibrational frequencies are sensitive to the size and shape of the cage. For methane in the more symmetric but more restricted small 5^{12} cages, only one peak at $75 \, cm^{-1}$ was observed. In comparison, the vibration of methane in the large elliptical $5^{12}6^2$ cages splits into two different vibrational modes corresponding to the motions along the minor and major axes (Figure 8c). The frequencies $(45 \, cm^{-1})$ are lower than those in the small cages. The observation of localized vibrations with energy lower than the maximum of the acoustic branches at the zone boundary is significant. It indicates that the localized rattling motion must intersect the strongly dispersive transverse and longitudinal acoustic (LA) translations inside the first Brilluion zone and, as a result, may help to promote Umklapp scatterings.

These theoretical predictions prompted the measurement of the methane VDOS in a deuterated lattice using IINS.[58] The agreement between the calculated and measured methane VDOS is excellent (Figure 8). The result confirmed unambiguously the existence of the low-frequency localized methane vibrations. The implication is significant. A factor group analysis of the symmetry of the lattice acoustic and guest translational motions[58] is summarized in Figure 9. For the cubic space group, the acoustic vibrations at the zone center transform as T_{1u} symmetry. The methane vibrations in both the large and small cavities also possess T_{1u} symmetry. Because the methane vibrations are found experimentally to be localized with small energy dispersions and within the acoustic vibrations, there must be symmetry avoided-crossings along certain phonon wave vectors between the zone center and the zone boundary. A schematic representation of this interaction is shown in Figure 9. The avoided-crossing of the guest and lattice phonon branches permits energy transfer between them and affects the phonon-scattering processes. This is referred to as the resonant phonon-scattering model.[59]

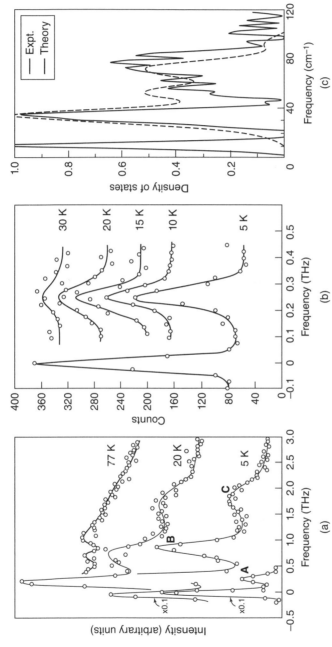

Figure 8 Inelastic incoherent neutron scattering spectra for methane hydrate (a) translational vibration; (b) quantum rotation; and (c) comparison with molecular dynamics calculations. Reprinted (adapted) with permission from *J. Phys. Chem. A*, **101**, 4491 (1997). Copyright 1997 American Chemical Society.

Cage	Site symmetry	Point group	Vibrational symmetry
Small, $[5^{12}]$	$m\bar{3}$	T_h	$T_{1u} + T_{2u}$
Large, $[5^{12}6^2]$	$\bar{4}m2$	D_{2d}	$A_{2g} + E_g + T_{1g} + 2T_{1u} + T_{2u} + T_{2g}$

Structure I hydrate: $G^s_2 \cdot G^L_6 \cdot 46H_2O$. Space group: $Pm3n$ (no. 223) O_h^3. $\Gamma_{guest} = 3(T_{1u}) + 2T_{2u} + T_{1g} + T_{2g} + A_{2g} + E_g$. $\Gamma_{acoustic} = T_{1u}$.

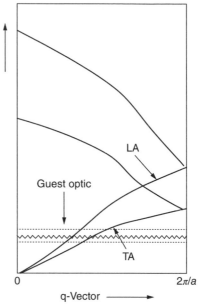

Figure 9   Factor group analysis of the symmetry of the guest vibrations and the schematic plot illustrating symmetry forbidden anti-crossing between localized guest vibrations and framework acoustic transverse (TA) and longitudinal (LA) vibrations.

IINS experiments revealed the quantum nature of the methane molecules at low temperature.[58] Below 30 K, a peak (A, in Figure 8a) at 8.3 cm^{-1} can be observed. This peak is assigned to the $J = 0 \rightarrow 1$ quantum rotational excitation inside the clathrate cages. The energy is remarkably close to the free rotor limit of 10.5 cm^{-1}. Moreover, from the correlation of the peak intensity variation with the neutron wave vector momentum transfer, it can also be concluded that the methane behaves almost like a free rotor. This indicated that the methane rotation is only slightly perturbed by the molecular field created by water molecules forming the cavity. Interestingly, unlike methane deposited in rare-gas matrices, the quantum rotational modes do not merge into the classical rotation-diffusion limit at high temperature (i.e., decreasing frequency with

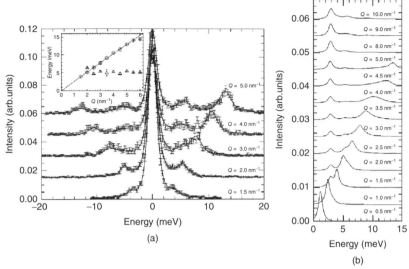

Figure 10 Experimental X-ray inelastic scattering spectra for methane hydrate (a). The inset shows the crossing of the localized methane vibrational band with the framework longitudinal acoustic branch. Comparison to theoretical prediction obtained from lattice dynamics calculations. *Phys. Rev. B*, **83**, 241403(R) (2011).

increasing temperature and eventually merges into the central elastic scattering peak, Figure 8a). In methane hydrate, the position of the quantum rotation peak remains constant but the linewidth increases substantially with increasing temperature and eventually disappears into the background when the temperature exceeds 35 K (Figure 8b). This observation is consistent with the translational motions being strongly hindered, and in this case, by the cage confinement. The IINS experiments on methane hydrate suggested the avoid-crossing of the localized guest vibrations with the water framework. Acoustic vibration branches provide a means for the exchange of energy between the two sub-systems, but the methane–water interaction remains weak. How the quantum tunnelling and anti-crossing of guest and the host phonon branches help to explain the low and glass-like behaviour of the thermal conductivity?

To investigate the proposed avoid-crossing scenario, which leads to resonant scattering inelastic X-ray scattering (IXS) experiments, a technique that probes the low-frequency and low-momentum transfer region of the phonon dispersion was used to study methane and xenon hydrates.[60] The IXS spectra of methane hydrate at several momentum transfers (Q) at 100 K are shown in Figure 10. Apart from the central elastic peak, the spectra display a well-defined dispersive mode and, from $q > 3\,nm^{-1}$, a second nondispersive peak. The dispersive excitation can be identified with the LA water–lattice phonon branch. The peak at 5 meV appearing at $q \approx 3\,nm^{-1}$ is close to the smallest size of the Brillouin zone in S-I clathrate ($Q_{min} = 3.8\,nm^{-1}$). This excitation is attributed to methane vibrations inside the large cage. These vibrations become visible in the spectrum after intersecting the LA lattice mode at $Q = 2.5\,\text{nm}^{-1}$, which is still inside the first Brillouin zone. This behaviour implies a coupling between the localized guest vibrations and the acoustic host lattice modes. The features observed in the IXS spectra can be reproduced correctly by theory.[60] In a one-phonon approach the scattering function $S(Q, \omega)$ at temperature T in an IXS experiment can be expressed as

$$S(Q, \omega) = G(Q, q, j) \cdot F(\omega, T, q, j) \qquad [21]$$

where q denotes the phonon wave vector in branch j, $G(Q, q, j)$ is the dynamical structure factor, and $F(\omega, T, q, j)$ is the response function. $G(Q, q, j)$ was calculated and averaged for the eigenvectors found in the lattice dynamics calculations for each of the 239 different directions chosen within the Q range of $0–10\,\text{nm}^{-1}$[60] As observed in the experiment, the calculated methane hydrate. IXS spectra display two distinct excitations: a dispersive LA host lattice mode and a nondispersive mode at $\omega \approx 3\,\text{meV}$ that becomes observable after crossing with the LA ($Q = 1.0\,\text{nm}^{-1}$) mode, corresponding to the methane vibrations inside the large cage.

For a harmonic crystal the phonon lifetime is infinite and there is no scattering of thermal phonons.[61] To understand the mechanism on how the guest–host interactions lead to the anomalous temperature dependence of the thermal conductivity, the lifetimes were calculated for phonon–phonon scatterings as a result of the anharmonic terms in the xenon-water potential of xenon hydrate in the small and large cage.[61] The inverse relaxation time (lifetime), τ^{-1}, of a lattice vibration with frequency $\omega_j(q)$ (j is the branch index and q is the direction of the momentum transfer) is related to the transition rate, W, of the lattice wave scattered from state $qj \rightarrow q'j'$ by a defect according to,[61]

$$\tau_{qj}^{-1} = \sum_{q'j'} W(qj \rightarrow q'j') \qquad [22]$$

The inverse relaxation time for the phonon scattering can be calculated by treating the guest (Xe) as an isolated point defect. The transition probability, W, can be computed from scattering theory using the Green's function method. The eigenfrequencies and eigenfunctions of an empty S-I hydrate were first obtained from a lattice dynamics calculation on a $2 \times 2 \times 2$ supercell. A Xe atom was then put in either the small cage or the large cage, and the scattering matrix, T, was computed from the retarded Green's function. The transition probability is related to the scattering T-matrix by

$$\delta T = \delta L + \delta L G \delta L \qquad [23]$$

where the matrix $\delta L = \delta L(\omega)$ describes the perturbation of the empty hydrate by an isolated defect (Xe). The results of the calculation for the inverse lifetime,

τ^{-1}, of the transverse (TA) and longitudinal (LA) acoustic phonons for a Xe atom in small and large cages are shown in Figure 11 ($1\,cm^{-1} = 3 \times 10^{10}\,s^{-1}$). If there is no scattering the inverse relaxation time will be zero. It is clear from Figure 11 that at the regions where the crossing of the Xe vibration with the TA or LA occurs, the inverse relaxation times become nonzero. Take the case of a Xe atom in the large cage; the lattice dynamics calculation (Figure 11a) shows that the localized Xe vibrations are located at 17 and $26\,cm^{-1}$. Furthermore, the lattice acoustic LA and TA branches intersect the lowest-energy localized band ($17\,cm^{-1}$) at $q = 0.23$ and 0.44 along the [1 1 1] direction, respectively. The q-values correspond nicely to the peak(s) in the calculated inverse lifetimes. For the LA phonon branch, three peaks in τ^{-1} at $q = 0.20$, 0.28 and 0.34 (Figure 11c) are observed, which can be attributed to the successive crossings of the LA branch with Xe vibration bands at 17, 23 and $32\,cm^{-1}$ in the experimental IINS spectrum (Figure 11b and 12),[62] respectively. A single strong peak is predicted for the TA mode related to the avoided-crossing at $q = 0.34$ with the lowest-energy Xe vibration band. The calculation also shows that the coupling of the TA and LA modes with the xenon in the large cage weakens substantially when the interactions are off resonance. The profile for the calculated inverse relaxation time for a Xe in the small cage can be rationalized in a similar manner. The resonance between Xe and the acoustic phonons now move to higher q due to the higher energy of the Xe-localized vibrations. The calculated magnitude for the inverse relaxation time of $10^{10}\,s^{-1}$ is in agreement with the value extracted from fitting the experimental THF hydrate data using the phenomenological resonant-scattering model.[50] The theoretical analysis shows that coupling of the lattice acoustic phonons with the localized low-frequency modes of the guest can lead to strong scattering of the lattice waves, and this is the cause of the low thermal conductivity in gas hydrates.

The phonon dispersion behaviour in xenon hydrate parallels the behaviour of electromagnetic waves propagating through a medium-containing localized ions resulting in the polariton and it can be explained by a phenomenological model. In the case of the gas hydrates, one can assume an elastic continuum, that is, sound waves representing the host lattice phonons interacting with the displacement field h of localized guest oscillators. The uncoupled elastic host field u of density ρ and sound velocity v along the wave vector k can be described by the Hamiltonian H_1,[62]

$$H_1 = \frac{p_1^2}{2\rho} + v^2 k^2 u^2 \qquad [24]$$

The Hamiltonian H_2 of the engaged xenon is that of a harmonic oscillator with mass m, frequency ω_0 and displacement h

$$H_2 = \frac{p_2^2}{2m} + m\omega_0^2 h^2 \qquad [25]$$

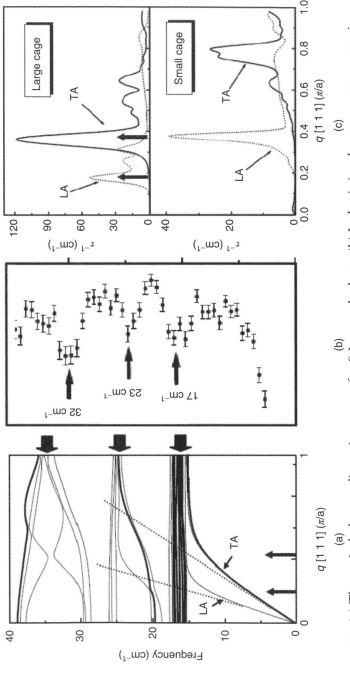

Figure 11 (a) Theoretical phonon dispersion curves for S-I xenon hydrate. (b) Inelastic incoherent neutron scattering spectrum of xenon hydrate. (c) Calculated phonon lifetime of Xe in the large (a) and small cage (b). Adapted from Ref. 61.

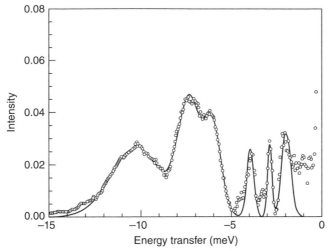

Figure 12 Experimental high-resolution inelastic incoherent neutron scattering spectrum of xenon hydrate. Reprinted with permission from *J. Chem. Phys.*, **116**, 3795 (2002). Copyright 2002 AIP Publishing, LLC.

If the host and the guest are coupled via the two displacement fields, u and h, the coupled Hamiltonian H_C is

$$H_C = H_1 + H_2 + \alpha u \cdot h \qquad [26]$$

where $u \cdot h$ is the coupling term and α is the coupling strength. A solution of the classical equations of motions can be found by assuming a wave-like behaviour of u and h as $e^{(i(\omega t - kx))}$. Solving the Eq. [26] results in the dispersion relation,

$$k^2 v^2 = \omega^2 + \frac{\alpha^2 \rho' \omega^2}{m(\omega_0^2 - \omega^2)} \qquad [27]$$

Equation [27] is sketched in Figure 13, and the dispersion relation is in qualitative agreement with dispersion curves calculated by lattice dynamics method for xenon hydrate (*cf.* Figure 11a).

The interactions between a localized vibrational state with the host elastic continuum can be described by the perturbation theory of Fano or in an equivalent way, with the second quantization field theoretical method of Anderson. The Anderson–Fano Hamiltonian appropriate for xenon hydrate is[63]

$$H = \omega_g \left(a_g^+ a_g + \frac{1}{2} \right) + \sum_k \left(\omega_l b_{l,k}^+ b_{l,k} + \frac{1}{2} \right) + \sum_{k,g} (b_{l,k}^+ a_g + a_g^+ b_{l,k}) \qquad [28]$$

where a_g^+ and $b_{l,k}^+$ are the creation and a_g and $b_{l,k}$ are the annihilation operators for the guest (g) and lattice (l) phonons. The first two terms are just the

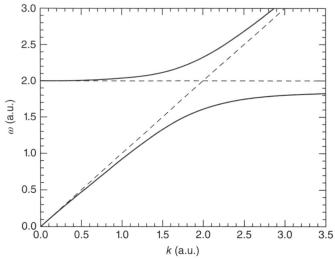

Figure 13   Calculated anti-crossing of the localized guest vibration and the lattice vibration using a classical continuum model. Reprinted with permission from *J. Chem. Phys.*, 116, 3795 (2002). Copyright 2002 AIP Publishing, LLC.

Hamiltonians for the zeroth order (isolated) state of the localized oscillator and lattice vibrations. The last term in Eq. [28] allows for the mixing of the lattice modes with the guest vibrations. The strength of the interaction is governed by the matrix element $M_{k,g}$. If $M_{k,g}$ is nonzero, there will be an exchange of energy between the lattice and the guest vibrational modes. The coupling term can be computed for a one-dimensional model system for the interaction of a localized state of energy ω_g with a single elastic (acoustic) lattice wave of energy ω_l. Representing the localized state by a harmonic Einstein oscillator $u_n(x)$ and the lattice wave as a plane wave with wave vector k, according to first-order perturbation theory, the matrix element is

$$M_{k \cdot g} \approx \langle u_n(x) \, | V(x) | e^{ikx} \rangle \qquad [29]$$

Considering only the lowest quadratic term in the polynomial expansion of the interaction potential $V(x)$ between the guest and the water lattice in the hydrate, and, that the harmonic oscillator is in the ground state $(n = 0)$, Eq. [29] reduces to the following integral,

$$|M_{k,g}| \sim \left| \frac{\sqrt{\pi}}{2} \int_{-\infty}^{\infty} e^{ikx} x^2 e^{-\alpha^2 x^2} dx \right|$$

$$= \left| \frac{\sqrt{\pi}}{2} \frac{e^{-\frac{k^2}{4\alpha^2}}}{\alpha^3} \left(\frac{1}{2} - \frac{k^2}{4\alpha^2} \right) \right| \qquad [30]$$

with α defined as $\sqrt{m_g \omega_g / 2\bar{h}}$, $\bar{h}$ is the Planck's constant and the energy of the plane wave, $E_{\omega l} = \bar{h}^2 k^2 / 2m_l$, and m_g and m_l are the masses of the guest and lattice waves, respectively. The latter can be approximated as the mass of a water molecule. Equation [30] has two interesting analytical properties. When $k^2 = 2\alpha^2$ the coupling matrix element vanishes. That is, the lattice plane wave and the harmonic oscillator are out of phase. Equating the first derivative of Eq. [30] with respect to the lattice wave vector k to zero, a maximum value is found when $k^2 = 6\alpha^2$. An expression relating the lattice and guest vibrational frequencies at maximum $|M_{k,g}|$ was found when $(m_l \omega_l / m_g \omega_g) = 3/2$. This simple relationship can be used to estimate the optimal lattice frequency that couples with a given guest vibration. For instance, the rattling motions of methane in the small and the large cages have frequencies of 75 and 42 cm$^{-1,52,58}$ respectively. Therefore, the coupling matrix element is predicted to maximize at lattice modes of 100 and 63 cm^{-1}. The predicted vibrational frequencies are well within the maxima of the transverse acoustic and transverse optic phonon branches of ice, which are about 70 and 110 cm^{-1}, respectively.

The thermal conductivity of a material can be calculated directly from equilibrium molecular dynamics (EMD) simulation based on the linear response theory Green–Kubo relationship.[64] The fluctuation–dissipation theorem provides a connection between the energy dissipation in irreversible processes and the thermal fluctuations in equilibrium. The thermal conductivity tensor, Λ, can be expressed in terms of heat current autocorrelation correlation functions (HCACFs), J_q,

$$\Lambda = \frac{1}{kT^2 V} \int_0^\infty J_q(t) \cdot J_q(0) dt \qquad [31]$$

The heat current is the amount of energy passing through a unit area. In the classical picture and if the interaction potential is pairwise additive, the energy (h_i) for an atom i in a N-particle system is

$$h_i = \frac{p_i^2}{2m_i} + \frac{1}{2} \sum_{j=1}^{N} V_{ij} \qquad [32]$$

The heat flux J_q is the time derivative of the work $r_i \cdot h_i$,

$$J_q(t) = \frac{d}{dt} \sum_{i=1}^{N} r_i \cdot h_i$$

$$= \sum_{i=1}^{N} \vec{v}_i h_i + \frac{1}{2} \sum_{i,j}^{N} \sum_{k,l}^{N} \vec{r}_{ik} \overrightarrow{F_{ij}^{kl}} \cdot \vec{v}_i \qquad [33]$$

where $\overrightarrow{F_{ij}^{kl}} = -\partial V_{kl} / \partial r_{ij}$.

For kinetic theory, the thermal conductivity is proportional to the thermal phonon mean free path.[64] Because the thermal conductivity of a disordered solid is very low, the mean free path is short. For a crystalline solid, the mean free path would be much longer. In an MD calculation, if the size of the model is too small, the time for a phonon to travel through the simulation cell is much shorter than the decay time of the current correlation function. This causes the phonons to be scattered more frequently than they would be in the infinite system. Therefore, it was expected that the calculation of the thermal conductivity would require a large model system and a very long simulation time. In practice, this was found not to be the case. According to the macroscopic law of relaxation and Onsager's postulate for microscopic thermal fluctuation, the asymptotic decay of the heat correlation function should be exponential. The kinetic coefficients depend mostly on long-time decay rate behaviour and because an exponential-decay behaviour is exhibited in a microscopic timescale, a medium-sized simulation cell can be used to extract the decay rate of the heat dissipation. This was demonstrated in a study on the thermal conductivity of cubic diamond where the mean free path and the thermal conductivity converge rapidly with the size of the model.[65]

The first study on the anomalous thermal conductivity of clathrate hydrates with MD simulation was reported in 1996.[66] EMD and nonequilibrium molecular dynamics (NEMD) simulations were performed on the xenon hydrate and compared to the results for the hypothetical empty hydrate and ice Ih. The results are summarized in Table 3. Both types of MD calculations reproduced the experimental trend that (i) the thermal conductivity of the gas hydrates is lower than ice at the same temperature; (ii) unlike ice the values decrease with temperature; (iii) the thermal conductivity of the empty hydrate is lower than that of ice but the temperature dependence is similar. There are significant differences in the EMD and NEMD results, however. For example, EMD predicted thermal conductivity for xenon hydrate of 0.49 W/m K at 100 K, which is almost three times higher than 0.16 W/m K obtained from NEMD. The calculated thermal conductivity for ice Ih is also significantly lower than the observed value.[66] These discrepancies may be partly attributed to the length of the simulation, to the small size of the models, and/or to the numerical techniques used. One unit cell consisting of 46 water molecules was used to model xenon hydrate. It was recognized only recently that truncations of long-range electrostatic interactions in charged systems may lead to serious errors in the stress tensor affecting the reliability of the heat flux autocorrelation function (HCACF).[67,68] Nonetheless, analysis of the normal modes shows strong anharmonicity in the xenon rattling motions particularly for the low-frequency vibrations in the large cages. It was then concluded that "the most likely mechanism of the low thermal conductivity and its anomalous temperature dependence seems to be the resonance scattering model".[66]

Table 3 Comparison of Theoretical and Experimental Thermal Conductivity (W/m K) of Xe Hydrate, Empty Hydrate, and Ice I*h*

System	Method	100 K	180 K	260 K
Xe hydrate	EMD	0.49		0.52
	NEMD	0.16	0.63	0.77
	Expt.	0.36	0.44	0.47
Empty hydrate	EMD	0.94		0.67
	NEMD	1.87		0.78
Ice I*h*	EMD	1.49		0.94
	NEMD	3.32		0.96
	Expt.	5.94	3.58	2.35

The calculation of the thermal conductivity of gas hydrate using EMD and the Green–Kubo linear response theory was repeated recently.[69] In that work, convergences of the relevant quantities were monitored carefully as a function of the model size. Subtleties in the numerical procedures were also carefully considered. The thermal conductivity of methane hydrate was found to converge within numerical accuracy for $3 \times 3 \times 3$ and $4 \times 4 \times 4$ supercells. In the calculation of the heat flux vector J_q there is an interactive term that is a pairwise summation over the forces exerted by atomic sites on one another. The species (i.e., water and methane) enthalpy correction term requires that the total enthalpy of the system is decomposed into contributions from each species. Because of the partial transformation from pairwise, real-space treatment to a reciprocal space form in Ewald electrostatics, it is necessary to recast the diffusive and interactive terms in this expression in a form amenable for use with the Ewald method using the formulation of Petravic.[70,71]

Results of the calculated thermal conductivity for ice I*h*, S-I methane hydrate and empty hydrate are depicted in Figure 14. The thermal conductivity of ice I*h* has improved, but the absolute value is still slightly smaller than the experiment. The calculations reproduced previous observation that the thermal conductivity of the hydrate is lower than ice I*h* and the empty hydrate.[66] Even though the empty hydrate has a lower thermal conductivity than ice I*h*, the crystalline temperature profile is similar. A surprising finding is the reversal in the thermal conductivity of methane hydrate at low temperature. From 250 to 100 K, the thermal conductivity decreases slightly. When the hydrate is cooled below 100 K, the conductivity increases and follows the trend as a crystal. This unusual temperature profile has indeed been observed in methane and xenon hydrates,[72,73] details of which will be deferred to a later part of this chapter. To unravel the thermal transport mechanism, various correlation functions were computed and the relaxation times analyzed.[74,75] The HCACF can be fitted to

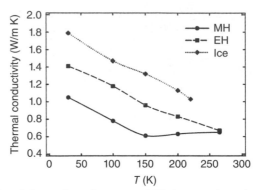

Figure 14   Calculated thermal conductivity of filled S-I methane hydrate (MH), hypothetical empty S-I hydrate (EH), and ice Ih. Reprinted with permission from N. J. English and J. S. Tse, *Phys. Rev. Lett.*, 103, 015901 (2009). Copyright 2009 American Physical Society.

sums of exponentially decay functions characterized by the respective relaxation time.

$$\text{HCACF}(t) = \sum_{i=1}^{n\_ac} A_i \exp\left(-\frac{t}{\tau_i}\right)$$

$$+ \sum_{j=1}^{n\_opt} \left(\sum_{k=1}^{n_o,j} B_{jk} \exp\left(-\frac{t}{\tau_{jk}}\right) \right) \cos \omega_{0,j} t + \sum_{j=1}^{n\_opt} C_j \cos \omega_0 t \qquad [34]$$

The first term in Eq. [34] accounts for the acoustic phonons (sum over n_ac). The second and third terms are related to the optical phonons and the oscillatory features in the HCACF because of the water optic modes with energy ω_0 (summed over n_opt).

The results of the fit are summarized in Table 4. In general, these relaxation times fall into two time regimes – a rapid initial drop ($\tau_{sh,ac}$) followed by a long decay ($\tau_{lg,ac}$).[64] This was the case for ice Ih, empty S-I and methane hydrate at 30 K. Between 100 and 250 K, the optimal fit to the HCACF for methane hydrate data required an additional term with a relaxation time intermediate ($\tau_{int,ac}$) between the long- and short-time regimes. The unique $\tau_{int,ac}$ is relatively insensitive to the temperature ($\sim$0.3–0.4 ps) and, incidentally, is comparable to the relaxation time determined from the analysis of the THF thermal conductivity with the resonant scattering model and the calculated lifetimes of the strongly scattered phonons in xenon hydrate (*vide infra*).[58,61] This is not a coincidence; it is a strong and unambiguous indication of the significant role of the guest in the scattering of heat carrying phonons.

To probe the heat transfer derived through guest–host interactions, energy correlation functions (ECFs) were computed. ECF_{WM} is the deviation

Table 4　Acoustic Relaxation Times and Contributions to the Overall Thermal Conductivity (W/m K) for Structure-I Methane Hydrate, Empty Structure Hydrate, and Ice I*h*

System	T (K)	$\tau_{\text{sh,ac}}$ (ps)	Short-Range Acoustic	$\tau_{\text{int,ac}}$ (ps)	Medium-Range Acoustic	$\tau_{\text{ilg,ac}}$ (ps)	Long-Range Acoustic	Optic	Total
Structure-I	30	0.38	0.21			4.4	0.69	0.15	1.5
	265	0.046	0.06	0.33	0.05	2.1	0.40	0.13	0.64
Empty	30	0.44	0.20			6.1	0.99	0.22	1.41
	265	0.057	0.08			2.2	0.43	0.16	0.67
Ice I*h*	30	0.47	0.24			8.2	1.31	0.24	1.79
	220	0.23	0.11			2.4	0.72	0.20	1.00

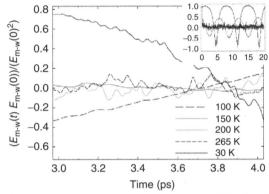

Figure 15　Normalized energy autocorrelation function (ACF) of methane–water interaction energies in S-I methane hydrate at various temperatures at a time slice of 1 ps (3–4 ps); (inset) the full 20 ps ACF. Reprinted with permission from N. J. English and J. S. Tse, *Phys. Rev. Lett.*, **103**, 015901 (2009). Copyright 2009 American Physical Society.

of the water–methane interaction energy from its long-time average value,[74,75]

$$\text{ECF}_{\text{WM}}(t) = \frac{\langle \Delta U_{\text{WM}}(t)\Delta U_{\text{WM}}(0)\rangle}{\langle \Delta U_{\text{WM}}(0)\Delta U_{\text{WM}}(0)\rangle} \qquad [35]$$

ECF$_{\text{WM}}$ were calculated from 30 to 265 K and shown in Figure 15. Damping is visible at and above 150 K. Below 150 K, the ECF of the methane–water interaction energy is almost harmonic with no damping thus indicating that heat dissipation is small. This agrees with the experimental observation that conductivity starts to increase when the temperature is lower than 90 K.[72,73] Above 150 K, strong ECF damping is observed.

The analysis of the theoretical results is consistent with the resonant scattering model. The nondispersive (localized) rattling guest vibrations intersect the acoustic branches of the lattice modes. The avoided-crossings provide the means to transfer energy from heat-carrying phonons to localized excitations of the guest vibrations. This process has a relaxation time of 0.3–0.4 ps and is largely independent of the temperature between 100 and 250 K. At sufficiently low temperature $(T < 3–100\,K)$ insufficient thermal energy exists $(kT = 69\,cm^{-1})$ to populate the guest vibration levels (reduced VDOS). The role of the resonant scattering is diminished and the thermal conductivity of the gas hydrate reverts back as a crystalline solid. The theoretical prediction is in agreement with experimental trend observed in Xe and CH_4 hydrates (Figure 16).[72,73] In fact, in the absence of rattling motions, that is, in the filled ice structure of methane, hydrate exists at high pressure or when the guest vibrational frequency is higher than the acoustic branches, for example, in hydrogen clathrate, a normal crystalline-like thermal conductivity is predicted.[76] It is noteworthy that the anomalous "glasslike" behaviour at higher temperatures has been exploited by the principle of phonon glass electron crystals for engineering the next generation of efficient thermoelectric materials.[77–79]

Thermodynamic Stability

In view of the computational simplicity of the vdW-P solid solution theory,[28] and apart from developing more reliable intermolecular potentials and methods for the evaluation of the configurational integrals (*vide supra*), there have been attempts to generalize the theory by relaxing restrictions that include: (i) a rigid water framework, (ii) the neglect of guest–guest interactions and (iii) single occupancy on each type of cage that is inappropriate for some gas hydrates. While including a guest with a large molecular size may not influence the overall hydrate structure, it may distort the local cage environment. Multiple occupancy has been reported in nitrogen[80] and hydrogen clathrate,[81,82] and this possibility has also been raised for the argon hydrate at high gas pressure.[29,83] The most comprehensive approach for generalizing the vdW-P theory was proposed by Tanaka and his colleagues.[84–96] The idea is to take into account the effect of the host lattice and guest vibrations on the coupling in consideration of thermodynamics stability. The free energy of the host lattice in the formation of a gas hydrate is decomposed into two contributions: the frequency modulation and the lattice distortions.[84] This enables calculations of the separate factors by knowing the water–water and guest–water intermolecular potentials.

The essences of the generalized vdW-P theory proposed by Tanaka follows. A system consisting of n_w unit cells is in equilibrium with a gas phase of the guest. For the S-I hydrate, there are $m_w = 46$ water molecules per unit cell. The total number of small cages (5^{12}) N_s is $2n_w$ and that of large cages $(5^{12}6^2)$ N_l is $6n_w$. Similarly, for S-II, $m_w = 136$, $N_s = 16n_w$ (5^{12}) and $N_l = 8n_w$ $(5^{12}6^4)$. The combined number of ways (probability, P) of placing j_s and j_l into the small and large cages in the hydrate structure is

$$P(j_s, N_s, j_l, N_l) = \binom{N_s}{j_s} \binom{N_l}{j_l}$$

[36]

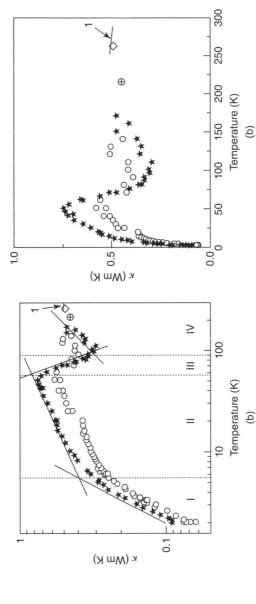

Figure 16 Experimental thermal conductivity of (a) xenon hydrate and (b) methane hydrate. Reprinted figure with permission from A. I. Krivchikov, B. Y. Gorodilov, O. A. Korolyuk, V. G. Manzhelii, O. O. Romantsove, H. Conrad, W. Press, J. S. Tse and D. D. Klug, *Phys Rev. B*, 73, 064203 (2006). Copyright 2006 American Physical Society.

It is straightforward to extend the probability factor to multiple occupancy. For example, the number of ways to choose j_l singly occupied and j_d doubly occupied large cage and j_s singly occupied small cages becomes

$$P(j_s, N_s, j_l, N_l) = \binom{N_s}{j_s} \binom{N_l}{j_d} \binom{N_l - j_d}{j_l} \qquad [37]$$

The canonical partition function Z_{j_s, j_l} is given by the free energy of the host lattice. The entropy arises from the proton disorder and from the occupation of the guests. The free energy is because of the guests' vibrations in the cages. For single occupancy,

$$\begin{aligned} Z_{j_s j_l} &= \binom{N_s}{j_s} \binom{N_l}{j_l} \sigma_w^{n_w} \exp[-\beta n_w (F + U)] \\ &= \binom{N_s}{j_s} \binom{N_l}{j_l} \sigma_w^{n_w} \exp[-\beta n_w (F_w^0 + U_w^0)] \times \exp[-\beta(j_s f_s + j_l f_l)] \qquad [38] \end{aligned}$$

where $\beta = 1/kT$, and $\sigma_w^{n_w}$ is the number of possible arrangements of protons for a single empty unit cell. F_w^0 and U_w^0 are the free energy of intermolecular vibrations and the internal energy of a geometry-optimized empty hydrate unit cell, respectively. F and U are those for the occupied clathrate hydrate. Hence, f_s and f_l are the free energy changes because of the enclathration of the guests. The guest–host interaction is incorporated implicitly into the free energy terms. The grand partition function, Ξ, of the system with respect to the guest molecules using the chemical potential of the guest species can be written as

$$\begin{aligned} \Xi &= \sigma_w^{n_w} \exp[-\beta n_w (F_w^0 + U_w^0) \\ &\quad \times \{1 + \exp[\beta(\mu - f_s)]\}^{N_s} \\ &\quad \times \{1 + \exp[\beta(\mu) - f_l]\}^{N_l} \qquad [39] \end{aligned}$$

The free energy of the system is given by the entropy arising from the occupancy of guest molecules and the free energy because of guests' vibrations inside the cavities.

The chemical potential of water (μ_w) can be obtained from the derivative of the partition function,

$$\mu_w == -kT \frac{\partial n\Xi}{\partial n_w} \qquad [40]$$

In a similar manner, the mean occupancy of the small $\langle N_s \rangle$ and large $\langle N_l \rangle$ cages are computed from the derivative of the partition function with respect to the free energy,

$$\langle N_{s,l} \rangle = \frac{\partial \ln \Xi}{\partial (\beta \mu)} = \frac{n_w \sum_{j_{s,l}=1}^{N_{s,l}} \binom{N_{s,l}}{j_{s,l}} \exp(\beta g_{j_{s,l}})}{\sum_{j_{s,l}=0}^{N_{s,l}} \binom{N_{s,l}}{j_{s,l}} \exp(\beta g_{j_{s,l}})}$$ [41]

All the quantities contributing to the partition function in Eqs [39–41] can be computed once the intermolecular potentials are known. The calculation proceeds as follows. Several proton-disordered hydrate structures with zero net dipole are chosen. The minimum energy of each gas hydrate structure is computed from a geometry optimization, usually using the steepest descent method. The vibrational entropies are from normal mode analysis of lattice dynamics results. The host and guest contributions are determined from the Hessian matrix V (second derivative of the intermolecular potential with respect to the displacements). The Hessian matrix can be divided into four sub-matrices, V_{ww}, V_{gg}, V_{gw}, and V_{wg},

$$V = \begin{pmatrix} V_{ww} & V_{wg} \\ V_{gw} & V_{gg} \end{pmatrix}$$ [42]

V_{gg} and V_{ww} are associated with the free energy changes f_g and f_w described above.

The assumption of a rigid water host framework can be eliminated and replaced by the following procedure. For a guest molecule with size comparable to the hydrate cage, the free energy (f_s or f_l) due to the guest molecule is approximated by

$$f_{s,l} = \Delta g + U$$ [43]

where U is the minimum value of the potential energy for the guest in the presence of the water surrounding it and Δg is the vibrational free energy change due to the guest binding in the cavity. The vibrational motions are assumed to be a collection of harmonic oscillators. Thus the vibrational free energy, g (empty or occupied hydrate) is computed according to the classical mechanical partition function for an isolated oscillator,

$$g = kT \int \ln(\beta \hbar \omega) h(\omega) d\omega$$ [44]

where $\hbar \omega$ is the VDOS and can be computed from the normal mode analysis. The difference in the vibrational free energy between empty and occupied hydrates, Δg, is computed taking into account the number of degrees of freedom for each system.

When the size of the guest is much smaller than the cavity, the free energy can be simplified to a free rotor,

$$f_{s,l} = kT \ln b \qquad [45]$$

b is the partition function of the isolated guest molecule and is given by

$$b = \left(\frac{mkT}{2\pi\hbar^2}\right)^{3/2} \int_{\text{cell}} \exp[\beta u(r)]dV \qquad [46]$$

where the integration spans the volume of the single cage, V_{cell}, m is the mass of the guest and $u(r)$ is the interaction potential between the guest and the surrounding water.

The theory can be improved further by including the anharmonic contribution to the free energy. This is accomplished by a thermodynamic integration method with a reference system of being a collection of harmonic oscillators. In the case of nonspherical molecules, the reference system is chosen to be the hydrate of spherical guest molecules. The free energy difference between the real and the reference system $\Delta\mu_A$ is given by

$$\Delta\mu_A = -kT \ln\langle\exp[-\beta(\Phi - \Phi_0)]\rangle \qquad [47]$$

where Φ and Φ_0 are the potential energies for the real and reference systems, respectively. The sum of the vibrational energies is simply the energy difference between the reference system and the potential energy of the reference system at its minimum structure, U_0,

$$\Phi_0 - U_0 = \sum_i \omega_i^2 q_i^2 \qquad [48]$$

and q_i is the normal mode coordinates. The average is taken over the reference harmonic oscillators using a Monte Carlo scheme weighed by the probability distribution,

$$P(q) = \prod_i^{6n_w-3} \left(\frac{\beta\omega_i^2}{2\pi}\right)^{1/2} \exp\left(-\frac{\beta\omega_i^2 q_i^2}{2}\right) \qquad [49]$$

The generalized vdW-P theory has been used to predict the thermodynamic stability of various gas hydrates having diverse characteristics, namely, atom and molecules, polar and nonpolar guests, spherical and nonspherical molecules, large molecules, and multiple occupancy in both small and large cages in S-I and S-II hydrates. A realistic description of guest–host interactions is crucial for guest molecules at a size comparable to the cavity. Large molecules, such as perfluoromethane, propane, and so on, may distort the host

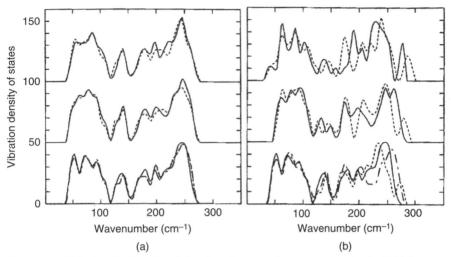

Figure 17 Calculated vibrational density of states of water molecules in S-I (a) xenon and (b) fluoromethane hydrate. Reprinted from H. Tanaka, The Stability of Xe and CF4 Clathrate Hydrates. Vibrational Frequency Modulation and Cage Distortion, *Chem. Phys. Lett.*, **202**, 345. Copyright 1993, with permission from Elsevier.

Table 5 Free energy of Intermolecular Vibration g at 273.15 K and the Potential Energy u at Structures of Minimum Potential Energy

System	Cage	$g + u$	u	$g_g' + u$	g_w'	$-RT \ln b$
Xenon	s	−36.52	−28.29	−39.46	+2.90	−38.90
	l	−45.35	−29.83	−46.36	+0.80	−44.74
CF$_4$	s	−8.68	−8,35	−14.86	+6.39	−14.64
	l	−35.81	−27.36	−38.21	+2,53	−37.72

The free energy and the potential energy are calculated for larger and smaller cage occupations, denoted by l and s, respectively. The free energy of a guest molecule is denoted by $-RT\ln b$. Energy is in kJ/mol.

environment and alter its lattice vibrations. The VDOS of S-I xenon and perfluoromethane hydrates[97] and the empty hydrate in the water translation vibration region are shown in Figure 17 and the relevant calculated free energies in Table 5. It is shown that the VDOS of xenon and the empty S-I hydrate are very similar. In comparison, including the large perfluoromethane in the large cages results in significant changes in the host lattice vibrations. In xenon hydrate the calculated difference of $g + U$ and $g' + U$ are within 3 kJ/mol for both the small and large hydrate cages. The corresponding differences are substantially larger in perfluoromethane, particularly in the small cage. Note that it is known from experiment there is little or no perfluoromethane occupying the small cages in the type I structure. Differences in the VDOS may be amplified by occupying the small cages with the perfluoromethane molecules. An improved calculation, employing the same theory, shows the small cages are not occupied.[95] In the same study, the predicted occupancy ratio of the small/large cages in xenon of 0.73 is in fair agreement with 0.77 ± 0.02 for a xenon deuterated hydrate determined from an NMR experiment.[35] The predicted dissociation pressures for argon and xenon hydrates over a temperature range 123.15–273.13 K are in good agreement with experiments; however, multiple occupancy of the large cage of the S-II argon hydrate has been suggested in a recent experimental study[98] where a new Raman peak emerges at 100 MPa as the intensity increases steadily with increasing pressure up to 400 MPa and attributed to double occupancy in the gas hydrate.[98] Calculations using the generalized vdW-P theory found that double occupancy becomes dominant over single occupancy when the equilibrium gas–hydrate pressure exceeds 270 MPa. Overall, the performance of the generalized vdW-P is satisfactory, but in some cases, it is still not quantitative. Eliminating the restrictions imposed by the original vdW theory has indeed improved agreement with experiment. However, the theoretical results are not yet quantitative. In selected cases, for example, xenon hydrate, discrepancy with experiments exists. Another unsettling problem is on the predicted chemical potential difference of water in ice and in empty clathrate hydrate (i.e., μ_w^0, see above).[85] The calculated μ_w^0 for S-I and S-II hydrates are 760 and 580 kJ/mol. Although empty S-II is predicted to be more stable than empty S-I, the chemical potentials differ significantly when compared to the currently accepted values of 1203 and 1077 kJ/mol for S-I and S-II, respectively. It is possible that using the empirical TIP4P water–water potential has contributed to the error. Improvements using more accurate water potentials may be needed to resolve the discrepancy.

Recall that encapsulation of guest molecules into cavities of a gas hydrate can be regarded as an adsorption process. In fact, a large Langmuir constant (Eq. [3]) is the main contributing factor for the stability of a gas hydrate. The grand canonical Monte Carlo (GCMC) technique is well suited for studying adsorption processes.[99] It is performed at constant chemical potential, volume and temperature, but the number of particles (molecules) can fluctuate. Hence, the adsorption isotherms can be determined directly from the simulation by evaluating the average number of adsorbed molecules. This method has been used extensively for investigating multicomponent adsorption in zeolites and mesoporous materials. In the case of a gas hydrate at a given pressure (chemical potential of the guests), the number of guest molecules that can be incorporated into the structure can be evaluated in the same way as the usual adsorption process using the GCMC method. Owing to its simplicity and its capability of handling gas mixtures, GCMC has steadily gained popularity as an alternative approach to obtain information on the gas content in gas hydrates.[100–102]

The GCMC protocol is similar to a conventional MC method except that particles are allowed to be inserted or deleted from the system. If insertion is chosen, a particle is placed into the system at a given probability density. If deletion is chosen, a randomly selected particle is removed. Insertions or deletions are accepted or rejected according to the Metropolis scheme. A sequence of trial moves generates a Markov chain.[99] GCMC was used to study the accommodation of propane in a hydrate structure, but the Metropolis method for selecting creation and deletion moves was replaced by a symmetrical algorithm[88,99] that works as follows. One of the cavities in the system is chosen randomly. If the cavity is empty, a molecule is inserted with its position and orientation assigned with a probability distribution $\phi(\Delta r, \Omega)$, where Δr is the position from the center of the cage and Ω is the associated set of Euler angles. The trial creation is accepted with a probability,

$$\min\left[1, zv \exp\left(\frac{-\beta U(\Delta r, \Omega)}{\phi(\Delta r, \Omega)}\right)\right] \qquad [50]$$

where z is the fugacity and v is the cage volume. $U(\Delta r, \Omega)$ is the interaction potential of the guest molecule with the host water framework and with other guests. The probability function is normalized to the volume of the cavity.

$$\iint (r,) drd = 4\pi^2 v \qquad [51]$$

If the cavity is occupied, the guest is deleted with a probability,

$$\min[1, \phi(\Delta r, \Omega)] \exp\left(-\frac{\beta U(\Delta r, \Omega)}{zv}\right) \qquad [52]$$

Using these rules, GCMC calculations were performed on propane hydrate in equilibrium with the gas using four proton-disordered lattice models.[88] Each calculation was equilibrated with 100,000 cycles of standard Metropolis MC steps. GCMC simulations were then continued for 600,000 cycles, in which the first 100,000 cycles were used to equilibrate the systems with respect to the occupation of guest molecules. Results of the simulations at 0.001, 0.005, 0.02, 0.1 and 0.2 MPa are compared with the direct generalized vdW-P theory in Figure 18. The agreement of the GCMC results with directly calculated cage occupation from the free energy is reasonable over the pressure range studied. The trend is reproduced correctly, but the deviation between the two set of calculations increases at lower pressures.

The GCMC technique has been extended to the isothermal–isobaric (*NPT*) ensemble for the study of multiple occupancy of hydrogen in the S-II hydrate.[95] Because the hydrogen clathrate is stable only under pressure, it is essential to calculate the cage occupancy at a given temperature and pressure. It is assumed that the hydrogen pressure on the hydrate is the same as in the fluid and accordingly the volume of the host lattice corresponding to the pressure is adjusted using the *NPT* MC scheme. The water molecules are sampled according to the usual Metropolis MC method. A hydrogen molecule is inserted into the hydrate system with N_g hydrogens placed initially at an

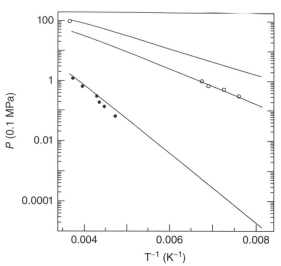

Figure 18 Dissociation pressure over a temperature range from 123.15 to 273.15 K. Solid and dashed lines show the calculated dissociation pressure for S-I argon and xenon clathrates, respectively. Dash–dot line shows the dissociation pressure for the argon hydrate. Open and black circles show the experimental results for argon and xenon clathrate hydrates, respectively. Reprinted by permission of Taylor & Francis Ltd, http://www.tandf.co.uk/journals from H. Tanaka and K. Nakanishi, The Stability of Clathrate Hydrates: Temperature Dependence of Dissociation Pressure in Xe and Ar Hydrate, Molecular Simulation, 1994.

arbitrary position and a random orientation. Each move is accepted with a probability

$$p = \exp[\beta(\mu' - U)]\left(\frac{2\pi mkT}{h^2}\right)^{3/2}\frac{V}{(N_g + 1)} \qquad [53a]$$

μ' is the chemical potential for hydrogen molecule excluding the contribution from the free-rotational motion. In the deletion move, a hydrogen molecule is chosen arbitrarily out of the system. The trial is accepted with a probability

$$p = N_g\left(\frac{2\pi mkT}{h^2}\right)^{-3/2}\exp\left(\frac{\beta\left(-\mu' + U'\right)}{V}\right) \qquad [53b]$$

where U' is the interaction energy of the chosen guest with all other molecules.

Results of these simulations performed at 170 and 273 K are shown in Figure 19.[95] The occupancies of the large cage in the S-II hydrogen hydrate depend on the pressure of hydrogen fluid. At 170 K and 100 MPa, single occupancy is a minor component. Double occupancy prevails in the low-pressure range (0–300 MPa) but decreases with increasing pressure. At high gas pressure, triple occupancy prevails. At even higher pressure, quadruple occupancy

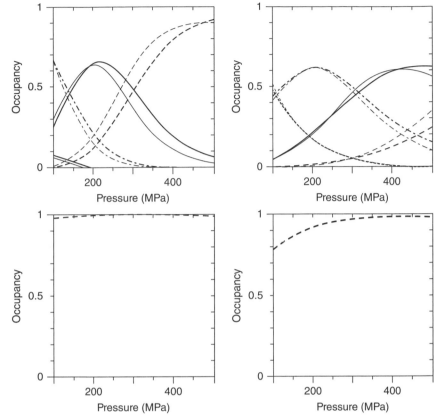

Figure 19 Occupancy of (a) large cage (dotted, single; dotted–dashed, double; solid, triple; dashed, quadruple occupancy) and (b) smaller cage (solid, single; dotted, double occupancy) obtained from GC/NPT and GC/NVT MC simulations at 170 K (heavy lines, GC/NPT; thin lines, GC/NVT). (c) and (d) same as above but calculation performed at 273 K. Reprinted with permission from *J. Chem. Phys.*, **127**, 044509 (2007). Copyright 2007 AIP Publishing LLC.

becomes dominant. No further incorporation of hydrogen is observed at 500 MPa. In comparison, the small cage contains only a single hydrogen molecule in the pressure range investigated. Only 0.2% of the smaller cage encapsulates two hydrogen molecules at 500 MPa. At 273 K, single occupancy of the smaller cage decreases and double occupancy increases slightly but it is still very low at only 0.7%, even at 500 MPa. For the large cage at 273 K, multiple occupancy is less noticeable compared to 170 K but it becomes significant at pressures exceeding 400 MPa.

GCMC simulations have also been used to explore the favorable thermodynamic conditions for exchanging carbon dioxide and methane in the S-I clathrate hydrates.[102] Of practical interest is the need to find a feasible thermodynamic pathway in the phase diagram consisting of methane, carbon

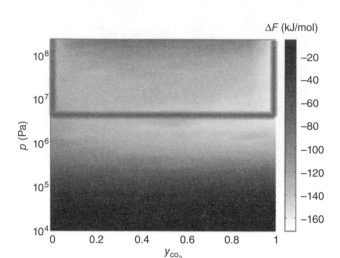

Figure 20   Helmholtz free energy difference (ΔF) (in gray scale) computed by GCMC of a carbon dioxide–methane gas mixture in the hydrate at 278 K as a function of the fugacity of each of the component. Reprinted (adapted) with permission from *J. Phys. Chem. B*, **116**, 3745 (2012). Copyright 2012 American Chemical Society.

dioxide and water, to convert a pure methane hydrate into a pure carbon dioxide hydrate. The theoretical approach is to obtain the Helmholtz free energy for a given configuration of a filled hydrate from the occupancy isotherms computed using GCMC simulation. In that study by Glavatsky et al., only the relative stabilities of the methane and carbon dioxide gas hydrates were evaluated. GCMC calculations were performed on a $2 \times 2 \times 2$ unit cell model of S-I hydrate. The TIP5PEw water potential[103] was used. The guest interactions were described by the TraPPE force fields.[104] Both flexible and rigid water frameworks were considered. Because the results are similar, only those obtained from the rigid framework were reported. Simulations were performed between 278 and 328 K with the pressure and fugacity varying between 10^4 and 10^9 Pa. The main results are summarized in Figure 20. The Helmholtz free energy differences (ΔF) relative to the empty hydrate for a methane and carbon dioxide mixture in the hydrate at 278 K as a function of the gas pressure and mole fraction of carbon dioxide are shown. The lighter shading indicates greater stability of the hydrate. The end points at 0 and 1 mole fraction of carbon dioxide correspond to the formation of stable methane and stable carbon dioxide hydrates at high pressure (e.g., 10^9 Pa), respectively. An important observation is that a low-energy path (light color) was found connecting the two stable hydrates (indicated by the thick lines) through a region of mixed methane and carbon dioxide hydrates. The theoretical results suggest that a thermodynamically feasible pathway exists for converting methane hydrate to carbon dioxide hydrate via the following steps.[103] First, the pressure of pure methane hydrate must decrease to $\sim 10^7$ Pa.

At this pressure, there is enough space in the large cages for carbon dioxide molecules to enter. Second, by increasing the concentration of carbon dioxide in the surrounding fluid the large cages fill with more carbon dioxide with a concomitant expulsion of the methane from the hydrate. At 10^7 Pa, there is no more methane in the mixture. The small cages are now empty and available for subsequent filling by carbon dioxide. Eventually the process leads to the formation of a pure carbon dioxide hydrate.

Hydrate Nucleation and Growth

It may be fair to say that flow assurance provided the major impetus for gas hydrate research in the past century.[10] Flow assurance is the prevention of pipeline blockages due to hydrate formation. Gas and water crystallize into an immobilized solid hydrate under high gas pressure and low temperature. This was recognized by Hammerschmidt in the 1930s[7] and has been a problem plaguing the oil and gas industry ever since. This is a predicament especially for deep water pipelines that transport natural gas. Providing an understanding of the thermodynamic and physical properties along with the mechanisms for nucleation and growth of gas hydrate are imperative to resolve and eliminate this problem. Studies on nucleation and growth can be loosely classified into two categories: hydrate crystallization and inhibition of hydrate formation. The study of nucleation of a complex liquid, like water, in a computer simulation is difficult because freezing is a stochastic event.[105] The disordered three-dimensional hydrogen-bonded network of water gives rise to a rugged PES and consists of many local minima thus making the direct observation of crystallization in an equilibrium MD simulation nontrivial. Very long simulation times on the order of hundreds of nanoseconds (ns) to microseconds (μs) are required to observe such rare events.

A trick to accelerate the simulation is to impose order parameters relevant to the description of the crystalline state to drive the system towards nucleation. This method was applied to the study of water nucleation and carbon dioxide hydrate formation using a conventional Monte Carlo method.[106,107] Steinhardt's bond-orientational order parameters[108] Q_l and W_l, based on quadratic and third-order invariants formed from bond spherical harmonics ($Y_{lm}(\theta, \varphi)$), were employed. These order parameters allow quantitative measures of the local symmetry in liquids and glasses. The rotationally invariant orientational order parameters are defined as[108]

$$Q_i = \left(\frac{4\pi}{2l + 1} \sum_{m=-l}^{+l} \left| \overline{Q}_{lm} \right|^2 \right)^{1/2} \qquad [54]$$

where

$$\overline{Q}_{lm} = \frac{1}{N} \sum_{i=1}^{N} Y_{lm}(\theta, \varphi) \qquad [55]$$

N is the number of nearest-neighbour contacts within a given radius cutoff. W_l is defined as

$$W_l = \sum_{\substack{m_1 m_2 m_3 \\ m_1+m_2+m_3=0}} \begin{bmatrix} l & l & l \\ m_1 & m_2 & m_3 \end{bmatrix} \overline{Q}_{lm_1} \overline{Q}_{lm_2} \overline{Q}_{lm_3} \qquad [56]$$

The coefficients $\begin{bmatrix} l & l & l \\ m_1 & m_2 & m_3 \end{bmatrix}$ are the Wigner 3-j symbols.

For water molecules an additional order parameter needed to measure the tetrahedricity, ζ[109] was used.

$$= \frac{1}{N}\sum_N \left[1 - \frac{3}{8}\sum_{i=1}^{3}\sum_{j=i+1}^{4}\left(\cos\varphi_{ij} + \frac{1}{3}\right)^2 \right] \qquad [57]$$

where N is the number of water molecules; the indices i, j run over the four nearest neighbours of a given water molecule; and φ_{Iij} is the angle between the nearest-neighbour bond associated with water i and j.

In the MC simulation using a TIP4P[110] water potential, the system was supercooled[107] to 180 K (the melting point of ice predicted by the TIP4P model is 250 K). Umbrella sampling was performed on the water molecules to bias the trial moves towards the bond order parameters Q_6 and ζ while Q_4 and W_4 were monitored. The simulation indeed generated an ice crystal. Note that the Markov chain generated by the biased MC did not generate a Boltzmann distribution.[100] Snapshots along the path of freezing together with the evolution of the Q_6 are shown in Figure 21.[107] The Gibbs energy profile shows a transition region at $Q_6 \sim 0.2$. The formation of a six-membered water ring configuration has been suggested as the initial step of nucleation in an MD calculation (*vide supra*). In contrast, the rings have already formed *prior* to the transition region in the MC calculations. The transition region is characterized by a locally ordered fluctuating "crystal-like" molecular cluster consisting of 210–260 water molecules embedded in the liquid phase.[106]

Similar MC calculations were used by Trout's group to study the carbon dioxide–liquid water interface at 220 K and 4 MPa near the phase boundary of a carbon dioxide hydrate (273 K and 4 MPa).[107] Nucleation was achieved by "seeding" the system with a cluster of carbon dioxide hydrate. It was found that a small cluster with diameter <9.6 Å dissolved into the solution readily. A hydrate crystal started to grow, however, when a hydrate cluster twice that size (19.3 Å) was implanted into the system. The crystal eventually spanned the whole system (Figure 22). Thus the critical nucleus size for hydrate nucleation is estimated to be about 19 Å consisting of approximately 200 water molecules. This is a considerably smaller number than that estimated from the local harmonic model of around 600 molecules. The theoretical results refuted the "labile cluster hypothesis."[111] This hypothesis speculates the agglomeration

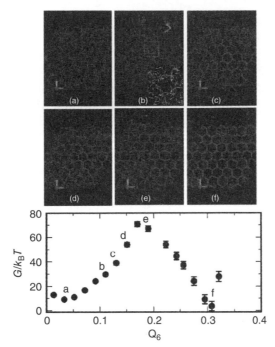

Figure 21 (Top) Snapshots generated from biased Monte Carlo simulation along the path of freezing showing the distribution of hydrogen bonds. (a)–(f) are at different sets of values of the four-order parameters (see text). The inset (dotted square) in (b) shows the formation of a hexagonal ringed structure. (Bottom) Average Gibbs free energy (*G*) along the path of nucleation corresponding to states (a)–(f) indicated above. Adapted with permission from *J. Am. Chem. Soc.*, **125**, 7743 (2003). Copyright 2003 American Chemical Society. (For a color version of this figure, please see plate 8 in color plate section.)

of a "labile" solvated guest that forms hydrate-like cavities (surrounded by 20–24 water molecules in its first coordination shell)[111,112] leading to the formation of the critical nucleus. The MC results show that a thermal fluctuation causes a group of the guest carbon dioxide molecules to arrange in a configuration similar to that in the clathrate phase. Another possibility is that the local ordering of the guest molecules induces ordering of the host molecules at the nearest- and next-to-nearest-neighbour shells, and thermodynamic fluctuations lead to the formation of the critical nucleus.[107]

It is preferable to observe nucleation under equilibrium conditions free of constraints. With the advent of computing technology in recent years, direct observation of crystallization in an equilibrium MD simulation became feasible. The nucleation and growth of an ice crystal have been reported from a canonical ensemble constant volume–constant temperature (*NVT*) MD simulation in 2002.[113] The calculations were performed on 512 water molecules using the TIP4P[110] water potential. The density of the water was

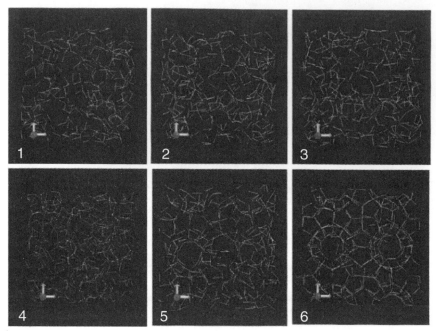

Figure 22   Six snapshots along the path of nucleation of carbon dioxide hydrate computed from Monte Carlo simulations. Reprinted with permission from *J. Chem. Phys.*, **117**, 1786 (2002). Copyright 2002 AIP Publishing LLC.

prepared at $0.96 \, \text{g/cm}^3$, close to the density of ice Ih ($0.92 \, \text{g/cm}^3$). The water was then supercooled to 230 K. The favorable simulation conditions (low density and low temperature) helped to promote ice nucleation, which was observed within 250–300 ns after thermodynamic equilibration. The MD calculations provide a molecular-level picture of the water freezing process that is not possible from phenomenological nucleation theory. The simulation revealed that ice nucleation occurs once a sufficient number of relatively long-lived hydrogen bonds develop spontaneously at the same location to form a fairly compact initial nucleus.[113] The initial nucleus then slowly changes shape and size until it reaches a stage that allows rapid expansion, resulting in crystallization of the entire system.

The mechanisms of nucleation followed by subsequent growth of the ice crystal discussed below are similar to that observed in the simulation of homogeneous nucleation of methane hydrate. More importantly, results of MD simulations show hydrate formation always initiate near the water/methane interface[114–116] where there is a large concentration gradient difference exists between the methane gas and the solution. This was observed in slab calculations that have a distinct water/methane interface and around the methane bubble in the spontaneous nucleation study. The nucleation prediction is confirmed by a recent X-ray scattering study on hydrate formation in a water/guest interface.[117] Finally, it was also shown that the time required for the initialization of hydrate nucleation depends on the water potential. Nucleation can be achieved much faster, in nanoseconds rather than in microseconds, using the six-site water model[118] instead of TIP4P.

An attempt has been made by Rodger et al.[114–116] to simulate nucleation under realistic experimental conditions at the methane (gas)/water interface. The model system consisted of two separated components – a liquid water slab undersaturated with methane and a gaseous methane region. Isobaric–isothermal (*NPT*) MD calculations using the SPC water model[57] and a one-site united atom Lennard-Jones potential for methane were performed at 250 K at 0.3 kbar. It is to be noted that the predicted melting point for ice I*h* with the SPC model is 190 K. It was found that a methane hydrate-like structure formed immediately at the water/methane interface when the MD calculation commenced,[115] a result that is possibly due to a large methane concentration difference at the interface. The temporal methane–methane radial distribution functions located within the water film and configurational snapshots of the clathrate-like clusters are plotted in Figure 23. Analysis reveals that structures of these clusters are inconsistent with the known bulk S-I or S-II structures but instead contain cages (5^{12}, $5^{12}6^2$, $5^{12}6^4$) from different clathrate structures. The formation of a clathrate-like structure arises from large-scale ordering of methane and water, which is the prelude to nucleation. The simulation, however, is too short to observe complete crystallization.

The first successful observation of spontaneous homogeneous nucleation and growth of methane hydrate by MD calculations was achieved in 2009.[119] That calculation was performed using the TIP4P/ICE water potential[120] on a model system whose initial configuration was obtained by melting 64 unit cells of S-I methane hydrate. This resulted in a liquid–vapor two-phase system with a mole fraction of methane in water of 0.0015. The initial system was then cooled to 250 K pressurized at 50 MPa. Canonical (*NVT*) ensemble MD calculations were performed up to 5 μs. The results were remarkable. The first several hundred nanoseconds of the simulation involve the dissolution of methane into water induced by high pressure and low temperature. The methane molecules aggregate forming a bubble at *ca.* 0.18 μs. At ~0.3 μs, small networks of water cages form around dissolved methane molecules, only to dissociate several nanoseconds later. At ~1.2 μs, sustainable clathrate cage clusters start to emerge and after ~1.3 μs the clathrate structure grows rapidly and spans the entire simulation cell (Figure 24). The resulting crystalline-like structure is not pure S-I but a mixture of S-I and S-II with the interface linked by novel $5^{12}6^3$ cages. In the inset of Figure 24, the temporal evolution of the potential energy and the F_4 order parameter are shown; $F_4 = \langle \cos 3\phi \rangle$ is a four-body order parameter monitoring the orientation of the nearest-neighbour water relative to a given water through the dihedral angle ϕ.[121] The formation of labile cages at 0.2 μs is accompanied by a simultaneous but small drop in the potential energy.

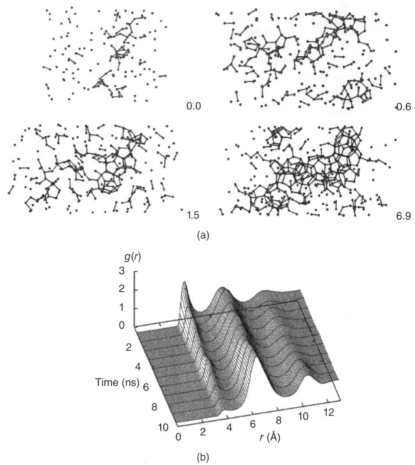

Figure 23 (a) Snapshots of clathrate clusters at four given times (ns). (b) Methane–methane radial distribution functions calculated from successive 0.9 ns time slice along the molecular dynamics trajectory. Reprinted (adapted) with permission from *J. Am. Chem. Soc.*, **125**, 4706, 2003. Copyright 2003 American Chemical Society.

The encapsulation of methane into a cavity is clearly favorable. A small drop in the potential with a concomitant increase of F_4 signifies the initialization of nucleation at 1.3 μs. The growth of the "crystalline" clathrate commences at *ca.* 23.0 μs. The cage-like structures formed around 1.3 μs are very mobile and they continuously break and form new H-bond connections (Figure 25). A well-formed and sustainable clathrate-like "crystal" was observed only at time longer than 3.0 μs.

The nucleation and growth stage of methane hydrate has been examined in detail by Vatamanu and Kusalik[122] in a recent MD study. They found that nucleation of a gas hydrate crystal involves a two-step process. It starts with the formation of disordered solid-like structures and then evolves spontaneously to

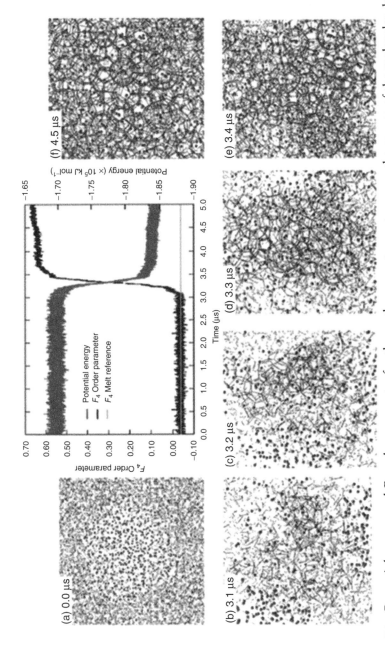

Figure 24 Potential energy and F_4 order parameter for the methane–water system over the course of the molecular dynamics simulation. The reference F_4 value (−0.04) for a fully melted hydrate (horizontal blue line) identifies the beginning of nucleation at ~1.2 μs. Snapshots (a) through (f) show the system evolution during the simulation. Adapted from Ref. 119. (For a color version of this figure, please see plate 9 in color plate section.)

361

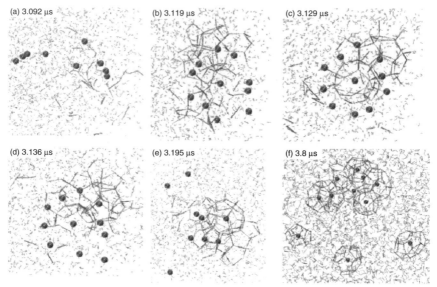

Figure 25   Figures illustrating the fluctuating nature of the nascent cages and methane molecules involved in hydrate nucleation from 5 µs simulation at 260 K and 45 MPa. Note the initial structure partially dissociates during the process of hydrate formation, as shown in (d) and (e), and the eventual fates of the methane molecules making up the adsorbed bowl are shown in (f). Adapted from Ref. 119.

more recognizable crystalline forms. At the early stage of the nucleation process, pentagonal and hexagonal water ring arrangements around methane were already evident. As time progresses, the formation of irregular-shaped cages became apparent. Cages of higher symmetry, such as those found in S-I and S-II hydrate, were then identified and, at that point in time, the nascent hydrate cluster persisted and became less susceptible to disordering fluctuations. The transition from mobile (liquid-like) to immobile (solid-like) behaviour is a gradual process lasting 10 ns with continuous structural fluctuations. In the initial solid-like region, the structure is disordered over a large length scale. A variety of cavities were observed (Figure 26). The disordered hydrate-like structure eventually evolves into more recognizable bulk-like S-I and S-II motifs.

The careful analysis by Kusalik[122] revealed similarities in the nucleation of methane hydrate among different MD studies. The first step of nucleation is the formation of labile cage-like structures around the methane molecules. Sometime later, these cage-like structures coalesce forming larger dynamic clathrate-like clusters. After a period of thermal fluctuation, a larger and less mobile clathrate precursor structure is observed. This structure consists of many closed and/or open cages that eventually develop into a more recognizable clathrate nucleus. It is likely that the earlier MD simulation of Rodger[114] provides only the incipient stage of nucleation.

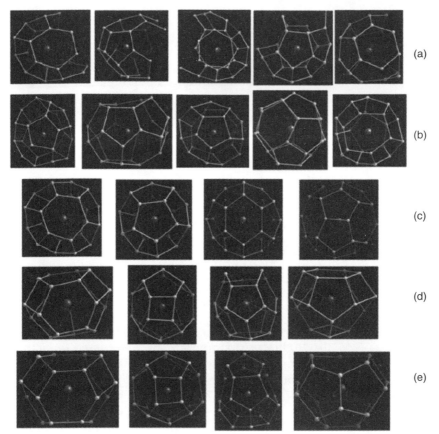

Figure 26 The diversity of cages formed from the initial stage of hydrate nucleation. (a) Partially formed regular or irregular cages, (b) irregular cages with low or no internal symmetry, (c) regular and symmetric cages that can be found in common MH structures, and (d) high symmetry cages not identified yet in the experimentally known hydrate crystals but found in the first-stage nucleated solid. (e) High symmetry cages identified in the annealed crystal. The blue and grey spheres represent the water and methane molecules identified to behave solid-like. Adapted from Ref. 122.

As just described during the final stage of forming a clathrate nucleus, the disordered cluster structure consists of many types of cages that must evolve into more recognizable bulk-like structures (Figure 26). It turns out the water and the water cages are highly fluxional: water can easily leave or add to these cages. Analysis of MD trajectories of the spontaneous nucleation reveals two basic types of cage–cage transformations.[123] (i) The first type is through water pair insertion/deletion. As shown in Figure 27a, when a water pair is inserted between the common edges shared by the n-polygon with the adjacent m and m' faces, the number of edges in the m and m' polygon must increase by one

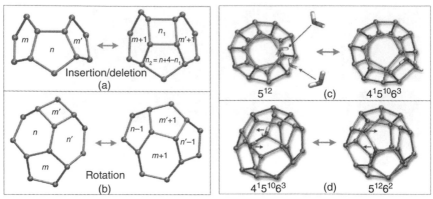

Figure 27   The two basic types of cage–cage transformations. (a) In water pair insertion/deletion moves, the n face is split, causing both m faces (m and m') to be increased by one. (b) In water pair rotation moves, the n and n' faces are reduced by one. (c) Insertion transformation: two water molecules split a pentagonal face of a 5^{12} cage to add one face and three edges to the cage, resulting in a $4^1 5^{10} 6^2$ cage. (d) Rotation transformation between a $4^1 5^{10} 6^3$ cage and a $5^{12} 6^2$ cage. Adapted from Ref. 123. (For a color version of this figure, please see plate 10 in color plate section.)

and, in addition, a square polygon is formed. An example is the transformation from a $5^{12} 6^2$ ($n = 6$, $m = m' = 5$) to $5^{12} 6^3$ by water insertion across the hexagonal face. The second type involves water pair rotation. In this case, the edges of the adjacent n and n' polygon are reduced by one while the m and m' faces are increased by one (Figure 27b). Thus, it is possible to transform a $4^1 5^{10} 6^4$ ($m = 5$, $m' = 4$, $n = n' = 6$) cage into a $5^{12} 6^3$ cage via such a rotation. Because the cages are connected in three dimension, transformations occurring in adjacent cages are coupled. Figure 27c shows two water molecules splitting a pentagonal face of a 5^{12} cage adding one face and three edges to that cage, resulting in a $4^1 5^{10} 6^2$ cage. Figure 27d shows how the rotation of a single water molecule can transform a $4^1 5^{10} 6^3$ cage to a $5^{12} 6^2$ cage. Cage interconversion via water pair insertion/removals and rotations can form solids of varying degrees of long-range order. Through a sequence of interconversion steps, the disordered cluster eventually transforms to a clathrate-like nucleus (Figure 28).

Knowledge gained on nucleation and growth processes from computer modelling has been applied to the design of inhibitors to prevent gas hydrate formation. Currently, there are two practical approaches to slow or stop the growth of gas hydrate in pipelines. The traditional approach is based on the thermodynamic consequence of the colligative property by adding methanol to reduce the chemical potential of water in the liquid phase mixture that lowers the freezing point of the solid hydrate phase. This method requires a large amount of methanol (up to 50% volume) and is estimated to cost the industry over US$500 million annually.[124] A more recent approach is to develop low-dosage hydrate inhibitors (LDHIs), which can operate in low quantity but are able to inhibit the formation and growth of gas hydrate

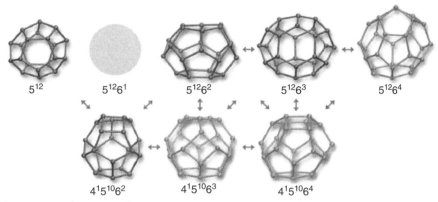

Figure 28 The seven dominant cages in incipient clathrate hydrate formation as observed from molecular dynamics simulations. The four most common cages to form in the incipient solids are shaded more prominently (5^{12}, $4^1 5^{10} 6^2$, $5^{12} 6^2$, $5^{12} 6^3$). Horizontal and diagonal arrows indicate a transformation via addition or removal of a pair of water molecules, while vertical arrows represent a transformation via rotation of a pair of water molecules. Adapted from Ref. 123.

kinetically. There are demonstrated successes using this approach. To improve the design, it is essential to understand the mechanisms of how LDHIs interact with a gas/water mixture and with a clathrate hydrate. Two mechanistic hypotheses currently exist: (i) the mere presence of a LDHI will hinder the rate of hydrate formation and (ii) physical interaction of the LDHI with incipient hydrate structure will slow down its growth into a solid clathrate hydrate.[125,126] These hypotheses have been examined by MD simulations on prototypical examples. Rodger et al.[125] starting from a methane/water (under-saturated with methane) model, prepared as described above in the hydrate formation study, inserted a test LDHI and then displaced methane molecules in the methane region at selected distances from the interface (Figure 29).[125] The initial structure was then relaxed by MD keeping the water positions fixed. The combined system was then allowed to equilibrate over a 100 ns time period. The formation/growth of hydrate-like structures in the water region was monitored using appropriate order parameters,[121] such as F_3 ($F_3 = \langle |\cos\theta| \cos\theta + \cos^2(109.47°) \rangle$; θ is the angle between a water oxygen and two of the water molecules in the first solvation shell) and F_4 (see above). Results of a well-known kinetic inhibitor and water-soluble polyvinylpyrrolidone (PVP) (Figure 29a) are depicted in Figure 29e.[125] Without PVP there is a steady growth in the number of hydrate-like waters. When PVP is added to the aqueous phase, the fraction of hydrate-like water clusters decreases substantially. It should be noted that the inhibition and even dissolution of small hydrate-like clusters in the aqueous phase was observed even without physical contact between the inserted PVP and the water interface. Obviously, the presence of PVP has suppressed the formation of hydrate-like clusters and

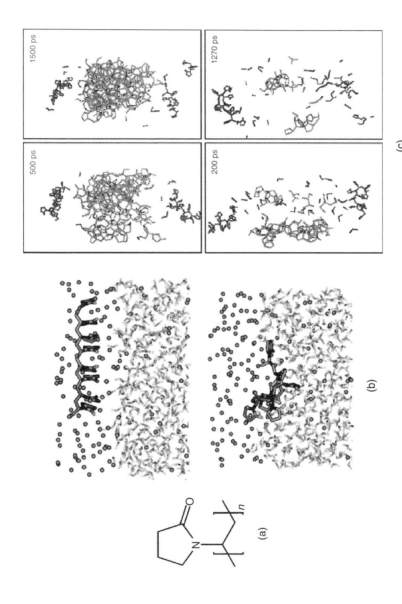

Figure 29 (a) Molecular formula for poly(vinylpyrrolidone) (PVP). (b) Location of PVP relative to the water/methane interface (top) initially and (bottom) after 1 ns. PVP is represented by thick lines, water with thin lines, and methane as spheres; N atoms are black, C dark grey, O light grey, and H white. (c) Snapshots of the hydrate network and PVP following insertion of PVP. Adapted from Refs. 121 and 127. (For a color version of this figure, please see plate 11 in color plate section.)

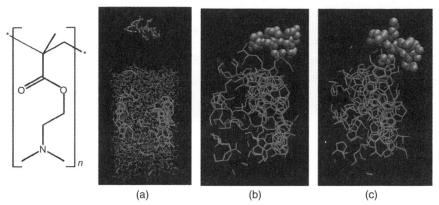

(a)　　　　　　　　(b)　　　　　　　　(c)

Figure 30 (a) Molecular formula of poly(dimethylaminoethylmethacrylate), PDMAEM. (b) Snapshot of the initial configuration of one of the atactic systems. Hydrate structure is shown in green. (c) Snapshots from helical isotactic insertion. Adapted from Ref. 128. (For a color version of this figure, please see plate 12 in color plate section.)

inhibits the formation of a critical nucleus. Similar MD calculations have been performed on the commercial LDHI poly(dimethylaminoethylmethacrylate) (PDMAEMA).[128] Two polymer conformations were considered, namely, a linear structure with all backbone dihedral angles *trans* (i.e., 180°) and a helical structure with the backbone dihedrals of 60°. The results for this polymer are very interesting. Unlike PVP, PDMAEMA adsorbs onto the outer boundary of the hydrate-like clusters in the water region (Figure 30). Depending on the conformation, the polymer is either partially engulfed in an incomplete cavity of the hydrate-like cluster or it is adsorbed onto the hydrate cluster via hydrogen bonds. More surprising, perhaps even counterintuitive, it is found that PDMAEMA actually helped to promote hydrate growth at a rate two to three times faster than observed in a simulation without the polymer![127] The polymer facilitates the adsorption of smaller hydrate-like clusters into the dominant cluster. The results are counterintuitive because one expects the kinetic inhibitor function for PDMAEMA. The apparent contradiction was rationalized by arguing that the growth sites on hydrate-like clusters are blocked by the inhibitor adsorbing onto the crystal. At present, it is not certain if this process would lead to the inhibition of hydrate crystal formation; much longer simulation times are required to address this issue.

The longtime (microseconds) and large size scale MD simulations described above have provided valuable insight to the hydrate nucleation and growth processes at the molecular level. The new finding is that the formation of a disordered ("amorphous") hydrate cluster serves as a precursor for the conversion into more recognizable crystalline hydrate structures. The nascent cavities are highly mobile and undergo continuous, dynamic interconversion through the addition and removal of water molecules. It is fair to say that a clathrate hydrate is the most extensively studied system concerning the crystallization mechanisms. Apart from helping to tackle the practical problem of blockage in natural gas pipelines, the results will benefit greatly our mechanistic understanding of homogeneous crystallization validating competing nucleation models. The results presented above are by no means conclusive in this regard, and more studies on this topic will certainly emerge in the near future.

Guest Diffusion Through Hydrate Cages

The possibility of guest migration between cages in solid hydrate has been the subject of much speculation. Diffusion has been proposed as a possible mechanism for forming gas hydrates under non-equilibrium conditions when ice is exposed to guest molecules at temperatures as low as 100–150 K, at low guest saturation pressure. It was postulated that the high-defect concentration in the host lattices, for example, Bjerrum defects (Figure 31)[130] and/or water vacancies may help guest migration from one cavity to another. More recently,

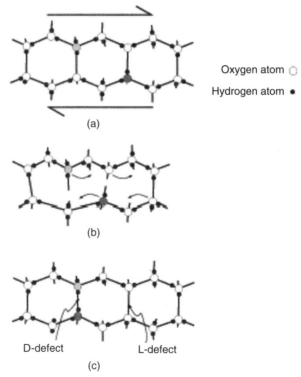

Figure 31 Calculated free energy barrier for the hopping of a carbon dioxide molecule between clathrate hydrate cages (a) containing no water defects; (b) water vacancy present between the cages; (c) through the shared hexagonal water face; and (d) through the shared pentagonal water face. Adapted from Ref. 129.

since the report on the synthesis of the hydrogen hydrate at high pressure, considerable research has appeared on the investigation of alternative ways to stabilize the hydrate at lower pressure for the purpose of hydrogen storage. Because the hydrogen molecule is smaller than the pentagonal and hexagonal faces linking the cavities, the possibility exists that it may diffuse out of the hydrate structure after it is formed and stored under low pressure. The current interest in the exchange between carbon dioxide and methane in solid hydrates and the injection of carbon dioxide into the deep ocean floor has added to the importance of this research topic.

No serious attempt to understand the diffusion process in the hydrate structure was made until the investigation of the diffusivity of carbon dioxide in the hydrate.[131,132] In this pioneering study,[132] it was realized that the activation energy for direct migration through the faces shared by two cavities is too high to be feasible. In fact, MD calculations of an ideal defect-free clathrate hydrate showed no evidence of carbon dioxide diffusion.[131] It was then assumed that diffusion occurs through neighbouring hydrate cages when a water vacancy site exists (i.e., one water molecule is removed from the polygon face). The free energy for CO_2 hopping from one cage to the other is reduced significantly from $20 kT$ to about $6-8 kT$ (Figure 32) depending on the face trespassed. When a water vacancy is introduced, hopping of both water and carbon dioxide were observed within 0.5 ns. The concentration of the water defect c was computed from thermodynamic integration on the free energy (ΔF) of an empty lattice (defect-free reference) to a structure with a single defect according to the Arrhenius equation, $c = \exp(-\Delta F/kT)$. From the number of hopping events observed in a given time window, the hopping rate can be determined (Table 6). It is informative to note that the hopping rates are similar for the four possible pathways, that is, large to large through pentagonal face, large to large through hexagonal face, small to large through pentagonal face and large to small through pentagonal face and all *ca.* 10^{10} hops/s. The diffusion coefficient (D) of carbon dioxide in the hydrate can be estimated by combining the following information,

$$D = \frac{1}{6}(d_{\text{cage}-\text{cage}}^2)\begin{pmatrix}\text{hopping}\\\text{rate}\end{pmatrix}\begin{pmatrix}\text{connectivity}\\\text{of cages}\end{pmatrix}\begin{pmatrix}\text{occupancy}\\\text{of cages}\end{pmatrix}$$
$$\times\begin{pmatrix}\text{concentration of}\\\text{water molecules}\end{pmatrix}\begin{pmatrix}\text{number of water}\\\text{in the water face}\end{pmatrix} \quad [58]$$

The average diffusion rate for carbon dioxide in the clathrate hydrate at 273 K is predicted to be *ca.* 1×10^{-12} m²/s (see Table 6).

It is reasonable to assume that guest diffusion in a hydrate will not be too facile without the presence of defects.[130] The Bjerrum D- and L-defects (Figure 31) are common in ice and hydrates. To test the effect of defects on the migration of guest molecules in S-I hydrate, nanosecond MD calculations employing empirical potentials were performed on model hydrogen sulfide and

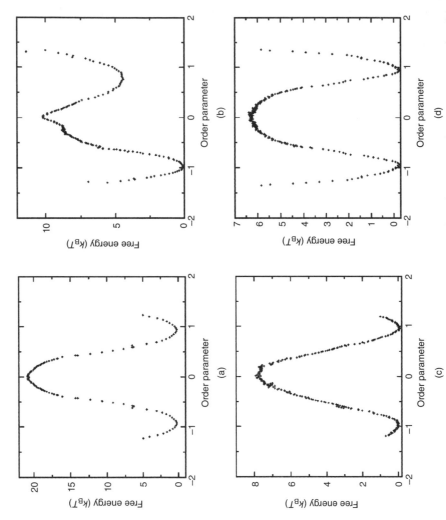

Figure 32 The formation of Bjerrum D- and L-defects via the transfer of a proton from a water molecule to adjacent water (a)–(d).

Table 6 CO_2 Hopping and Diffusion Rates and Between Clathrate Hydrate Cages with a Water Vacancy Between them at 273 K Determined from Monte Carlo Calculations

System	Rate (hops/s)	Diffusion (m^2/s)
Large-to-large through pentagonal face	6.2×10^{10}	1×10^{-12}
Large-to-large through hexagonal face	2.8×10^{10}	2×10^{-13}
Small-to-large through pentagonal face	9.1×10^{10}	
Large-to-small through hexagonal face	8.9×10^{9}	

ethylene oxide hydrates with L-defects.[129] A few guest-hopping events were observed, but they were not through these defects. Instead, a water molecule was found to jump from one of the cage walls into a neighbouring empty cage, leaving behind a vacancy in the host lattice. This defect pair often recombined within a few tens of picoseconds. However, the vacancy that was shared by several adjacent cages, occasionally, acted as a conduit for guest hopping from nearby filled to empty cages. The results are consistent with the assumption that water vacancies promote diffusional guest migration as already described above.[131] The most comprehensive investigation of guest diffusion in a gas hydrate involved a study using equilibrium path sampling combined with reactive flux and kinetic Monte Carlo simulations to estimate the self-diffusion of methane in the S-I hydrate.[133] In the path sampling methods, an ensemble of paths are generated by repeatedly generating a path from the previous path and then accepting the path according to a Metropolis rule for the path weight in the desired ensemble. The results affirm the vacancy hypothesis in which the barrier to migration is reduced if a defect is present. The collaborative and concerted motions of the water molecules forming the cages are important in the diffusion mechanism. The assumption of maintaining an undisrupted hydrate cage is invalid; if this scenario is assumed, the calculated free energy barrier for guest migration will always be much higher (see Figure 32).

Ab Initio Methods

In all of the computer modelling calculations described earlier, intermolecular potentials either derived from experimental data, constructed empirically, or fit to an *ab initio* quantum mechanical PES were used. In most cases, the results are satisfactory, but, for some properties of gas hydrates, subtleties in the electronic structure cannot be ignored. Examples where this is relevant are the breaking and formation of H-bonds, weak dispersive interactions and effects from quantum tunneling. Under these circumstances, appropriate electronic structure methods must be used.

NMR spectroscopy is a very sensitive technique that can distinguish subtle differences in the environment and determine the relative occupancy of the material by probing the nucleus of the guest atom (or molecule) in the hydrate structure.[134] The chemical shielding is affected directly by the electronic environment of the probing nucleus. So it is possible to identify the type of hydrate cage from the NMR chemical shifts of the encaged atom or molecule. Lineshape of the shielding tensor also provides information on the local symmetry. ^{1}H, ^{129}Xe and ^{13}C NMR have been employed routinely for the characterization of gas hydrates. An early example on the application of electronic structure calculations to a gas hydrate is the prediction of the isotropic NMR chemical shift of ^{129}Xe in the cavities of S-I and S-II hydrates.[135] ^{129}Xe chemical shielding surfaces in the 5^{12}, $5^{12}6^2$ and $5^{12}6^4$ cages of S-I and S-II hydrates have been generated from quantum mechanical calculations. The shielding surface of each cage type was constructed by computing the ^{129}Xe chemical shift at various nuclear configurations by the gauge invariant atomic orbital (GIAO) using both Hartree-Fock and density functional theory.[135,136] An almost completed atomic basis set consisting of 240 primitive Gaussian functions was used for the Xe. The effect of proton disorder was examined and found to be small. The Dunning split-valence basis sets augmented by a set of p and d polarization functions were used for the water molecule. The calculated shielding surface was fitted to a universal shielding function expanded in a power series of inverse distances (Xe—O and Xe—H) with even exponents from r^{-6} to r^{-12}. The universal Xe—O and Xe—H shielding functions were then used in subsequent canonical Monte Carlo simulations to compute the average isotropic Xe shielding values in the various cages. The theoretical and experimental results are compared in Table 7. The absolute agreement is satisfactory. The trend in the chemical shifts, in particular the ^{129}Xe chemical shielding in the small cage of S-II, is downfield from S-I by 7 ppm and is predicted correctly and comparable to the experimental observed shift of 17 ppm. Quantum mechanical calculations of chemical shift have since been applied to the study of proton shielding in S-II hydrate and ^{13}C NMR powder lineshape of linear guests in S-I clathrate hydrates.[137,138]

It was shown by Car and Parrinello that classical MD (CPMD, Car–Parrinello molecular dynamics) can be performed in the framework of *ab initio* electronic theory.[139] In CPMD, coefficients of the electronic basis functions (φ) are treated as dynamical variables with associate mass (μ). The trajectories of these fictitious particles representing the electronic wave function (Ψ_0) of the system can propagate simultaneously with the dynamics

Table 7 Average Isotropic Shielding of $\langle \sigma_{iso} \rangle$ and σ (Xe atom), ppm, in the Small and Large Cages of Structure I and Structure II Clathrate Hydrate Calculated at 275 K

System	Calculated $\langle \sigma_{iso} \rangle - \sigma$(atom)	Experiment $\langle \sigma_{iso} \rangle - \sigma$ (atom)
Small cage, structure-I	−213.98	−242
Large cage, structure-I	−146.93	−152
Small cage, structure-II	−206.7	−255
Large cage, structure-II	−104.67	−80

of the ions (M). The equation of motion for the coupled electron-ion MD is,[140,141]

$$L_{CP} = \sum_i \frac{1}{2} \mu_i \varphi_i | \varphi_i + \frac{1}{2} \sum_I M_I \dot{R}_I^2 - E(\Psi_0, \vec{R}) + \text{constraints} \qquad [59]$$

The second term given in Eq. 59 is the classical Newtonian equation of motion of the ions and can be written as

$$M_I \ddot{R}_I(t) = -\frac{\partial}{\partial R_{I0}} |H|_0 + \frac{\partial}{\partial R_I} (\text{constraints}) \qquad [60]$$

An alternate approach is to evaluate the ground state of the system at each time step and compute the forces on the atoms from the Hellman–Feynman theorem.

$$M_I \ddot{R}_I(t) = \left| \frac{\partial H}{\partial R_I} \right| \qquad [61]$$

In this way, the electronic ground state will always lie on the Born–Oppenheimer (BO) surface, and this method is known as Born–Oppenheimer molecular dynamics (BOMD).[142]

The first application of *ab initio* BOMD (AIMD) to study a gas hydrate involved the vibrational dynamics of methane in the cages of S-I hydrate.[143] That study was motivated by the anomalous experimental shift in the stretching vibrational frequencies of a guest in the hydrate when compared to the corresponding free molecule. For example, it was reported that in methane hydrate the C—H stretching vibrations in the large cage shifted to a lower frequency than those in the small cage.[144] This observation is inconsistent with the expectation that a guest–host interaction in the large cage should be weaker than in the small cage due to a larger free volume. The Charles–Pimentel "loose cage–tight cage" model[145] had been invoked to account for this novel phenomenon. This model suggests the existence of local minima in the cavity that enhance the guest–host interactions and lead to the weakening of the C—H vibrations. Canonical *NVT* AIMD calculations were performed on S-I methane hydrate with one unit cell using BOMD. For the carbon and oxygen atom, the core electrons were replaced by a pseudopotential. The atomic valence orbitals were expanded by localized split-valence plus p and d polarization function basis sets. Because the motions of the carbon and hydrogen atoms of methane are coupled, the methane vibrations can be examined from the hydrogen atom velocity autocorrelation function (VCF). The vibrations are clearly divided into three groups: translational $(0-300\,\text{cm}^{-1})$, H—C—H bend $(1000-1700\,\text{cm}^{-1})$ and C—H stretch $(2800-3200\,\text{cm}^{-1})$ (Figure 33). The VDOS of methane in small and large cages of the hydrate are shown in Figure 34. The calculations reproduced correctly the experimental observation that the frequencies of C—H stretch vibrations in the large cage (symmetric: $2967\,\text{cm}^{-1}$, asymmetric: $3152\,\text{cm}^{-1}$) are slightly lower than the corresponding

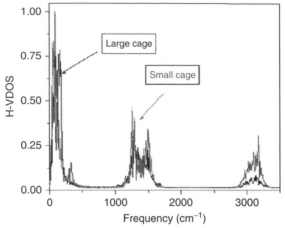

Figure 33 Total hydrogen atom vibrational density of states in the small and large cages of S-I methane hydrate obtained from *ab initio* molecular dynamics calculations. Reprinted (adapted) with permission from *J. Phys. Chem.*, **101**, 7371, 1997. Copyright 1997 American Chemical Society. (For a color version of this figure, please see plate 13 in color plate section.)

values ($3014 \, cm^{-1}$ and $3167 \, cm^{-1}$) in the small cage (Figure 34).[144] The reason for the shift in C—H stretching frequencies relative to that of the free molecule is due to a small lengthening of the C—H bond of the encapsulated methane. The calculated average C—H bond length of $1.1122 \, \text{Å}$ in the large cage is slightly longer than that in the small cage of $1.1107 \, \text{Å}$, and both are longer than the free molecule value of $1.1089 \, \text{Å}$. Elongation of the C—H bond is consistent with the observed lowering of the frequency. Incidentally, the predicted lengthening of C—H bonds has been confirmed experimentally.[146] The vibrational dynamics of methane in the hydrae has been repeated using CPMD.[147] In this calculation the C—H stretches were computed from both the H atom VCF and from the time evolution of the C—H bond vector VCF. The VDOS computed from the H atom VCF is similar to the earlier results.[143] The broad spectral feature in Figure 34 is due to coupling to the rotational modes. By specifically correlating the temporal difference in the C and H positions, the vibrational spectrum evaluated from the C—H bond correlation function are much sharper and clearly show the red shift of the vibrations in large cages relative to the small cages (Figure 34). The vibrations in S-I hydrogen sulfide hydrate have been studied by isobaric–isothermal (*NPT*) AIMD at several temperatures.[148] Once again the red shift of S-H stretching vibrations in the large cage is reproduced (see Table 8). In this case, instantaneous weak HSH…H_2O interactions were attributed to the softening of the S-H vibrations. The calculated coefficient of linear thermal expansion of 4×10^{-5} is in good agreement with the measured value of *ca.* 4–7×10^{-5} for S-I methane and ethylene oxide hydrates.[149,150] The AIMD calculations on methane and hydrogen sulfide hydrates refuted the

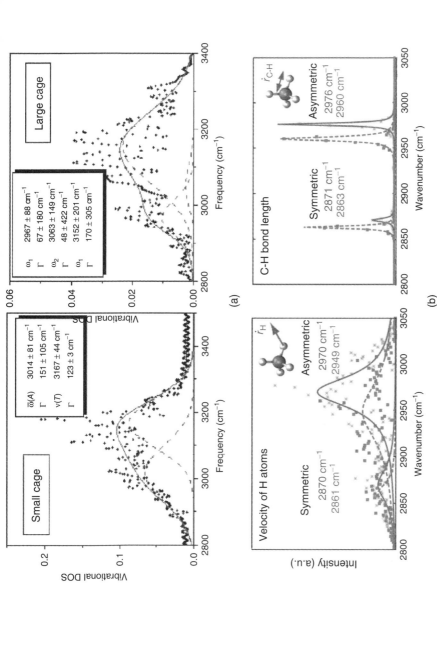

Figure 34 Hydrogen atom vibrational density of states of CH₄ in the small and large cages of S-I methane hydrate obtained from *ab initio* molecular dynamics calculations ((a), reprinted (adapted) with permission from *J. Phys. Chem.*, **101**, 7371 (1997). Copyright 1997 American Chemical Society) and CPMD ((b), reprinted (adapted) with permission from *J. Chem. Phys.*, **136**, 044508 (2011), American Institute of Physics). (For a color version of this figure, please see plate 14 in color plate section.)

Table 8 Comparison of Theoretical and Experimental Stretching and Bending Frequencies (cm^{-1}) for Hydrogen Sulfide Hydrate

Vibration	Cage	150 K	300 K	Expt
S—H	s	2604	2582, 2618	2591, 2604, 2616
	l	2591	2574	2532, 2548, 2556, 2570
H—S—H	s	1169	1136, 1176	1176
	l	1159	1156	1176

"loose cage–tight cage" Pimentel–Charles model that was often invoked to explain a lower frequency of the guest molecule stretching vibrations in the large cage compared to the small cage.[145] The mechanical, acoustic and thermal properties of methane hydrate in the S-I, S-II and S-H structure have also been calculated with *ab initio* electronic structure theory.

An interesting application of AIMD was used to explore the effect of pressure on the hydrate structure.[151,152] At room temperature, it was demonstrated that S-I methane hydrate undergoes a sequence of structural transformation: to S-II at 0.8 GPa and an ice-like structure (MH-III, Figure 35) at 1.9 GPa.[56,57] Variable cell *NPT* CPMD calculations have been performed to study the structural transitions in deuterated methane hydrate at room temperature.[153] No noticeable sign of phase transition was observed up to 4.5 GPa. However, the cage structure was found to break down with the small and large cages distorting gradually with increasing pressure and eventually vanishing at 4.5 GPa. To speed the recovery (crystallization) of the distorted structure, the fictitious mass of the cell was abruptly increased by a factor of 5 during the simulation, while the temperature was kept constant (by rescaling ionic velocities). The effect of doing this is twofold. First, it ensured an adiabatic separation of the cell motion and that of the remaining degrees of freedom. Second, because the fictitious kinetic energy of the cell were also increased by the same factor as the mass of the cell, this enabled a more efficient sampling of the configuration space. After 10 ps the fictitious mass was reset back to the original value. The reconstructed structure exhibited a water channel occupied by a chain of methane molecules. The channel structure resembles that observed in filling ice (Figure 35) where methane occupies the open hexagonal channels formed by the ice host structure. More interestingly, the calculated C-D stretch frequency shows a blue-shift relative to both the small and large cages of S-I methane hydrate. The theoretical prediction is consistent with Raman spectra of the transformation from S-I to MH-III.[150,151] Although the CPMD calculation did not reproduce the sequence of structural transformations or the structure of MH-III, it demonstrates the potential of *ab initio* electronic structure methods to survey energy landscapes and to search for candidate structures at high pressure.

It is well known that DFT approximations employing nonlocal functionals fail to reproduce the high-pressure phase diagram of ice.[154] In particular, the transition pressure between different forms of high-pressure ices are seriously

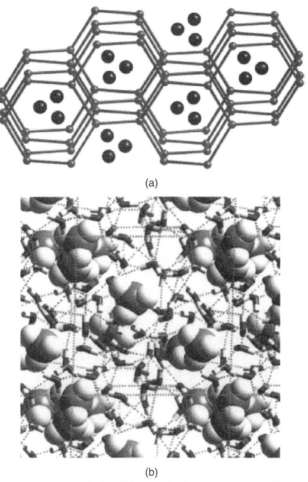

(a)

(b)

Figure 35 (a) Comparison of the filled-ice high-pressure crystalline polymorph of methane hydrate with (b) the structure obtained from the pressurization of the S-I hydrate using constant pressure *ab initio* molecular dynamics method showing methane molecules in the channels. Reprinted with permission from J. S. Loveday, R. J. Nelmes, M. Guthrie, D. D. Klug and J. S. Tse, *Phys. Rev. Lett.*, 87, 215501 (2001). Copyright 2001 American Physical Society.

overestimated. Recently, it was shown that at high pressure, the contribution to the lattice energy from the dispersive vdW[155] interaction increases and that from hydrogen bonding decreases, suggesting that dispersion has a substantial effect on the transition between the high-pressure ice phases.[156] This suggestion is consistent with the constant pressure CPMD calculations presented above where a structural transition to an MH-III-like phase was overestimated at 4.5 GPa while the observed pressure is 1.9 GPa. Therefore, the vdW interaction should also play a significant role in the structure and stability of gas

Table 9 Adsorption Energies (in eV per molecule) for Single CH_4, CO_2, and H_2 Molecules in One of the Different Cavities of Structures I and H Clathrates

Guest	Functional	Structure-I			Structure-H			
		5^{12}	$5^{12}6^2$	All	5^{12}	$4^3 5^6 6^3$	$5^{12}6^8$	All
CH_4	GGA	0.09	−0.07	−0.03	0.07	0.05	−0.12	0.04
	vdW	−0.52	−0.59	−0.51	−0.53	−0.54	−0.48	−0.55
CO_2	GGA	0.35	0.10	0.15	0.29	0.22	0.01	0.27
	vdW	−0.41	−0.56	−0.51	−0.41	−0.43	−0.38	−0.44
H_2	GGA	0.01	−0.01	−0.02	0.03	−0.01	0.00	0.03
	vdW	−0.20	−0.19	−0.21	−0.18	−0.22	−0.15	−0.18

"All" stands for one molecule inside each cavity. The adsorption energy is defined as the difference between the total energy of the clathrate, with the molecules inside, and that of the sum of empty clathrate and isolated molecules.

hydrates, particularly at high pressure. Correction to dispersive guest–host interaction has been made in *ab initio* calculations of gas hydrates in isolated cages. For example, Tse and Patchkovskii[42] noted the importance of dispersive energy on the calculated occupancy and Langmuir constant in multiple occupied hydrate cages. *Ab initio* vdW density functional (vdW-DF) theory has been used to study the adsorption energy of hydrogen, carbon dioxide and methane in the different cages of S-I and S-II hydrates.[157,158] Nonlocal gradient functionals that exclude vdW interactions predicted no stability when the gas molecules are enclathrated. In contrast, strong stabilization was found when the vdW-DF functional[156] is used. A summary of the calculated adsorption energies is tabulated in Table 9. It was further found that the energy barriers for the migration of gas molecules through the opening in the cage walls for a defect-free hydrate are very high, a not-so-surprising result in view of the discussion already presented.

Because the guest–host interactions for light guest molecules like hydrogen and methane are within the regime of vdW forces, the quantum nature of the dynamics, in particular at low temperature, cannot be neglected. These molecules behave almost like free rotors and because they are confined in the hydrate the translational and rotation (TR) motions are coupled strongly and the quantized energy levels are comparable to their zero-point energies. Quantum effects have been observed in the vibrational spectra of methane and hydrogen hydrate but there have been few research to incorporate quantum effects into dynamical calculations of clathrate hydrates. Even in these studies the quantum effect was mostly treated in a very approximate and simplified manner.[159] As described earlier, the vibrational spectrum obtained from IINS of methane hydrate below 30 K exhibits a distinct feature due to quantum rotation of the methane molecule.[58] It was found that the intrinsic linewidth is very broad. Analysis of the high-resolution spectrum obtained at 2 K using empirical atom–atom interactions shows the contribution of the vdW interaction to be one order of magnitude smaller than the electrostatic interaction for the frozen-in disorder of the water dipole moments in the cages.[160] The quantum effect cannot be ignored in hydrogen hydrate; including quantum effects in the calculation of the adsorption[42,159] energy of hydrogen in the gas hydrate lowers the value of the Langmuir constant by an order of magnitude.[159] Quantum tunneling has been considered using the Eckart model for determining the diffusion rates through walls of small and large cages of defect-free hydrogen gas hydrate.[161] Even with this correction, the diffusion barrier is still one order of magnitude higher than that observed in NMR experiments.[86,159]

Molecular hydrogen occurs in two nuclear spin states: with its two proton spins aligned either parallel (triplet *ortho*-H_2) or antiparallel (singlet *para*-H_2). The neutron–proton scattering cross section for *ortho*- and *para*-H_2 depends strongly on the neutron energy.[162] IINS has the ability to resolve quantum rotation excitations $\Delta J = 1$ for $J = 0 \to 1$ in *para*-H_2 and $J = 1 \to 2$ in *ortho*-H_2, which are forbidden in optical, infrared and Raman spectroscopy of H_2. IINS spectra for *ortho*- and *para*-hydrogen in the small cages of S-II tetrahydrofuran double hydrate has been reported.[161] The spectra are very complex and exhibit many distinctive features (Figure 36). To interpret the experimental results, a fully quantum methodology for calculating IINS spectra confined in a cavity has been developed.[163] The desirable quantity is the neutron double differential cross section i,

$$\frac{d^2\sigma}{dd\omega} = \frac{k'}{k} S(\vec{\kappa}, \omega)$$ [62]

where $S(\vec{\kappa}, \omega)$ is the energy-dependent dynamical structure factor, $\vec{\kappa} = \vec{k} - \vec{k'}$, the scattering vector.

$$S(\vec{\kappa}, \omega) = \sum_i p_i \sum_f |M_f^i|^2 \delta\left(\omega - \frac{(\varepsilon_f - \varepsilon_i)}{\hbar}\right)$$ [63]

p_i is the statistical weight of the initial state, and the matrix elements M_f^i are determined by the scattering power,

$$M_f^i = \sum_n \langle f | b_n \exp(i\vec{\kappa} \cdot r_n) | i \rangle$$ [64]

where b_n is the scattering cross section for nucleus n at r_n. The initial and final state wave functions $|i\rangle$ and $|f\rangle$ are products of the spin and spatial wave functions and are solutions of the Schrodinger equation for the TR Hamiltonian[164]

$$H_{TR} = -\frac{\hbar^2}{8\pi^2 m}\left(\frac{\partial^2}{\partial x^2} + \frac{\partial^2}{\partial y^2} + \frac{\partial^2}{\partial z^2}\right) + Bj^2 + V(x, y, z, \theta,)$$ [65]

In the case of hydrogen the potential V is the five-dimensional PES of the molecule inside the clathrate cage. Once the intermolecular potential is chosen, V can be calculated and Eq. [65] can be solved numerically. Then the matrix

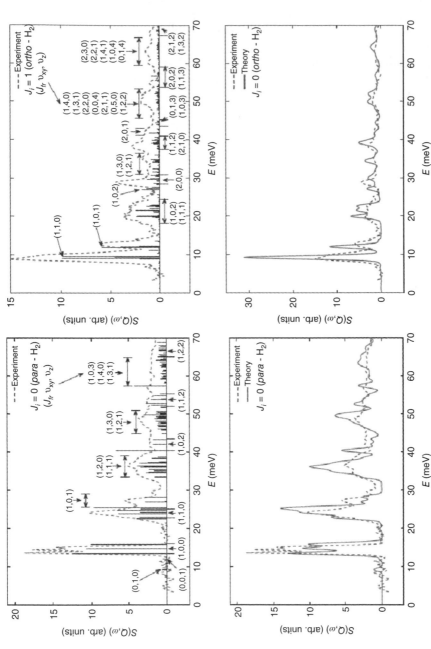

Figure 36 Calculated and measured IINS spectra of p-H_2 (left) and o-H_2 (right) in the small cage of S-II tetrahydrofuran clathrate hydrate. The top figures show the stick spectrum, while the one convolved with the instrumental resolution function is in the bottom figures. Reprinted figure with permission from M. Xu, L. Ulivi, M. Celli, D. Colognesi and Z. Bacci, *Phys. Rev. B*, **83**, 241403 (2011). Copyright 2011 American Physical Society. (For a color version of this figure, please see plate 15 in color plate section.)

elements in Eq. [64] are evaluated and the IINS cross section computed. The simulated spectra for *ortho* and *para*-H_2 are compared with the corresponding experimental spectra in Figure 36.[161,164] All spectral features are reproduced. Each rotational band is found to consist of a large number of unresolved individual TR transitions. Agreement between calculated and experimental IINS spectra are exceedingly good. The small differences can be attributed to deficiencies in the calculation of the potential surface V. Therefore, combining quantum dynamic calculations and high-resolution IINS spectra of hydrogen hydrate offers a new opportunity to determine an accurate hydrogen–water interaction potential. The multidimensional quantum dynamic methods can be extended to polyatomic molecules such as methane in the gas hydrate in which the 6D TR eigenstates have already been calculated.[165] It was found that the $(2j + 1)^2$ degeneracy of the rotational levels of methane in the gas phase is partially lifted by the anisotropy of the PES in the cages. This results in rather intricate patterns of rotational multiplet splitting. The splitting may also be a contributing factor to the large intrinsic linewidth observed in a high-resolution IINS experiment.[62]

OUTLOOK

Several computational methods as applied to the study of gas hydrates were discussed in this chapter. The topics were chosen with the intention to illustrate that the properties of gas hydrates can be reproduced, but more importantly, to demonstrate clathrate hydrates can be exploited as convenient models to advance our knowledge of related challenging research subjects such as thermal conductivity of disordered solids, nucleation processes in molecular crystals and quantum dynamics of isolated molecules. MD and Monte Carlo methods are the main modelling techniques for computational gas hydrates. Increasingly, empirical potentials have been replaced by more accurate *ab initio* electronic structure methods. Besides improving the accuracy and reliability, the application of advanced quantum mechanical methods is expected to provide new insights in the near future. This is particularly important, for example, in the study of the role of the Bjerrum defects on guest diffusion in the hydrate structure. The formation of the D- and L-defects can be described correctly only by quantum mechanics. Moreover, the migration of defects might have a significant effect on the rates of guest diffusion. In combination with other techniques for treating rare events, such as transition path sampling and defect formation is best studied by path integral quantum MD method. As mentioned, the elementary processes leading to the formation of a gas hydrate is probably the most extensive studied topic. Results of these studies are relevant to the foundation of current nucleation theory. Careful analysis of simulation results should shed light on the origin and thermodynamic characteristics of the hypothesized "critical nucleus." Finally, combined with advanced computational theory, the quantum dynamics of isolated, complex polyatomic molecules can now be studied in great detail by employing clathrate hydrates as matrices.

REFERENCES

1. J. Priestley, *Experiments and Observations on Different Kinds of Aid and Other Natural Branches of Natural Philosophy Connected with the Subject.* Vol. 3, Printed by T. Pearson, 1790.

2. H. Davy, *Phil. Trans. Roy. Soc. Lond.*, 100(231) (1810). Researches on the Oxymuriatic Acid, Its Nature and Combinations; And on the Elements of the Muriatic Acid. With Some Experiments on Sulphur and Phosphorus, Made in the Laboratory of the Royal Institution.

3. M. Faraday, *Quart. J. Sci.*, **15**, 71 (1823). On Hydrate of Chlorine.

4. M. v. Stackelberg and H. R. Muller, *Naturwiss*, **38**, 456 (1951). Feste Gashydrate.

5. M. v. Stackelberg and H. R. Milller, *J. Chem. Phys.*, **19**, 1319 (1951). On the Structure of Gas Hydrates.

6. L. Pauling, *Science*, **134**, 15 (1961). A Molecular Theory of General Anesthesia.

7. E. G. Hammerschmidt, *Am. Gas. Assoc. Monthly*, **18**, 273 (1936). Gas Hydrates.

8. K. C. Hester and P. G. Brewer, *Ann. Rev. Mar. Sci.*, **1**, 303 (2009). Clathrate Hydrates in Nature.

9. Y. Makogan, Deepwater Horizon Oil Spill, *Encyclopedia of Life Support System*. http://www.eoearth.org/article/Deepwater_Horizon_oil_spill?topic=50364, xxxx.

10. E. D. Sloan, *Nature*, **426**, 353 (2005). Fundamental Principles and Applications of Natural Gas Hydrates.

11. B. P. McGrail, H. T. Schaef, M. D. White, T. Zhu, A. S. Kulkami, R. B. Hunter, S. L. Patil, A. T. Owen and P. F. Martin, *Using Carbon Dioxide to Enhance Recovery of Methane from Gas Hydrate*, Report prepared for the U.S. Department of Energy under Contract DE-AC06-76RLO 1830 (2007).

12. R. W. Bradshaw, J. A. Greathouse, R. T. Cygan, B. A. Simmons, D. E. Dedrick, and E. H. Majzoub, *Desalination Utilizing Clathrate Hydrates*, Sandia Report, SAND2007-6565 (2008).

13. http://www.princeton.edu/grandchallenges/energy/research-highlights/desalination/, xxxx.

14. S. C. Hunt. *Gas Hydrate Thermal Energy Storage System*, U.S. patent 5140824 (1992).

15. C. Giavarini and K. Hester, *Gas Hydrates: Immense Energy Potential and Environmental Challenges*, Springer-Verlag, 2011.

16. J. P. Kennett, G. Cannariato, I. L. Hendy, and R. J. Behl, *Methane Hydrates in Quaternary Climate Change: The Clathrate Gun Hypothesis*, American Geophysica Union, Washington, DC, 2003.

17. D. W. Davidson, in *Water: A Comprehensive Treatise*, F. Franks (Ed.), Vol. 2, Plenum, 1979, p. 115. Clathrate Hydrates.

18. J. A. Ripmeester, C. I. Ratcliffe, D. D. Klug, and J. S. Tse, *Ann. New York Acad. Sci*, **715**, 161 (1994). Molecular Perspectives on Structure and Dynamics in Clathrate Hydrates.

19. E. D. Sloan and C. A. Koh, *Clathrate Hydrates of Natural Gases*, 3rd edition, CRC Press, Boca Raton, FL, 2007.

20. W. F. Claussen, *J. Chem. Phys.*, **19**, 259 (1951). Suggested Structures of Water in Inert Gas Hydrates; A Second Water Structure for Inert Gas Hydrate, *ibid*, **19**, 1425 (1951).

21. J. A. Ripmeester, J. S. Tse, C. I. Ratcliffe, and B. M. Powell, *Nature (London)*, **325**, 135 (1987). A New Clathrate Hydrate Structure.

22. K. A. Udachin, G. D. Enright, C. I. Ratcliffe, and J. A. Ripmeester, *J. Am. Chem. Soc.*, **119**, 11481 (1997). Structure, Stoichiometry, and Morphology of Bromine Hydrate.

23. H. Lu, Y. Seo, J. Lee, I. Moudrakovski, J. A. Ripmeester, N. R. Chapman, R. B. Coffin, G. Gardner, and J. Pohlman, *Nature*, 445, 303 (2007). Complex Gas Hydrate from the Cascadia Margin.

24. D. W. Davidson, Y. P. Handa, C. I. Ratcliffe, J. A. Ripmeester, J. S. Tse, J. R. Dahn, F. L. Lee, and L. D. Calvert, *Mol. Cryst. Liq. Cryst.*, 141, 141 (1986). Crystallographic Studies of Clathrate Hydrates. Part I.

25. D. W. Davidson, Y. P. Handa, C. I. Ratcliffe, J. S. Tse, and B. M. Powell, *Nature*, 31, 142 (1984). The Ability of Small Molecules to Form Clathrate Hydrates of Structure II.

26. J. H. van der Waals, *Trans. Faraday Soc.*, 52, 184 (1956). The Statistical Mechanics of Clathrate Compounds.

27. J. H. van der Waals and J. C. Platteeuw, *Rec. Trav. Chim.*, 75, 912 (1956). Thermodynamic Properties of Quinol Clathrates III.

28. J. H. van der Waals and J. C. Platteeuw, *Adv. Chem. Phys.*, 2, 1 (1959). Clathrate Solutions.

29. H. T. Lotz and J. A. Schouten, *J. Chem. Phys.*, 111, 10242 (1999). Clathrate Hydrates in the System H_2O–Ar at Pressures and Temperatures up to 30 kbar and 140 °C.

30. J. E. Lennard-Jones and A. F. Devonshire, *Proc. Roy. Soc., (London)*, A163, 53 (1937). Critical Phenomena in Gases I; Critical and Co-Operative Phenomena IV. A Theory of Disorder in Solids and Liquids and the Process of Melting, A165, 1 (1938).

31. V. McKoy and O. Sinanoğlu, *J. Chem. Phys.*, 38, 2946 (1963). Theory of Dissociation Pressures of Some Gas Hydrates.

32. G. D. Holder and D. J. Manganiello, *Chem. Eng. Sci.*, 37, 9 (1982). Hydrate Dissociation Pressure Minima in Multicomponent Systems.

33. J. W. Tester, R. L. Bivins, and C. C. Herrick, *AIChE J.*, 18, 1220 (1972). Use of Monte Carlo in Calculating the Thermodynamic Properties of Water Clathrates.

34. J. S. Tse and D. W. Davidson, Intermolecular Potential for Gas Hydrates, *Proc. 4th Can. Permafrost Conf.*, p.329, pubs.a.na.ucalgary.ca/ca (1982).

35. J. A. Ripmeester and D. W. Davidson, *J. Mol. Struct.*, 75, 67 (1981). [129]Xe Nuclear Magnetic Resonance in the Clathrate Hydrate of Xenon.

36. Z. Cao, J. W. Tester, and B. L. Trout, *J. Chem. Phys.*, 115, 2550 (2001). Computation of the Methane–Water Potential Energy Hypersurface via Ab Initio Methods.

37. Z. Cao, J. W. Tester, and B. L. Trout, *J. Phys. Chem. B*, 106, 7681 (2002). Sensitivity Analysis of Hydrate Thermodynamic Reference Properties Using Experimental Data and Ab Initio Methods.

38. Y. P. Handa and J. S. Tse, *J. Phys. Chem.*, 90, 5917 (1986). Thermodynamic Properties of Empty Lattices of Structure I and Structure II Clathrate Hydrates.

39. S. Ravipati and S. N. Punnathanam, *Ind. Eng. Chem. Res.*, 51, 9419 (2012). Analysis of Parameter Values in the van der Waals and Platteeuw Theory for Methane Hydrates Using Monte Carlo Molecular Simulations.

40. M. Z. Bazant and B. L. Trout, *Physica*, A300, 139 (2001). A Method to Extract Potentials from the Temperature Dependence of Langmuir Constants for Clathrate Hydrates.

41. B. J. Anderson, M. Z. Bazant, J. W. Tester, and B. L. Trout, *J. Phys. Chem. B*, 109, 8153 (2005). Application of the Cell Potential Method To Predict Phase Equilibria of Multicomponent Gas Hydrate Systems.

42. S. Patchkovskii and J. S. Tse, *Proc. Natl. Acad. Sci. U.S.A.*, 100, 14645 (2003). Thermodynamic Stability of Hydrogen Clathrates.

43. A. Martin, *J. Phys. Chem. B*, 114, 9602 (2010). A Simplified van der Waals-Platteeuw Model of Clathrate Hydrates with Multiple Occupancy of Cavities.

44. D. W. Davidson, in *Natural Gas Hydrates: Properties, Occurrence and Recovery*, J. L. Cox (Ed.), Butterworth, Boston, 1983; Chapter 1.

45. H. Kiefte, M. J. Clouter, and R. E. Gagnon, *J. Phys. Chem.*, 89, 3103 (1985). Determination of Acoustic Velocities of Clathrate Hydrates by Brillouin Spectroscopy.

46. R. D. Stoll and G. M. Bryan, *J. Geophys. Res.*, **84**, 1629 (1979). Physical Properties of Sediments Containing Gas Hydrates.

47. Y. P. Handa, *Can. J. Chem.*, **62**, 1659 (1984). Enthalpies of Fusion and Heat Capacities for $H_2^{18}O$ Ice and $H_2^{18}O$ Tetrahydrofuran Clathrate Hydrate in the Range 100–270 K.

48. R. G. Ross, P. Andersson, and G. Backstrom, *Nature*, **290**, 322 (1981). Unusual PT Dependence of Thermal Conductivity for a Clathrate Hydrate.

49. Y. P. Handa and J. G. Cook, *J. Phys. Chem.*, **91**, 6327 (1987). Thermal Conductivity of Xenon Hydrate.

50. J. S. Tse and M. A. White, *J. Phys. Chem.*, **92**, 5006 (1988). Origin of Glassy Crystalline Behavior in the Thermal Properties of Clathrate Hydrates: A Thermal Conductivity Study of Tetrahydrofuran Hydrate.

51. J. J. Freeman and A. C. Anderson, *Phys. Rev. B*, **34**, 5684 (1986). Thermal Conductivity of Amorphous Solids.

52. J. S. Tse and M. L. Klein, *J. Phys. Chem.*, **87**, 4198 (1983). Molecular Dynamics Studies of Ice Ic and the Structure I Clathrate Hydrate of Methane.

53. R. Berman, *Thermal Conduction in Solids*, Clarendon Press, Oxford, 1976.

54. F. Hollander and G. A. Jeffrey, *J. Chem. Phys.*, **66**, 4699 (1977). Neutron Diffraction Study of the Crystal Structure of Ethylene Oxide Deuterohydrate at 80°K.

55. G. Honjo and K. Shimaoka, *Acta Crystallogr.*, **10**, 710 (1957). Determination of Hydrogen Position in Cubic Ice by Electron Diffraction.

56. J. D. Bernal and R. H. Fowler, *J. Chem. Phys.*, **1**, 515 (1933). A Theory of Water and Ionic Solution, with Particular Reference to Hydrogen and Hydroxyl Ions.

57. H. J. C. Berendsen, J. P. M. Postma, W. F. van Gunsteren, and J. Hermans, in *Intermolecular Forces*, B. Pullman (Ed.), Reidal, Dordrecht, Holland, 1981, p. 331, Interaction Models for Water in Relation to Protein Hydration.

58. J. S. Tse, C. I. Ratcliffe, B. M. Powell, V. F. Sears, and Y. P. Handa, *J. Phys. Chem. A*, **101**, 4491 (1997). Rotational and Translational Motions of Trapped Methane. Incoherent Inelastic Neutron Scattering of Methane Hydrate.

59. P. Carruthers, *Rev. Mod. Phys.*, **33**, 92 (1961). Theory of Thermal Conductivity of Solids at Low Temperatures.

60. J. Baumert, C. Gutt, V. P. Shpakov, J. S. Tse, M. Krisch, M. Muller, H. Requardt, D. D. Klug, S. Janssen, and W. Press, *Phys. Rev. B*, **68**, 174301 (2003). Lattice Dynamics of Methane and Xenon Hydrate: Observation of Symmetry-Avoided Crossing By Experiment And Theory.

61. J. S. Tse, V. P. Shpakov, V. R. Belosludov, F. Trouw, Y. P. Handa, and W. Press, *Europhys. Lett.*, **54**, 354 (2001). Coupling of Localized Guest Vibrations with the Lattice Modes in Clathrate Hydrates.

62. C. Gutt, J. Baumert, W. Press, J. S. Tse, and S. Janssen, *J. Chem. Phys.*, **116**, 3795 (2002). The Vibrational Properties of Xenon Hydrate: An Inelastic Incoherent Neutron Scattering Study.

63. J. S. Tse, *J. Incl. Phenom. Mol. Recogn. Chem.*, **17**, 259 (1994). Localized Oscillators and Heat Conduction in Clathrate Hydrates.

64. A. J. C. Ladd, B. Moran, and W. G. Hoover, *Phys. Rev. B*, **34**, 5058 (1986). Lattice Thermal Conductivity: A Comparison of Molecular Dynamics and Anharmonic Lattice Dynamics.

65. J. Che, T. Cagin, W. Dong, and W. A. Goddard, III,, *J. Chem. Phys.*, **113**, 6888 (2000). Thermal Conductivity of Diamond and Related Materials from Molecular Dynamics Simulations.

66. R. Inoue, H. Tanaka, and K. Nakanishi, *J. Chem. Phys.*, **104**, 9569 (1996). Molecular Dynamics Simulation Study of the Anomalous Thermal Conductivity of Clathrate Hydrates.

67. N. Galamba, C. A. Nieto de Castro, and J. F. Ely, *J. Chem. Phys.*, **120**, 8676 (2004). Thermal Conductivity of Molten Alkali Halides From Equilibrium Molecular Dynamics Simulations.

68. N. Galamba, C. A. Nieto de Castro, and J. F. Ely, *J. Phys. Chem. B*, **108**, 3658 (2004). Molecular Dynamics Simulation of the Shear Viscosity of Molten Alkali Halides.

69. N. J. English and J. S. Tse, *Phys. Rev. Lett.*, **103**, 015901 (2009). Mechanisms for Thermal Conduction in Methane Hydrate.

70. N. J. English, J. S. Tse, and D. Carey, *Phys. Rev. B*, **80**, 134306 (2009). Mechanisms for Thermal Conduction In Various Polymorphs of Methane Hydrate.

71. J. Petravic, *J. Chem. Phys.*, **123**, 174503 (2005). Thermal Conductivity of Ethanol.

72. A. I. Krivchikov, B. Ya, O. A. Gorodilov, V. G. Korolyuk, H. C. Manzhelii, and W. Press, *Low Temp. Phys.*, **139**(693) (2005). Thermal Conductivity of Methane-Hydrate.

73. A. I. Krivchikov, B. Y. Gorodilov, O. A. Korolyuk, V. G. Manzhelii, O. O. Romantsova, H. Conrad, W. Press, J. S. Tse, and D. D. Klug, *Phys. Rev. B*, **73**, 064203 (2006). Thermal Conductivity of Xe Clathrate Hydrate at Low Temperatures.

74. J. H. McGaughey and M. Kaviany, *Int. J. Heat Mass Transf.*, **47**, 1783 (2004). Thermal Conductivity Decomposition And Analysis Using Molecular Dynamics Simulations.

75. J. H. McGaughey and M. Kaviany, *Int. J. Heat MassTransf.*, **47**, 1799 (2004). Thermal Conductivity Decomposition and Analysis using Molecular Dynamics Simulations: Part II. Complex Silica Structures.

76. N. J. English, P. D. Gorman, and J. M. D. MacElroy, *J. Chem. Phys.*, **136**, 044501 (2012). Mechanisms for Thermal Conduction in Hydrogen Hydrate.

77. G. A. Slack, in *Thermoelectric Materials – New Directions and Approaches*, T. M. Tritt, M. G. Kanatzidis, H. B. Lyon, and G. D. Mahan (Eds.), MRS Symposia, Proceedings No. 468, Materials Research Society, Pittsburgh, 1997, p. 47.

78. J. S. Tse and D. D. Klug, in *CRC Handbook of Thermoelectrics from Macro to Nano*, R. M. Rowe (Ed.), 2005, Recent Trends for the Design and Optimization of Thermoelectric Materials – A Theoretical Perspective.

79. G. S. Nolas, J. Sharp, and J. Goldsmid, *Thermoelectrics: Basic Principles and New Materials Developments*, Springer-Verlag, 2001.

80. W. F. Kuhs, B. Chazallon, P. G. Radaelli, and P. Pauer, *J. Inclusion Phenom. Mol. Recognit. Chem.*, **29**, 65 (1997). Cage Occupancy and Compressibility of Deuterated N2-Clathrate Hydrate by Neutron Diffraction.

81. W. L. Mao, H.-K. Mao, A. F. Goncharov, V. V. Struzhkin, Q. Guo, J. Shu, R. J. Hemley, M. Somayazulu, and Y. Zhao, *Science*, **297**, 2247 (2002). Hydrogen Clusters in Clathrate Hydrate.

82. K. A. Lokshin, Y. Zhao, D. He, W. L. Mao, H.-K. Mao, R. J. Hemley, M. V. Lobanov, and M. Greenblatt, *Phys. Rev. Lett.*, **93**, 125503 (2004). Structure and Dynamics of Hydrogen Molecules in the Novel Clathrate Hydrate by High Pressure Neutron Diffraction.

83. H. Itoh, J. S. Tse, and K. Kawamura, *J. Chem. Phys.*, **115**, 9414 (2001). The Structure and Dynamics of Doubly Occupied Ar Hydrate.

84. H. Tanaka, *Chem. Phys. Lett.*, **202**, 345 (1993). The Stability of Xe and CF$_4$ Clathrate Hydrates. Vibrational Frequency Modulation and Cage Distortion.

85. H. Tanaka and K. Kiyohara, *J. Chem. Phys.*, **98**, 4098 (1993). On the Thermodynamic Stability of Clathrate Hydrate. I.

86. H. Tanaka and K. Kiyohara, *J. Chem. Phys.*, **98**, 8110 (1993). The Thermodynamic Stability of Clathrate Hydrate. II. Simultaneous Occupation of Larger and Smaller Cages.

87. H. Tanaka, *J. Chem. Phys.*, **101**, 10833 (1994). The Thermodynamic Stability of Clathrate Hydrate. III. Accommodation of Nonspherical Propane and Ethane Molecules.

88. H. Tanaka, T. Nakatsuka, and K. Koga, *J. Chem. Phys.*, **121**, 5488 (2004). On the Thermodynamic Stability of Clathrate Hydrates IV: Double Occupancy of Cages.

89. Y. Koyama, H. Tanaka, and K. Koga, *J. Chem. Phys.*, **122**, 074503 (2005). On the Thermodynamic Stability and Structural Transition of Clathrate Hydrates.

90. H. Tanaka and M. Matsumoto, *J. Phys. Chem. B*, **115**, 14256 (2011). On the Thermodynamic Stability of Clathrate Hydrates V: Phase Behaviors Accommodating Large Guest Molecules with New Reference States.

91. H. Tanaka, *Fluid Phase Equilib.*, **104**, 331 (1995). A Novel Approach to the Stability of Clathrate Hydrate.

92. K. Koga, H. Tanaka, and K. Nakanishi, *J. Chem. Phys.*, **101**, 3127 (1994). Stability of Polar Guest-Encaging Clathrate Hydrates.

93. H. Tanaka, *Chem. Phys. Lett.*, **220**, 371 (1994). The Thermodynamic Stability of Clathrate Hydrate. Encaging Nonspherical Propane Molecules.

94. H. Tanaka and K. Nakanishi, *Mol. Sim.*, **12**, 317 (1994). The Stability of Clathrate Hydrates: Temperature Dependence of Dissociation Pressure in Xe and Ar Hydrate.

95. K. Katsumasa, K. Koga, and H. Tanaka, *J. Chem. Phys.*, **127**, 044509 (2007). On the Thermodynamic Stability of Hydrogen Clathrate Hydrates.

96. H. Tanaka, *Can. J. Phys.*, **81**, 55 (2003). On the Thermodynamic Stability of Clathrate Hydrates.

97. K. Sugahara, M. Yoshida, T. Sugahara, and K. Ohgaki, *J. Chem. Eng. Data*, **49**, 326 (2004). High-Pressure Phase Behavior and Cage Occupancy for the CF_4 Hydrate System.

98. H. Shimizu, S. Hori, T. Kume, and S. Sasaki, *Chem. Phys. Lett.*, **368**, 132 (2003). Optical Microscopy and Raman Scattering of a Single Crystalline Argon Hydrate at High Pressures.

99. M. P. Allen and D. J. Tildesley, *Computer Simulation of Liquids*, Oxford University Press, 1987.

100. S. J. Wierzchowski and P. A. Monson, *J. Phys. Chem. B*, **111**, 7274 (2007). Calculation of Free Energies and Chemical Potentials for Gas Hydrates Using Monte Carlo Simulations.

101. S. J. Wierzchowski and P. A. Monson, *Ind. Eng. Chem. Res.*, **45**, 424 (2006). Calculating the Phase of Gas-Hydrate-Forming Systems from Molecular Models.

102. K. S. Glavatsky, T. J. H. Vlugt, and S. Kjelstrup, *J. Phys. Chem. B*, **116**, 3745 (2012). Toward a Possibility To Exchange CO_2 and CH_4 in sI Clathrate Hydrates.

103. S. Rick, *J. Chem. Phys.*, **120**, 6085 (2004). A Reoptimization of the Five-Site Water Potential (TIP5P) for Use with Ewald Sums.

104. J. J. Potoff and J. I. Siepmann, *AIChE J.*, **47**, 1676 (2001). Vapor–Liquid Equilibria of Mixtures Containing Alkanes, Carbon Dioxide, and Nitrogen.

105. A. D. J. Haymet and T. W. Barlow, *Ann. New York, Acd. Sci.*, **715**, 549 (1994). Nucleation of Supercooled Liquids.

106. R. Radhakrishnan and B. L. Trout, *J. Am. Chem. Soc.*, **125**, 7743 (2003). Nucleation of Hexagonal Ice (I_h) in Liquid Water.

107. R. Radhakrishnan and B. L. Trout, *J. Chem. Phys.*, **117**, 1786 (2002). A New Approach for Studying Nucleation Phenomena Using Molecular Simulations: Application to CO_2 Hydrate Clathrates.

108. P. J. Steinhardt, D. R. Nelson, and M. Ronchetti, *Phys. Rev. B*, **28**, 784 (1983). Bond-Orientational Order in Liquids and Glasses.

109. P. I. Chau and A. J. Hardwick, *Mol. Phys.*, **93**, 511 (1998). A New Order Parameter for Tetrahedral Configurations.

110. W. L. Jorgensen, J. Chandrasekhar, J. D. Madura, R. I. Impey, and M. L. Klein, *J. Chem. Phys.*, **79**, 926 (1983). Comparison of Simple Potential Functions for Simulating Liquid Water.

111. E. D. Sloan and F. A. Fleyfel, *AIChE J.*, **37**, 1281 (1991). A Molecular Mechanism for Gas Hydrate Nucleation from Ice.

112. S. Swaminathan, S. W. Harrison, and D. L. Beveridge, *J. Am. Chem. Soc.*, **100**, 5705 (1976). Monte Carlo Studies on the Structure of a Dilute Aqueous Solution of Methane.

113. M. Matsumoto, S. Saito, and I. Ohmine, *Nature*, **416**, 409 (2002). Molecular Dynamics Simulation of the Ice Nucleation and Growth Process Leading to Water Freezing.

114. C. Moon, P. C. Taylor, and P. M. Rodger, *J. Am. Chem. Soc.*, **125**, 4706 (2003). Molecular Dynamics Study of Gas Hydrate Formation.

115. R. E. Westacott and P. M. Rodger, *J. Chem. Soc., Faraday Trans.*, **94**, 3421 (1998). A Local Harmonic Study of Clusters of Water and Methane.

116. R. W. Hawtin, D. Quigley, and P. M. Rodger, *Phys. Chem. Chem. Phys.*, **10**, 4853 (2008). Gas Hydrate Nucleation and Cage Formation at a Water/Methane Interface.

117. L. Boewer, J. Nase, M. Paulus, F. Lehmkühler, S. Tiemeyer, S. Holz, D. Pontoni, and M. Tolan, *J. Phys. Chem. C*, **116**, 8548 (2012). On the Spontaneous Formation of Clathrate Hydrates at Water–Guest Interfaces.

118. H. Nada and J. P. J. M. van der Eerden, *J. Chem. Phys.*, **118**, 7401 (2003). An Intermolecular Potential Model for the Simulation of Ice and Water Near the Melting Point: A Six-Site Model of H_2O.

119. M. R. Walsh, C. A. Koh, E. D. Sloan, A. K. Sum, and D. T. Wu, *Science*, **326**, 1095 (2009). Microsecond Simulations of Spontaneous Methane Hydrate Nucleation and Growth.

120. J. L. F. Abascal, E. Sanz, R. García Fernandez, and C. Vega, *J. Chem. Phys.*, **122**, 234511 (2005). A Potential Model for the Study Of Ices and Amorphous Water: TIP4P/Ice.

121. P. M. Rodger, T. R. Forester, and W. Smith, *Fluid Phase Equilib.*, **116**, 326 (1996). Simulations of the Methane Hydrate/Methane Gas Interface Near Hydrate Forming Conditions.

122. J. Vatamanu and P. Kusalik, *Phys. Chem. Chem. Phys.*, **12**, 15065 (2010). Observation of Two-Step Nucleation in Methane Hydrates.

123. M. R. Walsh, J. D. Rainey, P. G. Lafond, D.-H. Park, G. T. Beckham, M. D. Jones, K.-H. Lee, C. A. Koh, E. D. Sloan, D. T. Wu, and A. K. Sum, *Phys. Chem. Chem. Phys.*, **13**(19951) (2011). The Cages, Dynamics, and Structuring of Incipient Methane Clathrate Hydrates.

124. A. P. Mehta, P. B. Hebert; E. R. Cadena and J. P. Weatherman, *Offshore Technology Conference*, doi:10.4043/14057-MS. Houston, Texas (2002).

125. C. Moon, P. C. Taylor, and P. M. Rodger, *Can. J. Phys.*, **81**, 451 (2003). Clathrate Nucleation And Inhibition From a Molecular Perspective.

126. M. Storr, P. C. Taylor, J.-P. Monfort, and P. M. Rodger, *J. Am. Chem. Soc.*, **126**(1569) (2004). Kinetic Inhibitor of Hydrate Crystallization.

127. C. Moon, R. E. Hawtin, and P. M. Rodger, *Faraday Discussion*, **136**, 367 (2007). Nucleation and Control of Clathrate Hydrates: Insights from Simulation.

128. R. W. Hawtin and P. M. Rodger, *J. Mater. Chem.*, **16**, 1937 (2006). Polydispersity in Oligomeric Low Dosage Gas Hydrate Inhibitors.

129. V. Buch, J. P. Delvin, I. A. Monreal, N. Jagoda-Cwiklik, N. Uras-Aytemiz, and L. Cwiklik, *Phys. Chem. Chem. Phys.*, **11**, 10245 (2008). Clathrate Hydrates with Hydrogen-Bonding Guests.

130. N. Bjerrum, *Science*, **115**, 385 (1952). Structure and Properties of Ice.

131. B. J. Anderson, J. W. Tester, G. P. Borghi, and B. L. Trout, *J. Am. Chem. Soc.*, **127**, 17852 (2005). Properties of Inhibitors of Methane Hydrate Formation via Molecular Dynamics Simulations.

132. A. Demurov, R. Radhakrishnan, and B. L. Trout, *J. Chem. Phys.*, **116**, 702 (2002). Computations of Diffusivities in Ice and CO_2 Clathrate Hydrates via Molecular Dynamics and Monte Carlo Simulations.

133. A. Peters, N. E. Zimmermann, G. T. Backham, J. W. Tester, and B. L. Trout, *J. Am. Chem. Soc.*, **130**, 17342 (2008). Path Sampling Calculation of Methane Diffusivity in Natural Gas Hydrates from a Water-Vacancy Assisted Mechanism.

134. J. A. Ripmeester, C. I. Ratcliffe, and J. S. Tse, *J. Chem. Soc. Faraday Trans. I*, **84**, 3731 (1988). The NMR of [129]Xe Trapped in Clathrates and Some Other Solids.

135. B. Stueber and C. J. Jameson, *J. Chem. Phys.*, **120**, 1560 (2004). The Chemical Shifts of Xe in the Cages of Clathrate Hydrate Structures I and II.

136. R. Ditchfield, *Mol. Phys.*, **27**, 789 (1974). Self-Consistent Perturbation Theory of Diamagnetism.

137. S. Alavi, J. A. Ripmeester, and D. D. Klug, *J. Chem. Phys.*, **123**, 051107 (2005). NMR Shielding Constants for Hydrogen Guest Molecules in Structure II Clathrates.

138. H. Mohammadi-Manesh, S. Alavi, T. K. Woo, and B. Najafib, *Phys. Chem. Chem. Phys.*, **13**, 2367 (2011). Molecular Dynamics Simulation of NMR Powder Lineshapes of Linear Guests in Structure I Clathrate Hydrates.

139. R. Car and M. Parrinello, *Phys. Rev. Lett.*, **55**, 2471 (1985). Unified Approach for Molecular Dynamics and Density-Functional Theory.

140. J. S. Tse, *Ann. Rev. Phys. Chem.*, **53**, 249 (2002). Ab Initio Molecular Dynamics with Density Functional Theory.

141. D. Marx and J. Hutter, *Ab Initio Molecular Dynamics: Basic Theory and Advanced Methods*, Cambridge University Press, 2009.

142. M. C. Payne, M. P. Teter, D. C. Allan, T. A. Arias, and J. D. Joannopouos, *Rev. Mod. Phys.*, **64**, 1045 (1992). Iterative Minimization Techniques for Ab Initio Total-Energy Calculations: Molecular Dynamics and Conjugate Gradients.

143. J. S. Tse, *J. Supramolecular Chem.*, **2**, 429 (2002). Vibrations of Methane in Structure I Clathrate Hydrate – An Ab Initio Density Functional Molecular Dynamics Study.

144. A. K. Sum, R. C. Burruss, and E. D. Sloan, *J. Phys. Chem.*, **101**, 7371 (1997). Measurement of Clathrate Hydrates via Raman Spectroscopy.

145. G. C. Pimental and S. W. Charles, *Pure Appl. Chem.*, **7**, 111 (1963). Infrared Spectral Perturbations in Matrix Experiments.

146. B. C. Chakoumakas, C. J. Rawn, A. J. Rondinone, S. L. Marshall, L. A. Stern, S. Circone, S. H. Kirby, C. Y. Jones, B. H. Toby, and Y. Ishii, The Use of Rigid Body Constraints in Rietveld Refinements of Neutron Diffraction Data of Clathrate Hydrates, in *Proc. Fourth Int. Conf. Gas Hydrates*, Yokohama; 655 (2002).

147. M. Hiratsuka, R. Ohmura, A. K. Sum, and K. Yasuoka, *J. Chem. Phys.*, **136**, 044508 (2012). Molecular Vibrations of Methane Molecules in the Structure I Clathrate Hydrate from Ab Initio Molecular Dynamics Simulation.

148. N. J. English and J. S. Tse, *J. Phys. Chem. A*, **115**, 6226 (2011). Dynamical Properties of Hydrogen Sulphide Motion in its Clathrate Hydrate from Ab Initio and Classical Isobaric–Isothermal Molecular Dynamics.

149. J. S. Tse, W. R. McKinnon, and M. Marchi, *J. Phys. Chem.*, **91**, 4188 (1987). Thermal Expansion of Structure I Ethylene Oxide Hydrate.

150. K. C. Hester, Z. Huo, A. L. Ballard, C. A. Koh, K. T. Miller, and E. D. Sloan, *J. Phys. Chem. B*, **111**, 8830 (2007). Thermal Expansivity for sI and sII Clathrate Hydrates.

151. T. Ikeda and K. Terakura, *J. Chem. Phys.*, **119**, 6784 (2003). Structural Transformation of Methane Hydrate from Cage Clathrate to Filled Ice.

152. J. S. Loveday, R. J. Nelmes, M. Guthrie, S. A. Belmonte, D. R. Allan, D. D. Klug, J. S. Tse, and Y. P. Handa, *Nature (London)*, **410**(661) (2001). Existence of Stable Methane Hydrate Above 2 GPa and the Implication for Titan.

153. J. S. Loveday, R. J. Nelmes, M. Guthrie, D. D. Klug, and J. S. Tse, *Phys. Rev. Lett.*, **87**, 215501 (2001). Transition from Cage Clathrate to Filled Ice: The Structure of Methane Hydrate III.

154. O. Kambara, K. Takahashi, M. Hayash, and J. L. Kuo, *Phys. Chem. Chem. Phys.*, **14**, 11484 (2012). Assessment of Density Functional Theory to Calculate the Phase Transition Pressure of Ice.

155. M. Dion, H. Rydberg, E. Schröder, D. C. Langreth, and B. I. Lundqvist, *Phys. Rev. Lett.*, **92**, 246401 (2004). Van der Waals Density Functional for General Geometries.

156. B. Santra, J. Klimes, D. Alfe, A. Tkatchenko, B. Slater, A. Michaelides, R. Car, and M. Scheffler, *Phys. Rev. Lett.*, **107**, 185701 (2011). Hydrogen Bonds and van der Waals Forces in Ice at Ambient and High Pressures.

157. G. Roman-Perez, M. Moaied, J. M. Soler, and F. Yndurain, *Phys. Rev. Lett.*, **105**, 145901 (2010). Stability, Adsorption, and Diffusion of CH_4, CO_2 and H_2 in Clathrate Hydrates.

158. Q. Li, B. Kolb, G. Roman-Perez, J. M. Soler, F. Yndurain, L. Kong, D. C. Langreth, and T. Thonhauser, *Phys. Rev. B*, **84**, 153103 (2011). Ab Initio Energetics and Kinetics Study of H_2 and CH_4 in the SI Clathrate Hydrate.

159. S. Patchkovskii and S. N. Yurchenko, *Phys. Chem. Chem. Phys.*, **6**, 4152 (2004). Quantum and Classical Equilibrium Properties for Exactly Solvable Models of Weakly Interacting Systems.

160. C. Gutt, W. Press, A. Hüller, J. S. Tse, and H. Casalta, *J. Chem. Phys.*, **114**, 4160 (2001). The Isotope Effect and Orientational Potentials of Methane Molecules in Gas Hydrates.

161. M. Xu, L. Ulivi, M. Celli, D. Colognesi, and Z. Bacci, *Phys. Rev. B*, **83**, 241403 (2011). Quantum Calculation of Inelastic Neutron Scattering Spectra of a Hydrogen Molecule Inside a Nanoscale Cavity Based on Rigorous Treatment of the Coupled Translation-Rotation Dynamics.

162. J. Schwinger and E. Teller, *Phys. Rev.*, **52**, 286 (1937). The Scattering of Neutrons by Ortho- and Parahydrogen.

163. M. Xu and Z. Bacic, *Phys. Rev., B*, **84**, 195445 (2011). Inelastic Neutron Scattering Spectra of a Hydrogen Molecule in a Nanocavity: Methodology for Quantum Calculations Incorporating the Coupled Five-Dimensional Translation-Rotation Eigenstates.

164. M. Xu, Y. S. Elmatad, F. Sebastianelli, J. W. Moskowitz, and Z. Bačić, *J. Phys. Chem. B*, **110**, 24806 (2006). Hydrogen Molecule in the Small Dodecahedral Cage of a Clathrate Hydrate: Quantum Five-Dimensional Calculations of the Coupled Translation–Rotation Eigenstates.

165. I. Matanović, M. Xu, J. W. Moskowitz, J. Eckert, and Z. Bačić, *J. Chem. Phys.*, **131**, 224308 (2009). Methane Molecule Confined in the Small and Large Cages of Structure I Clathrate Hydrate: Quantum Six-Dimensional Calculations of the Coupled Translation-Rotation Eigenstates.

166. J. Cohn, G. S. Nolas, V. Fessatidis, T. H. Metcalf, and G. A. Slack, *Phys. Rev. Lett.*, **82**, 779 (1999). Glasslike Heat Conduction in High-Mobility Crystalline Semiconductors.

167. J. C. Owicki and H. A. Scheraga, *J. Am. Chem. Soc.*, **99**, 7413 (1977). Monte Carlo Calculations in the Isothermal-Isobaric Ensemble. 2. Dilute Aqueous Solution of Methane.

168. S. Alavi and J. A. Ripmeester, *Angew. Chem. Int. Ed.*, **46**, 6102 (2007). Hydrogen-Gas Migration through Clathrate Hydrate Cages.

169. T. Okuchi, I. L. Moudrakovski, and J. A. Ripmeester, *Appl. Phys. Lett.*, **91**, 171903 (2007). Efficient Storage of Hydrogen Fuel into Leaky Cages of Clathrate Hydrate.

CHAPTER 8

The Quantum Chemistry of Loosely-Bound Electrons

John M. Herbert

Department of Chemistry and Biochemistry, The Ohio State University, Columbus, OH 43210, USA

INTRODUCTION AND OVERVIEW

What Is a Loosely-Bound Electron?

By some measure, the title of this chapter could have been "The quantum chemistry of weakly-bound anions," because much of it will focus on how to describe the weak binding of an "extra" electron to a stable, neutral molecule using electronic structure theory. Electron binding energies in such cases may be quite small, typically less than the largest *atomic* electron affinities (EAs)[1] (3.4 eV for F and 3.6 eV for Cl), and even less than 0.1 eV in some cases. Unlike the case where a neutral molecule, M, is ionized, an electron separated from the anion M^- does not experience an attractive $-1/r$ potential at large separations,[2,3] but rather only charge–dipole and or higher order charge–multipole interactions. Cases where the electron affinity of M is $\lesssim 0.5$ eV are the signature of a short-range valence potential that is weakly attractive at best, such that electron binding in M^- results primarily from long-range electron–molecule, charge–multipole interactions. In such cases, one expects to find an unpaired electron in M^- that is radially diffuse, much more so than in F^- or Cl^-, for example. It is in this sense that the odd electron in M^- is "loosely-bound." Moreover, the preceding discussion assumes that M^- is bound at all, but in fact we will also consider cases in which the electron

Reviews in Computational Chemistry, Volume 28, First Edition.
Edited by Abby L. Parrill and Kenny B. Lipkowitz.

is adiabatically *unbound* (higher in energy than $M + e^-$), but where the species M^- can exist as a *temporary anion resonance*, trapped behind some energy barrier that often originates from the centrifugal potential required to conserve angular moment when removing an electron from an orbital with $\ell > 0$.

We will further broaden our definition of what constitutes a loosely-bound electron to include a discussion of *solvated electrons*, which one might define as cluster anions M_N^- where the odd electron is bound collectively by the solvent molecules, insofar as the anion M^- of a single solvent molecule is not a bound species. In a small cluster, M_N^- might be weakly-bound, but depending on the number (N) and nature of the solvent molecules, the electron binding energy of M_N^- can sometimes be quite large, up to a few eV in some cases. This is still much smaller than the ionization energy of the neutral cluster M_N, and in this sense the "extra" electron may still be considered to be weakly-bound. As a definite example, consider the case $M = H_2O$. Vertical electron binding energies in $(H_2O)_N^-$ clusters, as measured by photoelectron spectroscopy, can exceed 2 eV for $N \gtrsim 100$,[4,5] and the best estimates in the bulk limit ($N \to \infty$) lie in the range of 3.3–4.0 eV.[5–10] Insofar as H_2O^- is not bound,[11] however, the unpaired electron cannot be said to be associated to any one particular water molecule and is thus "loosely-bound."

As another example of what we have categorized, for the purpose of this chapter, as loosely-bound electrons, we will consider excited electronic states of anions that possess enough energy to access an electronic continuum, or in other words, excited states where the excitation energy is greater than the electron detachment energy. Such states exist, if at all, only as temporary, "auto-ionizing" resonances. Other temporary anion resonances (viz, shape resonances and Feshbach resonances) will be explained and discussed as well. These temporary anion resonances are chemically important in the context of *dissociative electron attachment* (DEA),[12] in which resonant attachment of low-energy electrons is followed by internal conversion to a state where covalent bond dissociation is energetically feasible. DEA provides a mechanism wherein energy barriers that would be thermodynamically insurmountable on the Born–Oppenheimer potential surface for $M + e^-$ are bypassed by means of nonadiabatic transitions, leading to molecular fragmentation in the presence of electrons whose kinetic energies are "just right."

Scope of This Review

Each of the aforementioned phenomena presents special challenges for quantum chemistry, and the methods required to meet these challenges are the primary topic of this chapter. In addition, this chapter provides a discussion of the basic quantum mechanical concepts that underlie the collection of phenomena that this author has categorized as "loosely-bound electrons." A limited discussion of some of the interesting chemical systems that fall under this moniker is provided as well, although this chapter is not intended to be a comprehensive review of the field of weakly-bound anions, solvated electrons, or anything else. Several excellent topical reviews have appeared in the past few years, to which the reader is referred for a thorough discussion of the chemistry. These include a comprehensive 2008 review of the whole field of molecular anions (experiment as well as theory),[2] and a more recent review focused on theoretical calculations.[3] Reactions induced by low-energy electrons have also been thoroughly reviewed in the past few years, including several general overviews of electron-induced reactions,[12-14] a review of biological radiation chemistry and the role of "presolvated" electrons in biological radiation damage,[13] and a review of electron attachment (EA) to DNA, focusing on electronic structure calculations.[14] Solvated electrons, both in clusters and in bulk liquids, have been reviewed recently from both a theoretical perspective[15,16] and an experimental perspective.[17-19]

This work is intended as an introduction to these topics and a tutorial guide to performing quantum chemistry calculations intended to model these types of molecular systems and phenomena. The focus here is on methods that are readily available in standard quantum chemistry software packages and thus, for example, we will discuss the calculation of resonance states using modifications of *bound-state* methodology,[20] since bound states are what one computes in traditional quantum chemistry. Alternative formalisms such as scattering theory[21-26] or the explicit treatment of the interaction of a discrete state with a continuum state[27,28] will not be discussed here. The use of complex absorbing potentials[29-31] is discussed only briefly.

Our discussion of electronic structure calculations is further limited to methods based on Gaussian basis sets rather than plane waves. This restriction is partly a matter of taste, and in fact one can make a good case that a plane-wave basis is better suited for representing the most diffuse parts of a weakly-bound anion's electron density, as compared to a basis comprised of localized, atom-centered functions. However, a more important concern (in this author's view) is the fact that Hartree–Fock exchange is prohibitively expensive to compute in a plane-wave basis. Although significant progress has been made in this respect,[32] the cost remains prohibitive unless the nonlocal exchange interaction is screened at long range.[33,34] As a result, the hybrid functionals that provide the best performance for many molecular properties are ordinarily not available in plane-wave density functional theory (DFT) calculations, which proves particularly egregious in the case of weakly-bound anions. Correlated, post-Hartree–Fock wave functions are similarly unavailable in most plane-wave codes, which is another reason to dismiss plane waves in the present context.

As the title of this chapter suggests, the discussion here is limited, for the most part, to *loosely* bound electrons, meaning that valence ionization is not considered to any significant extent. It is important to keep in mind, however, that "loosely-bound" need not mean "weakly-bound." As mentioned earlier, vertical electron detachment energies in excess of 1–2 eV are possible even in solvated-electron clusters where the "extra" electron is not strongly associated with any particular molecular unit or chemical moiety. That said, the quantum chemistry of weakly-bound, gas-phase anions is certainly discussed herein, at a somewhat more pedagogical level as compared to previous reviews on that subject.[2,11,35] An exception is that Refs. 2 and 37 describe the underlying theory behind various post-Hartree–Fock quantum chemistry models, material that is not covered here at all (it has been covered in previous chapters in this series[36,37]). This review assumes a basic familiarity with the nomenclature of quantum chemical models and basis sets, at the level of introductory textbooks[35,38,39] or previous chapters in this series.[36,40,41] The performance of a variety of quantum chemical models, as applied to weakly-bound anions, is addressed in detail, but the inner workings of these models is not discussed here.

A unique feature of this chapter, as compared to other reviews of anion quantum chemistry that have appeared in the past decade,[2,11,42] is a greater emphasis on treating larger systems, including anions in the condensed phase. In large systems, compromises must inevitably be made in terms of the theoretical methods that can be deployed. As such, density functional methods – which have hardly been discussed at all in previous reviews of anion quantum chemistry, except briefly in Ref. 2 and in a benchmarking capacity in Ref. 43 – are discussed at length here. Previous chapters in this series provide an introduction to DFT itself,[40,41] but the focus here is on the performance of the models, not their intimate details. There has been some controversy regarding the applicability of DFT to anions,[44–46] however, which will require delving into a bit of detail.

Finally, this chapter covers not just bound-state methods but also methods that can safely be applied to *metastable* anions (i.e., temporary anion resonances), which is a far less mainstream topic.[20] For metastable anions, the emphasis of this chapter is on those methods that have been implemented in standard quantum chemistry codes and are therefore widely available to the chemistry community.

Chemical Significance of Loosely-Bound Electrons

Weakly-bound anions with electron detachment energies < 0.1 eV can be produced and detected experimentally.[2,11] At some level, it is tempting to be cynical about the significance of an anion M^- whose electron detachment energy is that small, since excitation of any vibrational mode with $\tilde{v} \gtrsim 800$ cm^{-1}, or some combination of rotational and or vibrational excitations adding up to ~ 800 cm^{-1}, provides enough energy to detach the electron, depending on how the modes in question couple to the weakly-bound electron and modulate the electron binding energy.[47–49] However, there are other possible fates for M^- that are more interesting, such as a nonadiabatic transition into a different anion electronic state.

Anions that are formally unbound but metastable can be chemically important. These *temporary anion resonances* are discussed in detail later in this chapter, but a pictorial illustration of one important case is provided in Figure 1. In this particular example, the energy of the anion AB⁻ lies above that of neutral AB at the geometry of the latter. However, an incident electron with appropriate kinetic energy can attach (into a virtual orbital of AB) to form AB⁻, but the energy of the anion that is formed lies above the dissociative asymptote of AB⁻. This is an example of DEA.[12] Experimentally, this temporary anion resonance manifests as a yield of ions B⁻ and radicals A• for incident electron energies that lie within a certain range that is defined by the Franck–Condon envelope of AB's wave function projected onto the AB⁻ potential surface. The thermodynamic driving force for DEA is often the large electron affinity of the B moiety, as suggested in Figure 1.

From the standpoint of radiation chemistry, DEA reactions such as these are classified as low-energy electron-induced reactions,[12,13,50] since they are driven by "secondary" (subionizing) electrons with $E < 15$ eV, rather than the fast "primary" electrons that have enough kinetic energy to generate molecular ions directly. An excellent recent overview of low-energy electron-induced reactions, from an experimental perspective, can be found in Ref. 12. One key feature of electron-induced reactions is high selectivity for cleavage of specific bonds, which need not be (and often are not) the thermodynamically weakest

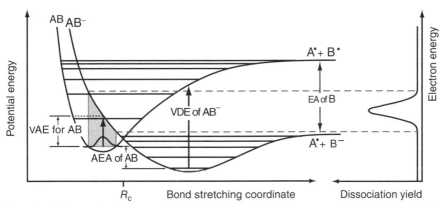

Figure 1 Schematic representation of a temporary anion resonance leading to dissociative electron attachment. Within the Franck–Condon envelope of the neutral molecule AB (shaded region), electron attachment is possible if the incident electron has an energy that matches the vertical attachment energy (VAE), equal to $E(AB^-) - E(AB)$. On the AB⁻ potential surface, the temporary anion possesses sufficient energy to dissociate to A• + B⁻, and this manifests as an observed yield of products (A• or B⁻) in the energy window defined by the Franck–Condon envelope. (The vertical energy scales at the left and right are the same, so that the potential surfaces map directly onto the observed dissociation yield.) Important energetic quantities are indicated, including the VAE and the adiabatic electron affinity (AEA) of the molecule AB, the vertical detachment energy (VDE) of AB⁻, and the electron affinity (EA) of species B. The last is the driving force, chemically speaking, that stabilizes the AB⁻ potential surface relative to that of AB. Adapted with permission from Ref. 12; copyright 2010 Elsevier.

bonds in a polyatomic molecule, owing to details of the nonadiabatic nature of the electron-induced reaction. This fact has been exploited in proof-of-concept experiments that use DEA to break selected chemical bonds with 100% selectivity,[51] and single-molecule specificity,[52,53] for molecules adsorbed on a solid substrate. (In these particular experiments, the reagent electrons were generated using the tip of a scanning tunneling microscope.) A longer list of examples and potential applications of reactions induced by low-energy electrons can be found in Ref. 12.

Chlorofluorocarbons (CFCs) tend to have large cross sections for DEA, owing to the large EAs of the halogen atoms, and have been common targets for laboratory studies of DEA. The DEA cross sections for CFCs are enhanced further still by several orders of magnitude relative to gas-phase values, for DEA occurring at the surface of water ice.[54-57] Laboratory experiments consist of adsorbing, for example, $CFCl_3$ at low-surface coverage onto a thin film of ice prepared on a metal substrate; solvated electrons at the ice/vacuum interface are subsequently generated by photoexcitation of the metal.[55,58,59] The reaction $CFCl_3 + e^- \rightarrow {}^{\bullet}CFCl_2 + Cl^-$ is observed even at very low surface coverage of $CFCl_3$,[55,59] suggesting a highly efficient reaction. Absent the CFC, the photogenerated electrons in these experiments are found to be stable for *minutes* at the ice/vacuum interface.[60] (For comparison, solvated electrons generated by radiolysis of bulk liquid water survive for only about 10 μs,[61] owing to fast diffusion and a variety of recombination reactions with other radiolysis byproducts.[62])

In view of this, it has been proposed that hydrated electrons generated on the surface of stratospheric ice crystals, via cosmic rays, could contribute to Cl^- formation via DEA of adsorbed CFCs.[54,56] Photodetachment of the chloride ions might then provide a mechanism to generate the Cl radicals that lead to ozone destruction. However, attempts to link these laboratory observations directly to stratospheric ozone chemistry have been strongly criticized,[63-71] although modeling does leave open the possibility that, at the very least, HCl destruction on ice crystals might be important for stratospheric chlorine chemistry.[67] More work is evidently needed to resolve this controversy.

A slight variant on the DEA picture introduced in Figure 1, but one worthy of mention in its own right, is the possibility of long-range electron capture in the gas phase into a Rydberg-type orbital of some molecule. Suppose that our molecule, M, consists of two different functional groups connected by a single bond, $M = R_1$—R_2. If an electron initially attaches to the molecule R_1—R_2 via some unoccupied (virtual) molecular orbital (MO) associated with the R_1 moiety, but subsequently tunnels into a σ^* orbital associated with the R_1—R_2 bond, then the net effect is to reduce the formal bond order for R_1—R_2. It may then be the case that the anion is no longer a bound state with respect to dissociation along the R_1—R_2 bond. The process in question can be written

$$R_1 - R_2 + e^- \rightarrow (R_1 - R_2)^- \rightarrow R_1^- + R_2^{\bullet} \qquad [1]$$

Assuming that the products $R_1^- + R_2^\bullet$ are lower in energy than the reactants ($R_1 - R_2$ plus an infinitely separated electron), then the intermediate species $(R_1 - R_2)^-$ is a temporary anion resonance, since the bound-state configuration of moieties R_1, R_2, and an extra electron is $R_1^- + R_2^\bullet$, not the anion $(R_1 - R_2)^-$. Whether reaction [1] will occur in practice depends sensitively on the barrier(s) and nonadiabatic couplings involved in the second step (nonadiabatic transition to the dissociative σ^* state), as well as the energy of the incident electron, which controls whether the temporary anion resonance can be accessed or not.

In the case of weakly-bound anions, we note that if the electron binding energy of M^- is very small, then the unpaired electron is bound very diffusely, so that it may, on average, be well-separated from the molecular core, M. This scenario is unlikely to occur except in the gas phase, where M^- is well separated from other molecules. Long-range electron capture into Rydberg orbitals, to form very weakly-bound gas-phase anions, is the mechanism that underlies the techniques known as *electron capture dissociation*[72] and *electron transfer dissociation*[73] that are used for protein sequence analysis via mass spectrometry. In these methods, capture of low-energy electrons by highly protonated, gas-phase polypeptides leads to highly specific fragmentation patterns, with cleavage of disulfide bonds and $N-C_\alpha$ bonds as the dominant fragmentation channels. Simons and coworkers[74–78] have performed quantum chemistry calculations in an attempt to understand the mechanism(s) behind this specificity, and the potential energy curves shown in Figure 2 illustrate the DEA phenomenon. In these calculations,[74] the molecule is actually a cation (consistent with the aforementioned experiments), namely, $M^+ = H_3CS—S(CH_2)_2NH_3^+$. EA directly to the $\sigma^*(S - S)$ orbital is energetically feasible at the parent cation's geometry, with a resulting potential curve that is indeed dissociative along the S—S coordinate. However, the cross section for this direct attachment process is found to be small. Instead, EA to excited Rydberg states of the $—NH_3^+$ moiety appears to offer a substantially enhanced cross section that also ultimately results in S—S dissociation, via a DEA mechanism.[77]

Moving from the gas phase into solution, there is evidence that very low-energy electrons can damage DNA,[79–81,13] via a DEA mechanism somewhat analogous to that discussed earlier.[82,83] In this context, "low energy" means electrons whose kinetic energies are less than the ~ 4 eV bond energies of the single bonds that are ruptured in DNA single-strand breaks. (In the context of radiation chemistry, electrons in the range 0–20 eV are typically classified as "low-energy" electrons, since this is the energy distribution of secondary electrons formed from primary ionization events,[84,13] but here we are focusing on the low end of this range.) Experimental studies of very low-energy ($E < 3$ eV) EA to gas-phase nucleobases show high site selectivity in the ion yields. For example, in thymine and uracil, there is a resonance around $E \approx 1$ eV corresponding to the dissociation of the N1—H bond (where the base would attach to the sugar in DNA) and a somewhat higher energy

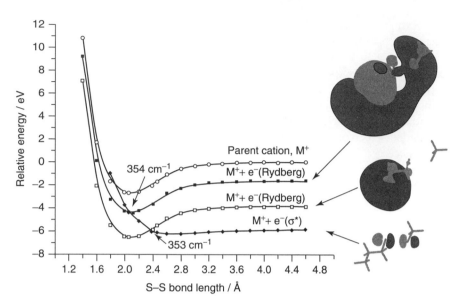

Figure 2   Potential energy curves along the S—S coordinate, for the protonated model peptide $M^+ = H_3CS—S(CH_2)_2NH_3^+$ and electron-attached states thereof. Shown are states in which the electron attaches to either of two different Rydberg orbitals localized on the $—NH_3^+$ moiety and the state that results from direct e^- attachment to the $\sigma^*(S—S)$ orbital. Orbital isosurfaces were computed at $R = 3.6$ Å and contain 60% of the orbital density in each case. Estimated nonadiabatic couplings between the Rydberg states and the σ^* state are shown as well. Adapted with permission from Ref. 74; copyright 2006 Elsevier.

resonance corresponding to N3–H dissociation. Notably, the former is absent when the N1 site is methylated,[85] or even deuterated,[86] suggesting that the lowest energy electrons might generate H atoms in DNA, but would not lead directly to the loss of the nucleobase.

A mechanism for single-strand breaks induced by electrons with $E \lesssim 2$ eV has been proposed by Simons[82] on the basis of theoretical calculations. The first step is formation of a temporary anion resonance involving electron capture by a π^* orbital of the nucleobase to form an anion radical, $(\text{nucleobase})^{\bullet-}$. Subsequently, the anion may undergo a nonadiabatic transition to a dissociative σ^* state involving a sugar–phosphate C—O bond,[82,14] and theoretical estimates of the nonadiabatic transition rate suggest that this process is feasible within the lifetime of the temporary anion resonance.[75,77,74,78] This mechanism is consistent with experimental results for gas-phase deoxyribose, in which near-zero kinetic energy electrons were found to dissociate C—O bonds of the sugar.[87] However, direct experimental identification of these putative $(\sigma^*)^-$ temporary anion resonances remains debatable,[21,88,23,26,89] because they tend to have shorter lifetimes as compared to $(\pi^*)^-$ resonances and are thus subject to a greater degree of lifetime broadening.

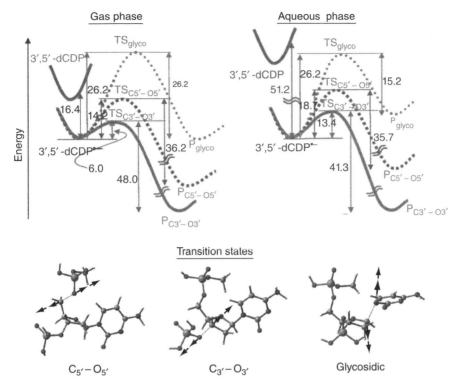

Figure 3 Pathways for three possible single-strand breaks in the radical anion of 2′-deoxycytidine-3′,5′-diphosphate, dCDP•−. The upper panels show computed minimum-energy pathways in the gas phase and in a polarizable continuum model of aqueous solvation, while the lower panels depict the calculated transition states corresponding to rupture of the 5′ sugar—phosphate C—O bond, the 3′ sugar—phosphate C—O bond, or the glycosidic C—N bond, with arrows to emphasize the direction of bond scission. Adapted with permission from Ref. 90; copyright 2010 Oxford University Press.

Figure 3 shows some computed minimum-energy pathways for single bond ruptures in the radical anion of the nucleoside deoxycytidine diphosphate, dCDP•−.[90] Pathways leading to dissociation of either the 3′ or the 5′ sugar–phosphate C—O bond, and also the glycosidic C—N bond, have been located, but the computed energetics suggest that the latter pathway is inaccessible with the energy available from forming dCDP•− from its dCDP precursor. The lower barrier of the 3′ pathway relative to the 5′ pathway and the high barrier to C—N bond cleavage are all qualitatively consistent with experimental studies in which DNA was bombarded by electrons in the 6–15 eV range.[80,81] In those experiments, the relative yield of 3′ cytidine strand breaks to 5′ strand breaks increased significantly as the energy of the incident electrons was decreased, and products corresponding to glycosidic bond cleavage were not detected.[80,81]

A possible precursor for DEA reactions occurring in condensed media is the solvated electron, and as such this species (in various solvents, but especially water) has been studied extensively.[91,17,18,15,16,13] Solvated electrons are generated in high yield by radiolysis of the solvent,[13] but are also generated by photoionization of common UV chromophores such as indole,[92] the chromophore in tryptophan. In that case, geminate recombination of the ion pair (a solvated electron and an indole cation radical) is slower than the diffusion limit,[92] such that these photogenerated solvated electrons may have a chemical role to play in solution-phase photochemistry. In the gas phase, finite cluster analogues of the solvated electron serve as interesting model systems for understanding how a solvent accommodates an excess charge.[93–95,16,19]

Challenges for Theory

Systems containing loosely-bound electrons pose special challenges for electronic structure calculations that are absent in calculations of cations, neutral molecules, or even strongly-bound valence anions. For an anion, separation of an electron from the molecular framework leaves behind a charge neutral molecule or radical, so that an "outgoing" electron does not feel a $-1/r$ Coulomb potential, as it would were an electron removed from a neutral molecule, leaving a cationic core. Rather, the long-range electron–molecule potential decays as $-1/r^2$ (charge-dipole) or faster.[2,3] The result, which is discussed quantitatively later in this chapter, is that the anion's electron density is significantly more diffuse as compared to that of a neutral molecule or cation. This is true for strongly-bound anions as well as weakly-bound ones, but the radial extent of the electron density increases exponentially as the electron binding energy decreases. This places special demands on the basis sets that are used to describe weakly-bound anions.

In addition, for a weakly-bound anion it may be necessary to consider the possibility that the molecule's vibrational motion could access regions of the potential energy surface where the anion is no longer thermodynamically stable with respect to electron ejection (autodetachment). For weakly-bound anions in the gas phase, autodetachment induced by rotational motion may also need to be considered.[2]

In terms of the level of electronic structure theory that is required, calculation of EAs, or, in other words, electron detachment energies for anions, tends to be more demanding than calculation of ionization potentials (IPs), if for no other reason than that the former tend to be smaller in magnitude than the latter. As evidence for this, one need look no further than the periodic table: atomic EAs are bounded above by that of chlorine, at 3.6 eV,[1] whereas atomic IPs range from 5.4 eV for Li up to 17.4 eV for fluorine and larger still for the noble gases.[96] A few examples for atoms and small molecules are shown in Table 1, illustrating that IPs tend to be $\gtrsim 8$–10 eV whereas EAs are typically ~ 3 eV. (Even for molecules, a comprehensive review of experimental EAs from photoelectron spectroscopy reveals only a very few examples larger than 4–5 eV.[43])

Table 1 Experimental Adiabatic Electron Affinities (AEAs) and Ionization Potentials (IPs) for Some Atoms and Small Molecules[a]

	IP (eV)	EA (eV)		IP (eV)	EA (eV)		IP (eV)	EA (eV)
C	11.26	1.26	OH	12.97	1.83	S_2	9.37	1.70
Cl	12.97	3.62	P	10.49	0.75	SH	10.36	2.31
Cl_2	11.50	2.41	PH	10.15	1.01	Si	8.15	1.38
O	13.61	1.46	PH_2	9.81	1.27			
O_2	12.09	0.47	S	10.36	2.08			

Reprinted with permission from Ref. 97; copyright 2003 American Chemical Society.
[a]This is the "EA13/03" database of Ref. 97 and estimated vibrational zero-point energies have been removed from the experimental data.

Moreover, EAs are *intensive* quantities, as are IPs, whereas both the total electronic energy and the correlation energy are extensive. Thus, as molecular size increases the EA that one is attempting to calculate represents a diminishing fraction of the total energy.[98,99,2] Simons[2] has used this fact to argue in favor of the so-called *equation-of-motion* (EOM) methods[98,99] (also known as *Green's function*[100] or *electron propagator*[101-104] methods), in which EAs and IPs are computed directly, in a single calculation, not evaluated as an energy difference. This is accomplished by means of a perturbative or cluster-type expansion of the EA or IP itself.

On the other hand, the correlation energy is *always* growing with system size, and therefore one *always* faces the problem that intensive energy differences such as bond dissociation energies or barrier heights are shrinking in comparison to the total correlation energy, as molecular size increases. The problem is intrinsic to large-molecule quantum chemistry and is the reason why only size-extensive methods such as coupled-cluster (CC) theory and many-body perturbation theory are appropriate for large systems. Methods that lack size-extensivity, such as truncated configuration interaction approaches, will recover a diminishing fraction of the correlation energy as the number of electrons increases. In the context of EA or IP calculations, the key is to use size-extensive quantum chemistry methods that are carefully calibrated to provide a balanced description of both the neutral and the ionized molecule.

On the topic of electron correlation and balanced approximations, another challenge in the application of quantum chemistry to weakly-bound anions is that the "zeroth order" estimate of the electron affinity, namely, the so-called *Koopmans' theorem* (KT) estimate[35] (EA $\approx -\varepsilon_{LUMO}$) is typically a worse approximation than is the analogous KT estimate for the IP (IP $\approx -\varepsilon_{HOMO}$). This has to do with an error cancellation in the latter estimate that is not present in the former, where errors arising from neglect of electron correlation and neglect of orbital relaxation have the same sign, whereas for IPs these errors have opposite signs. This underscores the need for high-level, correlated descriptions of molecular anions, which adds to the cost. Special problems in the density-functional description of anions, arising from self-interaction associated with the half-filled orbital of an extra-valence anion, complicate matters further, although the present situation is much better than it has been in the past, due to recent progress in functional development.

Finally, the calculation of temporary anion resonances poses a challenge for quantum chemistry because the metastable anion M^- lies higher in energy (at a fixed molecular geometry) as compared to $M + e^-$, where the electron is an outgoing plane wave. Attempts to compute the energy of M^- using standard, bound-state quantum chemistry methods are therefore susceptible to *variational collapse*, wherein the wave function for what is ostensibly M^- collapses to the wave function for $M + e^-$ as the basis set approaches completeness.[20] Finite-basis calculations thus afford a deceptively seductive but ultimately unrealistic description of M^-,[2] and modifications to standard quantum chemical methods are required to compute accurate energetics for the metastable anion.[20]

TERMINOLOGY AND FUNDAMENTAL CONCEPTS

This section defines and explains some basic chemical and quantum-mechanical concepts concerning anions. Bound anions (where M^- is lower in energy than M, at the minimum-energy geometry of the former) are considered first, and subsequently we discuss metastable anions, also known as temporary anion resonances. For easy reference, a list of acronyms is provided at the end of this chapter.

Bound Anions

Attachment and Detachment Energies

We begin with a careful exposition of the various energy differences associated with attaching or removing an electron. For negative ions in general, one of the key experimental observables that is directly accessible from *ab initio* calculations is the *vertical detachment energy* (VDE), sometimes called the *vertical electron binding energy*. This quantity is defined pictorially in Figure 1 as the energy gap between the ground-state energy of the anion (call it M^-) at its equilibrium geometry and the value of the neutral molecule's potential energy surface at the anion geometry. Note that the ground-state energy of the anion should properly include the anion's zero-point vibrational energy, as indicated in Figure 1. That said, a vertical transition from the anion's minimum-energy geometry need not land on a vibrational state of the neutral molecule, potentially leading to the appearance of a vibrational progression in the anion's photoelectron spectrum. In the example of Figure 1, the $v = 0$, $v = 1$, and $v = 2$ states of AB are likely to be accessed via photodetachment from the $v = 0$ state of AB^-.

Insofar as one is able to map out potential energy surfaces for both M and M⁻, one could compute such a vibrational progression by evaluating the appropriate Franck–Condon factors, $\langle \psi_{anion}(v = 0)|\psi_{neutral}(v)\rangle$. On the other hand, taking a classical-mechanical view of the nuclear motion, it is often convenient to define the VDE as a continuously varying function of molecular geometry:

$$VDE(\mathbf{R}) = E_{neutral}(\mathbf{R}) - E_{anion}(\mathbf{R}) \qquad [2]$$

Here, both energies are evaluated at the same geometry, $\mathbf{R}$. For $\mathbf{R} = \mathbf{R}_{anion}$, this definition coincides with the "experimental" definition of the VDE that is suggested in Figure 1, provided that zero-point corrections are included in $E_{anion}(\mathbf{R}_{anion})$, by computing vibrational frequencies for the anion. However, we allow for arbitrary $\mathbf{R}$ in the definition used here, since Eq. [2] can be used to sample the VDE along a molecular dynamics trajectory. In such a calculation, a histogram of the fluctuations in $VDE(\mathbf{R})$ provides a semiclassical explanation for the width of the photoelectron spectrum.[105] The semiclassical picture suggests that Eq. [2] can be used to find and delineate regions of the anion's potential energy surface where the electron would be expected to autodetach, that is, regions where $VDE(\mathbf{R}) \leq 0$.

Note that Eq. [2] defines the VDE, which excludes relaxation of the neutral species following electron detachment. Including that relaxation energy defines the *adiabatic* electron detachment energy, which is more often called the *adiabatic electron affinity* (AEA),

$$AEA = E_{neutral}(\mathbf{R}_{neutral}) - E_{anion}(\mathbf{R}_{anion}) \qquad [3]$$

The AEA is also depicted pictorially in Figure 1, as is the vertical *attachment* energy (VAE), which is defined analogously to the VDE but starting from the ground state of the neutral molecule:

$$VAE = E_{anion}(\mathbf{R}_{neutral}) - E_{neutral}(\mathbf{R}_{neutral}) \qquad [4]$$

One note of caution about terminology: the phrase "electron affinity" by itself (as opposed to AEA) is used somewhat ambiguously in the literature, in the sense that the EA in question might be the AEA, or it could be the EA for a vertical process. We caution against this ambiguous usage. In contrast, the vertical EA and the VDE are two different names for precisely the same energetic quantity, and the use of one term over the other is simply a matter of taste.

In principle, the AEA can be determined experimentally from the onset of the photoelectron spectrum.[43] In practice, however, large differences between the anion and neutral geometries, arising from electron penetration into antibonding orbitals of the neutral molecular framework, leading to reduced formal bond orders in the anion, can give rise to unfavorable Franck–Condon factors that make the AEA difficult to determine.[2] Indeed, the schematic potential energy surfaces in Figure 1 suggest that the M⁻($v = 0$) → M($v = 0$)

transition will be difficult to locate, experimentally. (The paucity of bound excited states for anions also limits the available experimental techniques to determine EAs.[2]) In such cases, theoretical calculations may be the only means to determine the AEA. In cases where the photoelectron spectrum cannot be vibrationally resolved and fit to a Franck–Condon progression, the VDE is usually taken to be the location of the maximum spectral intensity.

Classification of Molecular Anions

The diffuse nature of an anion's electron density can be understood by examining the asymptotic behavior of the potential energy function for removing an electron from either a neutral molecule, M, or else an anion, M$^-$. Such potential functions are illustrated schematically in Figure 4(a), and we consider the neutral case first. Ionization of the neutral molecule leaves behind a cationic core and thus asymptotically the potential energy function for this process looks like an attractive Coulomb potential between the outgoing electron and the cation, $V(r) \sim -1/r$. (At short range, the potential ultimately becomes repulsive due to the other valence electrons.) This deep potential well can support a large number of bound states, whose energies can be fit to the formula

$$E_n = -\frac{Z_{\text{eff}}R_H}{(n - \delta)^2} \qquad [5]$$

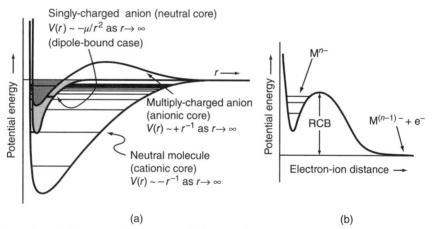

(a)	(b)

Figure 4   (a) Schematic depictions of the one-electron potential energy function for removing a valence electron from a neutral molecule, a singly-charged anion, or a (stable) doubly-charged anion. The bound-state energy levels represent Rydberg states, and although the figure is qualitative it correctly suggests that anions typically have very few bound Rydberg states. (b) Alternative potential for electron ejection from a multiply-charged anion M^{n-}, in the case where this species is metastable, being trapped behind a repulsive Coulomb barrier (RCB). Panel (a) is based on a similar figure in Ref. 2; copyright 2008 American Chemical Society. Panel (b) is based on a figure in Ref. 106; copyright 2000 American Chemical Society.

where Z_{eff} is some effective charge, $R_H = 13.6$ eV is the Rydberg constant, and δ is known as the *quantum defect*.[107] In analogy to the Rydberg series for the hydrogen atom, these states are known as *Rydberg states*, and it is clear from the figure that they constitute a series of increasingly diffuse excitations (as n increases) of one electron about a cationic core.

Removing an electron from M^- leaves a charge-neutral core. The monopole therefore vanishes in a multipole expansion of the electron–molecule $(M + e^-)$ Coulomb interaction, hence there is no strongly attractive $-1/r$ potential at long range, as there is in the $M^+ + e^-$ case. The long-range form of the potential for $M + e^-$ reflects higher order charge–multipole interactions, the longest range of which is the charge–dipole interaction, $V(r) \sim -\mu/r^2$, if M possesses a nonzero dipole moment, μ. [If M has no dipole moment but a nonzero quadrupole moment, Q, then the asymptotic form of the potential is a charge–quadrupole interaction, $V(r) \sim -Q/r^3$.] The $M + e^-$ potential well in Figure 4(a) is much shallower than that for $M^+ + e^-$, consistent with the observation that IPs for neutral atoms and molecules are large compared to EAs. As such, the anion typically possesses few (if any) bound Rydberg states.[2]

In view of this discussion, it is tempting to conceptualize the weakly-bound anions of polar molecules as *dipole-bound* anions.[11] Solution of the Schrödinger equation for an electron interacting with a point dipole reveals that bound states are obtained for dipole moments $\mu > 1.625$ debye,[108,109] with no further molecular structure required. (In practice this threshold should be modified to something like $\mu \gtrsim 2.4$ debye,[11] owing to the possibility of rotational-to-electronic energy transfer,[110–112] but the point remains that a sufficiently large dipole moment alone is enough to bind an electron.) As discussed in detail in Ref. 2, however, it is difficult to fully disentangle the long-range charge–dipole interactions from shorter-range valence-type interactions involving the Coulomb and exchange potentials established by the occupied MOs, which can also stabilize electron binding. A detailed mathematical analysis of these valence interactions has been given by Simons.[42,3]

As a result of these competing interactions, the distinction between a valence anion and a dipole-bound anion is sometimes ambiguous. For example, the acetonitrile (CH_3CN) molecule has a calculated dipole moment of 3.94 debye,[113] well above the threshold value, and an experimental VDE of 0.012 eV.[114] These values might suggest that electron binding will be extremely weak unless the dipole moment is extremely large, since CH_3CN already has a sizable dipole moment as small molecules go, yet barely binds an extra electron. At the same time, the $(BeO)_2^-$ anion has been classified as quadrupole-bound,[115] since the dipole moment of the D_{2h} $(BeO)_2$ framework vanishes by symmetry. One might therefore expect a smaller VDE for $(BeO_2)^-$ as compared to dipole-bound cases, yet the computed VDE for $(BeO)_2^-$ is 1.1 eV![115] Clearly, short-range valence attractions must contribute significantly to the stabilization of $(BeO)_2^-$.

Even in cases where it seems safe to classify M^- as a dipole-bound anion, the balance of long-range charge–dipole interactions and short-range valence interactions means that there is no clear correlation between the magnitude of the neutral molecule's dipole moment (call it μ_0) and the VDE of its anion. For example, whereas $\mu_0 = 3.94$ debye for CH_3CN^- (computed at the MP2 level),[113] and the experimental VDE of this anion is a mere 0.012 eV,[114] the water dimer anion exhibits a smaller value of μ_0 ($\lesssim 2.0$ debye at the MP2 level[116]), yet a *larger* experimental VDE (≈ 0.045 eV).[117,118]

The water dimer anion also provides a simple example of a system where vibrational motion can promote autodetachment. Figure 5 shows one-dimensional potential energy scans of $(H_2O)_2^-$ and $(H_2O)_2$ along the so-called "flap angle" that connects *cis* and *trans* isomers of the dimer. The *cis* isomer of the anion is stabilized by an enhanced dipole moment, but this isomer is destabilized in the neutral dimer owing to slightly larger steric repulsion. The ground vibrational state of $(H_2O)_2^-$ is bound by ≈ 0.045 eV, but the $v_{flap} = 1$ state is much closer to the neutral $v_{flap} = 0$ energy and might autodetach if nonadiabatic effects were considered. (The semiclassical picture is that motion along the flap angle coordinate may access the *trans* geometry, where autodetachment is more likely.)

Consider also the case of (uracil)$^-$. This anion exhibits both a diffuse, dipole-bound state and a valence anion state, the latter characterized by a

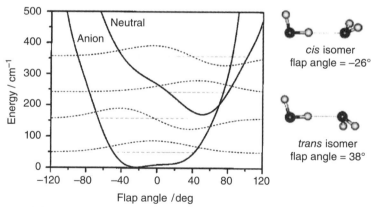

Figure 5   Solid curves: potential energy scans for $(H_2O)_2^-$ and $(H_2O)_2$ along the "flap angle" coordinate, with the other coordinates relaxed. (The *cis* and *trans* structures of the anion, shown at right, suggest the nature of the flap angle.) Calculations were performed at the CCSD(T) level with a large, diffuse basis set,[119] and harmonic zero-point corrections for all coordinates except the flap angle are included in these potentials. In the absence of zero-point corrections, both the *cis* and *trans* isomers of $(H_2O)_2^-$ are local minima, but the latter minimum (at 38°) disappears on zero-point correction. Broken curves: lowest two vibrational wave functions ($v_{flap} = 0$ and 1) for these one-dimensional potentials. Adapted with permission from Ref. 119; copyright 1999 American Institute of Physics.

half-filled π^* orbital that is bound fairly closely to the molecular framework. Calculations suggest that the two anion states are nearly isoenergetic, although the dipole-bound anion has a much smaller VDE.[120,121] This can be understood in terms of the schematic potential energy surfaces that are depicted in Figure 6(a). The dipole-bound state of (uracil)⁻ is classified as such because it has essentially the same geometry as neutral uracil, and the dipole-bound anion is characterized by a highly diffuse electron situated at the positive end of the neutral molecule's dipole moment, largely outside of the region of space occupied by neutral uracil's valence electrons. The photoelectron spectrum of (uracil)⁻ [Figure 6(b)] is typical of what is observed for a dipole-bound anion, namely, a single narrow peak, corresponding to the origin transition, with *much* weaker features at higher energies.[122] The lack of a significant Franck–Condon progression in this case is a consequence of the essentially identical geometries of uracil and dipole-bound (uracil)⁻. As such, the vibrational wave functions are nearly the same for both species, so when the anion is prepared in its ground

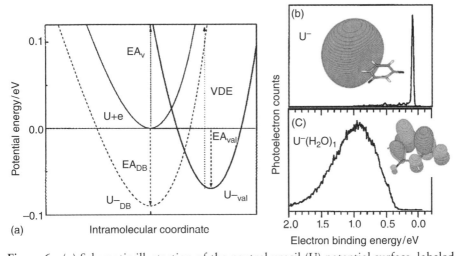

Figure 6 (a) Schematic illustration of the neutral uracil (U) potential surface, labeled U+e, the dipole-bound anion (U_{DB}^-) potential surface, and the valence anion (U_{val}^-) potential surface. The intramolecular coordinate involves out-of-plane displacement of the uracil ring atoms, leading to a nonplanar geometry for U_{val}^-. (b) Photoelectron spectrum of the dipole-bound state of U⁻, along with an isosurface plot of the singly occupied molecular orbital (SOMO) for U_{DB}^-. The nearly identical geometries for U and U_{DB}^- lead to a very narrow spectrum. (c) Photoelectron spectrum for [U(H₂O)]⁻ and isosurface plot of the SOMO for U_{val}^-. The spectrum illustrates both the larger binding energy of the valence anion, as compared to the dipole-bound state, as well as the much greater spectral width resulting from distortion of U⁻ away from a planar geometry. Panel (a) is reprinted with permission from Ref. 120; copyright 1998 American Chemical Society. Spectra in (b) and (c) are reprinted with permission from Ref. 122; copyright 1998 American Institute of Physics. Orbital isosurfaces plots are reprinted with permission from Ref. 123; copyright 2004 the PCCP Owner Societies.

vibrational state, only the origin transition is observed with any appreciable intensity.

In contrast, the valence anion character of the $(\pi^*)^-$ state can be deduced, computationally, from the nonplanar nature of the anion's geometry, which results from the lifting of aromaticity due to an odd number of π electrons. As a result of geometric distortion, the VDE of the valence anion is significantly larger than that of the dipole-bound state, as indicated in Figure 6(a), even though these two anions are very close in energy. Electron attachment to form the dipole-bound state, which can occur at or near the minimum-energy geometry of neutral uracil, has been suggested as the initial step in a DEA mechanism that involves subsequent internal conversion to the valence anion state, followed by bond cleavage.[124,125]

While theory predicts U_{DB}^- and U_{val}^- to be nearly isoenergetic, only the dipole-bound state is observed in the experiments reported in Ref. 122 [Figure 6(b)]. The thermodynamics of which isomers are stable provides no information about how these species are actually formed in a molecular beam experiment, and nonthermal ensembles are probably common in anion photoelectron spectroscopy.[126–129] In the experiments of Ref. 122, the valence anion state *was* observed (to the exclusion of the dipole-bound anion signal) in a complex with a single water molecule, $[(\text{uracil})(\text{H}_2\text{O})]^-$. The photoelectron spectrum of this complex [Figure 6(c)] is much broader, owing to the differences in the geometries of the anion and the neutral molecule that lead to nontrivial Franck–Condon factors, even if individual vibrational states of uracil cannot be resolved. Later, under different source conditions, the valence anion state *was* observed for bare uracil,[130] which highlights the fact that anion spectroscopy can be particularly sensitive to the source conditions of the molecular beam.

Finally, let us consider the case of multiply-charged anions, M^{n-}. One possible potential energy curve for a multiply-charged anion is sketched in Figure 4(a) for the case where M^{n-} is stable (lower in energy than $M^{(n-1)-} + e^-$). This potential is repulsive at medium-range distances since it correlates asymptotically to separating two negatively charged species. At short range, however, there may be stabilizing valence interactions leading to a local potential minimum. On the other hand, these stabilizing interactions may be insufficient to lower the potential well below the asymptotic $M^{(n-1)-} + e^-$ energy; Figure 4(b) depicts a case where they are not. In this case, the multiply-charged anion is metastable only. Insofar as M^{n-} can be formed, it exists only because it is trapped behind a repulsive Coulomb barrier (RCB). This species will persist only until such time as an electron is able to tunnel through the RCB.

Although multiply-charged anions are ubiquitous in polar solvents, where the internal Coulomb repulsion of M^{n-} is offset by highly favorable electrostatic and induction interactions with the solvent molecules, multiply-charged anions have historically been difficult to prepare in gas-phase experiments that could directly probe the $M^{(n-1)-} + e^-$ interaction potential. Recently, however, it has been demonstrated that such anions can be prepared using electrospray ionization,[131,132] and photoelectron spectra at different excitation energies place experimental bounds on the magnitude of the RCB. For the citrate trianion, $C_3H_5O(COO)_3^{3-}$, whose three negatively charged carboxylate moieties are separated by ≈ 6 Å, the result is 1.9 eV $<$ RCB < 2.5 eV.[133] Unbranched dicarboxylate dianions, $^-O_2C(CH_2)_nCO_2^-$, allow for systematic variation of the RCB, and these species are found to be stable on the timescale of a time-of-flight photoelectron spectroscopy experiment (~ 0.1 s) for $n \geq 3$.[134,135] This is consistent with back-of-the-envelope calculations suggesting that two negative charges separated by $\lesssim 4$ Å will be unstable;[2] in the $n = 3$ dicarboxylate, the two terminal carbon atoms are separated by ≈ 5 Å (assuming typical bond lengths), whereas in the $n = 2$ case the separation is < 4 Å. Interestingly, the $n = 2$ (succinate) dianion is rendered stable on complexation with just a single water molecule, and the photoelectron spectrum of this complex has been reported.[134,135] Calculations suggest that a complex containing 2–3 water molecules is necessary to stabilize HPO_4^{2-} but ~ 16 water molecules are required to stabilize PO_4^{3-}.[136]

Cluster Anions and Solvated Electrons

The examples of the succinate–H_2O complex and the uracil–H_2O valence anion state that were discussed earlier demonstrate that solvent molecules can play a critical role in anion binding, even in a gas-phase experiment, and thus cluster anion photoelectron spectroscopy[94,19] warrants some discussion in its own right. More dramatic examples of the role of solvent in electron binding belong to a broad class of systems known as *solvated electrons*. From a gas-phase point of view, one might define a solvated electron as any cluster anion M_N^- (or mixed cluster anion; the molecules need not all be the same) for which the molecular anion M^- is not a bound species, or at least is much more weakly-bound than is the cluster anion M_N^-. As such, electron binding is a collective phenomenon, and this collective binding can be quite strong for sizable clusters. For example, whereas $(H_2O)_2^-$ has a VDE of only 0.045 eV,[117,118] for $(H_2O)_N^-$ with $N \sim 80$–100, one can find isomers with VDEs > 2.0 eV.[4,5] Due to the collective nature of electron binding in clusters, the VDE is the primary observable of merit; AEAs are essentially impossible to determine for cluster anions, due to solvent reorganization on electron detachment. Indeed, it is not even clear whether the concept of an AEA is meaningful in such cases, since removing the solute (an electron) creates a qualitatively different system (a neat solvent cluster rather than a solvated-electron cluster).

Much of the interest in these clusters is driven by the importance of the aqueous electron in bulk water, e^-(aq). This species, along with $H^\bullet$ and $HO^\bullet$, is one of the primary radical intermediates that is formed on water radiolysis[84,17,50,18,13] and can be detected following radiolysis of ammonia, alcohols, and organic amines as well.[50] Figure 7 provides an overview of the sequence of events involved in water radiolysis, which is initiated either

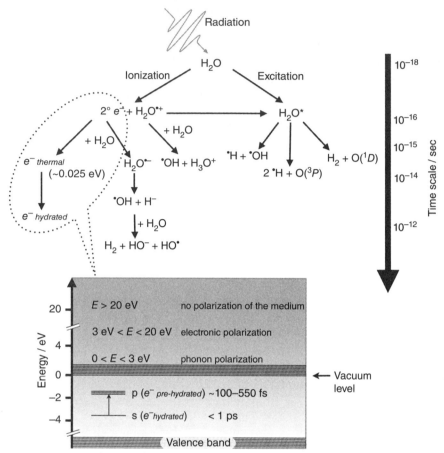

Figure 7 Upper scheme: schematic overview of water radiolysis, wherein ionizing radiation generates three primary radical intermediates, $e^-(aq)$, $H^\bullet$, and $HO^\bullet$. Lower panel: schematic energy diagram for an electron interacting with liquid water but modified to eliminate the conduction band of water in the lower panel, because other studies[137,138] indicate that the conduction band lies no more than ~ 0.1 eV below the vacuum level, whereas the original figure in Ref. 13 places the conduction band much lower in energy. (Adapted with permission from Ref. 13; copyright 2012 American Chemical Society.)

by electronically-excited water molecules or else water cation radicals that are generated either by absorption of ionizing radiation or else via collisions with high-energy "primary" electrons.[13] Following the track on the left in Figure 7, the initial ionization event leads to the formation of lower energy "secondary" (2°) electrons, which subsequently thermalize by dissipating energy into vibrational modes of the solvent and ultimately become solvated in a deep potential well. The depth of this well (relative to the vacuum level, corresponding to a free electron) can be estimated by extrapolation of VDEs for the highest-binding isomers of gas-phase $(H_2O)_N^-$ clusters. This extrapolation yields estimates of 3.4 eV[139] and ~4 eV,[5] depending on the experimental data set used. Direct experimental measurements of the VDE for e⁻(aq), using liquid microjet photoelectron spectroscopy,[140] afford values ranging from 3.3 to 3.7 eV.[7−10] A detailed theoretical calculation, including the effects of electronic relaxation on vertical detachment and extrapolating to an infinitely large simulation cell, affords a value of 3.7 eV.[141]

The lower panel of Figure 7 provides a rough energy-level structure for an "excess" electron in water that is consistent with the estimated VDEs quoted earlier. This diagram is labeled according to the conventional interpretation of the structure of e⁻(aq),[15,16,142,143] namely, that the ground state of the fully hydrated species can be conceptualized as a particle in a quasi-spherical solvent void, whose ground-state wave function exhibits pseudo-s symmetry. Within this picture, the states responsible for the strong electronic absorption at ~1.7 eV (720 nm), which is the most characteristic feature of e⁻(aq),[15] are quasi-degenerate p-type states, as suggested in the lower panel of Figure 7. Although this picture is quite entrenched and appears to be consistent with plane-wave DFT simulations performed in the liquid phase,[144−146] the cavity model has been questioned both historically[147−149] as well as recently.[150−156] Alternative structural models are a solvent–anion complex,[149] a HO⁻ ⋯ H_3O complex,[147,148] a hydronium radical ($H_3O^•$) that exhibits charge-separated biradical character on hydration,[150−153] and a delocalized wave function with a buildup of water density inside said wave function.[154,156] These alternative proposals have proved controversial,[157−162] and more work is needed to definitively resolve this question, because at present both the cavity and noncavity models explain *certain* features of the spectroscopy of the species called e⁻(aq), but each is inconsistent with other features.[15,162,156,163]

Cluster analogues of solvated electrons, M_N^-, have been studied extensively for a variety of polar and nonpolar solvents, as discussed in numerous recent reviews.[15,16,19,94,164] A key aspect of cluster ion spectroscopy is the attempt to determine how the solvent network accommodates the ionic solute, which in this case is an extra electron. Figure 8 shows VDE data, obtained from photoelectron spectroscopy, for some representative solvated-electron clusters. (Experimentally, the VDE is taken to be the energy at which the rather broad photoelectron spectrum is peaked.) Figure 8 employs the common practice of plotting cluster anion VDEs versus $N^{-1/3}$, where N denotes the cluster size. The reason for this convention is that the cluster radius, R, should be proportional to $N^{+1/3}$ for a spherical cluster, and by taking $R \propto N^{+1/3}$ in conjunction with a dielectric continuum treatment of the spherical solvent cluster, one obtains a VDE proportional to R^{-1} or $N^{-1/3}$.[167−170]

Let us consider this in a slightly more detail. Although the result $VDE(N) \propto N^{-1/3}$ can be derived from continuum dielectric theory under a variety of assumptions,[167−170] a simple model is the following. Consider a single point charge $-e$ centered in a cavity of fixed radius, a, representing the excluded volume of the solvent void inhabited by the solvated electron.

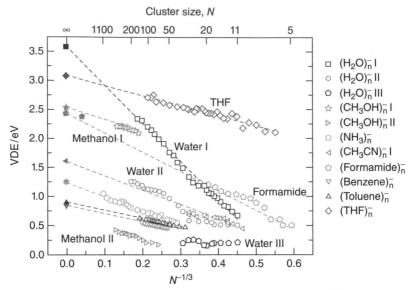

Figure 8   VDEs for various cluster anions M_N^- as a function of $N^{-1/3}$. For water, three distinct isomeric series are observed, which are labeled I, II, and III,[4] and for methanol, two isomers (I and II) are observed.[165] For acetonitrile, two isomers are observed[166] but only one is plotted here. Reprinted with permission from Ref. 19; copyright 2012 American Chemical Society.

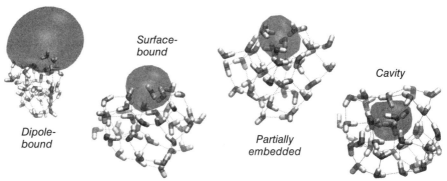

Figure 9   Illustration of the four distinct electron binding motifs identified in $(H_2O)_N^-$ clusters on the basis of one-electron QM MM simulations, for the case $N = 40$. The isosurfaces that are depicted encapsulate 70% of $|\psi|^2$. Reprinted from Ref. 95; copyright 2011 American Chemical Society.

[See, for example, the cavity isomer of $(H_2O)_{40}^-$, at the far right in Figure 9.] This cavity is assumed to be carved out of a spherical, homogeneous dielectric material whose radius is R and is characterized by a dielectric constant, ε_s. The solvation (free) energy for the $-e$ charge is the VDE for this model. The mathematical result, for $R \gg a$ and given in atomic units, is[170]

$$\text{VDE}(R) = \text{VDE}(\infty) - \frac{1}{2R} \left(1 + \frac{1}{\varepsilon_\infty} - \frac{2}{\varepsilon_s} \right) \qquad [6]$$

This leads immediately to the desired result:

$$\text{VDE}(N) = \text{VDE}(\infty) + AN^{-1/3} \qquad [7]$$

Experimentally, the parameter A corresponds to the slope of one of the best-fit extrapolations in Figure 8. It should be noted that the $N^{-1/3}$ proportionality in Eq. [7] can be derived *without* the assumption that the electron inhabits a cavity, provided only that the electron's wave function is spatially localized.[167–169]

The two dielectric constants in Eq. [6] warrant some discussion. The quantity ε_s, which is sometimes called simply "the dielectric constant" (often denoted ε instead of ε_s) is more precisely the *static* or zero-frequency dielectric constant. (Even more precisely, it is the scalar electric permittivity relative to that of vacuum,[171] and is therefore dimensionless.) This quantity includes the effects of both orientational and electronic polarization. For a vertical ionization process, however, the solvent's orientational degrees of freedom are frozen, but the electron densities of individual solvent molecules can adjust on the same timescale on which the ionization process occurs. Such considerations lead to a correction involving the optical (infinite-frequency) dielectric constant, $\varepsilon_\infty = n_{\text{refr}}^2$, where n_{refr} denotes the solvent's index of refraction.[172]

Values of ε_s vary widely from one solvent to the next, due to a particular solvent's ability (or lack thereof) to reorient permanent dipole moments of individual solvent molecules. Considering some examples at 25 °C, we have $\varepsilon_s = 78$ (water), 21 (acetone), 4.8 (chloroform), and 2.3 (benzene). Optical dielectric constants, on the other hand, are much more similar between different solvents: $\varepsilon_\infty = 1.78$ (water), 1.85 (acetone), 2.09 (chloroform), and 2.25 (benzene). From Eq. [6], one might anticipate that electronic polarization effects are most significant in polar solvents, for which $\varepsilon_s \gg \varepsilon_\infty$, and this is indeed the case. For electron solvation in bulk water, for example, continuum models more sophisticated than Eq. [6], as well as atomistic simulations using a polarizable solvent model, afford an estimate of ≈ 1.4 eV for the *electronic reorganization energy* that accompanies vertical detachment.[141] In other words, the predicted VDE is reduced by ≈ 1.4 eV (and brought into quantitative agreement with experiment[141]) when the solvent's electronic degrees of freedom are allowed to adjust to electron detachment, at fixed nuclear positions.

Returning to the VDE data in Figure 8, one can see that methanol cluster anions exhibit two distinct isomeric series. This implies that there are two peaks in the photoelectron spectra, whose peak positions exhibit different slopes, A, with respect to $N^{-1/3}$. (The relative intensities of the two peaks can be changed by modifying the backing pressure of the carrier gas, which effectively changes the temperature of the molecular beam, thus demonstrating that the two features are indeed distinct isomers.[4,19]) On the basis of theoretical calculations, these two isomeric series have been attributed to a surface-bound electron and a cavity-bound electron.[16] Note, however, that *both* the "methanol I" and "methanol II" data series in Figure 8 appear to be linear as a function of $N^{-1/3}$. In fact, one can construct a continuum electrostatics model such that, in large clusters, the VDE for a *surface-bound* electron is also proportional to $N^{-1/3}$.[170] Thus, one cannot directly infer structure from the linearity of the data in Figure 8.

Caution is warranted especially in the case of water cluster anions, as there has been much debate regarding the nature of the electron binding motifs in these systems.[4,5,93,95,173-177] Water cluster anions exhibit at least three[4] – and possibly four[5] – distinct isomeric series, depending on experimental conditions, and it is not altogether clear that the "water I" data in Figure 8 actually correspond to a single isomeric species.[5,95] In a recent study that relies on one-electron quantum mechanics/molecular mechanics (QM/MM) calculations using an electron–water pseudopotential model,[141,95] our group has suggested four distinct binding motifs,[95] examples of which are illustrated in Figure 9 for the case of $(H_2O)_{40}^-$. From left to right in the figure, the four categories of isomers that we have identified are the following:

- A weakly dipole-bound cluster anion that predominates when an extra electron is attached to a cold, equilibrated, neutral water cluster.
- A surface-bound electron with a somewhat larger VDE (and correspondingly less diffuse wave function) that can be reached from the dipole-bound state following modest rearrangement of some surface water molecules, so that several dangling O—H moieties coordinate to the e^- wave function.
- A considerably more strongly-bound "partially embedded" surface isomer. In this case, rearrangement of water molecules at the cluster surface has resulted in a partial solvation shell.
- A fully internalized, cavity-bound isomer.

VDEs computed on the basis of quantum classical molecular dynamics simulations at finite temperature correctly reproduce the N-dependent trends observed experimentally and are in semiquantitative agreement with the absolute VDEs determined from photoelectron spectroscopy[4,5] for cluster anions in the size range $20 \leq N \leq 200$.[95] These simulations suggest that the "water I" data in Figure 8 may actually represent a transition between a surface-bound and a partially embedded isomer, whereas somewhat higher-binding cavity isomers are observed (according to the calculations) only in certain experiments,[5] where the clusters are annealed in an ion trap prior to interrogation.[95] On the other hand, the predictions of one-electron models have been shown to be somewhat sensitive to the fine details of how the electron–molecule pseudopotential is constructed.[15,159-161] As such, it is probably too early to say that the identities of the isomers in $(H_2O)_N^-$ photoelectron spectroscopy have been definitely determined.

One final example of a solvated-electron cluster is that of acetamide cluster anions, $(CH_3CONH_2)_N^-$. For these clusters, photoelectron spectra reveal the presence of two isomers, and on the basis of calculations these are attributed to the two-electron binding motifs depicted in Figure 10.[178] One of these isomers consists of coplanar, ladder-like arrangements of acetamide units, with electron binding occurring at the free (non-hydrogen-bonded) N—H moieties. The other isomer consists of a folded form that one might begin to call a solvated electron. One might anticipate that the highly-ordered ladder-like isomers would be present in small clusters only, with the more globular structures prevailing in larger clusters, but precisely the opposite is observed in the experiments. For $n \geq 9$, the feature that is attributed to the folded isomer is entirely absent.[178]

All of these examples underscore the need for theoretical calculations to aid in the identification of various solvated-electron isomers and binding motifs.

Metastable (Resonance) Anions

The discussion earlier implicitly assumes that the anion M^- is bound, that is, lower in energy than the neutral molecule M at the geometry of M^-. At the heart of dissociative electron detachment, however, are temporary anion resonances that are metastable only with respect to autodetachment. This is the case, for example, when the anion M^- is formed at the neutral molecule's geometry in the example depicted in Figure 1. Here, the anion is higher in energy at the neutral molecule's most stable geometry.

To motivate the discussion of temporary anion resonances, we first discuss the basic quantum mechanics of the resonance phenomenon, using a piecewise constant potential that facilitates analytic results. This is a standard graduate-level quantum mechanics exercise, but the results should be qualitatively informative to readers who have not seen them.

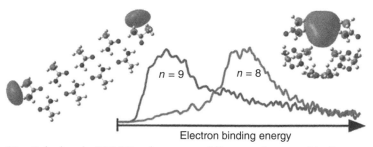

Figure 10 Calculated SOMOs for two different electron binding motifs in $(acetamide)_9^-$, on the left, and $(acetamide)_8^-$, on the right, superimposed with the corresponding photoelectron spectra. Reprinted from Ref. 178; copyright 2012 American Chemical Society.

Basic Quantum Mechanics

Consider the symmetric double square barrier potential that is plotted in Figure 11:

$$V(x) = \begin{cases} 0, & x < -\frac{1}{2}L - w \\ V_0, & -\frac{1}{2}L - w \le x \le -\frac{1}{2}L \\ 0, & -\frac{1}{2}L < x < \frac{1}{2}L \\ V_0, & \frac{1}{2}L \le x \le \frac{1}{2}L + w \\ 0, & x > \frac{1}{2}L + w \end{cases} \qquad [8]$$

For a particle incident from the left, the wave function should have the form

$$\psi_I(x) = A_I \exp(ik_I x) + B_I \exp(-ik_I x) \qquad [9]$$

in region I, and the outgoing wave function should have the form

$$\psi_V(x) = A_V \exp(ik_V x) \qquad [10]$$

in region V.

The one-dimensional *probability current density* is defined as[179]

$$j[\psi] = \frac{i\hbar}{2m} \left(\psi \frac{\partial \psi^*}{\partial x} - \psi^* \frac{\partial \psi}{\partial x} \right) \qquad [11]$$

For a wave function of the form $\psi(x) = A \exp(ikx)$, one obtains $j = |A|^2(\hbar k/m)$. As such, the *transmission probability*, T, for transit through the double-barrier in Figure 11 is appropriately defined as the ratio of the

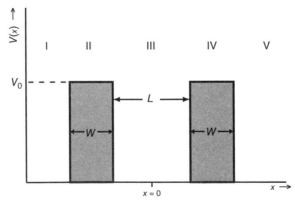

Figure 11 Potential energy function for a symmetric double square barrier. The regions labeled I–V represent the different piecewise constant values of $V(x)$.

probability flux exiting the rightmost extent of the potential ($x = L/2 + w$) to that entering the left-most extent of the potential ($x = -L/2 - w$):

$$T = \frac{j[A_V \exp(ik_V x)]}{j[A_I \exp(ik_I x)]} = \frac{k_V |A_V|^2}{k_I |A_I|^2} \tag{12}$$

This ratio is dependent on the energy of the incident particle, $E = (\hbar k_I)^2 / 2m$, and can be obtained by matching piecewise functions $\psi(x) = A \exp(\pm ikx)$ and their first derivatives in regions I–V, where each piecewise function is obligated to satisfy the time-independent Schrödinger equation. For the potential in Figure 11, and for solutions with $E < V_0$, the result is[180,181]

$$T(E) = \left[1 + \frac{V_0^2 M^2 \sinh^2 (\alpha w)}{4E^2 (V_0 - E)^2} \right]^{-1} \tag{13}$$

where

$$\alpha = \frac{1}{\hbar} \sqrt{2m(V_0 - E)} \tag{14}$$

is the inverse tunneling wavelength into the classically forbidden barrier regions and

$$M(E) = 2(\sqrt{E(V_0 - E)}) \cosh(\alpha w) \cos(kL) - (2E - V_0) \sinh(\alpha w) \sin(kL) \tag{15}$$

The incident wave vector, $k = (2mE)^{1/2}/\hbar$, is set by the energy of the incident particle, E.

The function $T(E)$ in Eq. [13] is plotted in Figure 12 for several values of w and L but a common value of V_0. When the barrier is wide ($w = 10$ bohr in Figure 12), there is very little transmission except around a sequence of narrow resonances, for which $T(E) \rightarrow 1$; these occur as $M(E) \rightarrow 0$. When the tunneling length scale α^{-1} is small compared to the barrier width w (which is the case for this particular set of parameters), then $\tanh(\alpha w) \approx 1$, and the condition $M(E) = 0$ becomes

$$\frac{2\sqrt{E(V_0 - E)}}{2E - V_0} = \tan(kL) \tag{16}$$

This is similar to the condition that defines the bound-state energy levels for a particle trapped in a finite square well of width L and depth V_0.[180] Indeed, if we narrow the barriers to $w = 4$ bohr, but leave the distance between them unchanged at $L = 20$ bohr, then the resonances in $T(E)$ broaden, due to additional tunneling through the narrower barriers, yet these resonances remain peaked around the same particle-in-a-square-well energy levels. On the other hand, decreasing L shifts these resonances to higher energies and results in a smaller number of them, consistent with the behavior of the square-well energy levels as a function of L, for a square-well potential with a finite binding energy, V_0.

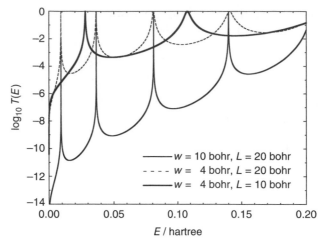

Figure 12 Transmission probability, $T(E)$, through the double-barrier potential in Figure 11, as a function of the incident particle's kinetic energy. The particle has the mass of an electron, and the barrier height is $V_0 = 0.2$ hartree.

This simple, analytically-solvable example captures the essence of the resonance phenomenon: incident particle energies that match, or nearly match, bound-state energy levels of a particular potential can become trapped, even if such states are asymptotically unstable. In a time-dependent picture, an incident particle whose kinetic energy matches a bound-state energy level can be captured for some finite period of time, before ultimately tunneling out of this potential. Importantly, the incoming particle's wave function e^{ikx} must be precisely matched to the wave function inside the potential well (or double-barrier, in this example), as shown in Figure 13.

Classification of Temporary Anion Resonances

Temporary anion resonances can be broadly classified according to two criteria. First, does the electron attach to the ground state of the molecule M, or is M excited in the process? If M remains in its ground state, then the resonance is classified as a *single-particle resonance*, since excitation of M's electrons can be ignored in a qualitative treatment. In contrast, a *core-excited* or *target-excited* resonance involves electronic excitation of M, for example,

$$e^- + M[(\pi)^2(\pi^*)^0] \rightarrow M^-[(\pi)^1(\pi^*)^2] \quad [17]$$

which provides a mechanism for the attachment of higher-energy electrons, since the M $\rightarrow$ M* excitation of the molecular "core" serves as a sink for electron kinetic energy. The second criterion is whether the separated species (M + e$^-$) are higher or lower in energy as compared to their complex (M$^-$). In other words, is the AEA of the molecule, M, positive or negative? If M$^-$ lies higher in energy, the resonance is classified as a *shape resonance* whereas if M$^-$ lies below

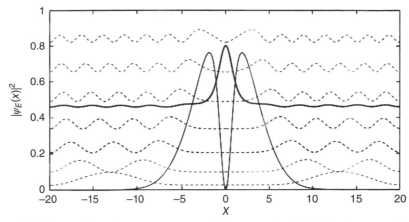

Figure 13 Illustration of a resonance state for a smooth double-barrier potential. At most energies $E = (\hbar k)^2/2m$, the continuum states $\psi_E(x)$ are affected by the presence of the barriers but not in a way that prevents significant transmission of the particle from left to right, or that results in significant accumulation of probability between the barriers. Such states are indicated by the dashed plots of $|\psi_E(x)|^2$. For certain incident energies, however, a resonance is observed (solid wave function). The vertical axis plots the particle's probability density, $|\psi_E(x)|^2$, but the plots are offset vertically according to the value of E. Reprinted from Ref. 182; copyright 2012 Elsevier.

$M + e^-$ it is classified as a *Feshbach resonance*. These definitions are most easily understood using simple one-dimensional potentials, as discussed later.

We first discuss the origin of anion shape resonances, which can be understood as follows. From the Schrödinger equation for the hydrogen atom, one learns that the electron experiences an effective radial potential of the form

$$U_{\text{eff}}(r) = U(r) + \frac{\hbar^2 \ell(\ell+1)}{2\mu r^2} \qquad [18]$$

where $U(r) = -1/r$ is the bare Coulomb potential (in atomic units), and the second term is a *centrifugal potential* arising from the conservation of angular momentum. The same form of the potential can be expected if we try to add an electron to a MO with angular momentum quantum number ℓ, except that in the molecular case the potential $U(r)$ would involve Coulomb and exchange operators for the valence electrons, averaged over MOs.[3] An effective electron–molecule potential with correct asymptotic behavior is sketched in Figure 14(a). A bound state of this potential with energy level ε is actually only metastable, in the sense that it lies above the asymptotic $(M + e^-)$ value of the potential but may be trapped temporarily by the centrifugal potential, if the attachment process involves an orbital with nonzero angular momentum [$\ell > 2$, in the example of Figure 14(a)]. Whether or not this occurs depends sensitively on the details of $U(r)$; for example, one may add an electron to a π^* orbital of O_2 to obtain a *bound* O_2^- anion, yet the lowest $(\pi^*)^-$ state of N_2^- is

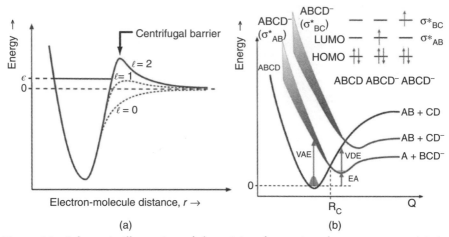

Figure 14   Schematic illustration of the origin of an anion shape resonance. (a) An electron–molecule interaction potential, illustrating the centrifugal barrier for various values of the angular momentum quantum number, ℓ. (b) Single-particle shape resonances arising from electron attachment to either the LUMO or LUMO+1 of molecule ABCD. Both orbitals are assumed to be of σ^* type, and the coordinate Q represents stretching of either the A—B or B—C bond. Shaded regions indicate energy widths arising from lifetime broadening. Adapted from Ref. 83; copyright 2008 Elsevier.

unbound, even though the centrifugal potential corresponds to $\ell = 1$ in either case.[2] As always, much of the beauty as well as the complexity of molecular physics lies in the fact that molecules are not all the same; subtle shifts in the energy-level structure afford qualitatively different properties.

Shape resonances are cases where the species in question (M^-) lies above its own continuum ($M + e^-$), but is trapped behind some barrier. [Note in Figure 14(b) that M^- is higher in energy than $M + e^-$ at the geometry of the neutral molecule, M.] Although we have stated this definition in terms of an anion shape resonance, cases not involving anions exist as well. For example, Ar_2 exhibits "orbiting" shape resonances, as do some simple atom–diatom (A + BC) scattering experiments.[183] In these cases, the rotational energy of the complex lies above the asymptotic dissociation threshold, yet the complex is trapped behind a centrifugal barrier arising from the orbital angular momentum.

For temporary anion resonances, VAEs can be measured experimentally by means of electron transmission spectroscopy,[2,184,185] in which an atomic or molecular sample is bombarded by a beam of electrons having well-defined kinetic energy. A change in current, due to attenuation of the electron beam, can then be detected as the kinetic energy of the electrons is tuned through a resonant VAE.

A simple one-dimensional picture of a single-particle anion shape resonance is depicted in Figure 14(b) for the hypothetical molecule ABCD. In this example, electron attachment to either the LUMO or the LUMO+1 of ABCD results in a temporary $(\sigma^*)^-$ anion. This species can decay either by autodetachment, which corresponds to the electron–molecule coordinate in Figure 14(a), or else by bond dissociation. Two different bond dissociation coordinates are mapped out in Figure 14(b), depending on which σ^* orbital captures the electron. Although qualitative, these sketches are realistic in the sense that the orbitals available for electron attachment in closed-shell molecules are typically antibonding orbitals with fairly small bond dissociation energies. In certain cases, these σ^* states might be dissociative,[82] meaning that they fail to bind any vibrational levels at all. Single-particle shape resonances tend to be found at low incident electron energies ($\lesssim 5$ eV), else M would likely be electronically excited by electron attachment. In addition, the presence of readily available exit channels, including both autodetachment and bond fission, means that shape resonances are typically very short-lived, with lifetimes on the order of $10^{-12} - 10^{-14}$ s,[12,83,185] and often closer to 10^{-14} s.[185]

As a result of these short lifetimes, electron attachment energies for anion shape resonances are considerably broadened by the time-energy uncertainty principle.[83,183] For a finite lifetime Δt, the resonance energy is subject to broadening according to[179]

$$(\Delta E)(\Delta t) \gtrsim h \qquad [19]$$

For $\Delta t = 1$ ps, this corresponds to a so-called "natural" line width $\Delta E \gtrsim 0.004$ eV, suggesting that a lifetime of ~ 1 ps constitutes an upper bound beyond which we need not worry too much about lifetime broadening. For $\Delta t = 10^{-14}$ s, however, the line width $\Delta E \gtrsim 0.4$ eV is certainly not negligible. As suggested by the diagrams in Figure 14(b), broadening increases as the separation between anion and neutral potential surfaces increases.

Figure 15 illustrates a pair of core-excited resonances, again involving the hypothetical molecule ABCD, which is now electronically excited on electron attachment. One of the temporary anion states suggested in the figure is a core-excited shape resonance and the other is a core-excited Feshbach resonance. The former is very much like the shape resonances discussed earlier, and thus we expect it to be short-lived. Since the resonance involves M*, however, the incident electron energies might be more like $\sim 5-15$ eV,[12] which can be understood as the $\sim 0-5$ eV of a single-particle shape resonance plus a typical molecular electronic excitation energy of something like $5-10$ eV. As a point of terminology, core-excited resonances involving electronic excitation are sometimes called *auto-ionizing* resonances, since M$^-$ in its excited electronic state is metastable with respect to autodetachment.

In the particular scenario illustrated in Figure 15, one obtains very different energetics depending on the orbital to which the electron attaches. For one particular electron-attached configuration, the anion ABCD*$^-$ lies below the neutral excited species, ABCD*, making this a Feshbach resonance. Because ABCD* is energetically inaccessible in this case, the autodetachment channel is closed and thus Feshbach resonances are sometimes called *closed-channel*

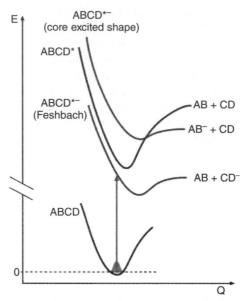

Figure 15   One-dimensional picture of a core-excited shape resonance and a core-excited Feshbach resonance. The two anion states differ in whether the electron attaches to an orbital associated with AB or with CD. Reprinted from Ref. 83; copyright 2008 Elsevier.

resonances. (Similarly, shape resonances are sometimes called *open channel resonances.*) Furthermore, the Feshbach resonance state cannot collapse to the ground state of neutral ABCD except via a two-electron process, whereas other decay mechanisms discussed so far for temporary anion resonances all involve one-electron processes. In a MO picture, the two-electron process requires the weak coupling between Slater determinants that arises from electron correlation, which is small in comparison to the determinantal energy levels themselves, hence the two-electron process is much slower. Feshbach resonances thus tend to be longer lived as compared to shape resonances.[186]

Next consider that a minimal set of requirements for DEA consists of the following.[12]

1. A resonance lifetime $\gtrsim 10^{-14}-10^{-12}$ s.
2. A transient negative ion state that is dissociative in the Franck–Condon region of the neutral species.
3. At least one fragment of the dissociation products that has a positive EA, so that the dissociation channel is energetically allowed.

These requirements are fulfilled by the Feshbach resonance illustrated in Figure 15. (The final requirement, energetic accessibility, is fulfilled relative to the excited state ABCD*.) These requirements are also fulfilled by the scenario depicted in Figure 1, which we may classify as a single-particle Feshbach resonance. Such states have sometimes been called *vibrational Feshbach resonances*. Note that in this case the anion is necessarily formed in a vibrationally excited state.

The distinction between a Feshbach resonance and a shape resonance for a negative ion can be stated succinctly in terms of whether M^- lies above (shape) or below (Feshbach) the energy of the neutral molecule, M. Stated differently: in the shape resonance case, M^- lies above its own continuum (that corresponding to $M + e^-$), while in the Feshbach case, M lies below this continuum. An example is shown in Figure 16, where we consider two isomeric forms of p-coumaric acid, which is a simplified chemical model of the chromophore in photoactive yellow protein.[187,188] For both the phenolate and the carboxylate isomers of this chromophore, the $S_1(\pi\pi^*)$ bright state lies above the adiabatic electron detachment threshold, hence the electronically excited state should be considered a resonance. However, the two isomers exhibit very different detachment processes. In the phenolate isomer (left side of Figure 16), the lowest detachment threshold corresponds to removing an electron from one of the π orbitals involved in the $\pi \to \pi^*$ ($S_0 \to S_1$) excitation, and in that sense the S_1 state lies above its own continuum and may be classified as a core-excited shape resonance. One expects the $^1\pi\pi^*$ state to be short-lived in this case. Meanwhile, for the carboxylate isomer (right side of Figure 16), the lowest

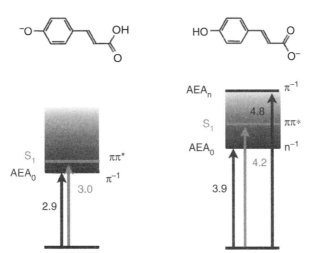

Figure 16 Schematic view of $S_0 \to S_1$ excitation in p-coumaric acid in its phenolate (a) or carboxylate (b) isomeric form. (Energies are given in electron volts.) In either case, the S_1 state lies above the anion's adiabatic detachment threshold and is thus embedded in a continuum of electron-detached states. The core-excited $\pi\pi^*$ resonance may be classified as a shape resonance (on the left) or a Feshbach resonance (on the right) depending on whether the low-lying continuum corresponds to detachment from the π system (a″ orbital) or from an a′ orbital that is not involved in the $\pi \to \pi^*$ excitation. Adapted with permission from Ref. 186; copyright 2013 American Institute of Physics.

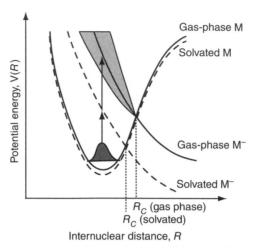

Figure 17 Qualitative depiction of how solvation might alter the solution-phase potentials (broken curves) of M and M⁻, relative to the corresponding gas-phase potentials (solid curves). The shaded region indicates the resonance width of the gas-phase anion. Adapted with permission from Ref. 187; copyright 2004 Elsevier.

detachment threshold corresponds to removing an electron from an orbital with a′ symmetry that is not involved in the $\pi \rightarrow \pi^*$ transition. In this case, e⁻ detachment from the π system (indicated as π^{-1} in Figure 16) lies above S_1. From S_1, electron detachment from the a′ orbital is a two-electron process, and the carboxylate case is an example of a core-excited Feshbach resonance.

Finally, it bears mention that where anions are involved, solvation effects can often have a qualitative impact on the basic physical picture, especially in polar solvents where M⁻ is likely to be dramatically stabilized with respect to M. Polar solvation might, for example, convert an open channel, core-excited resonance $(M^*)^-$ into a core-excited Feshbach resonance by dragging the $(M^*)^-$ potential curve below that of M^*. In such a case, the solution-phase anion resonance would be expected to have a significantly longer lifetime as compared to its gas-phase analogue.[186]

Even in cases where solvation is not enough to stabilize the anion with respect to the neutral molecule, the solution-phase environment can have important consequences. Figure 17 depicts schematic potential energy curves for M and M⁻ in both the gas phase and in solution, with the environment stabilizing M⁻ much more than it stabilizes M. On forming the anion M⁻ from the neutral molecule, an autodetachment channel is available for $R < R_c$. Taking a semiclassical view of electron attachment, the resulting anion spends less time in the autodetachment region $R < R_c$ in solution than it does in the gas-phase case. This enhances the lifetime of the solvated anion relative to the gas-phase species, and possibly allows it to escape the region $R < R_c$ without suffering autodetachment.[186]

QUANTUM CHEMISTRY FOR WEAKLY-BOUND ANIONS

In this section, we discuss the ways in which weakly-bound anions place special demands on quantum chemistry calculations. It is presumed, in this discussion, that the anion M^- is a bound species (VDE > 0) at the molecular geometry in question, such that the application of bound-state quantum chemistry methodology is appropriate. Referring to the situation in Figure 1, bound-state methods are appropriate for the description of the anion AB^- only for $R > R_c$. For $R < R_c$, the neutral molecule is lower in energy, and application of bound-state methods to M^- is not appropriate. Electronic structure methods for temporary anion resonances are discussed later in this chapter.

Gaussian Basis Sets

Atom-Centered Basis Sets

For calculations involving anions, one should use basis sets augmented with additional diffuse basis functions. However, the diffuse exponents for standard augmented basis sets were optimized to describe small molecular anions[189] (CH_3^-, NH_2^-, OH^-, etc.) and or atomic anions,[189,190] so while these basis sets may be appropriate for describing *valence* anions, they are inadequate for the description of dipole-bound or other loosely-bound electrons. This criticism applies to all of the standard, singly-augmented basis sets, including Pople-style basis sets such as 6-31+G*, 6-31++G**, and their triple $-\zeta$ analogues, as well as the Dunning-style correlation-consistent basis sets aug-cc-pVXZ, for X = D, T, Q, ... (The aug-cc-pVXZ basis set will sometimes be abbreviated aXZ in this work.) That these basis sets are inadequate to describe weakly-bound anions can be seen clearly in Figure 18(a), which depicts the convergence of calculated VDEs for an isomer of $(H_2O)_{12}^-$ for which the VDE ≈ 0 in the basis set limit. Setting aside the density-functional calculations, which strongly overbind the electron for reasons discussed later in this chapter, we see that the VDE converges to about zero only after *four* diffuse shells have been added.

In the calculations reported in Figure 18, exponents for the additional diffuse shells are chosen in an *even-tempered* manner, meaning they are arranged in a geometric progression (differing by a common scaling factor), starting from the smallest exponent in the standard 6-31++G* basis set. Use of a geometric progression is intended to reduce numerical linear dependencies that may hamper self-consistent field (SCF) convergence,[192] although such problems are ultimately inevitable as system size grows, especially when numerous diffuse shells are required. In cases where the anion SCF calculation proves difficult to converge, one may try either using the neutral molecule's MOs as an initial guess for the anion's MOs, or alternatively, converge the anion SCF calculation

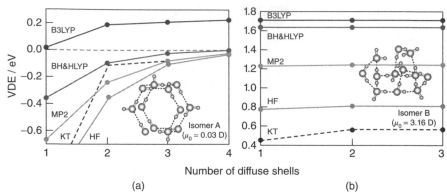

(a)   (b)

Figure 18 VDEs computed at various levels of electronic structure theory for two different isomers of $(H_2O)_{12}^-$, as a function of the number of diffuse shells included in the basis. The basis set is 6-31(1+,n+)G^*, meaning one set of diffuse sp functions for the oxygen atoms and n sets of diffuse exponents (with exponents comprising a geometric progression) for the hydrogen atoms. The quantity μ_0 is the dipole moment of the underlying neutral $(H_2O)_{12}$ cluster, and "KT" denotes the Koopmans' theorem result, $-\varepsilon_{SOMO}$. Reprinted with permission from Ref. 191; copyright 2005 American Chemical Society.

in a less diffuse basis set and then use these MOs (in conjunction with basis-set projection) as an initial guess for the MOs in the target basis set.

The valence basis sets that one wishes to augment are not always even-tempered themselves, so Schaefer and coworkers[193,43] recommend choosing the scaling factor, κ, according to

$$\kappa = \frac{1}{2}\left(\frac{\zeta_1}{\zeta_2} + \frac{\zeta_2}{\zeta_3}\right) \qquad [20]$$

where $\zeta_1 < \zeta_2 < \zeta_3$ are the smallest (most diffuse) s- and p-function exponents in the valence basis set. The diffuse exponents are then taken to be $\kappa\zeta_1, \kappa^2\zeta_1, \kappa^3\zeta_1, \ldots$. Often, the exponents contained in the valence basis set are such that κ is roughly 1/4–1/3. Taking oxygen atom as an example, the formula in Eq. [20] affords $\kappa^{-1} = 3.65$ for 6-31G^*, which is not so different from the ratio $\zeta_2/\zeta_1 = 3.20$ that is actually used for the diffuse exponent in 6-31+G^*, where ζ_1 was optimized for Hartree–Fock calculations on valence anions.[189] Skurski et al.[192] have studied the addition of even-tempered diffuse functions in a systematic way, for applications involving very weakly-bound anions, and they recommend scaling the orbital exponents by a factor of 1/3–1/5, that is, $\kappa^{-1} = 3-5$. In the Q-Chem program,[194,195] for example, a scaling factor of $\kappa^{-1} = 3.2$ is used, by default, for additional diffuse functions in Pople-style basis sets.

This leaves open the question of how many additional diffuse basis functions should be included, and clearly the convergence behavior with respect to diffuseness of the basis set is very different for the weakly-bound

$(H_2O)_{12}^-$ isomer in Figure 18(a) than it is for the more strongly-bound isomer in Figure 18(b). To understand just how diffuse the basis needs to be, consider that at large electron–molecule separation (when the loosely-bound electron is far from the nuclei), the asymptotic form of the wave function is $\psi(r) \sim e^{-r/\lambda}$ where

$$\lambda = \frac{\hbar}{\sqrt{2m \times \text{VDE}}} \qquad [21]$$

This is a rigorous and general result,[196,197] for both ground and excited states (provided that VDE is understood to be the detachment energy for the electronic state in question),[196] but can be understood using simple arguments. As $r \to \infty$, the electron tunnels out of whatever potential is responsible for its binding. For a simple model involving a square-well potential whose well depth is V_0, the characteristic length scale for the decay of the wave function into the classically forbidden region is

$$\lambda = \frac{\hbar}{\sqrt{2m(V_0 - E)}} \qquad \text{(square well)} \qquad [22]$$

and it seems reasonable to replace $V_0 - E$ with the VDE. Alternatively, consider that the asymptotic behavior of the Hartree–Fock wave function also has the form of a decaying exponential, with[198]

$$\lambda = \frac{\hbar}{\sqrt{2m(-\varepsilon_{\text{HOMO}})}} \qquad \text{(Hartree–Fock)} \qquad [23]$$

For a bound anion, the value $-\varepsilon_{\text{HOMO}} > 0$ furnishes the KT estimate of the VDE,[35] as discussed later in this chapter. Each of these arguments suggests that the VDE is directly related to the radial extent of the wave function.

Making a leap and replacing the wave function everywhere with its asymptotic form $\psi(r) \propto e^{-r/\lambda}$, with $\lambda = \hbar/(2m_e \times \text{VDE})^{1/2}$ as suggested above, one may then compute the expectation value $\langle r \rangle$ for this wave function. The value thus obtained should provide at least a crude estimate of the mean electron–molecule distance. The mathematical result is

$$\langle r \rangle_{\text{SOMO}} \approx \frac{3\hbar}{2\sqrt{2m_e \times \text{VDE}}} \qquad [24]$$

This is an exact result for $\psi(r) \propto e^{-r/\lambda}$ but is only a crude approximation to the actual anion's wave function. It should work best for the singly-occupied molecular orbital (SOMO), hence the notation $\langle r \rangle_{\text{SOMO}}$ in Eq. [24]. This result has been quoted previously,[2] albeit without the detailed justification presented herein. For what follows, we note that a computationally convenient form of Eq. [24] is

$$\langle r \rangle_{\text{SOMO}} / \text{Å} \approx \frac{2.928}{\sqrt{\text{VDE/eV}}} \qquad [25]$$

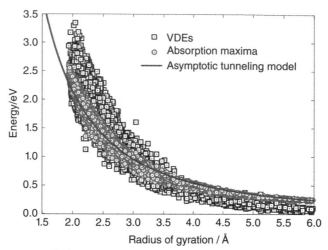

Figure 19   VDEs and electronic absorption maxima for $(H_2O)_N^-$ clusters ($20 \leq N \leq 200$), computed using a one-electron QM MM model[141] on a real-space grid. The "asymptotic model" denotes the VDE versus $\langle r \rangle$ relationship expressed in Eq. [25]. Adapted with permission from Ref. 95; copyright 2011 American Chemical Society.

The distance estimate in Eqs. [24] and [25] is derived only from the asymptotic form of the wave function, but data for a large number of $(H_2O)_N^-$ clusters suggest that this estimate is actually quite reasonable. Figure 19 plots VDEs for $(H_2O)_N^-$ clusters versus the *radius of gyration*, r_g, for a large number of cluster isomers ranging in size from $N = 20$ to $N = 200$.[95] The radius of gyration is defined as

$$r_g = \langle (\mathbf{r} - \langle \mathbf{r} \rangle) \cdot (\mathbf{r} - \langle \mathbf{r} \rangle) \rangle^{1/2} \qquad [26]$$

where the expectation value is with respect to some particular orbital, wave function, or density, as appropriate; r_g is a measure of the size of the probability distribution in question. The data in Figure 19 were computed using a one-electron QM MM model[141,95] rather than all-electron quantum chemistry (so r_g measures the size of the one-electron wave function), but this model has been shown to afford an accuracy of ~ 0.1 eV for VDEs as compared to MP2 benchmarks for clusters as large as $N = 32$.[141] Importantly, the one-electron wave function in these QM MM calculations is represented on a real-space grid as opposed to an atom-centered basis set, so there should be no question about whether a compact Gaussian basis set might skew the results for a diffusely-bound electron. The data in Figure 19 demonstrate that the quadratic relationship VDE $\propto r_g^{-2}$ that is suggested by Eqs [25] and [26] offers a fairly reasonable fit of the VDE data, especially for VDEs $\lesssim 1$ eV. (Note that the approximations underlying Eqs. [24] and [25] are expected to break down when the VDE is large, because in such cases it is no longer justified to replace the entire wave function with its asymptotic form.)

To understand the implications for the selection of diffuse Gaussian exponents, consider that the full width at half maximum (FWHM) of a primitive Gaussian function of the form $\exp(-\zeta r^2)$ is

$$\text{FWHM}(\zeta) = 2\left(\frac{\ln 2}{\zeta}\right)^{1/2} \approx \frac{1.665}{\zeta^{1/2}} \qquad [27]$$

Combining this with the estimated extent of the SOMO in Eq. [25] suggests that one ought to choose the smallest diffuse exponent such that

$$\frac{\zeta}{a_0^{-2}} \ll 0.09 \times (\text{VDE}/\text{eV}) \qquad [28]$$

where a_0 denotes the bohr radius. (Gaussian exponents are traditionally quoted in atomic units.) The diffuse s function on the oxygen atom has an exponent $\zeta = 0.0845\ a_0^{-2}$ for the 6-31+G* basis set and $\zeta = 0.07896\ a_0^{-2}$ in the case of aug-cc-pVDZ (abbreviated "aDZ" hereafter). These values correspond to FWHMs of 3.03 and 3.14 Å, respectively. According to Eq. [28], or reading from the plot in Figure 19, one would expect such a basis to be appropriate only if the VDE is $\gtrsim 1$ eV. The $(H_2O)_{12}^-$ isomer depicted in Figure 18(b), for example, exhibits an MP2-level VDE of ≈ 1.2 eV, and in fact this VDE does appear to be converged with only a single set of diffuse basis functions, that is, a standard, singly-augmented valence anion basis set.

For a species like $(H_2O)_2^-$ where the VDE is < 0.05 eV,[199–201] Eq. [28] suggests choosing $\zeta \ll 0.0045\ a_0^{-2}$. This suggests that three or perhaps four additional diffuse shells, constructed as suggested above, would be required for such a species, depending on the choice of κ. (One might also worry about the quality of the quadrature grid that is used in density-functional calculations that employ very diffuse basis functions, but this does not seem to be an issue, and the default grids in various software programs appear to be sufficient, even when three or four diffuse shells are included.[191,202]) This estimate of three to four diffuse shells turns out to be similar to a recommendation previously provided[191] on the basis of a systematic study of VDEs for small $(H_2O)_N^-$ cluster anions, using 6-31(m+,n+)G* basis sets that include m sets of diffuse sp functions on the oxygen atoms and n sets of diffuse s functions on the hydrogen atoms, using a scaling factor $\kappa^{-1} = 3$. In the study in Ref. 191 the 6-31(1+,3+)G* basis set was found to be sufficient to afford VDEs at the MP2 and CCSD(T) levels that were, in most cases, within ~ 0.01 eV of the VDEs obtained in more diffuse basis sets. An exception was when the VDE was very small ($\lesssim 0.05$ eV), in which case the accuracy was estimated to be 0.03–0.04 eV.

On the other hand, 6-31(1+,3+)G* is not appropriate for high-accuracy calculations of the most weakly-bound anions. To converge VDEs to within ~ 0.001 eV of the basis-set limit, Skurski et al.[192] have shown that seven diffuse shells (using $\kappa^{-1} = 3.2$) are required, for systems such as CH_3CN^- and $(H_2O \cdots NH_3)^-$ whose VDEs are both ≈ 110 cm^{-1} or 0.014 eV. This corresponds to smallest exponents on the order of 10^{-5} a_0^{-2}. Similar recommendations have been made by Gutowski et al.[113,203–205] For high-accuracy calculations, these authors employ aug-cc-pVDZ augmented further with either five or seven additional diffuse sp and diffuse d shells, depending on the value of κ, such that the smallest exponent is $\sim 10^{-5}$ a_0^{-2}. It is suggested that $\kappa^{-1} = 3.2$ is adequate if the neutral molecule's dipole moment, μ_0, is ≥ 6 debye, whereas $\kappa^{-1} = 5.0$ is more appropriate for $\mu_0 \sim 3.0$–4.5 debye.[205] The choice to augment aug-cc-pVDZ, as opposed to the roughly comparable 6-31++G** basis, originates in the observation that VDEs computed using Pople-style basis sets behave somewhat erratically as the number of diffuse shells is increased.[192,202]

Whether one chooses to apply this "enhanced augmentation" to 6-31+G* or to aug-cc-pVDZ, the fact that reasonable VDEs can be obtained using basis sets of double-ζ quality, using correlated wave functions, ultimately rests on a cancellation of errors, albeit a well-justified one. When VDEs for $(H_2O)_N^-$ clusters are computed at the MP2 level, for example, the differences between results obtained in the 6-31(1+,3+)G* and 6-311(1+,3+)G* basis sets are $\lesssim$ 0.01 eV.[206] No reasonable electronic structure theorist would expect such a cancellation in, say, a bond dissociation energy, or even an ionization energy for a closed-shell molecule computed at the same level of theory. The difference is the somewhat smaller correlation energy that is associated with the unpaired electron, which is better separated from the other electrons and therefore less strongly correlated. This is discussed in more detail later.

Although the convergence of VDEs provides a convenient means to evaluate basis set quality, one might object that if accurate VDEs are not the focus of a particular study, then the added computational cost of multiple diffuse shells is unwarranted. The danger in this reasoning – that compact basis sets might suffice, if only the computed VDEs can be ignored – is that overly compact basis sets *cannot* be trusted to describe the relative energetics of both dipole-bound anions *and* valence anions. Some molecules and clusters possess both types of states, at different geometries, and one must therefore consider whether a too-compact basis set – even one that might be quite reasonable for describing an anion whose VDE is large – may bias the calculation. Ignoring VDEs is also perilous because it leaves the user with no means to decide whether the anion in question is actually bound at the geometry in question, and whether that fact might change along a molecular dynamics trajectory, for example. If the anion is not bound, then the results of any bound-state quantum chemistry calculation are dubious at best.

As an example of how the lack of adequately diffuse basis functions can skew an entire potential energy surface, consider the case of $(H_2O)_2^-$. Calculations for this system on the basis of only singly-augmented basis sets suggest a stationary point corresponding to a $OH_2 \cdots e^- \cdots H_2O$ structure,[207–210] in which the two water molecules are arranged in either C_{2h} or D_{2d} symmetry, and the excess electron is apparently coordinated directly to all four O—H moieties. In an early paper that reported such a structure for $(H_2O)_2^-$, it was argued that inclusion of additional, more diffuse basis functions for the geometry optimization was not necessary, because the addition of such functions failed to change the contours of the SOMO in any qualitative way.[208] However, calculations using a more appropriate basis set, with three diffuse shells on each atom, later demonstrated that the $OH_2 \cdots e^- \cdots H_2O$ binding motif is unstable with respect to dissociation of the complex.[119,211] Using reasonable basis sets, the only stable structure identified for $(H_2O)_2^-$ corresponds to a dipole-bound $H_2O \cdots HOH \cdots e^-$ structure, with the electron bound to the positive-dipole end of a hydrogen-bonded water dimer.[211]

Considerations for the Condensed Phase

Convergence of the VDEs for $(H_2O)_{12}^-$ in Figure 18(b) suggests that when the VDE is larger, additional diffuse functions beyond the traditional set designed for valence anions (6-31++G*) do not alter the VDE by much. In condensed-phase systems, the presence of atomic centers is ubiquitous enough that one or a few diffuse shells may actually be enough to provide a converged wave function, even for solvated-electron systems where the unpaired electron resides in a solvent void (cavity) that lacks atoms, but which may be adequately covered via atom-centered basis functions. This is illustrated in Figure 20, which depicts a time-dependent density functional theory[215] (TD-DFT) calculation of the electronic absorption spectrum of $e^-(aq)$, using the 6-31+G* basis set.[212] Comparison to the experimental spectrum[139] shows nearly quantitative agreement, despite the use of an atom-centered basis set with only a single diffuse shell. The odd electron in these calculations inhabits a void in the solvent, and the ground-state orbital depicted in Figure 20(b) clearly spans this cavity, despite the lack of multiple diffuse shells. Careful convergence tests, with respect to the diffuseness of the atom-centered basis set as well as the size of the QM region in these QM MM calculations, reveal that the 6-31+G* basis is sufficient to converge the first few excited states (including, at least, the three 1s → 1p states), although basis-set effects are larger for more highly excited states.[162]

That the singly-augmented 6-31+G* basis appears to be sufficient for these $e^-(aq)$ calculations is fortunate, because the multiply-augmented basis sets discussed above increasingly suffer from numerical linear dependencies as the system size grows. This is problematic in larger clusters, as linear dependencies tend to hamper SCF convergence and may be catastrophic in calculations designed to model condensed-phase systems. Convergence problems can sometimes be mitigated using tight thresholds for shell-pair formation and integral evaluation, but the computational cost may increase substantially as a result.

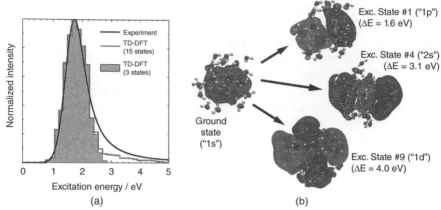

Figure 20   (a) Simulated absorption spectrum for e⁻(aq),[212] based on TD-DFT calculations at the LRC-μBOP/6-31+G* level, in comparison to the experimental spectrum
from Ref. 139 (The LRC-μBOP functional is described in more detail later in this
chapter; see Ref. 163 regarding the choice of functionals in TD-DFT). (b) Natural transition orbitals[213] (NTOs) for several excited states of this system,[15] with state labels 1s,
1p, and so on, corresponding to a "particle in a spherical box" model.[214] (For the particle in a spherical box eigenfunctions, the principal quantum number does not constrain
the angular momentum quantum number, as it does for the more familiar example of
the hydrogen atom, hence the lowest excited states of this system are 1p states, not
2p states.[214]) These plots demonstrate that the 6-31+G* basis adequately covers the
cavity formed in the solvent; in particular, a cavity-centered SOMO is obtained for
the ground state. Adapted with permission from Ref. 212; copyright 2010 American
Chemical Society.

    Gaussian basis sets designed specifically for condensed-phase calculations
do exist, in which basis-function parameters have been optimized using a metric designed to minimize the appearance of linear dependencies.[216] However,
these basis sets lack the diffuse functions needed to describe loosely-bound electrons. For *ab initio* molecular dynamics simulations of $(H_2O)_{32}^-$, Jungwirth and
coworkers[217] found it necessary to augment such condensed-phase basis sets
with an additional 1000 Gaussian functions centered on a $10 \times 10 \times 10$ Cartesian grid. The width of these functions was chosen to be ≈ 2.5 times the spacing
between their centers.[218]

    To appreciate another problem with highly diffuse basis sets, one must
first understand that it is common practice to use MOs from the neutral
molecule – call it M, as above – as an initial guess for the SCF calculation
on M⁻. This choice can greatly accelerate the anion calculation since the
closed-shell SCF cycles are 50% less expensive and often provide an excellent
guess if the anion is weakly-bound and only slightly perturbs the MOs of
the neutral molecule. Moreover, insofar as both the M and M⁻ calculations
are required to compute the VDE, nothing is lost by first computing the SCF
wave function for M. However, for $(H_2O)_N^-$ isomers with *very* small VDEs

($\lesssim$ 0.1 eV, in this author's experience), such a guess occasionally converges to a different SCF solution, as compared to an unbiased guess for $(H_2O)_N^-$!

In addition, attempts to optimize the geometry of M^- sometimes access geometries in which the anion is unbound, which may lead to convergence failure in the SCF calculation. The regions of the potential surface where this problematic behavior occurs can be expected to proliferate as the VDE of M^- approaches zero. Such problems are usually avoided – for all the wrong reasons – by the use of overly compact basis sets, which is perhaps why the literature is rife with weakly-bound anion calculations using inappropriate basis sets. The VDEs reported in these studies are certainly incorrect, and the structures may be bogus as well.

Floating Centers

The discussion up to this point has considered only atom-centered Gaussian basis sets, possibly augmented by an expensive Cartesian grid of Gaussian functions. To avoid problems with linear dependencies, at modest cost, while still providing highly diffuse basis sets that can be used, for example, to describe dipole-bound anions of small molecules, one can employ *floating-center* basis functions. Here, a standard augmented basis set is further augmented with a set of diffuse functions placed on some alternative center that is treated as an atom with zero nuclear charge (a so-called "ghost atom"). The floating-center approach is effective for small, weakly-bound anions such as CH_3CN^- and $(H_2O \cdots NH_3)^-$, although diffuse basis functions of d symmetry are required to converge VDEs.[192] Analytic models of electron binding to a point dipole suggest that d functions in a single-center expansion should be even more important for systems with larger dipole moments,[219] although calculations on $(H_2O)_N$ clusters show that diffuse, atom-centered d functions have very little impact on VDEs.[191] Presumably, this is because the greater asymmetry of these clusters, as compared to CH_3CN^- or a point dipole model, allows linear combinations of diffuse s and p functions to mimic the angular flexibility that would otherwise be provided by diffuse d functions. It stands to reason that in a single-center expansion, functions with higher angular momentum will be necessary, as compared to what is required when using a multicenter expansion.

When the size of the nuclear framework of the molecule or cluster is small compared to the extent of the diffuse basis functions, it probably does not matter much where the floating center is positioned. In applications to $(H_2O)_6^-$, for example, Sommerfeld et al.[128] place a single set of diffuse functions on one oxygen atom, using an even-tempered progression out to a maximum FWHM of 80 Å. It is reported that the VDE depends only weakly on which oxygen atom is chosen as the center of this expansion.[128]

In a large molecule or cluster, however, a single floating center cannot be expected to replace atom-centered diffuse basis functions. In the absence of these atom-centered functions, the floating-center approach is fundamentally unbalanced because some regions of space will be better covered by basis functions than other regions. Unless the electronic structure of the anion in question is well understood in advance, this technique is potentially dangerous for larger molecules of clusters because it necessarily biases the spatial location of the SOMO toward whatever region of space is best described by the floating center(s).[220] If using floating centers in larger systems, one should at the very least optimize their positions in any geometry optimization. (That is, the energy should be minimized with respect to the coordinates of *all* centers that support basis functions, including any ghost atoms.) In this context, it is worth noting that the presence of the floating centers will lead to artifactual vibrational frequencies, although these can be eliminated by increasing the (fictitious) mass of the ghost atom(s) so that their motion decouples from the actual vibrational degrees of freedom.[192]

Even so, the use of ghost atoms for larger molecules or clusters may present a biased description of the potential energy surface. It may be difficult to make a good *a priori* guess as to what the location of the floating center(s) should be, and geometry optimization only guarantees finding a *local* minimum with respect to the placement of these centers. A different initial guess for the position of the floating center(s) might easily lead to a prediction of an electron localized in a different region of space.

Orbital Isosurfaces

Given the extremely diffuse nature of the SOMO in a weakly-bound anion, an important but frequently overlooked consideration is how this orbital should be plotted for visual inspection and interpretation. Various software is available to render isocontour plots of MOs,[221−226] and an *isosurface* for the orbital ϕ is defined as the locus of points for which $\phi(\mathbf{r}) = c$, for some user-specified numerical value, c. A diligent author will faithfully report the value of c that was used to generate the isosurface, and while this does allow others to reproduce the same plot, it is basically meaningless from the standpoint of any physical interpretation. Moreover, any hand-waving arguments based on the spatial extent of the orbital thus plotted are dubious, insofar as the orbital can easily be made to appear larger or smaller by choosing a smaller or larger value of c, respectively.

While these comments pertain to orbital isosurface plots in general, the situation is particularly dire for diffuse electrons. In such cases, one should demand to know what fraction of the orbital density $|\phi|^2$ is encapsulated within a given isosurface plot. In other words, we need to know the fractional electron value,

$$f = \int_{\mathbf{r} \in V} d\mathbf{r} |\phi(\mathbf{r})|^2 \qquad [29]$$

that exists within the volume V defined by the orbital isosurface.

Figure 21 presents an example illustrating the danger of reporting isocontour plots with arbitrarily chosen contour values. Plotted in this example are the Hartree–Fock/6-31(1+,3+)G* SOMOs for an isomer of $(H_2O)_{20}^-$ and an isomer of $(H_2O)_{24}^-$ that both have VDEs of ≈ 1 eV (MP2 level),[15] but which exhibit very different electron binding motifs. In one case, the electron is bound at the surface of the cluster and its wave function extends mostly into vacuum, whereas in the other case the electron resides in cavity formed within the cluster, wherein H_2O molecules have reoriented to point their O—H moieties toward the excess electron. In the latter case, it is tempting to conceptualize the excess electron as a small, quasi-spherical "ball of charge" in water, and the surfaces in Figure 21 demonstrate that it is possible to choose a value of c such that an isocontour plot reinforces this "ball of charge" (or particle-in-a-box[15]) picture. However, such a plot encompasses no more than 75% of the total probability density $|\phi_{SOMO}|^2$. When > 90% of the probability density is plotted, it becomes apparent that the SOMO extends well beyond the first solvation shell, for the cavity-bound electron, whereas in the surface-bound case the SOMO extends a considerable distance out into vacuum but does not penetrate the interior of the cluster to any significant extent.[206] Condensed-phase calculations of $e^-(aq)$ in a solvent cavity also indicate that only $\sim 50\%$ of the spin density $(\rho_\alpha - \rho_\beta)$ is contained within the cavity.[146]

This example demonstrates that isocontour values c chosen with no knowledge of the corresponding value of f may present a highly distorted physical picture. Similarly, the ubiquitous practice of side-by-side comparison

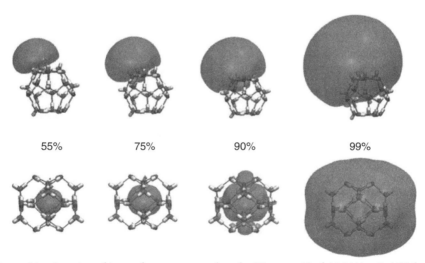

55%   75%   90%   99%

Figure 21 A series of isosurfaces computed at the Hartree–Fock/6-31(1+,3+)G* level that encompass ever-greater fractions of the SOMO for a surface-bound state of $(H_2O)_{20}^-$ top and a cavity-bound state of $(H_2O)_{24}^-$ bottom. The fraction f of the SOMO density, $|\phi|^2$, that is included within these surfaces is noted in the center. Both isomers exhibit VDEs of ≈ 1 eV at the MP2/6-31(1+,3+)G* level.[15] Reprinted with permission from Ref. 206; copyright 2008 American Chemical Society.

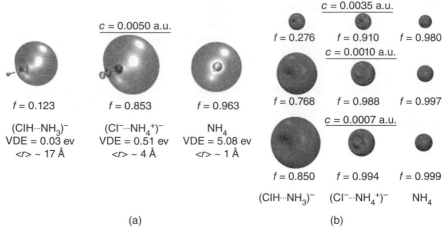

(a)   (b)

Figure 22   Isocontour plots of the SOMOs for three different loosely-bound electron systems studied in Ref. 227: the $(NH_3 \cdots HCl)^-$ dimer anion, the proton-transferred $(NH_4^+ \cdots Cl^-)^-$ isomer of the same dimer, and the charge-neutral NH_4 radical. (The last of these may be viewed as a Rydberg anion state of NH_4^+.) Panel (a) shows each SOMO plotted using a common isocontour value, $c = 0.005\ a_0^{-3/2}$. The corresponding fractions, f, of $|\phi_{SOMO}|^2$ are also listed, along with an estimate of the electron–molecule distance for an electron in the SOMO, as determined using Eq. [25]. In panel (b), these SOMOs are plotted for three different isocontour values, c, and the corresponding values of f are listed as well. Adapted with permission from Ref. 228; copyright 2008 American Chemical Society.

of different orbitals plotted using the same isocontour value may be an unfair comparison, if the radial extent of the orbitals in question is quite different. Because the radial extent is governed by the VDE, comparing common isocontours for systems with very different VDEs is not appropriate. An example is shown in Figure 22, which plots isocontours of the SOMO for three species with very different VDEs. When plotted side-by-side using a common value of c, the SOMO isosurfaces for these three systems can convey misleading information about which of the orbitals is the most diffuse. For example, choosing either $c = 0.0035\ a_0^{-3/2}$ or $c = 0.0050\ a_0^{-3/2}$ (see Figure 22), the SOMO for $(Cl^- \cdots NH_4^+)^-$, a species whose VDE is only 0.03 eV,[227] appears to be slightly larger than the SOMO for the proton-transferred $(ClH \cdots NH_3)^-$ isomer of the same cluster, whose VDE is 0.51 eV.[227] However, the fraction of the electron that is contained in the two isosurfaces is quite different: > 85% of the density for the more strongly-bound species but < 30% of the electron density for the weaker binding isomer.

In view of these examples, side-by-side comparison of orbital isosurfaces for different molecules, clusters, or orbitals is appropriate only when comparing common values of f, not c.[229,228,206,15] Unfortunately, this wisdom has yet to percolate into common practice. When plotting several MOs from the same molecule, this issue may be less pronounced because ε_{SOMO} controls the asymptotic decay of *all* occupied Hartree–Fock MOs, not just the SOMO (see Eq. [23]).[198] Nevertheless, best practice is to use consistent values of f whenever plotting orbitals side-by-side or when making "intuitive" arguments on the basis of the spatial extent of an orbital.

Whereas isocontour plots for arbitrary values of c are easy to generate from the output of most electronic structure programs, using readily available software,[221–226] plotting isoprobability contours for a specified value of f entails additional effort because volumetric data are required. Specifically, the function $\phi(\mathbf{r})$ must be computed on a grid and then integrated according to Eq. [29]. (In principle, this integration could be done internally within an electronic structure program, but to the best of this author's knowledge, no widely-used quantum chemistry package has yet implemented this feature.) Many electronic structure programs will output orbital and density data in the form of a so-called Gaussian cube file,[230] which has become something of an "industry standard" for storing volumetric data, insofar as this format can be read by a variety of visualization programs.[224,231,221] Nevertheless, the precise specification for "cube file" data does not appear to have been published in the literature.

This situation is rectified in Figure 23(a), which provides the format specification. The header portion of this file specifies the number of atoms (NAtoms) as well as their atomic numbers ($Z1$, $Z2$, ...), coordinates ($x1$, $y1$, $z1$, $x2$, ...), origin (XOrigin, YOrigin, ZOrigin), and axes of the volumetric grid. The number of cells (voxels) in each Cartesian dimension is specified as NVoxX, and so on, and the axes of each voxel must also be specified, so the values XAxisX, XAxisY, and XAxisZ, for example, determine the orientation of the first (X) axis of the voxel relative to the molecular frame. The length of these vectors specifies the size of the voxel. Often, the voxel axes are simply aligned with the molecular frame, in which case the three axis vectors constitute a 3×3 diagonal matrix whose diagonal entries represent the spacings between grid points. (Note that all distances are in units of bohr.) Following the atomic coordinates come the volumetric data, also in atomic units. The snippet of code in Figure 23(b) suggests how these data values are arranged; note in particular the line break after each batch of Z points, one of which can be seen in the penultimate line of Figure 23(a).

Next, given an appropriately formatted cube file, the freely available OpenCubMan program[228] can be used to convert between c (isocontour value) and f (fraction of the electron encapsulated by the specified isocontour), and vice versa. Figure 24, which plots the radial decay of the SOMOs for the two weakly-bound cluster anions from Figure 22, illustrates how the algorithm works. Given a cube file representing $\phi_{SOMO}(\mathbf{r})$ evaluated on a regular Cartesian grid, the algorithm first sorts the values of $\rho_{SOMO}(\mathbf{r}) = |\phi_{SOMO}(\mathbf{r})|^2$ into descending order, then numerically integrates ρ_{SOMO} starting from its maximum value, until the first point where the numerical integral equals or

(a)

```
Comment line #1
Comment line #2
NAtoms   XOrigin   YOrigin   ZOrigin
NVoxX    XAxisX    XAxisY    XAxisZ
NVoxY    YAxisX    YAxisY    YAxisZ
NVoxZ    ZAxisX    ZAxisY    ZAxisZ
Z1 0.00000 x1 y1 z1
Z2 0.00000 x2 y2 z2
         ...
ZN 0.00000 XN YN ZN
```

(b)

```
for (int i=0; i < Nx; i++){
    for (int j=0; j < Ny; j++){
        for (int k=0; k < Nz; k++){
            printf("%13.5E",data[i][j][k]);
            if (k % 6 == 5) printf("\n");
        }
        printf("\n");
    }
}
```

```
6.35606E-07   6.70235E-07   7.05477E-07   7.41220E-07   7.77344E-07   8.13717E-07
8.50200E-07   8.86644E-07   9.22893E-07   9.58785E-07   9.94153E-07   1.02883E-06
3.99687E-07   3.72104E-07   3.45449E-07   3.19779E-07   2.95144E-07   2.71584E-07
2.49128E-07   2.27797E-07
6.63832E-07   6.99866E-07   7.36503E-07   7.73622E-07   8.11090E-07   8.48765E-07
```

Figure 23 (a) Format specification for the "cube file" for storing volumetric data format. The file consists of a header that specifies the nuclei and the grid, followed by the data values at the grid points. All data are given in atomic units. (b) Inset: a snippet of C code that writes the data values in the appropriate order and format. This specification is consistent with that output of the Q-Chem electronic structure program[194,195] and has been tested for reading and visualization using the Visual Molecular Dynamics (VMD) program.[224] The column of zeros in the header file in panel (a) represents information that is not presently used by VMD.

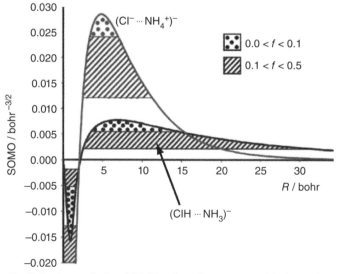

Figure 24 Radial plots of the SOMOs for the two weakly-bound anions from Figure 22, illustrating the $f \rightarrow c$ algorithm of Ref. 228. The dotted regions would be encapsulated by a 10% isoprobability contour and the hashed regions by a 50% isoprobability contour. Adapted with permission from Ref. 228; copyright 2008 American Chemical Society.

exceeds f. The value of $\phi_{SOMO}(\mathbf{r})$ at that point is reported as the contour value, c.[228]

With a bit of experimentation, this procedure can be used to plot an iso-probability contour corresponding to a desired value of f, provided that the volumetric data are output on a sufficiently fine grid. For coarse grids, one may find that the process $f \rightarrow c \rightarrow f$ or $c \rightarrow f \rightarrow c$ does not yield completely self-consistent results (i.e., the final value may be slightly different from the initial value), but such discrepancies should disappear as the volumetric grid density increases.

Wave Function Electronic Structure Methods

Importance of Electron Correlation

Simple textbook results for IPs and EAs serve as a starting point for thinking about quantum chemistry calculations. If we assume that the nonionized molecule is described by a Hartree–Fock wave function, and furthermore if we neglect any relaxation of the MOs on electron attachment or detachment – that is, if we use the neutral molecule's MOs to construct a Slater determinant for the ionized species, removing an electron from the HOMO or adding an electron to the LUMO – then the corresponding approximate IP and EA expressions are known as KT:[35]

$$EA \approx {}^{N}E^{N+1} - E_{LUMO} = -\varepsilon_{LUMO} \qquad [30]$$

$$IP \approx {}^{N-1}E_{HOMO} - {}^{N}E = -\varepsilon_{HOMO} \qquad [31]$$

Here, ${}^{N}E$ represents the Hartree–Fock energy for the N-electron system, and the energies of the determinants formed either by adding an electron to the LUMO or removing one from the HOMO are denoted as ${}^{N+1}E_{LUMO}$ and ${}^{N-1}E_{HOMO}$, respectively. The fact that these energy differences are equal to (minus) orbital energies is an exact result within the model described above; the approximations (with respect to actual EAs and IPs) involve neglect of orbital relaxation and neglect of electron correlation.

A recent review article[2] suggests that the accuracy of KT results is typically ± 0.5 eV for both IPs and EAs, but often the accuracy for IPs is better than that for EAs owing to favorable error cancellation in the case of IPs. To understand this, consider that neglect of orbital relaxation must necessarily destabilize the ionized species (relative to an exact calculation), since the Hartree–Fock method is variational. At the same time, the correlation energy is extensive and therefore tends to be larger for the species that contains more electrons. This correlation energy will probably stabilize the species in question, relative to the Hartree–Fock prediction, by allowing electrons to avoid one another more successfully. For the IP case, then, orbital relaxation would decrease (stabilize) ${}^{N-1}E_{HOMO}$, while electron correlation is expected to decrease ${}^{N}E$ to a greater extent than it decreases ${}^{N-1}E_{HOMO}$. The two errors thus partially cancel one another. Szabo and Ostlund[35] report small-molecule examples for which KT IPs afford values within 0.1 eV of experiment. For the DNA nucleobases, KT IPs evaluated using modest basis sets are also found to be within 0.15 eV of experimental values.[232]

In the case of EAs, however, the species where orbital relaxation is neglected (the anion) is also the species having the larger correlation energy, so if both effects are included then the net result is to move ^{N}E further away from $^{N+1}E_{LUMO}$. As such, errors in KT EAs tend to be much larger than errors in IPs, perhaps ~ 1 eV for EAs.[3] Empirical scaling of the KT EAs has been suggested,[232,124] although it is probably appropriate only across a narrow range of similar molecules. For example, the anion resonances in a series of substituted cyanoethylenes,[233] which can be determined experimentally by electron transmission spectroscopy,[184,185,2] are a good fit to the expression

$$VAE_{expt} = 0.74\varepsilon - 2.0 \text{ eV} \qquad [32]$$

Here, $\varepsilon > 0$ is an orbital energy (equal to the KT prediction of the VAE), computed in Ref. 233 at the Hartree–Fock/3-21G level. The VAEs measured experimentally and used to fit Eq. [32] range from 2 to 10 eV, meaning that the corresponding Hartree–Fock orbital energies lie in the range $\varepsilon \sim 5$–16 eV, according to Eq. [32]. In other words, the combination of orbital relaxation, electron correlation, and finite-basis effects (since 3-21G is far from the basis-set limit) modifies the KT prediction for the VAEs by 3–6 eV!

That said, and while KT EAs do still find some utility in stabilization calculations of temporary anion resonances (as discussed later in this chapter), for bound states of M$^-$ there is little reason to rely on KT since Hartree–Fock calculations are nowadays computationally facile on large molecules, often in large basis sets. It is therefore easy to compute a "ΔSCF" value for the EA, which includes the effects of orbital relaxation, simply by computing separately the Hartree–Fock energies of M and M$^-$, *assuming that the latter is bound.* (If it is not, then neither the KT nor the ΔSCF value of the EA is reliable.) This raises an important point, namely, that one obtains a positive EA from KT only when $\varepsilon_{LUMO} < 0$, and for very weakly-bound anions there may be no virtual orbitals with negative orbital eigenvalues. Simons[3] suggests that the orbital relaxation obtained in a ΔSCF calculation often amounts to ~ 0.5 eV, whereas EAs predicted using KT might have errors of ~ 1 eV, with the other 0.5 eV representing the effects of electron correlation. To do better than this for the calculation of EAs and/or VDEs, a correlated level of theory is required.

Much has been made of the critical role of electron correlation effects in the description of weakly-bound anions.[113,205,234,11,2,3] As with any electronic structure problem, electron correlation is always quantitatively important and occasionally qualitatively important. Cases where correlation is qualitatively important include certain anions with very small VDEs, where dipole binding effects alone (which might be reasonably well-described at the Hartree–Fock level) do not afford a positive VDE, even for anions whose existence can be confirmed experimentally. Such examples will be discussed in due course, but it should be noted at the outset that cases where electron correlation changes the sign of the VDE almost always correspond to cases where the VDE is very small (< 0.1 eV). In such cases, an estimate VDE $\approx$ 0 that might be obtained at the Hartree–Fock level represents a rather small absolute error, quantitatively speaking.

This distinction brings to mind the infamous example of the dipole moment of the carbon monoxide molecule, a historically important problem whose story is told in Ref. 235. To summarize: the experimentally-determined dipole moment is $\approx$ 0.1 D and exhibits $C^{\delta-}O^{\delta+}$ polarization, whereas Hartree–Fock calculations in reasonable basis sets predict $\mu \approx$ 0.3 D but with the opposite polarization.[236] Depending upon one's point of view, this is either an egregious, qualitative error in Hartree–Fock theory (wrong sign for μ!), or else a reasonable prediction that differs from experiment by only about 0.4 D. (Stated differently, Hartree–Fock theory *correctly* predicts that the CO dipole moment is small.) The VDEs of weakly-bound anions exist in a similar state of ambiguity. Since "chemical accuracy" (usually regarded as $\sim$ 1 kcal/mol or $\approx$ 0.04 eV) is achieved only with high-level treatments of electron correlation, one cannot expect that electron binding in any species whose VDE $\lesssim$ 0.04 eV will be described even qualitatively correctly in the absence of high-level treatments of electron correlation. On the other hand, an estimate of VDE $\approx$ 0 for such a species represents an error of $\lesssim$ 0.04 eV.

As pointed out previously in this chapter, VDEs are intensive quantities whereas the correlation energy is extensive. As such, it is sometimes argued that calculation of VDEs is especially difficult and sensitive to correlation effects.[2] Note, however, that other important chemical properties also do not scale with system size; reaction barriers, for example, tend to lie in the range 1–50 kcal/mol and depend on the local bonding environment more so than the overall system size. Similarly, electronic excitation energies for organic chromophores may be modulated in a chemically significant way by their environment, yet these excitation energies span a range of a few electron volts, largely independent of system size. In light of these facts, it is not altogether clear that calculation of VDEs is intrinsically more difficult than calculation of other chemically important energy differences.

Coupled-Cluster Theory

Rather than "walking up" the hierarchy of correlated wave function models, let us start from the top, with CC methods. For tutorial reviews of CC theory, see Refs. 38 and 39. Here, we simply note that the CCSD(T) method,[237] that is, CC theory with single and double excitations described self-consistently, and triple excitations treated perturbatively, is widely considered to be the "gold standard" of single-reference quantum chemistry. In conjunction with high-quality basis sets, this method often achieves "chemical accuracy" of $\lesssim$ 1 kcal/mol for a variety of molecular properties and energy differences, including difficult ones like atomization energies where one cannot rely on error cancellation in computing the energy difference.[238] A self-consistent treatment of the triple excitations (CCSDT) is usually a small correction ($\ll$ 1 kcal/mol).[238]

In defense of the assertion that EAs are fundamentally *not* significantly more challenging as compared to other energy differences in electronic structure theory, we first wish to make the case that the CCSD(T) method's reputation as an excellent benchmark level of theory is no less appropriate in the context of EAs and VDEs for weakly-bound anions. Considering EAs for the atoms H through Cl, where experimental values range from 0.28 eV for boron up to 3.61 eV for chlorine, complete-basis CCSD(T) results are in excellent agreement with experiment, with the largest deviation being 0.051 eV for Cl.[239] Notably, chlorine has the largest spin-orbit correction (estimated at -0.037 eV) and the largest relativistic correction (estimated at -0.015 eV) of any of the atoms H—Cl in the periodic table, and when these corrections are included, the discrepancy with experiment is reduced to < 0.0004 eV.[239] Errors in EAs for the 3d transition metal atoms, computed at the CCSD(T)/CBS level (where "CBS" denotes extrapolation to the complete basis set limit), are < 0.3 eV (or < 7 kcal/mol) in all cases. These errors are comparable in magnitude to the relativistic corrections.[240] Corrections for connected triple excitations ($E_{CCSDT} - E_{CCSD(T)}$) are no larger than 0.034 eV, and corrections for perturbative quadruple excitations ($E_{CCSDT(Q)} - E_{CCSDT}$) are < 0.01 eV.[240]

Although not directly relevant to CC theory *per se*, we mention in this context the performance of composite methods, including the W1 and W2 methods,[241,242] the Gaussian-3 (G3) method,[243–245] and the CBS-4 and CBS-Q methods.[246–249] These methods are "composites" in the sense that they combine the results of a variety of different levels of electronic structure theory, including vibrational zero-point energy and spin-orbit corrections, in an effort to design an overall computational strategy that can achieve $\lesssim$ 1 kcal/mol accuracy for equilibrium thermochemistry. (The W1 and W2 methods even target "spectroscopic accuracy" of $\sim$1 kJ/mol.[241,250]) For main group atoms and small molecules, the W1 and W2 methods predict AEAs with an accuracy of < 0.1 eV as compared to experiment.[242] The accuracy for AEAs predicted by the G3 method is 1.0–1.5 kcal/mol, or again better than 0.1 eV,[251,245] and the CBS-4 and CBS-Q methods perform similarly.[246] (Error statistics for several of these composite methods are summarized in Table 2.) The point of this brief digression about composite methods is to reinforce the idea that, as with CC methods, electronic structure models that are known to be accurate for thermochemistry are also accurate for AEAs.

Composite methods are designed for *equilibrium* thermochemistry, not for mapping out potential energy surfaces, so we now return to a discussion of CC methods and consider some cases involving prediction of VDEs for weakly-bound anions. Consider the case of $(H_2O)_2^-$, for example. The best available calculation for the VDE of the *trans* isomer (depicted in Figure 5)

Table 2 Error Statistics for Adiabatic EAs Computed by Various Composite Methods, for the 27 Atoms and Small Molecules in the G2-1 Data Set[a]

Method	Reference for Theory	Errors (eV)			
		Mean	MAD[b]	RMS[c]	Max[d]
G3[e]	243	0.031	0.049	0.065	0.182
G3(MP2)[e]	244	0.035	0.058	0.076	0.195
CBS-4M[e]	246	−0.006	0.111	0.136	0.312
CBS-QB3[e]	246	0.031	0.055	0.065	0.117
W1[f]	241	0.009	0.016	0.019	0.051
W2[f]	241	0.008	0.012	0.014	0.039

[a]252
[b]Mean absolute deviation (MAD).
[c]Root mean square (RMS) deviation.
[d]Maximum error for any of the 27 atoms.
[e]Error statistics taken from Ref. 246.
[f]Error statistics taken from Ref. 242.

is one performed at the CCSD(T)/aug-cc-pVDZ(6s6p6d) level, where the parenthetical indicates six additional diffuse shells on a floating center.[234] This calculation predicts a VDE of 0.039 eV, which should be compared to values of 0.045 ± 0.006 eV[199,200] and 0.030 ± 0.0004 eV[201] obtained from photoelectron spectroscopy. For the *cis* isomer, a CCSD(T)/aug-cc-pVDZ(5s5p5d) calculation affords a VDE of 0.013 eV,[234] in comparison to an experimental value of 0.017 eV,[253,254] the latter measured by field detachment of $(H_2O)_2^-$ in rare-gas clusters and extrapolated to the limit of zero rare-gas atoms. Table 3 provides several other examples, along with an energy decomposition that is discussed later in this chapter. Of the examples in Table 3, the largest difference between theory and experiment occurs in the case of $(HF)_2^-$, for which Jordan and Wang[11] cite "sizable discrepancies" between theory and experiment. CCSD(T)/aug-cc-pVDZ(5s5p5d) calculations for this dimer anion afford a VDE of 387 cm^{-1} (0.048 eV),[203] as compared to the experimental value of 508 cm^{-1} (0.063 eV).[255] These comparisons to experiment establish the limit of what it is possible to achieve with *ab initio* theory.

As with atomic EAs, comparisons to higher-level calculations suggest that correlation effects on VDEs beyond the CCSD(T) level are quite small for molecular anions.[113,203,204] Consider, for example, the notoriously challenging HNC$^-$ and HCN$^-$ anions,[234,256,257] whose binding energies are only ≈ 0.004 eV and ≈ 0.002 eV, respectively, with the VDE for HCN$^-$ arising almost entirely from electron correlation effects.[256] For these two species, VDEs computed at the CCSD(T) level and the CCSDT level agree to within 0.001 eV.[257] The $(H_2O)_6^-$ anion provides another example: here, the VDE computed at the CCSD(T) level[258] lies within the statistical error bars of a quantum Monte Carlo (QMC) calculation,[259] the latter of which is free of basis-set artifacts and does not require truncation of the excitation level.

Table 3 Decomposition of *Ab Initio* VDEs into Physically Meaningful Components[a]

Energy Component	Contributions to the VDE (cm^{-1})				
	CH$_3$CN^{-b}	C$_3$H$_2^{-b}$	(H$_2$O)$_2^{-c}$	(HF)$_2^{-d}$	(HF)$_3^{-e}$
KT	53	55	111	165	950
ΔE_{relax}	3	8	7	14	104
$\Delta E_{disp}^{(2)}$	57	70	114	177	625
$\Delta E_{non-disp}^{(2)}$	−38	5	−10	−73	−227
$\Delta E^{(3)}$	4	−38	0	−3	−24
$\Delta E^{(4)}$	8	34	20	27	93
ΔE_{HO}	21	39	100	81	145
VDE[CCSD(T)]^f	108	173	312	387	1666
VDE(expt)^g	93	171 ± 50	363 ± 48	508 ± 24	1613 ± 2420

[a]Reprinted with permission from Ref. 11 (copyright 2003 Annual Reviews); notation is explained in the text.
[b]From Ref. 113.
[c]From Ref. 234.
[d]From Ref. 203.
[e]From Ref. 204.
[f]Basis set is aDZ(7s7p8d) for CH$_3$CN$^-$ and C$_3$H$_2^-$, aDZ(5s5p5d) for (HF)$_2^-$, and aDZ(4s4p4d) for (HF)$_3^-$.
[g]References to the experimental literature can be found in the relevant theory papers.

Thus, CCSD(T) calculations with basis sets as described above are capable of reproducing experimental VDEs within the accuracy of the experiments, and further electron correlation effects are typically smaller than experimental precision, even in cases where the latter is quite good. The cost of CCSD(T) calculations, however, grows as $O(N^7)$ with respect to system size, N, which quickly becomes prohibitive. Therefore, it is important to understand how more modest treatments of electron correlation can be expected to perform. The convergence of VDEs with respect to the treatment of electron correlation, using basis sets that should be saturated with respect to diffuseness, is illustrated in Table 4 for several small (H$_2$O)$_N^-$ isomers. Two different isomers of (H$_2$O)$_6^-$ are considered, with VDEs of 0.42 and 0.78 eV at the CCSD(T) level.[260] Note that VDEs computed at the MP2 level for the same two isomers, which are also listed in Table 4, differ from the CCSD(T) values by only 0.06 eV (15%) and 0.03 eV (3%), respectively.

The very weakly-bound (H$_2$O)$_4^-$ isomer in Table 4 exhibits a larger MP2 error: 0.14 eV or 73%. That the MP2 error tends to be larger for less strongly-bound isomers is a general trend that will we see again later. From the KT results ($-\varepsilon_{LUMO}$), it is clear that this particular anion is nearly unbound at the Hartree–Fock level. According to Eq. [23], the asymptotic decay of the SCF wave function goes like $\sim \exp(-\varepsilon_{SOMO}r)$, so if $\varepsilon_{SOMO} > 0$ then the SCF

Table 4 VDEs for Several $(H_2O)_N^-$ Isomers Obtained Using Various Wave Function Models[a]

| | | VDE (eV) | |
| | | $(H_2O)_6^{-c}$ | |
Model	$(H_2O)_4^{-b}$	Isomer 1	Isomer 2
Koopmans' Theorem[d]	0.002	0.233	0.045
Hartree–Fock	0.003	0.259	0.254
MP2	0.051	0.361	0.750
CCSD	0.166	0.399	0.717
CCSD(T)	0.191	0.422	0.777
EOM-EA-CCSD	0.192	0.418	0.744
EOM-EA-CCSD(2)[e]	0.192	0.415	0.744
ADC(2)	0.192	0.400	0.748

[a]Reprinted with permission from Ref. 260; copyright 2012 American Chemical Society.
[b]Basis set is aug-cc-pVTZ(6s6p6d).
[c]Basis set is aug-cc-pVDZ(7s7p).
[d]VDE $= -\varepsilon_{SOMO}$.
[e]Uses MP2 amplitudes to construct the cluster operator.

wave function is not normalizable, and the application of perturbation theory is probably not appropriate, since the reference state is qualitatively wrong. In the particular example of the weakly-bound $(H_2O)_4^-$ isomer in Table 4, ε_{SOMO} remains slightly negative (bound), but clearly the Hartree–Fock determinant is approaching a regime where perturbation theory should be viewed with skepticism. CC wave functions, on the other hand, are far less sensitive to the choice of reference determinant,[261] and indeed the CCSD result (treating the singles and doubles self-consistently, as opposed to perturbatively) is a reasonable approximation to CCSD(T), much more along the lines of what is observed for the more strongly-bound $(H_2O)_6^-$ isomers.

Since the "(T)" correction is itself perturbative, one might reasonably question CCSD(T) results in cases where MP2 theory goes awry. Results quoted earlier for $(H_2O)_2^-$ and HCN$^-$ suggest that this is not an issue, however, presumably due to the superior reference state provided by the $\exp(\hat{T}_1)$ orbital rotation[261] in CCSD. A more direct assessment is possible using CC methods that do not rely on the weakly-bound anion as the reference state. One such method is the "electron attachment" version[262] of equation-of-motion (EOM) CC theory with single and double excitations.[263] This method, which goes by the acronym EOM-EA-CCSD, uses the CCSD wave function for the closed-shell neutral molecule, M, as a reference state to compute the cluster operator $\hat{T} = \hat{T}_1 + \hat{T}_2$, and is therefore free of any concern regarding whether or not ε_{SOMO} for M$^-$ is bound. The ground state of M$^-$ is then obtained by configuration interaction with respect to the similarity-transformed Hamiltonian,

$$\hat{\bar{H}} = e^{-\hat{T}} \hat{H} e^{\hat{T}} \qquad [33]$$

in a "two-particle, one-hole" (2p1h) basis.[262] For $(H_2O)_4^-$, EOM-EA-CCSD and CCSD(T) results are nearly identical (see Table 4), which suggests that the CCSD(T) method is a reasonable one even for weakly-bound anions. Evidently, for this particular example the $\exp(\hat{T}_1)$ orbital rotation operator is able to deal with any inadequacy in the reference state.

This *quantitative* agreement between EOM-EA-CCSD and CCSD(T) results may be fortuitous, however. An EOM-CCSD study of small-molecule EAs (in the range of 0.4–3.8 eV) found that the accuracy of EOM-EA-CCSD and EOM-IP-CCSD EAs is generally no better – and perhaps slightly worse – than the accuracy of a ΔCCSD approach, with the accuracy of the latter method being ≈ 0.3 eV.[262] This underscores a point made earlier: it is not completely clear whether EOM, Green's function, or electron propagator methods are intrinsically more accurate than standard methods that compute $E_{\text{neutral}} - E_{\text{anion}}$, assuming that: (i) each of the methods in question is extended to the same level of many-body theory, and (ii) size-extensive methods are employed for all calculations. Moreover, whereas the EOM-EA-CCSD calculations in Table 4 were performed in an effort to avoid a potentially problematic anion reference state,[260] it was noted in Ref. 262 that the EOM-IP-CCSD values in that study were in slightly better agreement with experiment as compared to the EOM-EA-CCSD results, by 0.05–0.10 eV on average. The EOM-IP-CCSD approach consists of a ground-state CCSD calculation on the anion, followed by an EOM calculation involving removal of an electron. Thus, the "IP" version of the method (EOM-IP-CCSD) uses orbitals optimized for the anion.

Similar accuracy (0.2–0.3 eV for valence ionization energies below ~ 20 eV) is available using propagator methods in conjunction with triple-ζ basis sets.[102,103] These propagator methods are known in the literature as the outer valence Green's function (OVGF) and partial-third order (P3) approximations.[102,103] With modern semidirect batching algorithms, these methods exhibit only fifth-order scaling (ov^4, where o and v denote the number of occupied and virtual orbitals, respectively, whereas EOM-CCSD methods scale as o^2v^4), although long calculation times (due to numerous batches) or else large amounts of memory (to reduce the number of batches) may be required.[101] That said, where point-group symmetry is available, molecules as large as fullerenes (C_{60} to C_{144}), porphyrins, and phthalocyanines [e.g., $(C_{32}H_{12}N_8NiO_{12}S_4)^{4-}$] have been considered.[264–266]

Perturbation Theory

When the VDE of M^- is small, changes in the treatment of electron correlation can have qualitative effects, analogous to the case of carbon monoxide's dipole moment.[235] A significant development in the study of (very) weakly-bound anions was the recognition, in the late 1990s, that electron correlation effects sometimes do play a qualitatively important role in electron binding,[113,204,205,234] in the sense that they may make the difference between electron binding or not binding. To interpret the physical origins of electron binding, Gutowski and coworkers[192,204,205] introduced a perturbative decomposition of the VDE, similar in spirit to the perturbative theory of intermolecular interactions[267,268] but with the loosely-bound electron serving as one of the "molecules." The method requires calculations involving triple excitations [MP4 and CCSD(T)] and is therefore applicable to small systems only, especially in view of the basis-set requirements for high-accuracy calculations. In a series of publications, Gutowski et al.[113,204,269,205,234] showed that (second-order) electron–molecule dispersion interactions – which are wholly electron correlation effects – make a significant contribution to the VDEs of a variety of dipole-bound anions, some of which are analyzed in Table 3.

Dispersion interactions are absent at the Hartree–Fock level and appear for the first time in second-order perturbation theory. Gutowski et al.[204,205] proposed that a subset of the terms in the MP2 correlation energy could be ascribed to dispersion interactions between the loosely-bound electron and the core molecular species, namely

$$\Delta E_{\text{disp}}^{(2)} = -\frac{1}{2}\sum_{i}^{\text{occ}}\sum_{ab}^{\text{virt}}\frac{|\langle\phi_a\phi_b||\phi_i\phi_{\text{SOMO}}\rangle|^2}{\varepsilon_a + \varepsilon_b - \varepsilon_i - \varepsilon_{\text{SOMO}}} \qquad [34]$$

This second-order dispersion correction represents the terms in the MP2 correlation energy that involve coupling of the transient dipoles induced by simultaneous excitation of the SOMO and one of the occupied MOs (ϕ_i) of the core molecular species. Because the electron repulsion integrals in Eq. [34] are anti-symmetrized, $\Delta E_{\text{disp}}^{(2)}$ includes second-order exchange-dispersion interactions as well, in the language of symmetry-adapted perturbation theory.[268]

For each of the anions in Table 3 except (HF)$_3^-$, this second-order dispersion energy is larger than the VDE predicted by KT, and in all of the cases examined in Table 3 it is larger than the remaining terms in the MP2 correlation energy ($\Delta E_{\text{non-disp}}^{(2)}$). Thus, with the possible exception of (HF)$_3^-$, one can say that electron–molecule dispersion is actually more important in stabilizing these particular weakly-bound anions than is the electron–dipole interaction, and as such, these are cases where correlation effects are *qualitatively* important. Much of this importance is captured already at second order in perturbation theory. The third- and fourth-order corrections to the VDE ($\Delta E^{(3)}$ and $\Delta E^{(4)}$ in Table 3) are fairly small, as is the "higher-order" correction, ΔE_{HO}, that is assessed as the difference between fourth-order perturbation theory and a benchmark VDE computed at either the CCSD(T) or CCSDT level.[11] For each of the weakly-bound anions in Table 3, the sum total of corrections to the VDE beyond second-order perturbation theory is no larger than 0.03 eV.

These higher-order effects are the largest in (HF)$_3^-$, which also has the largest VDE of any of the anions considered in Table 3. Much larger VDEs can be realized in larger (HF)$_N^-$ clusters,[11,270] and Table 5 shows the results of a perturbative energy decomposition for some examples. All of these are high-symmetry clusters in which the positive ends of the molecular dipoles have been oriented toward a common point, in an effort to construct small clusters with large VDEs. In the VDE decomposition, one important observation is that $\Delta E^{(2)}_{\text{disp}}$ is larger than the orbital relaxation correction (ΔE_{relax}) in every case, where the latter is defined as the difference between the KT estimate of the VDE and the ΔSCF value. Gutowski et al.[205] argue that ΔE_{relax} should approximate the second-order induction correction, that is, ΔE_{relax} is the leading-order polarization correction, and insofar as this is true, the ratio

$$\text{ratio} = \frac{\Delta E^{(2)}_{\text{disp}}}{\Delta E_{\text{relax}}} \qquad [35]$$

provides an estimate of the relative importance of dispersion versus induction. For the weakly-bound $(HF)_2^-$ isomer in Table 5, one obtains a ratio of 10.5, echoing the large ratios found in other weakly-bound anions such as $(H_2O)_2^-$ (ratio = 16.3) and CH_3CN^- (ratio = 19.0).[11] However, for the more strongly-bound $(HF)_N^-$ clusters in Table 5, this ratio is more like 1.5, meaning that both the induction and dispersion corrections are quite substantive.

These observations led to a significant change in the thinking about dipole-bound anions, whose very name implies electrostatic binding and summons notions of charge–dipole models.[271,219] On the other hand, the second-order dispersion and non-dispersion corrections have opposite signs and thus partially cancel, for all of the $(HF)_N^-$ clusters in Table 5 as well as many of the weakly-bound anions in Table 3. Moreover, as the VDE increases so, too, does the fraction of the VDE that comes from electrostatic and second-order induction effects ($\Delta E_{\text{relax}} - \varepsilon_{\text{LUMO}}$), and higher-order correlation effects constitute a smaller fraction of the VDE in larger clusters. In fact, the total fraction of the VDE arising from electron correlation gets smaller as the VDE gets larger, as shown in the left-most column of Table 5. To some extent, this reflects the intensive nature of the VDE versus the extensive nature of the correlation energy, even if the clusters in Table 5 are much too small to probe the thermodynamic limit.

Still, the correlation contribution to the VDE for $(HF)_6^-$, while only 7.6%, is hardly negligible at 0.2 eV. Note, however, that the third-order corrections and the "higher-order" correction have similar magnitudes but opposite signs, so that $|\Delta E^{(3)} + \Delta E^{(4+HO)}| \leq 0.015$ eV for each of the $(HF)_N^-$ clusters in Table 5, even for the dimer anion whose VDE is only 0.053 eV. This suggests that perhaps progress can be made in larger molecular and cluster anions by application of MP2 theory, even in cases where the VDE is rather small. In fact, MP2/6-31(1+,3+)G* VDEs for both $(H_2O)_N^-$ clusters[191,270] and $(HF)_N^-$ clusters[270] are consistently found to be only ~ 0.030 eV smaller than CCSD(T)/6-31(1+,3+)G* values. Considering also the basis-set error, it is suggested that the MP2 values are probably in error only by $\sim 0.03-0.05$ eV.[191,270,202]

Table 5 Decomposition of *ab Initio* VDEs for $(HF)_N^-$ Clusters into Physically Meaningful Components[a]

N	Point Group	Energy Components (eV)						VDE	% e^- Corr.[b]
		KT	ΔE_{relax}	$\Delta E^{(2)}_{disp}$	$\Delta E^{(2)}_{non\text{-}disp}$	$\Delta E^{(3)}$	$\Delta E^{(4+HO)}$		
	$D_{\infty h}$	0.012	0.004	0.042	−0.017	−0.000	0.013	0.053	70
	D_{3h}	0.244	0.159	0.427	−0.212	−0.017	0.032	0.633	36
	T_d	0.808	0.404	0.685	−0.427	−0.026	0.032	1.477	18
	D_{3h}	1.358	0.497	0.754	−0.513	−0.022	0.026	2.100	12
	O_h	1.931	0.556	0.797	−0.584	−0.018	0.010	2.691	8

[a] Adapted with permission from Ref. 11 (copyright 2003 Annual Reviews).
[b] Percentage of the VDE arising from electron correlation.

449

For $(H_2O)_N^-$ and $(HF)_N^-$ clusters, second-order many-body perturbation theory (MBPT2) based on Kohn–Sham (KS) orbitals, which we will call the MBPT2(KS) method, has been used to gauge the sensitivity of MP2 theory to the choice of reference determinant.[270,260] The functional used in Refs. 261 and 271 is Becke's "half and half" exchange functional[272] (BH&H) in conjunction with the Lee–Yang–Parr (LYP) correlation functional:[273]

$$E_{xc}^{BH\&HLYP} = 0.5E_x^{HF} + 0.5E_x^{B88} + E_c^{LYP}$$ [36]

Here, E_x^{HF} and E_x^{B88} denote the Hartree–Fock exchange energy and Becke's 1988 exchange functional,[274] respectively. The MBPT2(BH&H LYP) method has been applied to compute VDEs for a sizable database of $(H_2O)_N^-$ cluster isomers,[270] and in all but one case the canonical MP2 and the MBPT2(KS) values of the VDE were found to differ by no more than 0.15 eV. The outlier is a case where the Hartree–Fock determinant is unbound, and therefore the subsequent application of perturbation theory is ill-conceived.

Although it is not clear how general this MBPT2(KS) approach might be for other weakly-bound anions, for $(H_2O)_N^-$ clusters the bounds on the VDE that are obtained by MP2 calculations (lower bound) and MBPT2(BH&H LYP) calculations (upper bound) have recently been tested against benchmark results obtained using the ADC(2) method (second-order algebraic diagrammatic construction).[275,276] ADC(2) is a Green's function technique that, while based on second-order MBPT, is derived from a perturbative expansion of the VDE itself and should therefore be less circumspect than MP2 for very weakly-bound anions. Results for the $(H_2O)_4^-$ and $(H_2O)_6^-$ clusters considered previously (Table 4) show that VDEs computed at the ADC(2) level lie very close to EOM-EA-CCSD results and within 0.03 eV of CCSD(T) results. Table 6 compares ADC(2) and MBPT2 results for three isomers of $(H_2O)_{24}^-$. The Hartree–Fock SOMO eigenvalues for these three isomers are $\varepsilon_{SOMO} = +0.03$ eV (unbound), -0.05 eV (very weakly-bound), and -0.5 eV (moderately strong binding). ADC(2) results are shown in four high-quality basis sets, and variations in the VDE as a function of basis set are on the order of $0.05–0.08$ eV. Despite the wide range of SOMO eigenvalues exhibited by these three isomers, the ADC(2) results are bracketed in each case by the MP2 and MBPT2(BH&H LYP) values of the VDE computed in the 6-31(1+,3+)G* basis set. In other words, the ordering of the computed VDEs is

$$MP2 < ADC(2) < MBPT2(BH\&HLYP)$$ [37]

Since it is of interest to avoid the $O(N^6)$ cost of ADC(2) or EOM-EA-CCSD calculations in favor of the $O(N^5)$ cost of MBPT2 calculations, we note that the range of VDEs established by the MP2 and MBPT2(KS) results is ~ 0.2 eV for the two cases where the Hartree–Fock wave function is bound. These bounds can be improved by the empirical scaling

Table 6 VDEs (in eV) for Isomers of $(H_2O)_{24}^-$, Computed at Various Levels of Theory[a]

Method	Basis set	Isomer[b]		
		$4^{14}6^4$B	$5^{12}6^2$C	4^68^6B
$-\varepsilon_{SOMO}$ (Hartree–Fock)	6-31(1+,3+)G*	−0.027	0.054	0.544
MP2	6-31(1+,3+)G*	0.004	0.136	0.575
MP2, correlation, scaled[c]	6-31(1+,3+)G*	0.004	0.192	0.601
ADC(2)	aDZ(7s7p)	0.147	0.132	0.626
ADC(2)	aDZ(6s6p6d)	0.162	0.194	0.636
ADC(2)	aTZ′(7s7p)[d]	0.199	0.170	0.687
ADC(2)	aTZ′(6s6p6d)[d]		0.212	
$-\varepsilon_{SOMO}$(BH&HLYP)	6-31(1+,3+)G*	−0.136	0.027	0.136
MBPT2 (BH&HLYP)	6-31(1+,3+)G*	0.346	0.362	0.726
MBPT2 (BH&HLYP), scaled[e]	6-31(1+,3+)G*	0.302	0.316	0.632

[a]MP2 and MBPT2 results from Ref. 270; ADC(2) results from Ref. 260.
[b]Isomer labels from Ref. 270.
[c]MP2 correlation energies scaled by 1.053 (see Ref. 270).
[d]The aTZ′ basis (Ref. 260) removes the most diffuse shell from aug-cc-pVTZ (aTZ).
[e]MBPT2(KS) total energies scaled by 0.8715 (see Ref. 270).

of the MBPT2 energies based on CCSD(T) results for small cluster anions, as described in Ref. 270. Upon scaling in this manner, the spread between the MP2 and MBPT2 VDEs is only slightly larger than the magnitude of the ADC(2) basis-set effects.

The unbound Hartree–Fock case is an exception. Here, the range of VDEs established by MP2 and MBPT2 calculations is 0.3 eV even after scaling, although these results continue to bracket the ADC(2) results. Although the range of VDEs is only a bit larger than in the previous two cases, the SOMO is unbound at both the Hartree–Fock and BH&HLYP levels, and therefore application of perturbation theory is inappropriate for this particular isomer. It is also not surprising in this case that the SOMO obtained from a Hartree–Fock calculation differs qualitatively from the one that is obtained using BH&HLYP,[270] since both SOMOs are crude approximations to a plane wave.

The accuracy of the MP2 results rests on the fact that the *differential* correlation energy associated with the unpaired electron is reasonably small in the case that the unpaired electron is loosely-bound.[270,206,15] Let us denote this differential correlation energy for M^- as

$$\Delta E_{corr}(M^-) = E_{corr}(M) - E_{corr}(M^-) \quad [38]$$

Unlike the total correlation energy (E_{corr}), which is extensive, the quantity ΔE_{corr} is an *intensive* property. How big should one expect ΔE_{corr} to be? An oft-cited order-of-magnitude estimate of the correlation energy in a generic molecular system is 1 eV per electron pair. This estimate originates in accurate variational calculations for helium atom, for which $E_{corr} = 1.14$ eV.[277] An additional data point comes from an essentially exact calculation (in a triple-ζ basis set) for the equilibrium geometry of H_2O, which affords a correlation energy of 1.40 eV per electron pair.[278] Thus, one may expect at least 1 eV of correlation energy for electrons paired in valence MOs.

The differential correlation energy in Eq. [38] has been studied for $(H_2O)_N^-$ and $(HF)_N^-$ cluster anions,[270,206] where it is found that $\Delta E_{corr} \lesssim 0.3$ eV, independent of cluster size but increasing essentially linearly with the VDE. This behavior is demonstrated for $(H_2O)_N^-$ clusters in Figure 25. The increase in ΔE_{corr} with increasing VDE makes sense in light of the fact that the VDE is a measure of the mean electron–cluster distance, and the linear relationship that is observed in small clusters is preserved in surface-bound isomers of $(H_2O)_N^-$ that are at least as large as $N = 18$.[206,15] Only for cavity-bound isomers of $(H_2O)_N^-$ in larger clusters, where the SOMO has significant overlap with numerous H_2O molecules, does ΔE_{corr} begin to approach values of 0.5–0.6 eV.[206,15] Such values are closer to – but still smaller than – what one should anticipate for an electron in a doubly-occupied MO. This is consistent with the notion that spatial separation between the SOMO and the other occupied MOs reduces the magnitude of correlation contributions to the VDE, as compared to the correlation contribution to the IP of a closed-shell molecule.

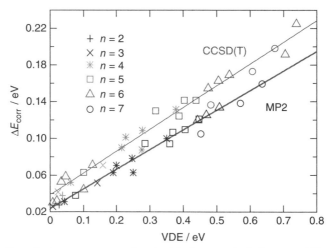

Figure 25   Correlation energy associated with the unpaired electron, ΔE_{corr} (Eq. [38]), computed at either the MP2/ or CCSD(T)/6-31(1+,3+)G* level for various isomers of $(H_2O)_N^-$. Reprinted with permission from Ref. 270; copyright 2006 the PCCP Owner Societies.

Large Systems and the Solution Phase

Our discussion of correlated wave function techniques demonstrates that accuracy sufficient to make quantitative comparisons with experiment is achievable using standard CC methods that work well in other contexts, albeit with some specialized basis-set requirements when weakly-bound anions are on the menu. Useful (if not quantitative) accuracy can be obtained at the MP2 level, provided that the VDE is $\gtrsim 0.05$ eV, the regime in which the Hartree–Fock determinant for the anion tends to be bound and thus normalizable. The reduced scaling of MP2 theory – $O(N^5)$ with respect to system size as compared to $O(N^6)$ for CCSD and $O(N^7)$ for CCSD(T) – offers the possibility of treating much larger systems. In 2008, this author[206] computed MP2/6-31(1+,3+)G*-level VDEs for cluster anions as large as $(H_2O)_{33}^-$ (≈ 1000 basis functions) using only modest computing resources (4 Gb of memory and 40 Gb of disk space per node, running in serial for systems with no symmetry). Today, MP2 calculations and even some higher-level methods are feasible in considerably larger systems,[264–266] by means of resolution-of-identity (RI) techniques and related methods,[279–281] which we shall briefly summarize.

At the heart of the RI-MP2 method is an expansion of the shell pairs $|\mu\nu)$ in an "auxiliary" basis of atom-centered Gaussians, $|K)$. This allows the four-index electron repulsion integrals required in MP2 theory to be expressed in terms of three-index integrals:[279]

$$(ia|jb) = \sum_{K,L}^{\text{auxiliary}} (ia|K)(K|L)^{-1}(L|jb) \qquad [39]$$

In this equation, which is exact if the auxiliary basis is complete, the quantity $(K|L)^{-1}$ is an element of the inverse overlap matrix (or some other metric matrix[279,282]) for the auxiliary basis, and the other quantities are electron repulsion integrals. While the formal scaling of the resulting RI-MP2 algorithm remains $O(N^5)$, the prefactor is reduced by up to a factor of ~ 20[282] such that the computational bottleneck is usually the iterative Hartree–Fock calculation rather than the post-SCF integral transformation. In practice, auxiliary basis sets $\{|K)\}$ are typically uncontracted and three to six times larger than the primary basis set, $\{|\mu)\}$, and extend to one unit higher in angular momentum. This ensures that the basis $\{|K)\}$ is sufficiently flexible to model the product functions $|\mu\nu)$. Using auxiliary basis sets that are specifically matched to – and optimized for – the primary basis set,[283,284] MP2 total energies are reproduced to within 30–60 μhartree per atom.[284,282] Relative energies and other energy differences are even more faithful to conventional MP2 results.

Timings for the Hartree–Fock part of the calculation can be dramatically reduced using either RI techniques (also known as *density fitting*[285,286]) to build the Coulomb and exchange matrices, or alternatively, dual-basis Hartree–Fock methods combined with RI-MP2.[287] The dual-basis RI-MP2 and density fitting RI-MP2 methods can routinely be extended to systems with ~ 50 atoms using high-quality basis sets, and to systems with ~100 atoms using smaller basis sets. As such, semiquantitative MP2-level VDE calculations should now be considered routine for systems of this size, although some care must be taken to modify and test the auxiliary basis sets for use with the highly diffuse primary basis sets that are required for weakly-bound anions. [Auxiliary basis sets designed for a floating-center treatment of $(H_2O)_6^-$ were developed in Ref. 288.] That said, and recognizing that efficient RI-MP2 gradient algorithms are also available,[289] for example, in Q-Chem,[194,195] geometry optimizations at the RI-MP2 level in systems with $\gtrsim 50$ atoms remain computationally demanding, and *ab initio* molecular dynamics simulations in this size regime are probably out of the question at present. For these tasks, one must rely on less expensive density-functional methods, which exhibit the same formal scaling as Hartree–Fock calculations and are amenable to the same density fitting and dual-basis acceleration techniques. The application of DFT to anions is considered later in this chapter.

The move toward larger systems allows one to consider solvation effects, by performing a quantum chemistry calculation involving the anion and one or more nearby solvent molecules. However, unless one is interested in relatively small gas-phase cluster anions (in which case the entire cluster might be described quantum mechanically), this sort of "microsolvation" approach is unlikely to describe solvation effects quantitatively or even semiquantitatively. For example, in an attempt to study $e^-(aq)$, this author has computed VDEs for $(H_2O)_N^-$ clusters that were extracted from a bulk-phase simulation of $e^-(aq)$, such that the cluster geometries are expected to be representative of electron binding motifs in bulk water. In these calculations, $N = 25$–30 water molecules were described using quantum chemistry while an additional 18,000 water molecules were described by classical point charges.[212,162] For the electronic absorption spectrum of $e^-(aq)$, this rather elaborate QM/MM setup results in *quantitative* agreement with the spectrum measured experimentally in bulk water.[212,162] Nevertheless, the computed VDEs were no larger than 2.0–2.5 eV, whereas experimental measurements for the bulk-phase VDE of $e^-(aq)$ range from 3.3 to 3.7 eV.[7–10] Evidently, the absorption spectrum is sensitive only to a correct treatment of the excess electron's wave function, which in turn depends on having a good description of the first two solvation shells into which this wave function penetrates. To predict the VDE, however, one must accurately describe the long-range Coulomb interactions in both charge states, which requires a much longer-range treatment of electronic reorganization.

Dielectric continuum solvation models can help to accelerate convergence toward the bulk limit. A simple, qualitative model for spherical solvent clusters was introduced in Eq. [6], in which the solvent's optical dielectric constant (ε_∞) provides a continuum correction for electronic reorganization. More general and more sophisticated dielectric continuum models exist, wherein the interface between the molecule and the continuum is allowed to be "molecule-shaped" rather than spherical (by means of a union of atom-centered van der Waals spheres, for example), and where the polarization of the continuum is iterated to self-consistency alongside the solute's wave function. In quantum chemistry, such methods are typically known as *polarizable continuum models* (PCMs),[290-292] or alternatively, *self-consistent reaction-field models*.[172,293] A complete discussion of these models is beyond the scope of this review, and the reader is referred to several recent reviews of the subject.[290,294,295] For a discussion of the theoretical connections between various treatments of continuum electrostatics [e.g., the COSMO, GCOSMO, IEF-PCM, and SS(V)PE models], see Refs. 293 and 294, and for a discussion of how these models perform relative to empirical continuum models such as SMx,[291] see the work of Cramer, Truhlar, and others.[291,296-299]

Here, we note only that the properties computed from these models are *extremely* sensitive to how the "molecular cavity" (solute/continuum interface) is constructed, and that nonelectrostatic solvation effects are typically (though not always[297,300]) neglected in the PCMs such as COSMO, GCOSMO, IEF-PCM, and SS(V)PE that are derived from Poisson's equation for continuum electrostatics. In contrast, such effects are built into the empirical SMx models and are crucial to accurate prediction of solvation free energies.[291,297,298] It is unclear how the neglect of nonelectrostatic effects might impact the calculation of anion VDEs; cavitation effect should cancel, but dispersion effects may not, as the anion is intrinsically more polarizable. One may hope that these effects will disappear if a sufficiently large number of explicit solvent molecules is included as part of the QM solute.

One aspect of PCMs that does warrant mention in the context of VDE calculations is the issue of equilibrium versus nonequilibrium solvation.[301,172,290,294] Traditionally, this distinction has been considered in the context of electronic excitation energies,[302-307] where the solute wave function is excited in the presence of a continuum description of the solvent. In such a situation, "equilibrium solvation" – in which both ground- and excited-state wave functions are equilibrated separately and self-consistently to a continuum solvent whose dielectric constant is ε_s – is not the appropriate way to proceed. This is because the static dielectric constant (ε_s) includes the effects of orientational averaging over the solvent molecules, but these orientational degrees of freedom are too slow to readjust on the timescale of a vertical electronic excitation. Rather, for the excited state one should include a correction term in which the excited-state electron density is equilibrated self-consistently with the "fast" part of the continuum polarization, meaning that generated by ε_∞, but also subject to the electrostatic potential arising from the "slow" (orientational) polarization from the ground state, which is generated by $\varepsilon_s - \varepsilon_\infty$.[301,307] Other versions of this nonequilibrium correction have been derived and implemented in electronic structure codes,[306] but only the one described in Ref. 307 (which is based on much older work by Marcus[301]) is correct in the high-dielectric ($\varepsilon_s \to \infty$) limit.

Table 7 Aqueous-Phase Ionization Energies Measured via Liquid Microjet Photoelectron Spectroscopy and Calculated Using Continuum Solvation Models[a]

| | | Vertical ionization energy (eV) | |
| | | MP2 + PCM | |
Molecule	Experiment	Equil.[b]	Nonequil.[c]
Cytosine		6.6	7.9
Cytidine	8.3	6.6	7.8
CMP^{-d}		6.7	7.8
CMP^{2-d}		6.6	7.7
Thymine		6.6	7.9
Deoxythymidine	8.3	6.7	7.8
dTMP^{-d}		6.7	7.7
dTMP^{2-d}		6.6	7.7
H$_2$PO$_4^-$	9.5	7.6	8.9

[a]Reprinted with permission from Ref. 308 (copyright 2009 Chemical Society), except for the experimental H$_2$PO$_4^-$ ionization energy, which is from Ref. 136
[b]Using an "equilibrium" PCM that incorporates ε_s only.
[c]Using a nonequilibrium PCM that incorporates electronic reorganization using ε_∞.
[d]CMP, cytidine monophosphate; dTMP, deoxythymidine monophosphate.

We conclude this discussion with some data demonstrating the importance of equilibrium versus nonequilibrium solvation for aqueous-phase ionization. Table 7 shows VDEs for several aqueous-phase nucleobases, nucleosides, and nucleotides, computed at the MP2 level using both equilibrium and nonequilibrium PCMs. In the former case, the parent and the ionized species are separately equilibrated to a solvent whose dielectric constant is $\varepsilon_s = 78$, whereas in the latter case the nonequilibrium version of TD-DFT + PCM is used to estimate a correction on electron detachment such that only $\varepsilon_\infty = 1.78$ is used to relax the solvent polarization. (See Ref. 309 for details.)

For the examples in Table 7, VDEs computed with the equilibrium PCM are 1.0–1.3 eV smaller than the nonequilibrium values, the latter of which lie closer to experimental results obtained from liquid microjet photoelectron spectroscopy.[309,136] The ≈ 0.5 eV discrepancies that remain between the nonequilibrium theory and the experimental VDEs probably have myriad origins, not the least of which is likely the inherent limitations of a continuum description of the solvent. This comparison between theory and experiment serves as evidence of the complexity of solution-phase VDE calculations.

Density Functional Theory

A pedagogical introduction to KS DFT can be found in a previous chapter in this series.[40] While DFT methods are highly appealing in terms of their low cost, the description of anions by DFT, even strongly-bound ones like F$^-$, has a controversial history that is summarized below. Until recently, the consensus view seemed to be that DFT methods were inappropriate for anions – especially in the context of VDE calculations – owing to problems associated with spurious self-interaction error (SIE).[46,310,308,311–313] A good historical overview of how the understanding of anion DFT calculations has evolved is given in a comprehensive review of anion DFT calculations by Schaefer and coworkers.[43] Summarizing briefly, there are two main objections to the application of DFT to anions.

1. DFT tends to yield *positive* HOMO eigenvalues, even for species with sizable EAs, suggesting that anions are unbound in DFT (in the KT sense).
2. SIE causes DFT to overstabilize half-filled orbitals, and in the context of electron attachment to a closed-shell molecule (forming a doublet radical anion) this means that the anion is overstabilized with respect to the neutral molecule, possibly drastically.

However, given the steady progress in functional development it is nowadays broadly recognized that DFT has an important role to play in anion electronic structure theory, even (with appropriate caveats, to be discussed) in the case of weakly-bound anions. At the same time, the literature is rife with egregious missteps and dubious conclusions because of ill-conceived DFT calculations for anions. In what follows, we attempt to sort this out and to address the two criticisms enumerated above.

Description of Anions in DFT
The effective one-electron potential in KS DFT is traditionally expressed as follows:

$$v_{KS}(\mathbf{r}) = v_{ne}(\mathbf{r}) + v_H[\rho](\mathbf{r}) + v_{xc}[\rho](\mathbf{r}) \qquad [40]$$

The three terms represent the sum of attractive nucleus–electron Coulomb potentials (v_{ne}), the Hartree potential

$$v_H[\rho](\mathbf{r}) = \int d\mathbf{r}' \frac{\rho(\mathbf{r}')}{|\mathbf{r} - \mathbf{r}'|} \qquad [41]$$

that represents classical electron–electron repulsion, and the exchange-correlation potential (v_{xc}), which is the rug under which all complexity is swept. For an electron that is well separated from any nucleus, $v_{ne}(r) \sim -Z/r$ in atomic units, where Z is the sum of all atomic numbers. Taking the lowest multipole moment of $\rho(\mathbf{r})$, we furthermore obtain $v_H(r) \sim N/r$ in the same asymptotic limit, where N is the total number of electrons. Finally, the asymptotic form of the exchange-correlation potential can be shown to be $v_{xc}(r) \sim -1/r$.[314] All together, the asymptotic KS potential is thus

$$v_{KS}(r) \sim -\frac{Z - N + 1}{r} \qquad [42]$$

For electron attachment to a neutral molecule $(Z = N - 1)$, one obtains $v_{KS}(r) \to 0$ asymptotically, consistent with previous remarks that there is no long-range $-1/r$ potential for electron attachment to a neutral molecule.

In Hartree–Fock theory, the nonlocal exchange operator $-\hat{K}$ replaces v_{xc} in Eq. [40], and the nonlocality recovers the correct form, $v_{xc}(r) \sim -1/r$, for an electron that is well separated from the other $N - 1$ electrons. The density-functionals in common use, however, are based on *local* exchange and correlation approximations, for which v_{xc} falls off with the density, which is to say, exponentially with distance. Note that this includes not just the local density (homogeneous electron gas) approximation, LDA, but also generalized gradient approximations (GGAs) that are still local potentials in the mathematical sense. (Sometimes GGAs are termed *semilocal* approximations, to distinguish them from LDA, but v_{KS}^{GGA} is still a local, scalar potential.)

The result is that, in practice, $v_{KS}(r) \sim -(Z - N)/r$ for a well-separated electron subject to a local KS potential. The basic problem is that $v_H(r)$ is the classical electrostatic potential for N electrons rather than $N - 1$, and approximate, local exchange potentials fail to fully cancel this spurious "self-interaction." As a result, an electron that is well separated from the molecule feels a repulsive potential generated by N electrons rather than $N - 1$. Setting $Z = N - 1$, the SIE leads to an anomalous $v_{KS}(r) \sim +1/r$ asymptotic potential for electron attachment to a neutral molecule. The electron–molecule potential is therefore repulsive at long range in approximate DFT!

As a result of SIE and the repulsive long-range potential that it engenders, anions that should be bound states in the real world are actually metastable resonances in the universe described by most approximate density-functionals. Figure 26(a) shows an example for the case of Li^-. On the basis of an accurate QMC density for Li^-, the Kohn–Sham equation can be inverted to determine the potential, $v_{KS}(r)$, whose ground-state orbitals reproduce the QMC density.[46] This may be considered the exact KS potential for this particular system, and indeed the asymptotic form of this potential is found to be $-1/r$, as it should be. This potential also binds an energy level corresponding to a stable Li^- ion with a $(1s)^2(2s)^2$ electron configuration. However, when the QMC density ρ is used to evaluate the KS potential

$$v_{KS}[\rho](r) = \frac{\delta E[\rho]}{\delta \rho} \qquad [43]$$

using an approximate density-functional $E[\rho]$, the potential rises above zero at intermediate distances and decays only very slowly back to zero [see Figure 26(b)].[46] Solution of the KS equations affords an energy level $\varepsilon = +0.80$ eV, which represents a metastable resonance since $v_{KS}(r) \to 0$ as $r \to \infty$. This point has been raised in various places, and used to suggest that DFT (with approximate functionals containing SIE) is not appropriate for the study of negative ions.[315,316,45] Notably, the density in KS DFT is represented in terms of a Slater determinant, so the considerations discussed

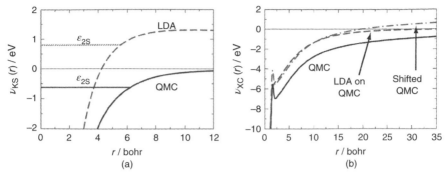

Figure 26 (a) The Kohn–Sham potential, $v_{KS}[\rho](r)$, for Li⁻. The solid curve is the potential obtained from an accurate QMC density, for which the KS equations are inverted to obtain the corresponding KS potential. The broken curve is the potential obtained from the LDA functional, evaluated using the QMC density. (b) The exchange-correlation part of v_{KS} for Li⁻, demonstrating that in the asymptotic region the LDA result (evaluated using the QMC density) differs from the QMC result by a roughly constant overall shift of the potential. Adapted with permission from Ref. 46; copyright 2010 American Chemical Society.

previously regarding positive orbital exponents in Hartree–Fock calculations are pertinent here as well: if $\varepsilon_{SOMO} > 0$ for a doublet radical anion, then in the limit of a complete basis set, the density for the "anion" M⁻ should converge to that of the neutral molecule M superimposed on a free electron.

The LDA version of $v_{KS}(r)$ that is plotted in Figure 26(a) has an outer classical turning point at $r = 17$ Å,[46] and one could therefore argue that in typical atom-centered basis sets, the lack of basis functions at such large values of r effectively forces the potential to go to infinity in those regions, and therefore the state that is metastable in the basis-set limit is transformed into a bound state in a finite-basis calculation. (In other words, the electron is trapped behind a very wide barrier, but the finite basis set does not allow one to notice this fact or to sample regions beyond the barrier.) *Extremely* diffuse basis sets may be required to detect this behavior. For example, DFT calculations on F⁻ show that ε_{HOMO} appears to converge to a positive value, even when the aug-cc-pV5Z basis is further augmented with an even-tempered progression of diffuse p functions out to $\zeta = 2 \times 10^{-5} \, a_0^{-2}$, corresponding to a FWHM of 197 Å.[44] However, with diffuse functions out to $\zeta = 10^{-10} \, a_0^{-2}$ (FWHM $> 9 \times 10^4$ Å), the HOMO eigenvalue converges to zero from above.[317] In the latter calculation, only a fraction (14%) of an electron is transferred in the asymptotic region, so that the converged DFT solution describes the system as $F^{0.86-} + 0.14 \, e^-$.[317]

At the same time, these extra diffuse functions change the energy (and therefore the computed EA) only modestly, if at all, and DFT values for EAs are often not *disastrously* wrong,[43] even if the accuracy may be insufficient for the study of weakly-bound anions. (The accuracy of approximate DFT methods is discussed in more detail below.) This apparent paradox between predicted

"ΔSCF" EAs that are often quite reasonable, but HOMO energy levels that are unbound, can be explained by noting that the LDA potential $v_{KS}[\rho](r)$, evaluated using the QMC density, differs by a roughly constant shift from the QMC potential itself, at least outside of the core region [Figure 26(b)].[46] Note that the KS potential is arbitrary up to a constant. Moreover, orbital energies play different roles in Hartree–Fock theory (where the Slater determinant is intended as a genuine approximation to the wave function, and the HOMO–LUMO gap approximates the "fundamental gap", IP – EA) as compared to Kohn–Sham theory (where the Slater determinant represents a fictitious reference system, and the HOMO–LUMO gap approximates the "optical gap," or in other words the lowest electronic excitation energy).[318] Together, these observations suggests that a positive HOMO eigenvalue is not automatically fatal in KS DFT, nor is it a sign that the DFT result must be rejected out of hand.[46] That said, if the basis set extends beyond the turning point in the potential, then the density *will* begin to resemble an unbound electron. In the case of F⁻, the potential barrier is high enough and/or wide enough that this does not occur except in ludicrously diffuse basis sets. For weakly-bound anions, however, one cannot be certain that the situation will be equally favorable.

Performance of Standard Functionals

The situation described above implies that one cannot trust the approach to the basis-set limit for most anion DFT calculations. (In fact, the complete-basis limit may not be completely well-defined for anion DFT,[313] as it may correspond to fractional electron transfer to the continuum.[317]) One solution to the lack of a well-defined basis-set limit in anion DFT calculations is to compute orbitals using Hartree–Fock theory, where the anion's HOMO tends to be bound (except in cases of very weak electron binding) and then use the Hartree–Fock density to evaluate energy by means of a density-functional containing both exchange *and* correlation.[46,313] The results of such a procedure, as applied to the G2-1 set[252] of EAs, are shown in Figure 27. (The G2-1 set includes EAs for atoms and small molecules that range from ∼0 up to 3.6 eV, with most of the EAs in the data set being > 1 eV.) The figure compares self-consistent B3LYP predictions for the EAs to predictions obtained by computing the B3LYP energy of the Hartree–Fock density. Unfortunately, one cannot say that, on the whole, the composite procedure affords any clear improvement over self-consistent B3LYP calculations; the advantage is that the HF-B3LYP procedure has a well-defined basis-set limit that should bind the electron.

In light of previous discussion, positive HOMO eigenvalues do not appear to be fatal for DFT anion calculations, and thus another approach is simply to charge ahead but benchmark thoroughly. To this end, Schaefer and coworkers[43] have benchmarked the performance of numerous density-functionals and basis sets against a data set consisting of 91 AEAs that have been determined, using photoelectron spectroscopy, to an accuracy of better than ±0.09 eV. (On the

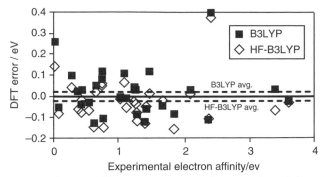

Experimental electron affinity/ev

Figure 27 Errors in electron affinities (as compared to experiment) for self-consistent B3LYP/aug-cc-pVTZ calculations, and also for cases where the B3LYP energies for the neutral molecule and the anion were evaluated using Hartree–Fock densities (HF-B3LYP). The dashed horizontal lines are mean errors, which include considerable cancellation between over- and underestimation of the experimental EAs. The data set is G2 1,[252] except that the CN molecule is removed because of spin contamination in the Hartree–Fock determinant. Adapted with permission from Ref. 313; copyright 2011 American Institute of Physics.

Table 8 Error Statistics for DFT DZP++ Calculations of Adiabatic EAs[a]

	B3LYP	B3P86	BH&HLYP	BLYP	BP86	LSDA
Mean abs. error	0.14	0.59	0.24	0.14	0.18	0.68
	(0.16)	(0.60)	0.25	(0.15)	(0.19)	(0.67)
Max. abs. error	0.71	1.04	0.87	0.67	0.62	1.23
	(0.76)	(1.14)	(0.87)	(0.71)	(0.66)	(1.01)
Std. deviation	0.14	0.16	0.17	0.13	0.13	0.16
	(0.17)	(0.18)	(0.18)	(0.13)	(0.15)	(0.14)
% of AEAs that are overest'd	71	99	25	46	87	100
	(68)	(99)	(29)	(45)	(86)	(100)

Reprinted with permission from Ref. 43; copyright 2002 American Chemical Society.
[a]The data set is 91 atoms and molecules for which experimental uncertainties are no worse than ±0.09 eV, excluding SF_6. Values in parenthesis are vibrational zero-point energy contributions.

basis of the comprehensive review of 1,101 experimentally-determined AEAs in Ref. 43, this data set consists of essentially *all* of the accurately determined AEAs, as of 2002.) Error statistics for a subset of these functionals are reported in Table 8.

In general, the species in the data set are not weakly-bound anions, and the basis set used in these calculations (DZP++) reflects that. Although this basis does contain diffuse s and p functions on the heavy atoms, with different exponents, the ratio of those exponents is only, for example, 1.185 for carbon and 1.264 for oxygen,[43] rather than the scaling factors of ≈ 3.5 that

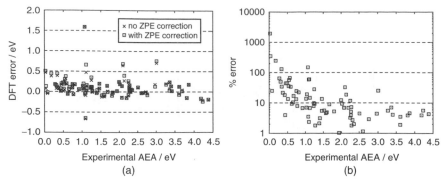

Figure 28   (a) Errors in B3LYP/DZP++ calculations of adiabatic EAs, either with or without a correction for the vibrational zero-point energy (ZPE), for 91 atoms and molecules whose experimental AEAs are known to be better than ±0.09 eV. (The outlier, with an error of 1.59 eV, is SF_6, which was excluded from the error statistics in Table 8.) (b) Percentage errors in the ZPE-corrected values, on a logarithmic scale. Both plots are based on data provided in Ref. 43.

were suggested in the discussion of basis sets presented earlier in this chapter. As such, the DZP++ basis set is effectively a singly-augmented one. (A variety of other singly-augmented basis sets were tested in Ref. 43 but did not afford results that were improved in any statistically meaningful way relative to DZP++ results.) The B3LYP/DZP++ and BH&HLYP/DZP++ results are fairly good, with mean absolute errors of 0.14 and 0.24 eV, respectively, although the maximum errors are 0.7–0.9 eV. The error for SF_6 is larger, and this molecule is therefore excluded from the statistics in Table 8. The authors of Ref. 43 attribute this larger error to the lack of f-type polarization functions in the DZP++ basis set.

The actual B3LYP/DZP++ errors are plotted in Figure 28(a) as a function of the magnitude of the experimental AEA. Somewhat surprisingly, the errors do not appear to be any larger for cases where the AEA is small (say, < 0.5 eV) as compared to cases where the AEA is several eV. Certainly, the *percentage* error is quite large for the more weakly-bound cases, and an error of ≈ 0.5 eV (toward overbinding) is fairly egregious for an anion that, experimentally, is almost unbound. Figure 28(b) shows that while B3LYP/DZP++ errors are *mostly* < 10% in cases where the AEA is larger than 1.5 eV, errors of 50–100% or more are not uncommon when the AEA is less than 1 eV. It is also worth noting that zero-point corrections have almost no effect on the error statistics, according to results in Table 8. Detailed examination of the B3LYP data from Ref. 43 shows that the difference between zero-point corrected and uncorrected AEAs is no larger than 0.05 eV for any molecule in the data set. Although one expects the zero-point corrections to be small (possibly vanishing) for dipole-bound anions, these are not well represented in the data set, hence the smallness of the zero-point corrections is perhaps something of a surprise. It may be attributable to the relatively small sizes of the molecules in the data set (organic molecules no larger than tetracene, $C_{18}H_{12}$, and inorganic compounds no larger than SF_4).

The comprehensive benchmark study of Schaefer and coworkers,[43] which recommended B3LYP/DZP++ for EA calculations, is already more than a decade old at the time of this writing. Although B3LYP continues to enjoy widespread use in computational chemistry, considerable progress in functional development has transpired since the writing of Ref. 43 to the point that B3LYP is arguably no longer state-of-the-art. However, more modern functionals have not yet been tested exhaustively for EA and VDE calculations. Truhlar and coworkers[319–322] have compared a great many functionals, both new and old, against a small database, consisting of 13 AEAs for atoms and small molecules for which the data range from 1.3 to 3.6 eV.[97] (This is the data set that is listed in Table 1 of this chapter.) In compiling this data set, experimental AEAs were adjusted by subtracting out a harmonic DFT estimate of the zero-point energy, to obtain "zero-point exclusive" experimental AEAs.[97]

Mean unsigned errors (MUEs) for the AEAs in this data set, computed using a wide range of functionals, are shown in Figure 29. (Only functionals affording a smaller MUE than B3LYP are considered.) For context, the MUE for Hartree–Fock AEAs in this same data set is 1.2 eV,[320] an order-of-magnitude larger than the DFT errors, consistent with a value of $\Delta E_{corr} \sim 1$ eV for the differential correlation energy in Eq. [38]. This is larger than the values of ΔE_{corr} that were observed for $(H_2O)_N^-$ and $(HF)_N^-$ cluster anions, but many of the AEAs in the data set (Table 1) are quite sizable and correspond to placing an electron into a valence orbital, for which one should expect ΔE_{corr} to be closer to 1 eV.

In examining the error statistics in Figure 29, one should keep in mind that the MUE for B3LYP of Table 1 and Ref. 97 is 0.10 eV, as compared to the MUE of 0.16 eV that is obtained in the larger data set of Ref. 43. This comparison affords some estimate of how these MUEs might shift around if a different set of molecules was considered. With this in mind, the only functionals that really stand out in comparison to B3LYP are M11[325] and M08-HX,[326] which have MUEs of 0.04 and 0.06 eV, respectively, for the small data set. It is worth mentioning, however, that all 13 AEAs in this data set are included in the training set that is used to parameterize both of these functionals. (They are also included in the training set for the SOGGA11-X functional,[327] which according to Figure 29 also performs reasonably well for these AEAs.) Errors in AEAs for these functionals, using a data set on which the functionals have not been trained, do not appear to have been reported in the literature.

Let us now turn our attention back to weakly-bound anions. The average error in B3LYP/6-31(1+,3+)G* VDEs for $(H_2O)_N^-$ isomers was quoted as 0.24 eV in Ref. 202 on the basis of CCSD(T) benchmarks for $N \leq 6$. In truth, this value is not much larger than the average B3LYP error of 0.16 eV that is reported in Table 8, although absolute errors of this magnitude constitute a

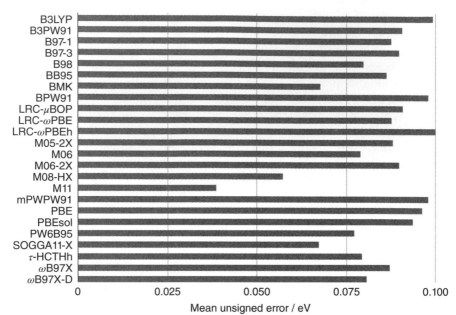

Figure 29 Mean unsigned errors for the database of 13 AEAs given in Table 1, with zero-point corrections removed from the experimental data,[97] and considering only functionals that perform at least as well as B3LYP. Note that the Minnesota (M) functionals and SOGGA11-X were parameterized based in part on this same set of AEAs. The basis set is G3(MP2)large,[244] but with diffuse functions removed from the hydrogen atoms, a basis that Truhlar and coworkers call MG3S. Data for the three long-range corrected (LRC) functionals are reported here for the first time; the parameters in these functionals are taken from Ref. 323 (LRC-ωPBE), Ref. 324 (LRC-ωPBEh), and Ref. 202 (LRC-μBOP). The remaining data are taken from Refs. 319–322.

much larger *fraction* of the VDE for a very weakly-bound anion, where even a 0.1–0.2 eV error might constitute 50–100% of the actual VDE. Moreover, the basis sets used to compile the error statistics in both Table 8 and Figure 29 are all singly-augmented, and thus inadequate for weakly-bound anions. Looking at the most diffuse exponent on the carbon and hydrogen nuclei in these basis sets and using Eq. [28] to convert the exponent into a minimum VDE suggests that these basis sets are appropriate only for VDEs $\gg 0.4$–0.5 eV. For the more electronegative oxygen atom (which has a correspondingly smaller diffuse exponent), the same estimate suggests that these singly augmented basis sets are appropriate when the VDE is > 0.9 eV.

This is especially troubling in view of the fact that SIE, which leads to *overestimates* of VDEs for doublet radical anions in the complete-basis limit (because the SIE preferentially stabilizes the anion, with its half-filled orbital), can be substantially cancelled by the incompleteness of the basis set. (Diffuse functions preferentially stabilize the anion, so their omission *destabilizes* the anion.) The weakly-bound $(H_2O)_{12}^-$ isomer in Figure 18(a) is a good example. In an overly compact basis set such as 6-31++G*, which is $n = 1$ in Figure 18(a), the B3LYP functional predicts VDE ≈ 0, which is in fact the right answer, based on MP2 calculations with very diffuse basis sets. However, the B3LYP VDE increases steadily as the quality (diffuseness) of the basis set is improved, such that when a more appropriate basis like 6-31(1+,3+)G* is used, the B3LYP prediction is a VDE of 0.2 eV. As such, it is possible (though certainly not advisable!) to use benchmark calculations or experimental data to "tune" the diffuseness of the basis set such that a given functional may afford an accurate VDE. However, there is no way to tell whether that particular combination of functional and basis set has any robustness or generality, and there is every reason to suspect that it does not.

Another hallmark of self-interaction problems, and another opportunity for ill-advised tweaking of DFT parameters, is the fact that DFT errors in VDEs are tunable, over a fairly wide range, as a function of the fraction of Hartree–Fock exchange that is included in the functional.[191] Of the commonly used functionals (as of 2005), BH&HLYP was the only choice found to yield useful results,[191] which is why this functional was selected for tests of the MBPT2(KS) approach that was discussed previously. VDEs predicted using BH&HLYP were typically ~ 0.03 eV larger than CCSD(T) predictions,[191] and this is borne out in the benchmark VDE comparison for $(H_2O)_N^-$ clusters that is shown in Figure 30(a). This comparison suggests a tractable way of putting error bars on computed VDEs, namely, by comparing the MP2 and BH&HLYP values,[191,270] although this approach would need to be tested against other methods on a wider variety of systems before it could convincingly be deployed to study weakly-bound anions other than the water and hydrogen fluoride cluster anions studied in Refs. 191 and 271.

Self-Interaction Corrections
Rather than tuning the combination of functional and basis set, and benchmarking against wave function methods for similar systems, another strategy is to try to eliminate the SIE that is a major source of error in VDE calculations. A simple self-interaction correction (SIC) was proposed in 1981 by Perdew and Zunger (PZ),[328] in which the self-interaction is subtracted orbital-by-orbital from the KS energy:

$$E_{PZ}[\{\rho_{i\sigma}\}] = E_{KS}[\rho_\alpha, \rho_\beta] - E_{SIC}[\{\rho_{i\sigma}\}] \qquad [44]$$

Here, $\rho_{i\sigma} = |\phi_{i\sigma}|^2$ is the density associated with an occupied MO, and $\sigma = \alpha$ or β is a spin index (thus $\rho_\sigma = \sum_i \rho_{i\sigma}$). The SIC is

$$E_{SIC}[\{\rho_{i\sigma}\}] = \sum_{\sigma=\alpha,\beta} \sum_i^{occ} (E_{xc}[\rho_{i\sigma}, 0] + J[\rho_{i\sigma}]) \qquad [45]$$

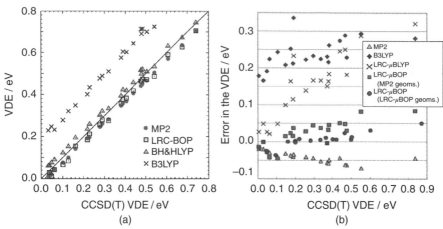

Figure 30   Comparison of VDEs for $(H_2O)_N^-$, $N = 2–7$, versus benchmark CCSD(T) results. In (a), cluster geometries were optimized at the B3LYP/6-31(1+,3+)G* level. MP2 and CCSD(T) calculations employ the 6-31(1+,3+)G* basis set while DFT calculations use the aDZ+diff basis from Ref. 202 which contains one more diffuse shell as compared to aug-cc-pVDZ, generated using a scaling factor $\kappa = 1/8$. In (b), all calculations use the aDZ+diff basis and MP2 geometries are used, except where indicated for the LRC-μBOP calculations. Panel (a) is adapted with permission from Ref. 15; copyright 2009 Taylor and Francis. Panel (b) is adapted with permission from Ref. 202; copyright 2008 American Chemical Society.

Using the notation of Hartree–Fock theory,[35]

$$J[\rho] = \int d\mathbf{r} \; v_H(\mathbf{r})\rho(\mathbf{r}) \qquad [46]$$

is the so-called Hartree (or Coulomb) functional. The Perdew–Zunger SIC in Eq. [45] failed to find widespread use, however, because it spoils the invariance of the energy with respect to unitary transformations of the occupied MOs and is thus difficult to implement self-consistently within the standard KS formalism. Systematic tests of this SIC were not reported until 2004,[329] and it was then determined that this SIC does *not* improve the accuracy of DFT thermochemistry, except in the case of the LDA functional. (Enthalpies of formation were the only properties examined in Ref. 329.)

As a simpler alternative for doublet radical anions, one might try to eliminate the SIE from the SOMO only, since that is likely the main source of error in the VDE. Jungwirth and coworkers[330,331,218,217] have reported *ab initio* molecular dynamics simulations on $(H_2O)_{32}^-$, using the BLYP and PBE functionals plus an *ad hoc* SIC of this form, which had been proposed previously.[332,333] This SIC is based on the reasoning that, if the exact E_{xc} were known, then it would cancel the self-interaction in $J[\rho]$ exactly. In view of this consideration, a sensible SIC for doublet radicals is[332]

$$\Delta E_{SIC}[\rho_\alpha, \rho_\beta] = -J[m(\mathbf{r})] - E_{xc}[m(\mathbf{r}), 0] \qquad [47]$$

where $m(\mathbf{r}) = \rho_\alpha(\mathbf{r}) - \rho_\beta(\mathbf{r})$ is the spin density. For self-consistency, one should implement ΔE_{SIC} by taking its functional derivatives with respect to density, and adding this as a correction to $v_{KS}(\mathbf{r})$,[332] although this is not always done in practice. Equation [47] can be viewed as an *ad hoc* modification of the original Perdew–Zunger procedure and is appropriate only for systems with a single unpaired electron. Unlike the more general Perdew–Zunger scheme, however, the correction in Eq. [47] is inexpensive to evaluate and does not spoil orbital invariance among the doubly occupied MOs. Unfortunately, this simplified SIC *does* lead to unphysical distortions of these MOs, such that $\phi_{i\alpha}$ becomes very different from $\phi_{i\beta}$. To prevent this, the slightly more complicated[333] restricted open shell Kohn–Sham formalism[334–336] is required,[332] in which $\phi_{i\alpha} = \phi_{i\beta}$ (by construction) for each doubly-occupied MO.

In the context of simulating the aqueous hydroxyl radical using plane-wave DFT, VandeVondele and Sprik suggested a modified form of Eq. [47] that introduces two adjustable parameters, a and b:

$$\Delta E_{SIC}[\rho_\alpha, \rho_\beta] = -aJ[m(\mathbf{r})] - bE_{xc}[m(\mathbf{r}), 0] \qquad [48]$$

These authors then took $b = 0$ by fiat, citing studies of hemibonded cation radical systems in which the SIE, which arises primarily from an overly delocalized cation hole, was found to be roughly proportional to $J[m(\mathbf{r})]$.[337,338] The value $a = 0.2$ was then chosen by comparison to CCSD(T) calculations for some cation dimer radicals. Whether this rationale extends to anions is unclear, and in fact very different parameters ($a = 0.8$ and $b = 0.5$) have been suggested on the basis of studies of other, non-hemibonded cation radicals.[339] Nevertheless, the parameters $a = 0.2$ and $b = 0$ were adopted in the aforementioned $(H_2O)_{32}^-$ calculations,[217] whereas in *ab initio* molecular dynamics simulations of $e^-(aq)$ in bulk water, the value $a = 0.3$ (with b again fixed at zero) was found to provide better agreement with the experimental absorption spectrum.[145] In Ref. 330, isosurface plots of $m(\mathbf{r})$ are presented for one particular isomer of $(H_2O)_{32}^-$, and the result obtained at the SIC-PBE level is seen to be qualitatively similar to the MP2 result. However, the SIC-PBE and RI-MP2 VDEs are rather different, and these differences do not appear to be systematic.[330] (On the other hand, the RI-MP2/6-311G* benchmarks in Ref. 330 could certainly be improved, in terms of the diffuseness of the basis set.) Moreover, the SIC is found to have *qualitative* effects on reactivity; the aqueous-phase reaction $H^+ + e^- \rightarrow H$, simulated inside of a water cluster, proceeds readily with the SIC but not without it.[340] In view of these issues, it seems that careful, systematic benchmark studies of SIC functionals for weakly-bound anions are probably warranted.

An alternative to explicit SICs is the use of long-range corrected (LRC) functionals,[341–344] also known as range-separated hybrid functionals,[345] which have garnered significant interest over the past decade. The LRC idea, as put forward by Hirao and coworkers[341] on the basis of earlier ideas by Savin and coworkers,[346,347] is to partition the electron–electron Coulomb operator, r_{12}^{-1}, into short- and long-range components, which are then described using different theoretical models. The most common partition uses the error function, erf(x):

$$\frac{1}{r_{12}} = \underbrace{\frac{1 - \text{erf}(-\mu r_{12})}{r_{12}}}_{SR} + \underbrace{\frac{\text{erf}(-\mu r_{12})}{r_{12}}}_{LR} \tag{49}$$

The first term on the right is the short-range component, and decays to zero on a length scale $\sim \mu^{-1}$, where μ is taken as an adjustable *range separation parameter*. Consider an exchange-correlation functional of the form $E_{xc} = E_x + E_c$, and furthermore separate the exchange functional into a nonlocal Hartree–Fock component and a local GGA component:

$$E_{xc} = E_c + C_{HF}E_x^{HF} + (1 - C_{HF})E_x^{GGA} \tag{50}$$

Here, C_{HF} represents the coefficient of Hartree–Fock exchange. For example, $C_{HF} = 0.2$ for B3LYP, $C_{HF} = 0.5$ for BH&HLYP, and $C_{HF} = 0$ for BLYP. Then the corresponding LRC functional is

$$E_{xc}^{LRC} = E_c + C_{HF}E_x^{HF,SR} + (1 - C_{HF})E_x^{GGA,SR} + E_x^{HF,LR} \tag{51}$$

The labels "SR" and "LR" indicate whether each component is evaluated using the short-range or the long-range component of r_{12}^{-1}, as defined in Eq. [49].

The terms $E_x^{HF,SR}$ and $E_x^{HF,LR}$ in Eq. [51] can be handled using modified electron repulsion integrals,[348] but $E_x^{GGA,SR}$ requires fundamental modifications to the GGA exchange functional. Two different ways to perform these modifications have been suggested: one on the basis of a modification to the GGA exchange enhancement factor,[341] and another on the basis of a modified exchange hole.[349] The papers developing the former *ansatz* have mostly been written by Hirao and coworkers,[341,342,350,351,343] who consistently use the symbol μ to represent the range separation parameter, as in Eq. [49], whereas Scuseria and coworkers[33,352,353,349,345] consistently use the symbol ω instead. As such, a notation for LRC functionals is the following:[354] LRC-μGGA is the LRC version of a particular GGA functional (e.g., GGA = PBE or BLYP) that is constructed according to the modified exchange enhancement factor introduced by Hirao's group.[341] LRC-ωPBE, on the other hand, denotes the LRC functional that uses the short-range ωPBE functional from the modified exchange hole introduced by Scuseria and coworkers.[349] (Whereas Hirao's construction can be applied to both the PBE and Becke88 exchange functionals, affording short-range exchange functionals that we denote as μPBE and μB88, respectively, the exchange-hole formalism has been applied only to PBE, affording the ωPBE short-range exchange functional.) This notation, introduced in Ref. 354, is consistent with how these functionals are designated in the Q-Chem electronic structure program.[194,195] Note, however, that the LRC-ωPBE functional has also been called LC-ωPBE,[349] and the functional discussed below that is called LRC-μBOP in Q-Chem has also been called LC-BOP.[202] One must also take care that different electronic structure programs may have different default values for the μ or ω parameter, and computed properties can be very sensitive, in some cases, to the precise value that is used.[355,356,324]

For well-separated electrons, the exchange-correlation functional in Eq. [51] consists of 100% Hartree–Fock exchange, since E_c is local. Hartree–Fock theory is free of self-interaction, at least in the sense that this term has historically been defined, namely, $E_{xc} = 0$ for a one-electron system. (More recently, this traditional definition has been called the "one-electron" SIE.[357,311]) As such, one might expect LRC functionals to mitigate some of the problems associated with the DFT description of VDEs, especially in cases where the SOMO is largely separated from the valence MOs, and this is indeed the case with *certain* LRC functionals.[202,358] This is demonstrated in Figure 30, where VDEs for a variety of $(H_2O)_N^-$ clusters are computed using LRC-μBOP and LRC-μBLYP and compared to CCSD(T) benchmarks. The LRC-μBOP functional with $\mu = 0.33$ bohr^{-1} is found to perform just as well as MP2 theory and slightly better than MP2 if the geometries are optimized at the same level of theory that is used to compute the VDEs. In the latter case, the mean absolute deviations relative to CCSD(T) are 0.014 eV for LRC-μBOP versus 0.044 eV for MP2.[202]

On the other hand, LRC-μBLYP results are only moderately better than B3LYP results; see Figure 30(b). Both LRC-μBLYP and LRC-μBOP use a short-range version of Becke88 exchange[274] (μB88) for the $E_x^{GGA,SR}$ term in Eq. [51], and the same value of the range separation parameter, μ. Thus, the only difference between the two is the correlation functional: the well-known LYP functional[273] in one case, versus the "one-parameter progressive" (OP) functional[359] in the other. Yagi et al.[202] suggest that the superior performance of LRC-μBOP originates in the fact that the OP correlation functional satisfies the exact constraint that the correlation energy density should vanish in the limit of a rapidly varying density,[359] whereas the LYP functional violates this constraint.

As mentioned above, predicted properties can be sensitive, in some cases, to the value of the range separation parameter.[356,355,324] In this context, it is worth noting that the original suggestion to use $\mu = 0.33\ a_0^{-1}$ comes not from fitting anion VDEs but rather was optimized to reproduce bond lengths for second-row diatomic molecules.[342] In other LRC functionals, the μ (or ω) parameter has been optimized to reproduce various experimental data.[353,343,349,324,323] A less empirical approach has been advocated[344] and may have particular appeal for the calculation of anion VDEs and excitation spectra, but to understand this approach we must first digress to discuss a particular theorem in exact DFT, on which the method is based.

It is known that if the exact functional $E[\rho]$ were employed in a KS DFT calculation, then the value $-\varepsilon_{HOMO}$ obtained from that calculation would be *exactly* equal to the system's smallest IP.[197,360,361] This result is in sharp contrast to the Hartree–Fock case, where $-\varepsilon_{HOMO}$ is the KT *approximation* to the IP. On the other hand, for typical approximate density functionals, the value of $-\varepsilon_{HOMO}$ is often a very crude approximation to the actual IP, much worse than the KT approximation obtained from a Hartree–Fock calculation. For example, a recent study[362] compared IPs for atoms and small molecules computed at the CCSD(T)/aQZ level to $-\varepsilon_{HOMO}$/aQZ values for several functionals and found mean differences of 1.5 eV for M05-2X, 3.5 eV for B3LYP, and 5.3 eV for BOP, as compared to 0.8 eV for Hartree–Fock theory. As such, the recommended way to compute an IP using DFT is *not* to use $-\varepsilon_{HOMO}$, but rather to use a ΔSCF approach. LRC functionals, however, do a much better job of achieving $-\varepsilon_{HOMO} \approx$ IP, with a mean error of only 0.2 eV for LRC-μBOP (with $\mu = 0.47\ a_0^{-1}$) as compared to benchmark CCSD(T)/aQZ results.[362]

Motivated by the aforementioned theorem (IP = $-\varepsilon_{HOMO}$), Baer et al.[344] have suggested tuning μ (or ω) in a system-specific way to satisfy this condition. Taking the IP to be defined by a ΔSCF calculation then provides some degree of self-consistency to this optimization procedure and provides an asymptotically correct exchange-correlation potential for the system of interest. Although Baer and coworkers have mostly been interested in ionization of neutral molecules,[363,344] in keeping with the spirit of this chapter we will state this condition in terms of the IP (= VDE) of the anion M^-:

$$E_M(\mu) - E_{M^-}(\mu) = -\varepsilon_{HOMO,M^-}(\mu)$$ [52]

The left-hand side of this equation is simply the ΔSCF value of the VDE for M^-, evaluated at a particular value of μ. The "tuning" procedure of Baer et al.[344] consists in locating a (system-specific) value of μ that satisfies Eq. [52]. While there is no theoretical guarantee that such a value must exist, some commonly used LRC functionals approximately satisfy this condition already, with values of μ that were obtained by minimizing the statistical errors with respect to some set of experimental data. This suggests that exceptional cases,[364] where no value of μ that satisfies Eq. [52] can be found, may be just that: exceptions.

Figure 31 shows an example of this tuning procedure for one particular isomer of $(H_2O)_6^-$.[162] This system was chosen because its VDE has been reported on the basis of a large-basis CCSD(T) calculation and a QMC calculation, and the two benchmark VDEs agree to within the statistical error of the QMC result.[259] For this system, the condition in Eq. [52] is satisfied when $\mu = 0.25\ a_0^{-1}$, and at this value of μ the error in the VDE (with respect to the benchmark) is 0.08 eV. In contrast, a ΔSCF calculation of the VDE agrees with

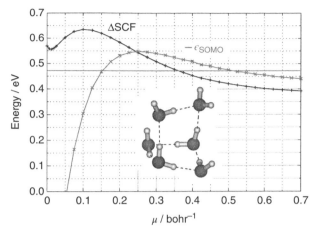

Figure 31 VDE for the "OP1-AA" isomer[128] of $(H_2O)_6^-$, computed at the LRC-μBOP/aDZ+diff level, as a function of the range separation parameter, μ. Results from a traditional ΔSCF calculation of the VDE are plotted along with the value $-\varepsilon_{SOMO}$. The solid horizontal line represents the CCSD(T) VDE from Ref. 259, which is in good agreement with quantum Monte Carlo results. Reprinted with permission from Ref. 162; copyright 2011 American Chemical Society.

the benchmark for $\mu = 0.35\ a_0^{-1}$. At this larger value of μ, the condition in Eq. [52] is violated only by about 0.05 eV, which suggests that LRC-μBOP with $0.25\ a_0^{-1} \lesssim \mu \lesssim 0.35\ a_0^{-1}$ is a good method for this system. Note, however, that the tuned value of μ often varies strongly with system size, such that a rather different value may be optimal for larger $(H_2O)_N^-$ clusters.[163]

These considerations lend some credence to the use of LRC functionals for moderate accuracy (~ 0.1 eV) VDE calculations in weakly-bound anions, which is especially promising for studies of cluster anions and other condensed-phase species, where large system sizes are required to obtain a realistic model. It has also been found that LRC functionals avoid the spurious transfer of fractional electrons from negatively to positively charged moieties on molecules that contain both.[317,365] At the same time, even for LRC functionals one is not free of the dependence of the result on the choice of functional, hence benchmarking against wave function methods remains advisable. The "tuning" criterion for μ in Eq. [52] suggests how this parameter can be validated, or perhaps modified, if some other LRC density-functional were to be used in place of LRC-μBOP.

QUANTUM CHEMISTRY FOR METASTABLE ANIONS

The traditional quantum chemistry methods discussed up to this point are designed to find the lowest Born–Oppenheimer energy and wave function (or density) for a given arrangement of the nuclei. For the AB^- system described in Figure 1, the lowest-energy state for $R > R_c$ does indeed correspond to an anion: either the molecular anion AB$^-$ (around the minimum-energy geometry), or else the radical A$^\bullet$ plus the anion B$^-$ (in the exit channel as $R \rightarrow \infty$). For $R < R_c$, however, the lowest-energy form of AB$^-$ corresponds (in the basis-set limit) to a neutral AB molecule and a free electron. As such, bound-state quantum chemistry methods are not appropriate for $R < R_c$. In that regime, bound-state quantum chemistry applied to AB$^-$ will afford an *orthogonalized discretized continuum* (ODC) solution,[366,233,367−369] consisting of ψ_{AB} for the neutral molecule superimposed on a poor quality, finite-basis approximation of a free electron. In such a case, the VDE should converge to zero in the limit of an infinitely-diffuse basis set.

Stated differently, while it is possible to use standard quantum chemistry methods to compute *an* energy for the unbound ($R < R_c$) state of AB$^-$, in general the computed wave function will not match the correct boundary conditions to qualify as a proper resonance state. In terms of the one-dimensional example in Figure 13, not every particle that is incident from the left yields a resonant enhancement of the probability density inside of the potential well. Only for a narrow range of incident particle energies is the phase matching "just right" to afford a resonance. As Simons puts it,[2] "An arbitrarily chosen basis, even with diffuse functions included, will yield but an arbitrary energy for the metastable anion rather than the correct resonance." This has not prevented the appearance of numerous studies of purportedly stable molecular anions in which what was really being examined is the structure of a neutral molecule in the field of a $-e$ charge that is smeared out over the most diffuse functions that are included in the basis set.

An example is depicted in Figure 32, which shows the results of two different calculations aimed at understanding whether electrons with near-zero kinetic energy can initiate DEA reactions within the DNA backbone.[370−372] The hypothesis was put forward that e$^-$ attachment to P–O π^* orbitals of a phosphate moiety could subsequently lead to a rupture of a sugar–phosphate C—O bond.[371] Using a charge stabilization method that is discussed later (as an appropriate technique for computing potential curves for metastable anions), Simons and coworkers[370] obtained the C—O potential energy scans shown in Figure 32(a). These potentials do suggest the existence of a $(\pi^*_{CO})^-$ anion resonance and also a dissociative $(\sigma^*_{CO})^-$ resonance. Electron attachment to form the $(\pi^*_{CO})^-$ state is therefore capable of causing C—O bond dissociation, according to the calculation. However, the calculations also suggest that the $(\pi^*_{CO})^-$ resonance lies > 2.5 eV above the energy of the neutral species in its equilibrium geometry. Bond rupture by ~ 0 eV electrons is therefore excluded, according to this calculation.[370,372]

A wholly different picture emerges when traditional bound-state quantum chemistry methods are applied, as shown in Figure 32(b).[371] A B3LYP/6-31+G* calculation of the C—O moiety affords "anion" potential energy curves that

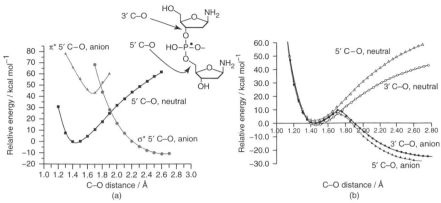

Figure 32 One-dimensional potential energy scans for a section of DNA backbone, addressing the question of whether attachment of zero-energy electrons to P–O π^* orbitals can subsequently cause rupture of a sugar—phosphate C—O bond. (a) Potential energy curves for the neutral 5' C—O bond and two-electron-attached states thereof, the latter obtained using stabilization methods that can correctly describe metastable anion resonances. Qualitatively similar curves are obtained for the 3' C—O bond but are not shown, for clarity. An electron whose energy matches the neutral-to-$(\pi^*)^-$ energy gap could potentially attach to form a temporary negative ion, which might subsequently undergo internal conversion to the $(\sigma^*)^-$ state, resulting in C—O bond dissociation. (b) Potential energy curves along both the 3' and the 5' C—O bond, obtained from a conventional B3LYP/6-31+G* calculation of the sugar—phosphate—sugar moiety and its anion. In both panels, the equilibrium value of the neutral C—O bond is used to define the zero of energy. Panel (a) is reprinted with permission from Ref. 370; copyright 2004 American Chemical Society. Panel (b) and the chemical structure drawing are reprinted with permission from Ref. 371; copyright 2003 American Chemical Society.

are essentially isoenergetic with the neutral curves, in the vicinity of the equilibrium geometry that both the neutral and anionic species reportedly share. After passing over a small barrier, the anion's potential energy curves become dissociative at larger C—O bond lengths, which was taken as evidence in favor of the hypothesis that very low-energy electrons could induce single-strand breaks.[371] However, the isoenergetic minima and similarly-shaped potential curves are also features that one would expect for a neutral molecule that is subject to the weak field arising from a smeared-out one-electron density trapped a few Angstroms away by a set of diffuse (but atom-centered) Gaussian functions. Larger bond lengths lead to a larger C—O bond dipole, such that the extra electron is probably stabilized into a σ^*_{CO} orbital as the bond is stretched. In a sense, this calculation captures the correct physics at large C–O bond lengths, yet the description of the anion states in the vicinity of the neutral molecule's equilibrium geometry is almost certainly wrong.[370]

The remainder of this section is devoted to methods that can get such systems *right*. As mentioned in the introductory remarks to this chapter, the discussion here is limited to methods that are based on relatively straightforward modifications to traditional (bound-state) quantum chemistry methods, since these are the only theoretical approaches that are widely accessible to the chemistry community. Such methods have also been reviewed in Ref. 20.

Maximum Overlap Method

Let us consider a slightly different case first, namely, the protonated model peptide system that was discussed in the context of electron transfer dissociation spectroscopy (see Figure 2). The goal of the calculations reported in Figure 2 is to obtain electron-attached Rydberg states, two of which are shown in the figure, and then to determine whether internal conversion to a dissociative $(\sigma^*)^-$ state is feasible.[77,74] At the equilibrium geometry of the parent species (the peptide cation), the lowest of the electron-attached Rydberg states is also the lowest Born–Oppenheimer electronic energy for a system composed of the parent molecule and an extra electron. As such, this state can be determined using traditional quantum chemistry, but the higher-lying Rydberg-attached state cannot be, nor can the $(\sigma^*)^-$ state.

Instead, the higher lying states were found using the following procedure.

1. First, compute MOs for the parent molecule (which happens to be a cation, in the calculations of Refs. 74 and 77 but might be a neutral molecule in other applications).
2. Second, compute the electron-attached Hartree–Fock determinant for the state of interest, using the neutral molecule's MOs as an initial guess but altering the initial occupancies such that an electron is added not to the LUMO but to some higher-lying virtual MO.

Following this procedure, the SCF calculation *may* converge (in the sense of finding a stationary point in the space of MO coefficients) to a legitimate solution of the SCF equations, but one that is *not* the lowest-energy solution. If so, then this state represents a Hartree–Fock approximation to an excited state (a higher-lying Rydberg-attached state, or the $(\sigma^*)^-$ state in the present example) including full orbital relaxation upon electron attachment. Electron correlation can be included subsequently, simply by using this excited Hartree–Fock determinant as the reference state for a correlated wave function. (The curves in Figure 2 were computed in this way at the MP2 level.[74])

On the other hand, this procedure is by no means guaranteed to work. In many cases, the "non-*aufbau*" guess for the initial MO occupancies may suffer variational collapse as the SCF iterations proceed, such that despite the elaborate initial guess, the final, converged SCF solution is simply the lowest energy anion state. In the calculations reported in Figure 2,[74] it proved possible to locate the excited Rydberg-attached state in this way, but attempts to determine the σ^*-attached state suffered variational collapse, and the potential curve for that state was ultimately computed in an entirely different way, using stabilization methods that are described later in this chapter.

To avoid variational collapse, it is probably advisable to use an SCF convergence algorithm that is based on direct minimization[373] rather than extrapolation methods such as direct inversion in the iterative subspace (DIIS)[374] and related methods,[375] which are the default convergence algorithms in most quantum chemistry programs. Direct minimization, while often very slow to reach convergence, is more likely to converge to the desired *local* minimum in the space of MO coefficients.

Nevertheless, even direct minimization remains vulnerable to variational collapse, since the newly-occupied MO of the anion is subject to a different potential as compared to virtual MOs that might be nearby in energy. Subsequent SCF iterations can therefore modify the energetic ordering of the MOs, and in such cases, it is unclear which MOs should be the occupied ones at the next SCF iteration. The *maximum overlap method* (MOM)[376,377] offers a possible solution to this problem, and a more refined version of the orbital relaxation technique.

Originally implemented in Q-Chem[194,195] as a way to assist SCF convergence in cases of near degeneracies, MOM has more recently been used as a means to compute electronically excited states using ground-state SCF technology,[376,377] and extension to electron-attached states is straightforward. The basic idea is that once the initial MO occupancies have been specified (with one or more holes below the Fermi level), the Fock matrix is constructed and diagonalized to obtain new MOs, but then the occupancies of these MOs are chosen, not according to the *aufbau* principle but rather by determining which orbitals exhibit maximum overlap with the orbitals from the previous SCF iteration.[376] Although not immune to variational collapse (especially, in our experience, when large, diffuse basis sets are employed, which may be a problem), this method has been applied successfully to the challenging problem of computing core-level excitation spectra.[377] In other words, SCF solutions that correspond to the promotion of an electron from a core orbital into a low-lying virtual orbital have been converged successfully, without collapse of the core hole. In the future, MOM may provide a more robust way to perform orbital relaxation for electron-attached states, although the algorithm has not yet been tested extensively in this capacity.

Note that the states that are generated by MOM are *diabatic* states, in the sense that the character of the wave function changes smoothly as each state passes through the interaction region with another state (see Figure 2). In a one-dimensional, two-state diabatic model, we can write the Hamiltonian in the form

$$\mathbf{H} = \begin{pmatrix} E_1(R) & V(R) \\ V(R) & E_2(R) \end{pmatrix}$$

[53]

Here, $E_1(R)$ and $E_2(R)$ are the diabatic potential energy curves obtained, for example, using the MOM procedure. The adiabatic (Born–Oppenheimer) potential energy curves are the eigenvalues of this matrix:

$$E_\pm = \frac{1}{2}(E_1 + E_2) \pm \frac{1}{2}\sqrt{(E_1 - E_2) + 4V^2} \qquad [54]$$

The quantities E_1, E_2, and V each depend on molecular geometry, R, but this is suppressed in Eq. [54], for brevity. At the point R_x where the two diabatic curves cross ($E_1 = E_2$), the adiabatic states are split by $\Delta E = E_+(R_x) - E_-(R_x) = 2V(R_x)$. Having obtained the point R_x from plotting the diabatic potential curves, if one can compute *adiabatic* energies at R_x for the two states in question, then one can evaluate the coupling, $V(R_x)$. (In fact, this procedure could be followed at any value of R, using Eq. [54] to map out the coupling $V(R)$, but we assume for simplicity that this coupling changes little in the vicinity of the crossing point.) Calculating the Born–Oppenheimer adiabatic energies $E_\pm(R)$ requires performing some type of electronic structure calculation that is capable of computing excited states, and this procedure has been used by Simons and coworkers[74–78] to evaluate $V(R_x)$ for systems such as the protonated peptide electron capture problem that is depicted in Figure 2.

With the coupling matrix element $V = \langle \psi_1^{\text{diabatic}} | \hat{H} | \psi_2^{\text{diabatic}} \rangle$ in hand, the semiclassical Landau–Zener formula[378,379] can be used to estimate the rate of nonadiabatic transitions. According to Landau–Zener theory, the probability (call it p_{LZ}) of making a transition from state 1 to state 2 is

$$p_{LZ} = 1 - \exp\left(-\frac{2\pi V^2}{\hbar \dot{R} |\Delta F|}\right) \qquad [55]$$

where $\dot{R}$ is the speed at which the nuclei are passing through the interaction region and ΔF is the difference in the forces (slopes) on the two diabatic potential energy curves. (Both of these ideas are semiclassical in that they take a classical view of the nuclear motion.)

To use Eq. [55], one must estimate the nuclear speed $\dot{R}$ along the reaction coordinate in question. A harmonic approximation for the reaction coordinate (e.g., the S–S stretching coordinate for the example in Figure 2) affords a simple means to do this,[74] and one that is consistent with the semiclassical nature of Landau–Zener theory. Having computed the (linear) harmonic frequency v for the mode in question, using some flavor of quantum chemistry, one may compute the classical turning points of the harmonic potential:

$$x_\pm = \pm\sqrt{\frac{\left(n + \frac{1}{2}\right)h}{4\pi^2 m v}} \qquad [56]$$

where m is the reduced mass for the mode in question, and $n = 0, 1, \ldots$ is the vibrational quantum number. Taking twice the distance between the two turning points as a measure of the distance traversed during a single vibrational period, one obtains the estimate $\dot{R} \approx 4x_+ v$ (speed = distance × frequency), which works out to be

$$\dot{R} \approx \frac{2}{\pi} \left[\frac{\left(n + \frac{1}{2} \right) h v}{m} \right]^{1/2} \tag{57}$$

Finally, one may estimate the nonadiabatic transition rate according to[74]

$$\text{rate} = (\text{frequency}) \times (\text{transition probability}) = v p_{LZ} \tag{58}$$

Simons and coworkers[74–78] have used this procedure to estimate electron capture and DEA rates for problems such as the one in Figure 2.

Landau–Zener theory is applicable in the regime where the nuclear kinetic energy is large compared to the spacing between the adiabatic potential energy surfaces in the crossing region.[380] The latter may be estimated as $\Delta E \approx 2V$ (from Eq. [54] with $E_1 = E_2$), whereas the nuclear kinetic energy is $m\dot{R}^2/2$. These considerations suggest that Landau–Zener theory is appropriate in the limit

$$|V| \ll \frac{\left(n + \frac{1}{2} \right) h v}{\pi^2} \tag{59}$$

Assuming a typical disulfide harmonic frequency $hv \approx 515$ cm^{-1},[381] the couplings $V \approx 350$ cm^{-1} for the system considered in Figure 2 lie outside of this limit. However, they are at least small compared to hv, so that Landau–Zener theory can be used to estimate the nonadiabatic transition rate, even if the theory is not completely rigorous in this particular system.

Complex Coordinate Rotation

As compared to the MOM approach discussed earlier, other methods for treating metastable states are somewhat more involved, and understanding them requires a few concepts that go beyond bound-state quantum mechanics. One idea that is needed is the notion of analytic continuation of the bound-state energy levels into the complex plane. A heuristic explanation of why this is necessary goes as follows.[382] In some ways, a temporary anion resonance resembles a stationary state of the molecular potential, at least in the sense that the probability distribution is relatively localized around the molecule (see Figure 13). At the same time, however, the resonance has a finite lifetime and will ultimately tunnel out of the potential that is responsible for it. In view of these facts, it seems reasonable that the simplest possible mathematical description of a resonance that is localized in space at time $t = 0$ might be the usual stationary-state time evolution formula,

$$\psi(\mathbf{r}, t) = e^{-iEt/\hbar}\, \psi(\mathbf{r}, 0) \qquad [60]$$

but with a complex energy

$$E = E_R - \frac{1}{2}i\Gamma \qquad [61]$$

(The quantities E_R and Γ are real.) Putting these two equations together, we have

$$\psi(\mathbf{r}, t) = e^{-t\Gamma/2\hbar}\, [e^{-iE_Rt/\hbar}\, \psi(\mathbf{r}, 0)] \qquad [62]$$

The quantity in square brackets looks like an ordinary stationary state, suggesting that the real part of the energy (E_R) is the resonance energy. The imaginary part of the energy $(-\Gamma/2)$ contributes an envelope function that decays exponentially on a timescale $\tau \sim \hbar/\Gamma$. This is consistent, up to factors of order unity, with the time-energy uncertainty principle in Eq. [19], if we take $\Delta E \sim \Gamma$. It is therefore not surprising that the quantity Γ is known as the *resonance width*. The lifetime of the metastable resonance is $\tau \sim \hbar/\Gamma$.

Scattering states are non-normalizable (and thus not in L^2, the Hilbert space of square-integrable functions), since they remain nonzero as $r \to \infty$. Moreover, one cannot obtain a complex energy as the eigenvalue of any self-adjoint operator. As such, the resonances cannot be computed directly using the machinery of bound-state quantum chemistry, but *can* be computed as eigenfunctions and eigenvalues of a modified, non-Hermitian Hamiltonian. The mathematical basis of this statement is a theorem by Balslev and Combes,[383] which we shall motivate below using a simple example and then summarize in a pedagogical way.

The underlying idea behind the *complex coordinate rotation* (CCR) method[382,384] that is suggested by the Balslev–Combes theorem is a complex scaling of the Cartesian coordinates in the Hamiltonian operator, each by the same complex phase factor: $x \to xe^{i\theta}$. This transformation defines a new, complex-scaled Hamiltonian, $\hat{H} \to \hat{H}(\theta)$. In one dimension (for simplicity), the complex-scaled Hamiltonian is

$$\hat{H}(\theta) = -\frac{\hbar^2 e^{-2i\theta}}{2m}\frac{\partial^2}{\partial x^2} + U(xe^{i\theta}) \qquad [63]$$

This idea is readily extended to the Born–Oppenheimer electronic Hamiltonian by noting that $x \to xe^{i\theta}$ implies that interparticle coordinates should be scaled as $r \to re^{i\theta}$. For $\theta \neq 0$, the operator $\hat{H}(\theta)$ is non-Hermitian and therefore admits complex eigenvalues. In its simplest form, the CCR method consists of determining these eigenvalues.

Before stating the general Balslev–Combes theorem, let us first consider two simple examples.[382] Up to constants and in atomic units ($\hbar = m_e = 1$), the 1s wave function for the hydrogen atom is $\psi(r) = e^{-r}$, which becomes $\psi(re^{i\theta}) = \exp(-re^{i\theta})$ upon complex scaling. The latter is still in L^2, provided that $|\theta| \leq \frac{1}{2}\pi$, else the function is not single-valued. (We will see that the CCR method uses only quadrant IV of the complex plane, corresponding to rotation angles $0 \geq \theta \geq -\frac{1}{2}\pi$.) To determine the energy of the state $\psi(re^{i\theta})$, we compute the expectation value

$$\langle E \rangle = \frac{\langle \psi(re^{i\theta}) | \hat{H}(\theta) | \psi(re^{i\theta}) \rangle}{\langle \psi(re^{i\theta}) | \psi(re^{i\theta}) \rangle} . \qquad [64]$$

In more detail, this expectation value is

$$\langle E \rangle = \frac{\int_0^\infty e^{-re^{i\theta}} \left[\left(-\frac{e^{-2i\theta}}{2} \frac{1}{r^2} \frac{\partial}{\partial r} r^2 \frac{\partial}{\partial r} - \frac{1}{re^{i\theta}} \right) e^{-re^{i\theta}} \right] e^{3i\theta} r^2 dr}{\int_0^\infty (e^{-re^{i\theta}})^2 e^{3i\theta} r^2 dr} \qquad [65]$$

Two points warrant explanation here. First, $e^{3i\theta} r^2 dr$ is simply the radial volume element following complex scaling. Second, and more subtle, is the fact that the CCR factor of $e^{i\theta}$ does *not* get complex-conjugated, that is, $[\psi(re^{i\theta})]^* = \exp(-re^{i\theta})$. The reason is that this factor ultimately results from an integration contour in the complex plane, taking advantage of Cauchy's residue theorem; see Ref. 382 for a more detailed explanation of this point. The factor in square brackets in Eq. [65] represents the action of $\hat{H}(\theta)$ on $\psi(re^{i\theta})$ and can be evaluated directly; the result is $-\frac{1}{2}\psi(re^{i\theta})$. Evaluating the integrals in Eq. [65] then affords $\langle E \rangle = -\frac{1}{2}$ in atomic units, meaning that the energy of this bound-state solution is unchanged on complex scaling.

The scattering wave functions, on the other hand, will behave something like e^{ikr} at long range. On scaling $r \to re^{i\theta}$, these continuum functions will not remain finite as $r \to \infty$ unless $k \to ke^{-i\theta}$,[382] which also makes sense in terms of the inverse relationship between r (position) and $\hbar k$ (momentum). A continuum state is characterized by its kinetic energy, and therefore the energy $E = (\hbar k)^2 / 2m$ is transformed upon complex scaling into $E = e^{-2i\theta} (\hbar k)^2 / 2m$. It has been rotated into the complex plane by an angle of -2θ.

With these examples in hand, a pedagogical version of the Balslev–Combes theorem[383] can be stated as follows.[382]

1. Bound-state eigenvalues of the original Hamiltonian are equal to bound-state eigenvalues of $\hat{H}(\theta)$ and are independent of θ, provided that $|\theta| \leq \frac{1}{2}\pi$.
2. Segments of the continuum beginning at a given scattering threshold are rotated by -2θ into quadrant IV of the complex plane.

3. Resonance states appear as *discrete* (albeit complex) eigenvalues of $\hat{H}(\theta)$, having L^2 eigenfunctions. The quantities E_R and Γ are obtained from the real and imaginary parts of the complex energy, respectively, according to Eq. [61].

4. The discrete resonances are independent of θ so long as they are isolated from the continuum states, appearing or disappearing as the continuum states rotate past.

A pictorial illustration is provided in Figure 33. It is important to emphasize that the theorem is rigorously valid for *exact* eigenfunctions of $\hat{H}(\theta)$ in L^2. For approximate solutions, including finite-basis solutions, these results are not exact, and in particular the complex resonance energies acquire a θ-dependence in finite basis sets.[187] It is suggested in practice to compute $E(\theta)$ and to use the value where $|dE/d\theta|$ is minimized as a best estimate of the resonance energy.[385,187]

The CCR idea has been around for a long time, as reviewed in Refs. 389 and 391, and many applications to temporary anion resonances have been reported. Nevertheless, this technique has remained somewhat specialized. Within the context of electronic structure theory, what is required for a CCR calculation is to combine the complex-scaled Hamiltonian in Eq. [63] with the usual wave function *ansätze*, and this involves extending quantum chemistry codes to handle complex-valued wave functions and energies and non-Hermitian matrices. CCR implementations of the Hartree–Fock,[386,387] configuration interaction,[386] and multiconfigurational SCF (MCSCF) models[387] have been reported but are not available in standard

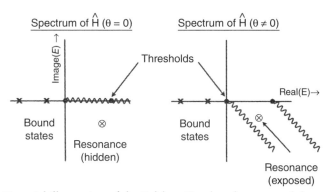

Figure 33 Pictorial illustration of the Balslev–Combes theorem and the complex coordinate rotation method. Horizontal and vertical axes represent the real and imaginary parts of the complex energy, respectively. Application of the complex-scaling transformation $x \rightarrow xe^{i\theta}$ rotates the continuum by an angle of -2θ in the complex plane, leaving the resonances "exposed" as discrete states with square-integrable wave functions and complex energies. Bound states remain on the real axis. Adapted with permission from Ref. 187; copyright 2013 American Institute of Physics.

quantum chemistry codes. In any case, the Hartree–Fock version of CCR is incapable of describing auto-ionizing Feshbach resonances, since these are two-electron processes (see the right side of Figure 16, for example). Hartree–Fock theory is obviously not suitable to describe such states, and a time-dependent DFT (TD-DFT) description of a true two-electron excitation requires the use of a frequency-dependent exchange-correlation kernel.[388] Such kernels so far exist only at the proof-of-concept level.[388,389]

Very recently, however, a CCR implementation of the EOM-CCSD method has been reported[187] in the Q-Chem program,[194,195] which promises to make such methods more routinely available to a wider variety of researchers. As it is based on a correlated wave function, this approach is capable of describing auto-ionizing, core-excited Feshbach resonances. So far, only atomic anion resonances have been considered, and the generation of molecular potential energy surfaces is complicated by the fact that the mathematical proof of the Balslev–Combes theorem requires the nuclear coordinates to be complex-scaled as well. Special techniques are therefore required to extract potential energy surfaces from CCR calculations.[390-394] Alternatively, the EOM-CCSD method has also been implemented in conjunction with a complex absorbing potential,[31,395] $\hat{H}_\xi = \hat{H} - i\xi W$, where the potential W turns on at large electron–molecule separation. Historically, the resonance energies thus obtained have been sensitive to the details of this potential, including most notably the coupling strength ξ, but recent progress has been made toward analyzing the ξ-dependent results in a way that minimizes this dependence.[31]

It is emphasized in Ref. 187 that very flexible basis sets are required to deal with the finite-basis θ-dependence of the complex eigenvalues in CCR methods, and in particular to converge the resonance widths. However, the term "very flexible" is used in comparison to standard basis sets for valence anions, and in fact good results are obtained for auto-ionizing resonances of He, H^-, and Be using the aug-cc-pVTZ+[3s3p] basis,[187] which includes three even-tempered diffuse s and p shells. This is not much different from the basis sets recommended here for proper description of loosely-bound electrons in general gas-phase calculations.

Since this methodology is fairly new to mainstream quantum chemistry, it is impossible to provide a broad overview of its performance at this time, so let us close this section instead with a provocative example,[187] illustrating the fundamental importance of such methods. Earlier in this chapter, we introduced *p*-coumaric acid as a molecule with an auto-ionizing resonance, in the sense that its $S_1(\pi\pi^*)$ bright state lies above its adiabatic electron detachment energy. (The molecule in question was introduced previously in Figure 16, but its structure is reproduced here in Figure 34.) This particular molecule carries significant chemical interest insofar as it is a simplified model of the chromophore in photoactive yellow protein.[188] In Figure 34, we show a large number of excitation energies computed for this molecule,[187] in a sequence of increasingly diffuse basis sets, at the level of singles configuration interaction (CIS). By the time one has put a single set of diffuse functions on all atoms (6-31++G**), the $^1\pi\pi^*$ excited state – which can be identified either by inspecting the orbitals involved in the transition, or more easily by examining oscillator strengths – is *not* the lowest energy excitation! That honor goes instead to an excitation that transfers an electron from the π system into the diffuse orbitals, with very little oscillator strength. Adding more and more diffuse shells results in an increasingly large density of states corresponding to excitations of the latter type, to the point where these π-to-diffuse excitations start to mix strongly with the $\pi \rightarrow \pi^*$ excitations, gradually bleeding the oscillator strength out of the latter. (This is reminiscent of the manner in which spurious, low-energy charge-transfer states predicted by TD-DFT with asymptotically incorrect exchange-correlation functionals can form a dense manifold that steals intensity from the true bright states.[397])

The explanation of the behavior in Figure 34 is the following. The auto-ionizing nature of the $S_1(\pi\pi^*)$ state means that it is *not* the lowest energy excited state at the S_0 geometry; an electron-detached state is lower. As the finite basis set extends farther into space, more and more ODC states appear

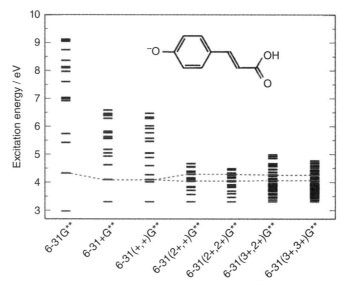

Figure 34   CIS calculations, in a sequence of increasingly diffuse basis sets, for the phenolate isomer of *p*-coumaric acid, a chemical model of the chromophore in photoactive yellow protein. (The energy level structure of this molecule is depicted in Figure 16.) Dashed lines connect the $\pi \rightarrow \pi^*$ transition(s) obtained in each basis, which mix with the continuum states in the more diffuse basis sets. The lowest-energy level in each basis set represents $-\varepsilon_{HOMO}$, which is the threshold energy for the onset of the continuum in the basis-set limit.[396] The other levels denote CIS excitation energies, most of which correspond to discretized continuum states. Adapted with permission from Ref. 187; copyright 2013 American Institute of Physics.

at energies below the $^1\pi\pi^*$ state. This type of behavior is expected whenever the lowest electronic excitation energy lies above $-\varepsilon_{HOMO}$, which represents the threshold energy for the onset of the continuum states in a complete-basis CIS (or complete-basis TD-DFT) calculation.[396,187] This type of behavior may well be endemic for electronically excited states of anions (it is well known, for example, in the case of the aqueous halide anions[398]), suggesting that methods appropriate for *metastable* states may be necessary in many such cases. To date, these considerations have not been widely recognized.

Moreover, in aqueous solution there is always a solvated-electron energy level sitting ~ 3.5 eV below vacuum level, and the $S_0 \rightarrow S_1$ excitation of an aqueous chromophore may be enough energy to access such a state. Common aromatic chromophores have photoionization thresholds corresponding to ultraviolet wavelengths (e.g., 4.35 eV for indole[399]), so that these solvated-electron states may be very much in play in solution-phase photochemistry and photophysics.[92] The molecular-level details of the electron ejection process, and the structure of the initially formed solvated electron, remain poorly understood.

Stabilization Methods

As compared to the complex-scaling method described earlier, the methods discussed in this section have been used more widely in quantum chemistry, although the *proper* computation of temporary anion resonances (as opposed to dubious calculation of random ODC states) continues to be a niche market. Historically, stabilization methods were introduced as an alternative to quantum scattering methods,[400] by means of which resonances could be located by solving a Schrödinger equation for bound-state energy levels only. In the present context, this means that the well-developed machinery of bound-state quantum chemistry can be deployed to compute temporary anion resonances. The basic idea (stated here using the language of a temporary anion resonance, although the technique is more general[400]) is to first stabilize an otherwise metastable anion M$^-$ by placing it in confining potential and thereby converting it into a bound state, whose energy can be computed using standard methods. Then, by examining how that bound-state energy level varies as a function of the spatial extent of the confining potential, the resonance energy can be extracted as described below. (With some additional effort, resonance widths can be extracted as well.) In effect, we aim to place the molecule into a box that is large enough so that its presence does not perturb the energy levels of the neutral molecule. Assuming this is so, then the effect of the box (confining potential) is to modulate the kinetic energy of the extra electron, whereas the molecular energy levels should not be significantly perturbed.

Exponent Stabilization and Analytic Continuation

In the context of quantum chemistry, a finite Gaussian basis set provides a natural confining potential, because any region of space that is not supported by basis functions is in effect subject to a potential of $+\infty$. In what is sometimes known as the *exponent stabilization method*,[369,401] one first selects a Gaussian basis set that is appropriate to describe the molecule, M, then constructs an adjustable confining potential simply by adding additional diffuse basis functions and scaling their orbital exponents, $\zeta \rightarrow \eta\zeta$. This is tantamount to varying the size of the box in which M^- is confined. The additional diffuse functions should include s, p, d, ... angular momentum functions as appropriate to describe any centrifugal barrier that might exist in the electron–molecule potential (see Figure 14).

The exponent stabilization method consists of a sequence of bound-state calculations in which various energy levels of M^- are computed as a function of the scaling parameter, η. Properly speaking, one should compute actual electronic states of M^-. At a simple level, however, this might consist of attaching the electron to different orbitals of the neutral molecule, using KT estimates of the VAE, or perhaps by performing a proper SCF calculation to include orbital relaxation. The MOM can be used in the case that the electron is attached to an orbital of M other than the HOMO. (To date, most applications of the exponent stabilization method have been to small molecules where orbital symmetry may prevent collapse to the HOMO.) Figure 35(a) provides a schematic depiction of the results,[2] which are η-dependent energy levels for M^-, while Figure 35(b) and (c) provide examples from actual calculations.[402,368] A fixed value of η in these plots corresponds to a particular set of orbital exponents, and a vertical slice through the stabilization graph therefore represents the energies computed for various states of M^- (measured, for convenience, relative to the energy of neutral M). It is important to bear in mind that these states are generally ODC states, none of which need represent a true temporary anion resonance, but rather an electron that is artificially confined by a finite basis set. Techniques to extract the resonance energies and widths from these plots are discussed below.

The true resonance energy can be extracted by examining the η-dependence of these pseudo-bound-state energies. To understand how, let us first understand the behavior of the curves sketched in Figure 35(a). Note that the action of the radial kinetic energy operator, $\hat{T}_r$, on a Gaussian basis function $\exp(-\eta\zeta r^2)$, is

$$\hat{T}_r \, e^{-\eta\zeta r^2} = \left[-\frac{\hbar^2}{2mr^2} \frac{\partial}{\partial r} \left(r^2 \frac{\partial}{\partial r} \right) \right] e^{-\eta\zeta r^2} = -\frac{\hbar^2\eta\zeta}{m}(2\eta\zeta r^2 - 3)e^{-\eta\zeta r^2} \qquad [66]$$

For an anion that is bound only by the confining potential, one expects the largest contribution to the wave function to come from the Gaussian(s) with the smallest exponent, and Eq. [66] suggests that the energy of such a state

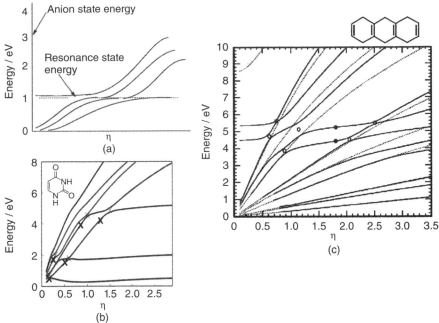

Figure 35 (a) Schematic illustration of the stabilization method, wherein anion energies are computed as a function of a dimensionless parameter, η, that is used to scale the diffuse basis function exponents. (b) Data for the a″ orbitals [$(\pi^*)^-$ electron-attached states] of (uracil)⁻, in which crosses mark the locations of avoided crossings. ODC states that are unaffected by exponent scaling have been removed from this plot. (c) Data for electron attachment to a_u orbitals of 1,4,5,8,9,10-hexahydroanthracene (whose structure is shown at the top of the figure), where the open circles indicate avoided crossings and the two filled circles are located on relatively stable plateaus at $\eta = 1.8$. The dashed curves are states that are not affected by interactions with any resonance state(s). Panel (a) is reprinted with permission from Ref. 2; copyright 2008 American Chemical Society. Panel (b) is adapted with permission from Ref. 368; copyright 2011 American Chemical Society. Panel (c) is reprinted with permission from Ref. 402; copyright 2000 American Institute of Physics.

should vary as η^2 for large r. At certain energies, however, the extra electron's wave function is precisely matched – in magnitude, slope, and phase – for resonant enhancement by the molecular potential, and the result is that the $\sim \eta^2$ dependence of the energy levels vanishes as the electron's wave function becomes localized on the molecule. The result is a plateau in $E(\eta)$, as depicted in Figure 35(a), and avoided crossings that can be understood in terms of the interaction between a bound state of the molecule and a continuum (ODC) state. [Some of these avoided crossings are indicated in Figure 35(b).] The location of the avoided crossing(s) identifies the true resonance energy or energies. A free electron with these special kinetic energies can be captured by the molecular potential, so that the energy of the resulting anion is only weakly dependent on η in the vicinity of the resonance.

As suggested in the schematic example of Figure 35(a), one sometimes observes a "plateau" in the $E(\eta)$ curves, or in other words a range of η over which the energy is approximately independent of η. The origin of this behavior is that the electron's wave function has localized on the molecule, and thus its size is controlled by the spatial extent of the *molecular* potential rather than the artificial confining potential introduced by the finite basis set. In such cases, one may safely identify the resonance energy as the energy of the plateau in $E(\eta)$. Alternatively, one may estimate the resonance energy by finding the point of the closest approach (call it η_0) of two $E(\eta)$ curves and simply taking the average of the two $E(\eta)$ values:

$$E_R \approx \frac{1}{2}[E_1(\eta_0) + E_2(\eta_0)] \qquad [67]$$

This is the procedure suggested by the dashed lines in Figure 35(a). It has been called the *midpoint method*,[233] and is basically equivalent to setting E_R to be the energy of the "plateau," in the case where the plateau exists and the avoided crossings are well separated.

Unfortunately, the complexity of the real stabilization graphs is often significantly greater than that of the idealized example in Figure 35(a). In Figure 35(c), for example, two plateau-like regions can be identified, and the point $\eta = 1.8$ is singled out in this figure because it was found to lie in the middle of the plateau for other, similar molecules.[402] For this particular molecule, however, the "plateaus" in Figure 35(c) are constant to no better than ~ 0.5 eV. Moreover, there are numerous avoided crossings and some states (the dashed curves) that lack any avoided crossings at all. The latter are ODC states that, for one reason or another, do not interact strongly with the resonances. Data for (uracil)$^-$ [Figure 35(b)] show hardly any plateaus at all, and the value of $E(\eta)$ also changes significantly as one moves through the region of the avoided crossing.

For these more complicated cases, more sophisticated methods for extracting the resonance energy are required. Such methods are based on fitting $E_1(\eta)$ and $E_2(\eta)$ in the region near η_0 and then analytically continuing the energy into the complex plane, taking advantage of the fact that the complex energy in Eq. [61] should correspond to a stationary point, $\partial E / \partial \bar{\eta} = 0$, for complex $\bar{\eta}$.[403–406] (The notation $\bar{\eta}$ will be used whenever the real scaling parameter η has been analytically continued into the complex plane.) While these approaches are more complicated, they are also more rigorous, and provide the resonance width Γ in addition to its energy. Two variants of the analytic continuation method are described later.

To understand these methods, first consider the underlying physics of the avoided crossings in the stabilization graphs. These arise due to the interaction between a bound state and a continuum state, with a noncrossing rule in one dimension as the parameter η is varied. This suggests a two-state model,[403,405]

$$\mathbf{H} = \begin{pmatrix} H_1(\eta) & V(\eta) \\ V(\eta) & H_2(\eta) \end{pmatrix} \qquad [68]$$

in which H_1 represents a slowly-varying resonance root and H_2 is a continuum root that couples to the resonance, with a coupling V. This two-state Hamiltonian has eigenvalues

$$E_\pm(\eta) = \frac{1}{2}H_+(\eta) \pm \frac{1}{2}\{[H_-(\eta)]^2 + 4[V(\eta)]^2\}^{1/2} \qquad [69]$$

where $H_\pm = H_1 \pm H_2$. The two branches of the avoided crossing are to be understood as data sets representing the functions $E_+(\eta)$ and $E_-(\eta)$.

One possible analytic continuation procedure is as follows.[405] Since the data plotted in the stabilization graphs are $E_\pm(\eta)$, one can express the three unknown functions $H_1(\eta)$, $H_2(\eta)$, and $V(\eta)$ as polynomials whose coefficients are determined by fitting to one or both branches of the $E_\pm(\eta)$ data, using the two-state model of Eq. [69]. Once these polynomial coefficients are fixed, one then analytically continues the functions H_1, H_2, and V into the complex plane by solving for the *complex* value $\bar{\eta}_0$ such that $E_+(\bar{\eta}_0) = E_-(\bar{\eta}_0)$. Thus, $\bar{\eta}_0$ is the solution to the equation

$$[H_-(\bar{\eta}_0)]^2 + 4[V(\bar{\eta}_0)]^2 = 0 \qquad [70]$$

Having found $\bar{\eta}_0$, one then expands the two eigenvalues $E_\pm(\bar{\eta})$ in the vicinity of $\bar{\eta}_0$ and searches for a stationary point where $\partial E/\partial \bar{\eta} = 0$. This stationary point is the complex energy in Eq. [61], from which one obtains E_R and Γ.

An alternative way to perform the analytic continuation is based on generalized Padé approximations.[403,406,368] In this approach, the energy E is taken to be a solution to the polynomial equation

$$E^3 P_3 + E^2 P_2 + E P_1 + P_0 = 0 \ , \qquad [71]$$

in which each P_k is itself a polynomial in η:

$$P_0(\eta) = c_{0,0} + c_{0,1}\eta + \cdots + c_{0,m_0}\eta^{m_0} \qquad [72]$$

$$P_1(\eta) = c_{1,0} + c_{1,1}\eta + \cdots + c_{1,m_1}\eta^{m_1} \qquad [73]$$

$$P_2(\eta) = c_{2,0} + c_{2,1}\eta + \cdots + c_{2,m_2}\eta^{m_2} \qquad [74]$$

$$P_3(\eta) = 1 + c_{3,1}\eta + \cdots + c_{3,m_3}\eta^{m_3} \ . \qquad [75]$$

The cubic polynomial for the energy in Eq. [71] allows for the possibility that three states might be strongly interacting in the vicinity of the avoided crossing.[406] If only two states are strongly interacting, as assumed in the analytic continuation procedure based on the two-state model in Eqs. [68] and [69], then one may set $P_3 \equiv 0$ in Eq. [71] and set $c_{2,0} = 1$ in Eq. [74]. As earlier, the data points for determining the polynomial coefficients are points (η, E) along the stabilization graphs. Given $m_0 + m_1 + m_2 + m_3 + 3$ such data points, all coefficients in Eq. [35] can be determined exactly, since there are exactly as many (η, E) data points as there are parameters to define the polynomials P_k. (In the two-state case, only $m_0 + m_1 + m_2 + 2$ data points are required.)

Once the polynomial coefficients in Eq. [35] are determined, they are taken to be fixed parameters, but η is allowed to become complex, and Eq. [71] for the energy is extended into the complex plane. One searches for complex roots E that are also stationary, $\partial E / \partial \bar{\eta} = 0$. The quality of the results is sensitive to the quality of the (η, E) data and also to the order of the polynomials in Eq. [35]; in recent applications, the quadratic (two-state) version of the procedure with $m_0 = m_1 = m_2 = 3$, 4, or 5 has been used.[368] It should be noted that this procedure can produce spurious roots[368] (which tend to appear far from the real line), just as it is not clear *a priori* which root in Eq. [69] is the physically correct one.[405] If the (η, E) data are accurate, the $E(\eta)$ fits are good, and the nonlinear search for stationary points is thorough, then it is claimed that these spurious solutions should be easy to identify,[404,405,368] as they are sensitive to the number and choice of the points (η, E) used in the analytic continuation procedure.[368]

In principle one may construct a stabilization graph by computing the energies of various electronic states (including excited states) of M^- at various values of η, each corresponding to a different set of orbital exponents. In practice, the calculation of proper electronic excited states is usually unnecessary, and sufficient accuracy to assign experimental VAEs (obtained from electron transmission spectroscopy[184,185,2]) can be achieved using simple modifications of the KT VAEs, which are equal to $-\varepsilon_{\text{virtual}}$. Often, electron transmission spectra are assigned on the basis of empirical shifting and scaling of the virtual MO eigenvalues to match known data in similar molecules,[233,407,125] as in Eq. [32] for example. In doing so, it is important to use relatively compact basis sets, as the low-lying virtual orbitals will mix with ODC states as one approaches the basis-set limit, and any correspondence between virtual MOs and actual electronic states of M^- will be lost.

From an *ab initio* perspective, a more satisfying procedure, yet one that is still enormously simpler than computing proper anion electronic states, is to use a stabilized version of KT.[406,366,408,401,367,409,368] This approach amounts to the approximation $E(\eta) \approx -\varepsilon_{\text{virtual}}(\eta)$, that is, the virtual MO eigenvalues are used to construct the stabilization graphs, from which resonance energies can be extracted either using the midpoint method (Eq. [67]) or else by proper analytic continuation. Recently, Chen and coworkers have explored this technique using virtual MOs from KS DFT, with good results obtained when LRC functionals with correct asymptotic behavior are used.[401,409,368] There is a formal objection along the lines that incorporating Hartree–Fock exchange into these functionals serves to push the Kohn–Sham eigenvalue spectrum (which is fundamentally an approximate electronic excitation spectrum) toward the Hartree–Fock spectrum (where the HOMO–LUMO gap approximates IP – EA, but the occupied → virtual transitions are poor approximations to electronic excitations),[318] thereby defiling the meaning of the KS DFT orbital energies. For VDE calculations, however, this choice may be justified.

Table 9 shows an example, for the case of $(\pi^*)^-$ and $(\sigma^*)^-$ resonances of (uracil)$^-$. Stabilized KT (S-KT) results are shown for the LRC functional ωB97X-D;[410] the "double hybrid" functional B2PLYP-D,[411] which mixes MP2 and DFT correlation as well as Hartree–Fock and DFT exchange; and the meta-GGA functional M06-HF.[320] Experimentally, only the $(\pi^*)^-$ resonances have been measured by electron transmission spectroscopy,[407] as the $(\sigma^*)^-$ resonances are shorter-lived. The S-KT results for all three functionals, which are based on stabilization graphs resembling the one in Figure 35(b), are in reasonable agreement with the lowest two $(\pi^*)^-$ VAEs measured experimentally, with differences of 0.2–0.3 eV that are comparable to the differences between the resonance energies (E_R) obtained with the three different functionals. In contrast, when Hartree–Fock virtual orbital energies are used [S-KT(HF)], the accuracy is unacceptable. Even for the DFT results, the agreement is less good for the third $(\pi^*)^-$ resonance, but at least good enough to assign the experimental spectrum. Note also that the resonance energies obtained using the midpoint method are nearly identical to those obtained using analytic continuation based on Padé approximations, although only the latter method affords resonance widths. Both the resonance energies and widths are in good agreement with the results of several quantum scattering calculations,[23,89,26] although there are other scattering calculations in the literature that place these resonances 1–2 eV higher in energy.[368] Overall, this (uracil)$^-$ example demonstrates the limits of how accurately one can expect to obtain resonance energies via the exponent stabilization method.

Extrapolation into the Metastable Domain

The advantage of the exponent stabilization method is that all necessary calculations can be performed using standard quantum chemistry codes, without modification. This makes such calculations readily accessible to the average chemist, and in addition various levels of theory can be brought to bear to compute the $E(\eta)$ stabilization curves. That said, the procedure is somewhat more complicated as compared to ordinary bound-state quantum chemistry calculations, because multiple states of M^- must be calculated, the stabilization graphs must be fit to analytic functions in the avoided crossing region(s), and finally these functions must be analytically continued and stationary points located

Table 9 Resonance Energies and Widths (Both in eV) for the Lowest Temporary Anion Resonances of Uracil

Method	Ref.	π*(a'') Resonances						σ*(a') Resonances					
		E_R	Γ	E_R	Γ	E_R	Γ	E_R	Γ	E_R	Γ	E_R	Γ
Shifted and scaled $\varepsilon_{virtual}$[a]	125, 407	0.22		1.61		5.01		2.26		3.67		3.87	
S-KT(HF), midpt. method[b]	368	2.07		4.21		7.59		6.03		6.29		7.60	
S-KT(ωB97X-D), midpt. method[c]	368	0.40		1.74		4.53		2.45		4.32		4.79	
S-KT(ωB97X-D), 3rd-order Padé[c,d]	368	0.36	0.04	1.69	0.17	4.58	0.27	2.35	1.94	4.35	0.77	4.82	0.68
S-KT(ωB97X-D), 5th-order Padé[c,e]	368	0.36	0.05	1.75	0.10	4.52	0.23	2.20	1.62	4.28	0.86	4.68	0.67
S-KT(B2PLYP-D), midpt. method[c]	368	0.11		1.51		4.30		1.57		3.75		4.11	
S-KT(B2PLYP-D), 3rd-order Padé[c,d]	368	0.06	0.08	1.52	0.15	4.33	0.33	1.52	1.53	3.86	0.81	4.11	0.53
S-KT(M06-HF), midpt. method[c]	368	0.44		2.01		5.26		1.81		4.29		4.45	
S-KT(M06-HF), 3rd-order Padé[c,d]	368	0.46	0.08	2.06	0.24	5.29	0.27	1.84	1.90	4.15	0.89	4.50	0.64
Quantum scattering	23	0.32	0.02	1.91	0.16	5.08	0.38	1.45		4.5		4.5	
Quantum scattering	26	0.13	0.00	1.94	0.17	4.95							
Quantum scattering	89	0.33		1.70		3.50							
Experiment	407	0.22		1.58		3.83							

[a]Scaled Hartree–Fock/6-31G* virtual orbital shifted and scaled to match the lowest experimental resonance.[407]

[b]Basis set is 6-31G* augmented with a set of sp diffuse functions.

[c]Basis set is aug-cc-pVDZ with the most diffuse exponent on each atom scaled by η.

[d]Padé approximation with $m_0 = m_1 = m_2 = 3$ (Eq. [35]).

[e]Padé approximation with $m_0 = m_1 = m_2 = 5$ (Eq. [35]).

in the complex plane. In the end, however, one obtains not just the resonance energy but also the resonance width.

An alternative *charge stabilization method* has been suggested for computing resonance energies (only),[412,413] which is simpler insofar as computing multiple states of M^- is not required. (This method has also been called Z-extrapolation, for reasons that will become clear.) The idea is to identify the likely binding site for the extra electron and then artificially increase the atomic numbers (Z) of one or more nearby nuclei. For sufficiently large values of Z, the anion M^- should be converted to a bound state treatable with standard quantum chemistry methods. Calculations are performed at a variety of nuclear charges, and the actual resonance energy is approximated as

$$E_R = \lim_{Z \to Z_0} E(\eta) \qquad [76]$$

where Z_0 denotes the actual atomic number. The success of this method depends critically on using values of Z that do indeed transform M^- into a bound state.

An example is shown in Figure 36 for the case of SO_4^{2-}.[414] This species is an example of a metastable dianion of the sort suggested in Figure 4(b), where the dianion can exist only when trapped behind the RCB to e^- detachment. As such, the lowest-energy state of gas-phase SO_4^{2-} in a finite basis set corresponds

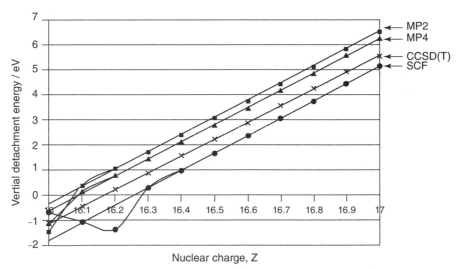

Figure 36 Variation of the VDE for the sulfate dianion, VDE $= E(SO_4^{2-}) - E(SO_4^-)$, as a function of the atomic number, Z, that is assigned to the sulfur atom. In cases where VDE < 0, the scaling of Z is insufficient to convert SO_4^{2-} into a bound state; these data points should be discarded. Reprinted with permission from Ref. 414; copyright 2002 American Institute of Physics.

to an ODC state as discussed above, and the energetics from bound-state quantum chemistry calculations should not be trusted for this species. However, an increase of the sulfur atom's atomic number by 0.1–0.3 (depending on the level of theory) is sufficient to obtain a bound state whose energetics *can* be trusted and can be extrapolated back to $Z_0 = 16$. Plots of the VDE as a function of Z (Figure 36) are remarkably linear except close to $Z = 16$, where the dianion is unbound. These unbound data points should not be used in the extrapolation.

Figure 37 shows an example of the charge stabilization method as applied to several different electronic states of O_2^- computed at a high level of theory, CCSD(T)/aug-cc-pVTZ. (Each state has a different symmetry and thus can be computed from a reference determinant of appropriate symmetry, as if it were the ground state.) Excellent agreement with spectroscopic potentials is obtained, demonstrating the feasibility of this method. Other studies have shown that this Z-extrapolation method affords resonance energies in good

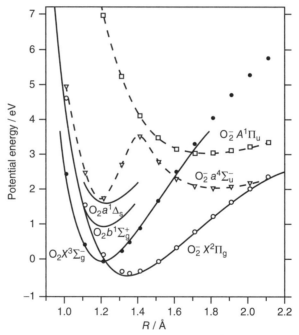

Figure 37   Potential energy curves for the ground state ($X^3\Sigma_g^-$) of O_2 (solid circles) along with three different states of O_2^-: the $X\ ^2\Pi_g$ state (open circles), the $a^4\Sigma_u^-$ state (inverted triangles), and the $A^2\ \Pi_u$ state (squares). The solid curves are derived from experimental spectroscopic parameters, whereas the data points come from theory. All points where O_2^- is not the lowest-energy species were obtained using the charge stabilization method at the CCSD(T)/aug-cc-pVTZ level. Reprinted with permission from Ref. 415; copyright 2003 American Chemical Society.

agreement with those obtained from methods based on complex absorbing potentials, provided that one extrapolates the VAE rather than the total energy.[416] (This is implicit in the stabilization graphs shown in Figure 36.) It was also recommended in Ref. 416 to scale the Gaussian exponents by a factor of $(Z/Z_0)^2$, so as to provide a balanced basis set for each value of Z. (Recall that the radial wave functions for a one-electron atom decay as e^{-Zr/a_0}, so the corresponding Gaussian basis function looks like $e^{-Z^2 \zeta r^2}$.)

A clever alternative to Z-extrapolation has been introduced by Cederbaum and coworkers,[416] for cases where one-dimensional potentials are desired along some relatively simple distance coordinate. The proposed "R-extrapolation" technique is based upon the observation that the attachment energy VAE(R) is often nearly linear in R, or at least the curvature is such that VAE(R) can easily be fit to a low-order polynomial, unlike potential energy curves. This is shown in Figure 38(a) for N_2^- and CO_2^-. The solid VAE(R) curves have been computed in the region where the anion is lower in energy than the neutral molecule, then extrapolated into the region $R \gtrsim 1.4$ Å where the anion is metastable. Alternative methods based on complex absorbing potentials (open symbols in the figure) can be performed on both sides of the metastability threshold $R_c \approx 1.4$ Å, and agree quite well with the R-extrapolated results.[416] Adding the VAE(R) curve (in both the stable and metastable domains) to potential energy curves for the neutral molecule then affords a reasonable potential energy curve for the anion, on both sides of the metastability threshold. This is shown in Figure 38(b).

In both the Z- and R-extrapolation examples, only resonance energies were computed, not resonance widths. To date, that seems to be the state of things: resonance widths have been extracted only on the basis of the exponent stabilization technique, not by Z- or R-extrapolation. However, we note that the analytic continuation procedure discussed above in the context of exponent stabilization has also been used in conjunction with an alternative stabilization technique in which the anion is stabilized by placing it inside of a large spherical array of positive point charges,[417,418,406,366,408] which will convert the anion into a bound state if the charges are sufficiently large. Stabilization graphs, complete with avoided crossings, are then generated by varying either the magnitude of the charges or the radius of the spherical array, and analytic continuation that is based on Padé approximations is applied. The fact that this is a form of *charge* (rather than *exponent*) stabilization suggests that it might be possible to perform similar techniques on the basis of Z-extrapolation, if sufficiently diffuse basis functions were included, so that ODC states that might interact with the resonance states would appear in the calculations. In Ref. 416, it was demonstrated in the case of CO_2^- that avoided crossings could be generated as Z was varied, although no attempt was made to locate these quantitatively or to apply analytic continuation.

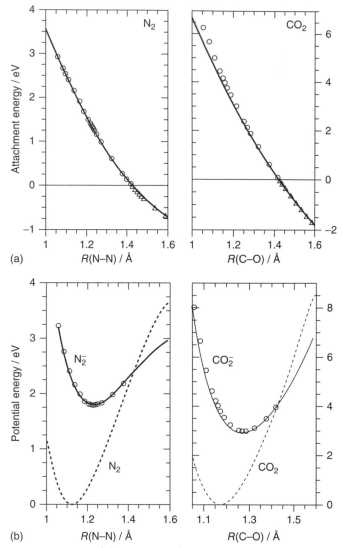

Figure 38   Demonstration of the R-extrapolation technique for N_2^- and CO_2^-. (a) Solid curves show the VAE for the anion, computed where the anion is stable and then extrapolated to larger bond lengths where the anion is metastable (VAE < 0). The open symbols represent calculations performed using a method that can be applied on both sides of the metastability threshold, which occurs around $R \approx 1.4$ Å. (A complex absorbing potential is used in the metastable region; see Ref. 416 for details.) (b) Potential energy curves, where the anion's potential was obtained by adding the VAE to the neutral molecule's potential energy curve. Reprinted with permission from Ref. 416; copyright 2004 American Institute of Physics.

CONCLUDING REMARKS

The loosely-bound electron, orphaned from any molecular unit and ranging far outside of the valence region as it is bound only by weak charge–multipole interactions, nevertheless manages to play a role in a variety of important chemical phenomena. These range from DEA reactions in the gas phase – where a weakly-bound anion may serve as a "doorway" to molecular dissociation – to putative DEA reactions in condensed phases. Solvated electrons *may* play a role in condensed-phase DEA, but in any case the solvated electron is undisputedly present as a reactive intermediates following solvent radiolysis. This species can actually be quite strongly bound, in the sense of its VDE, despite the fact that it is not strongly associated with any particular molecular unit and is bound only collectively by solvent molecules.

Quantum chemistry calculations for loosely-bound electrons come with some specialized demands in terms of the one-particle basis set, which must be "ultra-diffuse" in weakly-bound cases. (On the other hand, highly diffuse but double-ζ basis sets often suffice for the calculation of vertical detachment energies.) With that caveat about basis-set requirements, however, standard high-accuracy *ab initio* methods such as CCSD(T) continue to perform well, even for weakly-bound anions. Perturbative methods such as MP2 also work reasonably well, except in cases where the electron binding energy is exceedingly small ($\lesssim 0.05$ eV), such that the Hartree–Fock reference determinant is unbound or nearly so. DFT methods can also achieve semiquantitative accuracy (~ 0.1–0.3 eV), especially when SICs are employed. A promising route to correcting the SIE for weakly-bound anions is the use of LRC functionals. These appear to eliminate most of the SIE associated with the unpaired electron in doublet radical anions, especially when they are "tuned" to achieve a system-specific asymptotic correction, according to the criterion $\varepsilon_{\text{HOMO}} = -\text{IP}$.

While CCSD(T), MP2, DFT, and so on are appropriate for *bound* anions, theoretical description of *metastable* anions requires specialized techniques. Many of these techniques are well-established but have seen far less use as compared to bound-state quantum chemistry. In this chapter, we have discussed a variety of techniques (the maximum overlap method, CCR, and stabilization methods) that are all based, at some level, on modifications to bound-state quantum chemistry that can be implemented as reasonably straightforward modifications of standard bound-state quantum chemistry codes. It is this author's hope that this review of such methods for temporary anion resonances will prompt renewed and increased interest in these techniques.

ACKNOWLEDGMENTS

The author's own work on negative ions (primarily solvated electrons) has been supported over the years by the National Science Foundation (grant nos. CHE-0748448 and CHE-1300603, and a Mathematical Sciences Postdoctoral Fellowship), the Alfred P. Sloan Foundation, the Henry and

Camille Dreyfus Foundation, and through generous allocations of computing time from the Ohio Supercomputer Center (project nos. PAS-0291 and PAA-0003).

APPENDIX A: LIST OF ACRONYMS

ADC(2)	second-order algebraic diagrammatic construction
aDZ	aug-cc-pVDZ basis set
AEA	adiabatic electron affinity
aQZ	aug-cc-pVQZ basis set
aTZ	aug-cc-pVTZ basis set
CBS	complete basis set
CCSD	coupled-cluster theory with single and double excitations
CCSDT	coupled-cluster theory with single, double, and triple excitations
CCSD(T)	CCSD plus perturbative (noniterative) triple excitations
CIS	configuration interaction with single excitations
DEA	dissociative electron attachment
DFT	density-functional theory
EA	electron affinity
EOM-CCSD	equation-of-motion coupled-cluster theory
EOM-EA-CCSD	equation of motion coupled-cluster theory with electron attachment
EOM-IP-CCSD	equation of motion coupled-cluster theory with electron detachment
FWHM	full width at half maximum
GGA	generalized gradient approximation
HF	Hartree–Fock
HOMO	highest occupied molecular orbital
IP	ionization potential
KS	Kohn–Sham
KT	Koopmans' theorem
LDA	local density approximation
LRC-DFT	long-range-corrected density-functional theory
LUMO	lowest unoccupied molecular orbital
MBPT2	second-order many-body perturbation theory
MO	molecular orbital
MOM	maximum overlap method
MP2	second-order Møller–Plesset perturbation theory
MUE	mean unsigned error
ODC	orthogonalized discretized continuum
PCM	polarizable continuum model
QM/MM	quantum mechanics molecular mechanics
QMC	quantum Monte Carlo

RCB	repulsive Coulomb barrier
RI	resolution of identity
SCF	self-consistent field
SIC	self-interaction correction
SIE	self-interaction error
S-KT	stabilized Koopmans' theorem
SOMO	singly occupied molecular orbital
TD-DFT	time-dependent density-functional theory
VAE	vertical attachment energy
VDE	vertical detachment energy

REFERENCES

1. H. Hotop and W. C. Lineberger, *J. Phys. Chem. Ref. Data*, **14**, 731–750 (1985). Binding Energies in Atomic Negative Ions: II.

2. J. Simons, *J. Phys. Chem. A*, **112**, 6401–6511 (2008). Molecular Anions.

3. J. Simons, *Annu. Rev. Phys. Chem.*, **62**, 107–128 (2011). Theoretical Study of Negative Molecular Ions.

4. J. R. R. Verlet, A. E. Bragg, A. Kammrath, O. Cheshnovsky, and D. M. Neumark, *Science*, **307**, 93–96 (2005). Observation of Large Water-Cluster Anions with Surface-Bound Excess Electrons.

5. L. Ma, K. Majer, F. Chirot, and B. von Issendorff, *J. Chem. Phys.*, **131**, 144303:1–6 (2009). Low Temperature Photoelectron Spectra of Water Cluster Anions.

6. J. V. Coe, S. T. Arnold, J. G. Eaton, G. H. Lee, and K. H. Bowen, *J. Chem. Phys.*, **125** 014315: 1–11 (2006). Photoelectron Spectra of Hydrated Electron Clusters: Fitting Line Shapes and Grouping Isomers.

7. K. R. Siefermann, Y. Liu, E. Lugovoy, O. Link, M. Faubel, U. Buck, B. Winter, and B. Abel, *Nat. Chem.*, **2**, 274–279 (2010). Binding Energies, Lifetimes and Implications of Bulk and Interface Solvated Electrons in Water.

8. Y. Tang, H. Shen, K. Sekiguchi, N. Kurahashi, T. Mizuno, Y. I. Suzuki, and T. Suzuki, *Phys. Chem. Chem. Phys.*, **12**, 3653–3655 (2010). Direct Measurement of Vertical Binding Energy of a Hydrated Electron.

9. A. T. Shreve, T. A. Yen, and D. M. Neumark, *Chem. Phys. Lett.*, **493**, 216–219 (2010). Photoelectron Spectroscopy of Hydrated Electrons.

10. A. Lübcke, F. Buchner, N. Heine, I. V. Hertel, and T. Schultz, *Phys. Chem. Chem. Phys.*, **12**, 14629–14634 (2010). Time-Resolved Photoelectron Spectroscopy of Solvated Electrons in Aqueous NaI Solution.

11. K. D. Jordan and F. Wang, *Annu. Rev. Phys. Chem.*, **54**, 367–396 (2003). Theory of Dipole-Bound Anions.

12. C. R. Arumainayagam, H.-L. Lee, R. B. Nelson, D. R. Haines, and R. P. Gunawardane, *Surf. Sci. Rep.*, **65**, 1–44 (2010). Low-Energy Electron-Induced Reactions in Condensed Matter.

13. E. Alizadeh and L. Sanche, *Chem. Rev.*, **112**, 5578–5602 (2012). Precursors of Solvated Electrons in Radiobiological Physics and Chemistry.

14. J. Gu, J. Leszczynski, and H. F.Schaefer, III, *Chem. Rev.*, **112**, 5603–5640 (2012). Interaction of Electrons with Bare and Hydrated Biomolecules: From Nucleic Acid Bases to DNA Segments.

15. J. M. Herbert and L. D. Jacobson, *Int. Rev. Phys. Chem.*, **30**, 1–48 (2011). Nature's Most Squishy Ion: The Important Role of Solvent Polarization in the Description of the Hydrated Electron.

16. L. Turi and P. J. Rossky, *Chem. Rev.*, **112**, 5641–5674 (2012). Theoretical Studies of Spectroscopy and Dynamics of Hydrated Electrons.

17. B. C. Garrett, D. A. Dixon, D. M. Camaioni, D. M. Chipman, M. A. Johnson, C. D. Jonah, G. A. Kimmel, J. H. Miller, T. N. Rescigno, P. J. Rossky, S. S. Xantheas, S. D. Colson, A. H. Laufer, D. Ray, P. F. Barbara, D. M. Bartels, K. H. Becker, K. H.Bowen, Jr., S. E. Bradforth, I. Carmichael, J. V. Coe, L. R. Corrales, J. P. Cowin, M. Dupuis, K. B. Eisenthal, J. A. Franz, M. S. Gutowski, K. D. Jordan, B. D. Kay, J. A. LaVerne, S. V. Lymar, T. E. Madey, C. W. McCurdy, D. Meisel, S. Mukamel, A. R. Nilsson, T. M. Orlando, N. G. Petrik, S. M. Pimblott, J. R. Rustad, G. K. Schenter, S. J. Singer, A. Tokmakoff, L.-S. Wang, C. Wittig, and T. S. Zwier, *Chem. Rev.*, **105**, 355–389 (2005). Role of Water in Electron-Initiated Processes and Radical Chemistry: Issues and Scientific Advances.

18. M. Mostafavi and I. Lampre, in *Radiation Chemistry: From Basics to Applications in Material and Life Sciences*, M. Spotheim-Maurizot, M. Mostafavi, J. Belloni, and T. Douki (Eds.), EDP Sciences, 2008; chapter 3, pp. 33–52, The Solvated Electron: A Singular Chemical Species.

19. R. M. Young and D. M. Neumark, *Chem. Rev.*, **112**, 5553–5577 (2012). Dynamics of Solvated Electrons in Clusters.

20. K. D. Jordan, V. K. Voora, and J. Simons, *Theor. Chem. Acc.*, **133**, 1445:1–15 (2014). Negative Electron Affinities from Conventional Electronic Structure Methods.

21. F. A. Gianturco and R. R. Lucchese, *J. Chem. Phys.*, **120**, 7446–7455 (2004). Radiation Damage of Biosystems Mediated by Secondary Electrons: Resonant Precursors for Uracil Molecules.

22. S. Tonzani and C. H. Greene, *J. Chem. Phys.*, **122**, 014111:1–8 (2005). Electron–Molecule Scattering Calculations in a 3D Finite Element *R*-Matrix Approach.

23. C. Winstead and V. McKoy, *J. Chem. Phys.*, **125**, 174304:1–8 (2006). Low-Energy Electron Collisions with Gas-Phase Uracil.

24. C. Winstead and V. McKoy, *J. Chem. Phys.*, **125**, 244302:1–7 (2006). Interaction of Low-Energy Electrons with the Purine Bases, Nucleosides, and Nucleotides of DNA.

25. C. Winstead, V. McKoy, and S. d'Almeida Sanchez, *J. Chem. Phys.*, **127**, 085105:1–6 (2007). Interaction of Low-Energy Electrons with the Pyrimidine Bases and Nucleosides of DNA.

26. A. Dora, J. Tennyson, L. Bryjko, and T. van Mourik, *J. Chem. Phys.*, **130**, 164307:1–8 (2009). *R*-Matrix Calculation of Low-Energy Electron Collisions with Uracil.

27. U. Fano, *Phys. Rev.*, **124**, 1866–1878 (1961). Effects of Configuration Interaction on Intensities and Phase Shifts.

28. D. Chen and G. A. Gallup, *J. Chem. Phys.*, **93**, 8893–8901 (1990). The Relationship of the Virtual Orbitals of Self-Consistent-Field Theory to Temporary Negative Ions in Electron Scattering from Molecules.

29. R. Santra and L. S. Cederbaum, *J. Chem. Phys.*, **117**, 5511–5521 (2002). Complex Absorbing Potentials in the Framework of Electron Propagator Theory. I. General Formalism.

30. S. Feuerbacher, T. Sommerfeld, R. Santra, and L. S. Cederbaum, *J. Chem. Phys.*, **118**, 6188–6199 (2003). Complex Absorbing Potentials in the Framework of Electron Propagator Theory. II. Application to Temporary Anions.

31. T.-C. Jagau, D. Zuev, K. B. Bravaya, E. Epifanovsky, and A. I. Krylov, *J. Phys. Chem. Lett.*, **5**, 310–315 (2014). A Fresh Look at Resonances and Complex Absorbing Potentials: Density Matrix-Based Approach.

32. W. A. Al-Saidi, E. J. Walter, and A. M. Rappe, *Phys. Rev. B*, **77**(075112), 1–10 (2008). Optimized Norm-Conserving Hartree-Fock Pseudopotentials for Plane-Wave Calculations.

33. J. Heyd, G. E. Scuseria, and M. Ernzerhof, *J. Chem. Phys.*, **118**, 8207–8215 (2003). Hybrid Functionals Based on a Screened Coulomb Potential.

34. J. Paier, M. Marsman, K. Hummer, G. Kresse, I. C. Gerber, and J. G. Ángyán, *J. Chem. Phys.*, **124**, 154709:1–13 (2006). Screened Hybrid Density Functionals Applied to Solids.

35. A. Szabo and N. S. Ostlund, *Modern Quantum Chemistry*, Macmillan, New York, 1982.

36. R. J. Bartlett and J. F. Stanton, in *Reviews in Computational Chemistry*, K. B. Lipkowitz and D. B. Boyd (Eds.), Vol. 5, VCH Publishers, Inc., New York, 1993, pp. 65–169, Applications of Post-Hartree–Fock Methods: A Tutorial.

37. T. D. Crawford and H. F. Schaefer, III, in *Reviews in Computational Chemistry*, K. B. Lipkowitz and D. B. Boyd (Eds.), Vol. 14, VCH Publishers, Inc., New York, 2000, pp. 33–136, An Introduction to Coupled Cluster Theory for Computational Chemists.

38. F. Jensen, *Introduction to Computational Chemistry*, Wiley, 2006.

39. I. N. Levine, *Quantum Chemistry, Prentice Hall*, 6th edition, 2008.

40. F. M. Bickelhaupt and E. J. Baerends, in *Reviews in Computational Chemistry*, K. B. Lipkowitz and D. B. Boyd (Eds.), Vol. 15, VCH Publishers, Inc., New York, 2000, pp. 1–86, Kohn-Sham Density Functional Theory: Predicting and Understanding Chemistry.

41. P. Elliott, F. Furche, and K. Burke, in *Reviews in Computational Chemistry*, K. B. Lipkowitz and D. B. Boyd (Eds.), Vol. 26, VCH Publishers, Inc., New York, 2009, pp. 91–165, Excited States from Time-Dependent Density Functional Theory.

42. J. Simons, *J. Phys. Chem. A*, **114**, 8631–8643 (2010). One-Electron Electron–Molecule Potentials Consistent with ab Initio Møller–Plesset Theory.

43. J. C. Rienstra-Kiracofe, G. S. Tschumper, H. F. Schaefer III, S. Nandi, and G. B. Ellison, *Chem. Rev.*, **102**, 231–282 (2002). Atomic and Molecular Electron Affinities: Photoelectron Experiments and Theoretical Computations.

44. J. M. Galbraith and H. F. Schaefer III, *J. Chem. Phys.*, **105**, 862–864 (1996). Concerning the Applicability of Density Functional Methods to Atomic and Molecular Ions.

45. N. Rösch and S. B. Trickey, *J. Chem. Phys.*, **106**, 8940–8941 (1997). Comment on "Concerning the Applicability of Density Functional Methods to Atomic and Molecular Negative Ions" [J. Chem. Phys. 105, 862 (1996)].

46. D. Lee, F. Furche, and K. Burke, *J. Phys. Chem. Lett.*, **1**, 2124–2129 (2010). Accuracy of Electron Affinities of Atoms in Approximate Density Functional Theory.

47. J. Simons, *J. Am. Chem. Soc.*, **103**, 3971–3976 (1981). Propensity Rules for Vibration-Induced Electron Detachment of Anions.

48. P. K. Acharya, R. A. Kendall, and J. Simons, *J. Am. Chem. Soc.*, **106**, 3402–3407 (1984). Vibration-Induced Electron Detachment in Molecular Anions.

49. G. Chalasinski, R. A. Kendall, H. Taylor, and J. Simons, *J. Phys. Chem.*, **92**, 3086–3091 (1988). Propensity Rules for Vibration–Rotation-Induced Electron Detachment of Diatomic Anions: Application to $NH^- \rightarrow NH + e^-$.

50. G. V. Buxton, in *Radiation Chemistry: From Basics to Applications in Material and Life Sciences*, M. Spotheim-Maurizot, M. Mostafavi, J. Belloni, and T. Douki (Eds.). EDP Sciences, 2008; chapter 1, pp. 3–16, An Overview of the Radiation Chemistry of Liquids.

51. R. Balog and E. Illenberger, *Phys. Rev. Lett.*, **91**, 213201:1–4 (2003). Complete Chemical Transformation of a Molecular Film by Subexcitation Electrons (< 3 eV).

52. G. Dujardin, R. E. Walkup, and P. Avouris, *Science*, **255**, 1232–1235 (1992). Dissociation of Individual Molecules with Electrons from the Tip of a Scanning Tunneling Microscope.

53. S.-W. Hla, L. Bartels, G. Meyer, and K.-H. Rieder, *Phys. Rev. Lett.*, **85**, 2777–2780 (2000). Inducing All Steps of a Chemical Reaction with the Scanning Tunneling Microscope Tip: Towards Single Molecule Engineering.

54. Q.-B. Lu and L. Sanche, *Phys. Rev. Lett.*, **87**, 078501:1–4 (2001). Effects of Cosmic Rays on Atmospheric Chloroflurocarbon Dissociation and Ozone Depletion.

55. S. Ryu, J. Chang, H. Kwon, and S. K. Kim, *J. Am. Chem. Soc.*, **128**, 3500–3501 (2006). Dynamics of Solvated Electron Transfer in Thin Ice Film Leading to Large Enhancement in Photodissociation of $CFCl_3$.

56. Q.-B. Lu, *Phys. Rep.*, **487**, 141–167 (2010). Cosmic-Ray-Driven Electron-Induced Reactions of Halogenated Molecules Adsorbed on Ice Surfaces: Implications for Atmospheric Ozone Depletion and Global Climate Change.

57. J. Stähler, C. Gahl, and M. Wolf, *Acc. Chem. Res.*, **45**, 131–138 (2012). Dynamics and Reactivity of Trapped Electrons on Supported Ice Crystals.

58. J. Stähler, U. Bovensiepen, M. Meyer, and M. Wolf, *Chem. Soc. Rev.*, **37**, 2180–2190 (2008). A Surface Science Approach to Ultrafast Electron Transfer and Solvation Dynamics at Interfaces.

59. M. Bertin, M. Meyer, J. Stähler, C. Gahl, M. Wolf, and U. Bovensiepen, *Faraday Discuss.*, **141**, 293–307 (2009). Reactivity of Water–Electron Complexes on Crystalline Ice Surfaces.

60. U. Bovensiepen, C. Gahl, J. Stähler, M. Bockstedte, M. Meyer, F. Baletto, S. Scandolo, X.-Y. Zhu, A. Rubio, and M. Wolf, *J. Phys. Chem. C*, **113**, 979–988 (2009). A Dynamic Landscape from Femtoseconds to Minutes for Excess Electrons at Ice–Metal Interfaces.

61. F. P. Sargent and E. M. Gardy, *Chem. Phys. Lett.*, **39**, 188–190 (1976). Direct ESR Detection of the Solvated Electron in Pulse Irradiated Liquid Water.

62. G. V. Buxton, C. L. Greenstock, W. P. Helman, and A. B. Ross, *J. Phys. Chem. Ref. Data*, **17**, 513–886 (1988). Critical Review of Rate Constants for Reactions of Hydrated Electrons, Hydrogen Atoms and Hydroxyl Radicals ($^\bullet$OH/$^\bullet$O) in Aqueous Solution.

63. N. R. P. Harris, J. C. Farman, and D. W. Fahey, *Phys. Rev. Lett.*, **89**, 219801 (2002). Comment on "Effects of Cosmic Rays on Atmospheric Chlorofluorocarbon Dissociation and Ozone Depletion".

64. Q.-B. Lu and L. Sanche, *Phys. Rev. Lett.*, **89**, 219802 (2002). Reply to "Comment on Effects of Cosmic Rays on Atmospheric Chlorofluorocarbon Dissociation and Ozone Depletion".

65. P. K. Patra and M. S. Santhanam, *Phys. Rev. Lett.*, **89**, 219803 (2002). Comment on "Effects of Cosmic Rays on Atmospheric Chlorofluorocarbon Dissociation and Ozone Depletion".

66. Q.-B. Lu and L. Sanche, *Phys. Rev. Lett.*, **89**, 219804 (2002). Reply to "Comment on Effects of Cosmic Rays on Atmospheric Chlorofluorocarbon Dissociation and Ozone Depletion".

67. R. Müller, *Phys. Rev. Lett.*, **91**, 058502:1–4 (2003). Impact of Cosmic Rays on Stratospheric Chlorine Chemistry and Ozone Depletion.

68. R. Müller, *J. Chem. Phys.*, **129**, 027101:1–2 (2008). Comment on "Resonant Dissociative Electron Transfer of the Presolvated Electron to CCl_4 in Liquid: Direct Observation and Lifetime of the CCl_4^{*-} Transition State" [J. Chem. Phys. 128, 041102 (2008)].

69. C.-R. Wang, K. Drew, T. Luo, M.-J. Lu, and Q.-B. Lu, *J. Chem. Phys.*, **129**, 027102:1–2 (2008). Response to "Comment on 'Resonant Dissociative Electron Transfer of the Presolvated Electron to CCl_4 in Liquid: Direct Observation and Lifetime of the CCl_4^{*-} Transition State" [J. Chem. Phys. 129, 027101 (2008)]".

70. R. Müller and J.-U. Grooß, *Phys. Rev. Lett.*, **103**, 228501:1–4 (2009). Does Cosmic-Ray-Induced Heterogeneous Chemistry Influence Stratospheric Polar Ozone Loss?

71. J.-U. Grooß and R. Müller, *Atmos. Environ.*, **45**, 3508–3514 (2011). Do Cosmic Ray-Driven Electron-Induced Reactions Impact Stratospheric Ozone Depletion and Global Climate Change?

72. R. A. Zubarev, N. L. Kelleher, and F. W. McLafferty, *J. Am. Chem. Soc.*, **120**, 3265–3266 (1998). Electron Capture Dissociation of Multiply Charged Protein Cations. A Nonergodic Process.

73. J. E. P. Syka, J. J. Coon, M. J. Schroeder, J. Shabanowitz, and D. F. Hunt, *Proc. Natl. Acad. Sci. U.S.A.*, **101**, 9528–9533 (2004). Peptide and Protein Sequence Analysis by Electron Transfer Dissociation Mass Spectrometry.

74. M. Sobczyk and J. Simons, *Int. J. Mass. Spectrom.*, **253**, 274–280 (2006). Distance Dependence of Through-Bond Electron Transfer Rates in Electron-Capture and Electron-Transfer Dissociation.

75. I. Anusiewicz, J. Berdys-Kochanska, and J. Simons, *J. Phys. Chem. A*, **109**, 5801–5813 (2005). Electron Attachment Step in Electron Capture Dissociation (ECD) and Electron Transfer Dissociation (ETD).

76. I. Anusiewicz, J. Berdys-Kochanska, P. Skurski, and J. Simons, *J. Phys. Chem. A*, **110**, 1261–1266 (2006). Simulating Electron Transfer Attachment to a Positively Charged Model Peptide.

77. M. Sobczyk and J. Simons, *J. Phys. Chem. B*, **110**, 7519–7527 (2006). The Role of Excited Rydberg States in Electron Transfer Dissociation.

78. D. Neff and J. Simons, *J. Phys. Chem. A*, **114**, 1309–1323 (2010). Analytical and Computational Studies of Intramolecular Electron Transfer Pertinent to Electron Transfer and Electron Capture Dissociation Mass Spectrometry.

79. B. Boudaffa, P. Cloutier, D. Hunting, M. A. Huels, and L. Sanche, *Science*, **287**, 1658–1661 (2000). Resonance Formation of DNA Strand Breaks by Low-Energy (3 to 20 eV) Electrons.

80. Y. Zheng, P. Cloutier, D. J. Hunting, L. Sanche, and J. R. Wagner, *J. Am. Chem. Soc.*, **127**, 16592–16598 (2005). Chemical Basis of DNA Sugar–Phosphate Cleavage by Low-Energy Electrons.

81. L. Sanche, in *Radical and Radical Ion Reactivity in Nucleic Acid Chemistry*, M. M. Greenberg (Ed.), Wiley, 2010, pp. 239–293, Low Energy Electron Interaction with DNA: Bond Dissociation and Formation of Transient Anions, Radicals, and Radical Anions.

82. J. Simons, *Acc. Chem. Res.*, **39**, 772–779 (2006). How Do Low-Energy (0.1–2 eV) Electrons Cause DNA Strand Breaks?

83. I. Bald, J. Langer, P. Tegeder, and O. Ingólfsson, *Int. J. Mass Spectrom.*, **277**, 4–25 (2008). From Isolated Molecules through Clusters and Condensates to the Building Blocks of Life.

84. G. V. Buxton, in *Charged Particles and Phonon Interactions with Matter*, A. Mozumder and Y. Hatano (Eds.), Marcel Dekker, New York, 2004; chapter 12, pp. 331–364, The Radiation Chemistry of Liquid Water.

85. S. Ptasinkska, S. Denifl, P. Scheier, E. Illenberger, and T. D. Märk, *Angew. Chem. Int. Ed. Engl.*, **44**, 6941–6943 (2005). Bond- and Site-Selective Loss of H Atoms from Nucleobases by Very-Low-Energy Electrons (<3 eV).

86. A. M. Scheer, C. Silvernail, J. A. Belot, K. Aflatooni, G. A. Gallup, and P. D. Burrow, *Chem. Phys. Lett.*, **411**, 46–50 (2005). Dissociative Electron Attachment to Uracil Deuterated at the N_1 and N_3 Positions.

87. S. Ptasinkska, S. Denifl, P. Scheier, and T. D. Märk, *J. Chem. Phys.*, **120**, 8505–8511 (2004). Inelastic Electron Interaction (Attachment Ionization) with Deoxyribose.

88. P. D. Burrow, *J. Chem. Phys.*, **122**, 087105:1–2 (2005). Comment on "Radiation Damage of Biosystems Mediated by Secondary Electrons: Resonant Precursors for Uracil Molecules" [J. Chem. Phys. 120, 7446 (2004)].

89. F. A. Gianturco, F. Sebastianelli, R. R. Lucchese, I. Baccarelli, and N. Sanna, *J. Chem. Phys.*, **128**, 174302:1–8 (2008). Ring-Breaking Electron Attachment to Uracil: Following Bond Dissociations via Evolving Resonances.

90. J. Gu, J. Wang, and J. Leszczynski, *Nucleic Acids Res.*, **38**, 5280–5290 (2010). Electron Attachment-Induced DNA Single-Strand Breaks at the Pyrimidine Sites.

91. E. J. Hart and M. Anbar, *The Hydrated Electron*, Wiley, New York, 1970.

92. J. Peon, G. C. Hess, J.-M. L. Pecourt, T. Yuzawa, and B. Kohler, *J. Phys. Chem. A*, **103**, 2460–2466 (1999). Ultrafast Photoionization Dynamics of Indole in Water.

93. N. I. Hammer, J. W. Shin, J. M. Headrick, E. G. Diken, J. R. Roscioli, G. H. Weddle, and M. A. Johnson, *Science*, **306**, 675–679 (2004). How Do Small Water Clusters Bind an Excess Electron?

94. O. T. Ehrler and D. M. Neumark, *Acc. Chem. Res.*, **42**, 769–777 (2009). Dynamics of Electron Solvation in Molecular Clusters.

95. L. D. Jacobson and J. M. Herbert, *J. Am. Chem. Soc.*, **133**, 19889–19899 (2011). Theoretical Characterization of Four Distinct Isomer Types in Hydrated-Electron Clusters, and Proposed Assignments for Photoelectron Spectra of Water Cluster Anions.

96. W. C. Martin, A. Musgrove, S. Kotochigova, and J. E. Sansonetti. Technical report, National Institute of Standards and Technology, Gaithersburg, MD, 2013. Ground Levels and Ionization Energies for the Neutral Atoms.

97. B. J. Lynch, Y. Zhao, and D. G. Truhlar, *J. Phys. Chem. A*, **107**, 1384–1388 (2003). Effectiveness of Diffuse Basis Functions for Calculating Relative Energies by Density Functional Theory.

98. J. Simons, *Collect. Czech. Chem. C.*, **70**, 579–604 (2005). Equations of Motion Theory for Electron Affinities.

99. J. Simons, *Adv. Quantum Chem.*, **50**, 213–233 (2005). Response of a Molecule to Adding or Removing an Electron.

100. D. Danovich, *WIREs Comput. Mol. Sci.*, **1**, 377–387 (2011). Green's Function Methods for Calculating Ionization Potentials, Electron Affinities, and Excitation Energies.

101. J. V. Ortiz, in *Computational Chemistry: Reviews of Current Trends*, J. Leszczynski (Ed.) Vol. 2, World Scientific, Singapore, 1997, pp. 1–61, The Electron Propagator Picture of Molecular Electronic Structure.

102. R. Flores-Moreno and J. V. Ortiz, in *Practical Aspects of Computational Chemistry*, J. Leszczynski and M. K. Shukla (Eds.), Springer, 2009, chapter 1, pp. 1–17, Efficient and Accurate Electron Propagator Methods and Algorithms.

103. R. Flores-Moreno, J. Melin, O. Dolgounitcheva, V. G. Zakrezewski, and J. V. Ortiz, *Int. J. Quantum Chem.*, **110**, 706–715 (2010). Three Approximations to the Nonlocal and Energy-Dependent Correlation Potential in Electron Propagator Theory.

104. J. V. Ortiz, *WIREs Comput. Mol. Sci.*, **3**, 123–142 (2013). Electron Propagator Theory: An Approach to Prediction and Interpretation in Quantum Chemistry.

105. J. M. Herbert and M. Head-Gordon, *Proc. Natl. Acad. Sci. U.S.A.*, **103**, 14282–14287 (2006). First-Principles, Quantum-Mechanical Simulations of Electron Solvation by a Water Cluster.

106. J. Simons, P. Skurski, and R. Barrios, *J. Am. Chem. Soc.*, **122**, 11893–11899 (2000). Repulsive Coulomb Barriers in Compact Stable and Metastable Multiply Charged Anions.

107. T. P. Hezel, C. E. Burkhardt, M. Ciocca, L.-W. He, and J. J. Leventhal, *Am. J. Phys.*, **60**, 329–335 (1992). Classical View of the Properties of Rydberg Atoms: Application of the Correspondence Principle.

108. R. F. Wallis, R. Herman, and H. W. Milnes, *J. Mol. Spectrosc.*, **4**, 51–74 (1960). Energy Levels of an Electron in the Field of a Finite Dipole.

109. O. H. Crawford, *Proc. Phys. Soc. London*, **91**, 279–284 (1967). Bound States of a Charged Particle in a Dipole Field.

110. W. R. Garrett, *Chem. Phys. Lett.*, **5**, 393–397 (1970). Critical Binding of an Electron to a Non-Stationary Electric Dipole.

111. W. R. Garrett, *Phys. Rev. A*, **3**, 961–972 (1971). Critical Binding of an Electron to a Rotationally Excited Dipolar System.

112. O. H. Crawford and W. R. Garrett, *J. Chem. Phys.*, **66**, 4968–4970 (1977). Electron Affinities of Polar Molecules.

113. M. Gutowski, P. Skurski, A. I. Boldyrev, J. Simons, and K. D. Jordan, *Phys. Rev. A*, **54**, 1906–1909 (1996). Contribution of Electron Correlation to the Stability of Dipole-Bound Anionic States.

114. C. Desfrançois, H. Abdoul-Carime, N. Khelifa, and J. P. Schermann, *Phys. Rev. Lett.*, **73**, 2436–2439 (1994). From $1/r$ to $1/r^2$ Potentials: Electron Exchange between Rydberg Atoms and Polar Molecules.

115. M. Gutowski and P. Skurski, *Chem. Phys. Lett.*, **303**, 65–75 (1999). Theoretical Study of the Quadrupole-Bound Anion $(BeO)_2^-$.

116. H.-Y. Chen and W.-S. Sheu, *J. Chem. Phys.*, **110**, 9032–9038 (1999). Dipole-Bound Anion of Water Dimer: Theoretical *ab-Initio* Study.

117. J. V. Coe, G. H. Lee, J. G. Eaton, S. T. Arnold, H. W. Sarkas, K. H. Bowen, C. Ludewigt, H. Haberland, and D. R. Worsnop, *J. Chem. Phys.*, **92**, 3960–3962 (1990). Photoelectron Spectroscopy of Hydrated Electron Cluster Anions, $(H_2O)_{n=2-69}^-$.

118. G. H. Lee, S. T. Arnold, J. G. Eaton, H. W. Sarkas, K. H. Bowen, C. Ludewigt, and H. Haberland, *Z. Phys. D*, **20**, 9–12 (1991). Negative Ion Photoelectron Spectroscopy of Solvated Electron Cluster Anions, Water $(H_2O)_n^-$ and Ammonia $(NH_3)_n^-$.

119. J. Kim, S. B. Suh, and K. S. Kim, *J. Chem. Phys.*, **111**, 10077–10087 (1999). Water Dimer to Pentamer with an Excess Electron: *Ab Initio* Study.

120. C. Desfrançois, V. Periquet, Y. Bouteiller, and J. P. Schermann, *J. Phys. Chem. A*, **102**, 1274–1278 (1998). Valence and Dipole Binding of Electrons to Uracil.

121. O. Dolgounitcheva, V. G. Zakrzewski, and J. V. Ortiz, *Chem. Phys. Lett.*, **307**, 220–226 (1999). Structures and Electron Detachment Energies of Uracil Anions.

122. J. H. Hendricks, S. A. Lyapustina, H. L. de Clercq, and K. H. Bowen, *J. Chem. Phys.*, **108**, 8–11 (1998). The Dipole Bound-to-Covalent Anion Transformation in Uracil.

123. R. A. Bachorz, J. Rak, and M. Gutowski, *Phys. Chem. Chem. Phys.*, **7**, 2116–2125 (2005). Stabilization of Very Rare Tautomers of Uracil by an Excess Electron.

124. A. M. Scheer, K. Aflatooni, G. A. Gallup, and P. D. Burrow, *Phys. Rev. Lett.*, **92**, 068102:1–4 (2004). Bond Breaking and Temporary Anion States in Uracil and Halouracils: Implications for the DNA Bases.

125. P. D. Burrow, G. A. Gallup, A. M. Scheer, S. Denifl, S. Ptasinkska, T. Märk, and P. Scheier, *J. Chem. Phys.*, **124**, 124310:1–7 (2006). Vibrational Feshbach Resonances in Uracil and Thymine.

126. F. Wang and K. D. Jordan, *J. Chem. Phys.*, **119**, 11645–11653 (2003). Parallel-Tempering Monte Carlo Simulations of the Finite-Temperature Behavior of $(H_2O)_6^-$.

127. T. Sommerfeld and K. D. Jordan, *J. Phys. Chem. A*, **109**, 11531–11538 (2005). Quantum Drude Oscillator Model for Describing the Interaction of Excess Electrons with Water Clusters: An Application to $(H_2O)_{13}^-$.

128. T. Sommerfeld, S. D. Gardner, A. DeFusco, and K. D. Jordan, *J. Chem. Phys.*, **125**, 174301:1–7 (2006). Low-Lying Isomers and Finite Temperature Behavior of $(H_2O)_6^-$.

129. A. DeFusco, T. Sommerfeld, and K. D. Jordan, *Chem. Phys. Lett.*, **455**, 135–138 (2008). Parallel Tempering Monte Carlo Simulations of the Water Heptamer Anion.

130. R. A. Bachorz, W. Klopper, M. Gutowski, X. Li, and K. H. Bowen, *J. Chem. Phys.*, **129**, 054309:1–10 (2008). Photoelectron Spectrum of Valence Anions of Uracil and First-Principles Calculations of Excess Electron Binding Energies.

131. T. Waters, X.-B. Wang, and L.-S. Wang, *Coordin. Chem. Rev.*, **251**, 474–491 (2007). Electrospray Ionization Photoelectron Spectroscopy: Probing the Electronic Structure of Inorganic Metal Complexes in the Gas-Phase.

132. X.-B. Wang and L.-S. Wang, *Annu. Rev. Phys. Chem.*, **60**, 105–126 (2009). Photoelectron Spectroscopy of Multiply Charged Anions.

133. X.-B. Wang, C.-F. Ding, and L.-S. Wang, *Phys. Rev. Lett.*, **81**, 3351–3354 (1998). Photodetachment Spectroscopy of a Doubly Charged Anion: Direct Observation of the Repulsive Coulomb Barrier.

134. L.-S. Wang, C.-F. Ding, X.-B. Wang, and J. B. Nicholas, *Phys. Rev. Lett.*, **81**, 2667–2670 (1998). Probing the Potential Barriers and Intramolecular Electrostatic Interactions in Free Doubly Charged Anions.

135. C.-F. Ding, X.-B. Wang, and L.-S. Wang, *J. Phys. Chem. A*, **102**, 8633–8636 (1998). Photoelectron Spectroscopy of Doubly Charged Anions: Intramolecular Coulomb Repulsion and Solvent Stabilization.

136. E. Pluharřová, M. Ončák, R. Seidel, C. Schroeder, W. Schroeder, B. Winter, S. E. Bradforth, P. Jungwirth, and P. Slavček, *J. Phys. Chem. B*, **116**, 13254–13264 (2012). Transforming Anion Instability into Stability: Contrasting Photoionization of Three Protonation Forms of the Phosphate Ion upon Moving into Water.

137. J. V. Coe, A. D. Earhart, M. H. Cohen, G. J. Hoffman, H. W. Sarkas, and K. H. Bowen, *J. Chem. Phys.*, **107**, 6023–6031 (1997). Using Cluster Studies to Approach the Electronic Structure of Bulk Water: Reassessing the Vacuum Level, Conduction Band Edge, and Band Gap of Water.

138. J. V. Coe, *Int. Rev. Phys. Chem.*, **20**, 33–58 (2001). Fundamental Properties of Bulk Water from Cluster Ion Data.

139. J. V. Coe, S. M. Williams, and K. H. Bowen, *Int. Rev. Phys. Chem.*, **27**, 27–51 (2008). Photoelectron Spectra of Hydrated Electron Clusters Vs. Cluster Size.

140. B. Winter and M. Faubel, *Chem. Rev.*, **106**, 1176–1211 (2006). Photoemission from Liquid Aqueous Solutions.

141. L. D. Jacobson and J. M. Herbert, *J. Chem. Phys.*, **133**, 154106:1–19 (2010). A One-Electron Model for the Aqueous Electron that Includes Many-Body Electron-Water Polarization: Bulk Equilibrium Structure, Vertical Electron Binding Energy, and Optical Absorption Spectrum.

142. L. Kevan, *Acc. Chem. Res.*, **14**, 138–145 (1981). Solvated Electron Structure in Glassy Matrices.

143. P. J. Rossky and J. Schnitker, *J. Phys. Chem.*, **92**, 4277–4285 (1988). The Hydrated Electron: Quantum Simulation of Structure, Spectroscopy, and Dynamics.

144. M. Boero, M. Parrinello, K. Terakura, T. Ikeshoji, and C. C. Liew, *Phys. Rev. Lett.*, **90**, 226403:1–4 (2003). First-Principles Molecular-Dynamics Simulations of a Hydrated Electron in Normal and Supercritical Water.

145. M. Boero, *J. Phys. Chem. A*, **111**, 12248–12256 (2007). Excess Electron in Water at Different Thermodynamic Conditions.

146. F. Uhlig, O. Marsalek, and P. Jungwirth, *J. Phys. Chem. Lett.*, **3**, 3071–3075 (2012). Unraveling the Complex Nature of the Hydrated Electron.

147. G. W. Robinson, P. J. Thistlewaite, and J. Lee, *J. Phys. Chem.*, **90**, 4224–4233 (1986). Molecular Aspects of Ionic Hydration Reactions.

148. H. F. Hameka, G. W. Robinson, and C. J. Marsden, *J. Phys. Chem.*, **91**, 3150–3157 (1987). Structure of the Hydrated Electron.

149. T. R. Tuttle, Jr. and S. Golden, *J. Phys. Chem.*, **95**, 5725–5736 (1991). Solvated Electrons: What is Solvated?

150. A. L. Sobolewski and W. Domcke, *Phys. Chem. Chem. Phys.*, **4**, 4–10 (2002). Hydrated Hydronium: A Cluster Model of the Solvated Electron?

151. A. L. Sobolewski and W. Domcke, *J. Phys. Chem. A*, **106**, 4158–4167 (2002). Ab Initio Investigation of the Structure and Spectroscopy of Hydronium–Water Clusters.

152. A. L. Sobolewski and W. Domcke, *Phys. Chem. Chem. Phys.*, **9**, 3818–3829 (2007). Computational Studies of Aqueous-Phase Photochemistry and the Hydrated Electron in Finite-Size Clusters.

153. S. Neumann, W. Eisfeld, A. Sobolewski, and W. Domcke, *Phys. Chem. Chem. Phys.*, **6**, 5297–5303 (2004). Simulation of the Resonance Raman Spectrum of the Hydrated Electron in the Hydrated-Hydronium Cluster Model.

154. R. E. Larsen, W. J. Glover, and B. J. Schwartz, *Science*, **329**, 65–69 (2010). Does the Hydrated Electron Occupy a Cavity?

155. B. Abel, U. Buck, A. L. Sobolewski, and W. Domcke, *Phys. Chem. Chem. Phys.*, **14**, 22–34 (2012). On the Nature and Signatures of the Solvated Electron in Water.

156. J. R. Casey, R. E. Larsen, and B. J. Schwartz, *Proc. Natl. Acad. Sci. U.S.A.*, **110**, 2712–2717 (2013). Resonance Raman and Temperature-Dependent Electronic Absorption Spectra of Cavity and Noncavity Models of the Hydrated Electron.

157. F. Muguet, M.-P. Bassez, and G. W. Robinson, *J. Phys. Chem.*, **92**, 7262–7263 (1988). Reply to the Comment "Aquated Electrons, H_2O^- Anions, and OH^- H_3O Units".

158. M. J. Tauber and R. A. Mathies, *J. Am. Chem. Soc.*, **125**, 1394–1402 (2003). Structure of the Aqueous Solvated Electron from Resonance Raman Spectroscopy: Lessons from Isotopic Mixtures.

159. L. Turi and A. Madarász, *Science*, **331**, 1387 (2011). Comment on "Does the Hydrated Electron Occupy a Cavity?".

160. L. D. Jacobson and J. M. Herbert, *Science*, **331**, 1387 (2011). Comment on "Does the Hydrated Electron Occupy a Cavity?".

161. R. E. Larsen, W. J. Glover, and B. J. Schwartz, *Science*, **331**, 1387 (2011). Response to Comment on "Does the Hydrated Electron Occupy a Cavity?".

162. J. M. Herbert and L. D. Jacobson, *J. Phys. Chem. A*, **115**, 14470–14483 (2011). Structure of the Aqueous Electron: Assessment of One-Electron Pseudopotential Models in Comparison to Experimental Data and Time-Dependent Density Functional Theory.

163. F. Uhlig, J. M. Herbert, M. P. Coons, and P. Jungwirth, *J. Phys. Chem. A*, **118**, 7507–7515 (2014). Optical Spectroscopy of the Bulk and Interfacial Hydrated Electron from ab Initio Calculations.

164. D. M. Neumark, *Mol. Phys.*, **106**, 2183–2197 (2008). Spectroscopy and Dynamics of Excess Electrons in Clusters.

165. A. Kammrath, J. R. R. Verlet, G. B. Griffin, and D. M. Neumark, *J. Chem. Phys.*, **125**, 076101:1–2 (2006). Photoelectron Spectroscopy of Large (Water)$_n^-$ ($n = 50$–200) Clusters at 4.7 eV.

166. M. Mitsui, N. Ando, S. Kokubo, A. Nakajima, and K. Kaya, *Phys. Rev. Lett.*, **91**, 153002:1–4 (2003). Coexistence of Solvated Electrons and Solvent Valence Anions in Negatively Charged Acetonitrile Clusters, $(CH_3CN)_n^-$ ($n = 10$–100).

167. R. N. Barnett, U. Landman, C. L. Cleveland, and J. Jortner, *J. Chem. Phys.*, **88**, 4429–4447 (1988). Electron Localization in Water Clusters. II. Surface and Internal States.

168. J. Jortner, U. Landman, and R. N. Barnett, *Chem. Phys. Lett.*, **152**, 353–357 (1988). Optical Absorption Spectra of $(H_2O)_n^-$.

169. R. N. Barnett, U. Landman, G. Makov, and A. Nitzan, *J. Chem. Phys.*, **93**, 6226–6238 (1990). Theoretical Studies of the Spectroscopy of Excess Electrons in Water Clusters.

170. G. Makov and A. Nitzan, *J. Phys. Chem.*, **98**, 3459–3466 (1994). Solvation and Ionization near a Dielectric Surface.

171. S. E. Braslavsky, *Pure Appl. Chem.*, **79**, 292–465 (2007). Glossary of Terms Used in Photochemistry, 3rd Edition (IUPAC Recommendations 2006).

172. C. J. Cramer and D. G. Truhlar, *Chem. Rev.*, **99**, 2161–2200 (1999). Implicit Solvation Models: Equilibria, Structure, Spectra, and Dynamics.

173. L. Turi, W.-S. Sheu, and P. J. Rossky, *Science*, **309**, 914–917 (2005). Characterization of Excess Electrons in Water-Cluster Anions by Quantum Simulations.

174. J. R. R. Verlet, A. E. Bragg, A. Kammrath, O. Cheshnovsky, and D. M. Neumark, *Science*, **310**, 1769–1769 (2005). Comment on "Characterization of Excess Electrons in Water-Cluster Anions by Quantum Simulations".

175. L. Turi, W.-S. Sheu, and P. J. Rossky, *Science*, **310**, 1769–1769 (2005). Response to Comment on "Characterization of Excess Electrons in Water-Cluster Anions by Quantum Simulations".

176. J. R. Roscioli, N. I. Hammer, and M. A. Johnson, *J. Phys. Chem. A*, **110**, 7517–7520 (2006). Infrared Spectroscopy of Water Cluster Anions $(H_2O)_{n=3-24}^-$ in the HOH Bending Region: Persistence of the Double H-Bond Acceptor (AA) Water Molecule in the Excess Electron Binding Site of the Class I Isomers.

177. K. R. Asmis, G. Santabrogio, J. Zhou, E. Garand, J. Headrick, D. Goebbert, M. A. Johnson, and D. M. Neumark, *J. Chem. Phys.*, **126**, 191105, 1–5 (2007). Vibrational Spectroscopy of Hydrated Electron Clusters $(H_2O)_{15-50}^-$ via Infrared Multiple Photon Dissociation.

178. T. Maeyama, K. Yoshida, and A. Fujii, *J. Phys. Chem. A*, **116**, 3771–3780 (2012). Size-Dependent Metamorphosis of Electron Binding Motif in Cluster Anions of Primary Amide Molecules.

179. C. Cohen-Tannoudji, B. Diu, and F. Laloë, *Quantum Mechanics*, Vols. I and II, Wiley, New York, 1977.

180. H. Yamamoto, *Appl. Phys. A*, **42**, 245–248 (1987). Resonant Tunneling Condition and Transmission Coefficient in a Symmetrical One-Dimensional Rectangular Double-Barrier System.

181. H. Xu, Y. Wang, and G. Chen, *Phys. Stat. Sol. B*, **171**, K9–K12 (1992). Shape of the Transmission Spectrum in Rectangular Double-Barrier Structures.

182. S. Klaiman and I. Gilary, *Adv. Quantum Chem.*, **63**, 1–31 (2012). Chapter 1 – On Resonance: A First Glance in the Behavior of Unstable States.

183. J. Simons, in *Resonances in Electron-Molecule Scattering, van der Waals Complexes, and Reactive Chemical Dynamics*, D. Truhlar (Ed.), Vol. 263 of, *ACS Symposium Series*, American Chemical Society, 1984; chapter 1, pp. 3–16, Roles Played by Metastable States in Chemistry.

184. K. D. Jordan and P. D. Burrow, in *Photon, Electron, and Ion Probes of Polymer Structure and Properties*, ACS Symposium Series Vol. 162, American Chemical Society, 1981; chapter 1, pp. 1–10, Resonant Electron Scattering and Anion States in Polyatomic Molecules.

185. K. D. Jordan and P. D. Burrow, *Chem. Rev.*, **87**, 557–588 (1987). Temporary Anion States of Polyatomic Hydrocarbons.

186. R. Balog, J. Langer, S. Gohlke, M. Stano, H. Abdoul-Carime, and E. Illenberger, *Int. J. Mass Spectrom.*, **233**, 267–291 (2004). Low Energy Electron Driven Reactions in Free and Bound Molecules: From Unimolecular Processes in the Gas Phase to Complex Reactions in a Condensed Environment.

187. K. B. Bravaya, D. Zuev, E. Epifanovsky, and A. I. Krylov, *J. Chem. Phys.*, **138**, 124106:1–15 (2013). Complex-Scaled Equation-of-Motion Coupled-Cluster Method with Single and Double Substitutions for Autoionizing Excited States: Theory, Implementation, and Examples.

188. D. Zuev, K. B. Bravaya, T. D. Crawford, R. Lindh, and A. I. Krylov, *J. Chem. Phys.*, **134**, 034310:1–13 (2011). Electronic Structure of the Two Isomers of the Anionic Form of p-Coumaric Acid Chromophore.

189. T. Clark, J. Chandrasekhar, G. W. Spitznagel, and P. v. R. Schleyer, *J. Comput. Chem.*, **4**, 294–301 (1983). Efficient Diffuse Function-Augmented Basis Sets for Anion Calculations. III. The 3-21+G Basis Set for First-Row Elements, Li–F.

190. R. A. Kendall, T. H. Dunning, Jr. and R. J. Harrison, *J. Chem. Phys.*, **96**, 6796–6806 (1992). Electron Affinities of the First-Row Atoms Revisited. Systematic Basis Sets and Wave Functions.

191. J. M. Herbert and M. Head-Gordon, *J. Phys. Chem. A*, **109**, 5217–5229 (2005). Calculation of Electron Detachment Energies for Water Cluster Anions: An Appraisal of Electronic Structure Methods, with Application to $(H_2O)_{20}^-$ and $(H_2O)_{24}^-$.

192. P. Skurski, M. Gutowski, and J. Simons, *Int. J. Quantum Chem.*, **80**, 1024 (2000). How to Choose a One-Electron Basis Set to Reliably Describe a Dipole-Bound Anion.

193. T. J. Lee and H. F. Schaefer, III, *J. Chem. Phys.*, **83**, 1784–1794 (1985). Systematic Study of Molecular Anions within the Self-Consistent-Field Approximation: OH^-, CN^-, C_2H^-, NH_2^-, and CH_3^-.

194. A. I. Krylov and P. M. W. Gill, *WIREs Comput. Mol. Sci.*, **3**, 317–326 (2013). Q-Chem: An Engine for Innovation.

195. http://www.q-chem.com

196. J. Katriel and E. R. Davidson, *Proc. Natl. Acad. Sci. U.S.A.*, **77**, 4403–4406 (1980). Asymptotic Behavior of Atomic and Molecular Wave Functions.

197. J. P. Perdew, R. G. Parr, M. Levy, and J. L.Balduz, Jr., *Phys. Rev. Lett.*, **49**, 1691–1694 (1982). Density-Functional Theory for Fractional Particle Number: Derivative Discontinuities of the Energy.

198. N. C. Handy, M. T. Marron, and H. J. Silverstone, *Phys. Rev.*, **180**, 45–48 (1969). Long-Range Behavior of Hartree-Fock Orbitals.

199. K. H. Bowen and J. G. Eaton, in *The Structure of Small Molecules and Ions*, R. Naaman and Z. Vager (Eds.), Plenum, New York, 1989, pp. 147–169, Photodetachment Spectroscopy of Negative Cluster Ions.

200. S. T. Arnold, J. G. Eaton, D. Patel-Misra, H. W. Sarkas, and K. H. Bowen, in *Ion and Cluster Ion Spectroscopy and Structure*, J. P. Maier (Ed.), Elsevier, Amsterdam, 1989, pp. 417–472, Continuous Beam Photoelectron Spectroscopy of Cluster Anions.

201. Y. Bouteiller, C. Desfrançois, H. Abdoul-Carime, and J. P. Schermann, *J. Chem. Phys.*, **105**, 6420–6425 (1996). Structure and Intermolecular Motions of the Water Dimer Anions.

202. K. Yagi, Y. Okano, T. Sato, Y. Kawashima, T. Tsuneda, and K. Hirao, *J. Phys. Chem. A*, **112**, 9845–9853 (2008). Water Cluster Anions Studied by Long-Range Corrected Density Functional Theory.

203. M. Gutowski and P. Skurski, *J. Chem. Phys.*, **107**, 2968–2973 (1997). Theoretical Study of the Dipole-Bound Anion $(HF)_2^-$.

204. M. Gutowski and P. Skurski, *J. Phys. Chem. B*, **101**, 9143–9146 (1997). Dispersion Stabilization of Solvated Electrons and Dipole-Bound Anions.

205. M. Gutowski, K. D. Jordan, and P. Skurski, *J. Phys. Chem. A*, **102**, 2624–2633 (1998). Electronic Structure of Dipole-Bound Anions.

206. C. F. Williams and J. M. Herbert, *J. Phys. Chem. A*, **112**, 6171–6178 (2008). Influence of Structure on Electron Correlation Effects and Electron–Water Dispersion Interactions in Anionic Water Clusters.

207. T. Clark and G. Illing, *J. Am. Chem. Soc.*, **109**, 1013–1020 (1987). Ab Initio Localized Electron Calculations on Solvated Electron Structures.

208. H. Tachikawa and M. Ogasawura, *J. Phys. Chem.*, **94**, 1746–1750 (1990). Ab Initio Molecular Orbital Study on Water Dimer Anions.

209. H. Tachikawa, *Chem. Phys. Lett.*, **370**, 188–196 (2003). Electron Capture Dynamics of the Water Dimer: A Direct ab Initio Dynamics Study.

210. H. Tachikawa, *J. Chem. Phys.*, **125**, 144307, 1–8 (2006). Electron Hydration Dynamics in Water Clusters: A Direct *ab Initio* Molecular Dynamics Approach.

211. J. Kim, J. Y. Lee, K. S. Oh, J. M. Park, S. Lee, and K. S. Kim, *Phys. Rev. A*, **59**, R930–R933 (1999). Quantum-Mechanical Probabilistic Structure of the Water Dimer with an Excess Electron.

212. L. D. Jacobson and J. M. Herbert, *J. Am. Chem. Soc.*, **132**, 10000–10002 (2010). Polarization-Bound Quasi-Continuum States are Responsible for the "Blue Tail" in the Optical Absorption Spectrum of the Aqueous Electron.

213. R. L. Martin, *J. Chem. Phys.*, **118**, 4775–4777 (2003). Natural Transition Orbitals.

214. J. S. Townsend, *Modern Approach to Quantum Mechanics*, University Science Books, Sausalito, CA, 2000.

215. A. Dreuw and M. Head-Gordon, *Chem. Rev.*, **105**, 4009–4037 (2005). Single-Reference ab Initio Methods for the Calculation of Excited States of Large Molecules.

216. J. VandeVondele and J. Hutter, *J. Chem. Phys.*, **127**, 114105:1–9 (2007). Gaussian Basis Sets for Accurate Calculations on Molecular Systems in Gas and Condensed Phases.

217. O. Marsalek, F. Uhlig, J. VandeVondele, and P. Jungwirth, *Acc. Chem. Res.*, **45**, 23–32 (2012). Structure, Dynamics, and Reactivity of Hydrated Electrons by ab Initio Molecular Dynamics.

218. O. Marsalek, F. Uhlig, and P. Jungwirth, *J. Phys. Chem. C*, **114**, 20489–20495 (2010). Electrons in Cold Water Clusters: An ab Initio Molecular Dynamics Study of Localization and Metastable States.

219. S. Ronen, *Phys. Rev. A*, **68**, 012106:1–7 (2003). Electron Structure of a Dipole-Bound Anion Confined in a Spherical Box.

220. J. M. Herbert and M. Head-Gordon, *J. Am. Chem. Soc.*, **128**, 13932–13939 (2006). Charge Penetration and the Origin of Large O–H Vibrational Red-Shifts in Hydrated-Electron Clusters, $(H_2O)_n^-$.

221. http://www.iqmol.org, xxxx.

222. G. Schaftenaar and J. H. Noordik, *J. Comput.-Aided Mol. Design*, **14**, 123–134 (2000). Molden: A Pre- and Post-Processing Program for Molecular and Electronic Structures.

223. B. M. Bode and M. S. Gordon, *J. Mol. Graphics Mod.*, **16**, 133–138 (1998). MacMolPlt: A Graphical User Interface for GAMESS.

224. W. Humphrey, A. Dalke, and K. Schulten, *J. Molec. Graphics*, **14**, 33–38 (1996). VMD – Visual Molecular dynamics.

225. C. B. Hübschle and P. Luger, *J. Appl. Crystallogr.*, **39**, 901–904 (2006). MolIso – A Program for Colour-Mapped Iso-Surfaces.

226. C. B. Hübschle and B. Dittrich, *J. Appl. Crystallogr.*, **44**, 238–240 (2011). MoleCoolQt – A Molecule Viewer for Charge-Density Research.

227. S. N. Eustis, D. Radisic, K. H. Bowen, R. A. Bachorz, M. Haranczyk, G. K. Schenter, and M. Gutowski, *Science*, **319**, 936–939 (2008). Electron-Driven Acid–Base Chemistry: Proton Transfer from Hydrogen Chloride to Ammonia.

228. M. Haranczyk and M. Gutowski, *J. Chem. Theory Comput.*, **4**, 689–693 (2008). Visualization of Molecular Orbitals and the Related Electron Densities.

229. A. Rauk and D. A. Armstrong, *Int. J. Quantum Chem.*, **95**, 683–696 (2003). Potential Energy Barriers for Dissociative Attachment to HF.HF and HCl.HCl: Ab Initio Study.

230. http://www.gaussian.com/g_tech/g_ur/u_cubegen.htm

231. K. Momma and F. Izumi, *J. Appl. Crystallogr.*, **41**, 653–658 (2008). VESTA: A Three-Dimensional Visualization System for Electronic and Structural Analysis.

232. M. D. Sevilla, B. Besler, and A.-O. Colson, *J. Phys. Chem.*, **99**, 1060–1063 (1995). Ab Initio Molecular Orbital Calculations of DNA Radical Ions. 5. Scaling of Calculated Electron Affinities and Ionization Potentials to Experimental Values.

233. P. D. Burrow, A. E. Howard, A. R. Johnston, and K. D. Jordan, *J. Phys. Chem.*, **96**, 7570–7578 (1992). Temporary Anion States of Hydrogen Cyanide, Methyl Cyanide, and Methylene Dicyanide, Selected Cyanoethylenes, Benzonitrile, and Tetracyanoquinodimethane.

234. M. Gutowski and P. Skurski, *Recent Res. Dev. Phys. Chem.*, **3**, 245–260 (1999). Electronic Structure of Dipole-Bound Anions.

235. K. A. Peterson and T. H.Dunning, Jr., *J. Mol. Struct. (THEOCHEM)*, **400**, 93–117 (1997). The CO Molecule: The Role of Basis Set and Correlation Treatment in the Calculation of Molecular Properties.

236. R. K. Nesbet, *J. Chem. Phys.*, **40**, 3619–3633 (1964). Electronic Structure of N_2, CO, and BF.

237. K. Raghavachari, G. W. Trucks, J. A. Pople, and M. Head-Gordon, *Chem. Phys. Lett.*, **157**, 479–483 (1989). A Fifth-Order Perturbation Comparison of Electron Correlation Theories.

238. T. Helgaker, P. Jørgensen, and J. Olsen, *Molecular Electronic-Structure Theory*, Wiley, New York, 2000.

239. G. de Oliveira, J. M. L. Martin, F. de Proft, and P. Geerlings, *Phys. Rev. A*, **60**, 1034–1045 (1999). Electron Affinities of the First- and Second-Row Atoms: Benchmark *Ab Initio* and Density-Functional Calculations.

240. N. B. Balabanov and K. A. Peterson, *J. Chem. Phys.*, **125**, 074110:1–10 (2006). Basis Set Limit Electronic Excitation Energies, Ionization Potentials, and Electron Affinities for the 3d Transition Metal Atoms: Coupled Cluster and Multireference Methods.

241. J. M. L. Martin and G. de Oliveira, *J. Chem. Phys.*, **111**, 1843–1866 (1999). Towards Standard Methods for Benchmark Quality *ab Initio* Thermochemistry – W1 and W2 Theory.

242. S. Parthiban and J. M. L. Martin, *J. Chem. Phys.*, **114**, 6014–6029 (2001). Assessment of W1 and W2 Theories for the Computation of Electron Affinities, Ionization Potentials, Heats of Formation, and Proton Affinities.

243. L. A. Curtiss, K. Raghavachari, P. C. Redfern, V. Rassolov, and J. A. Pople, *J. Chem. Phys.*, **109**, 7764–7776 (1998). Gaussian-3 (G3) Theory for Molecules Containing First and Second-Row Atoms.

244. L. A. Curtiss, P. C. Redfern, K. Raghavachari, V. Rassolov, and J. A. Pople, *J. Chem. Phys.*, **110**, 4703–4709 (1999). Gaussian-3 Theory Using Reduced Møller-Plesset Order.

245. L. A. Curtiss and K. Raghavachari, *Theor. Chem. Acc.*, **108**, 61–70 (2002). Gaussian-3 and Related Methods for Accurate Thermochemistry.

246. J. A. Montgomery, M. J. Frisch, J. W. Ochterski, and G. A. Petersson, *J. Chem. Phys.*, **112**, 6532–6542 (2000). A Complete Basis Set Model Chemistry. VII. Use of the Minimum Population Localization Method.

247. J. A. Montgomery, J. W. Ochterski, and G. A. Petersson, *J. Chem. Phys.*, **101**, 5900–5909 (1994). A Complete Basis Set Model Chemistry. IV. An Improved Atomic Pair Natural Orbital Method.

248. J. W. Ochterski, G. A. Petersson, and J. A. Montgomery, *J. Chem. Phys.*, **104**, 2598–2619 (1996). A Complete Basis Set Model Chemistry. V. Extensions to Six or More Heavy Atoms.

249. J. A. Montgomery, M. J. Frisch, J. W. Ochterski, and G. A. Petersson, *J. Chem. Phys.*, **110**, 2822–2827 (1999). A Complete Basis Set Model Chemistry. VI. Use of Density Functional Geometries and Frequencies.

250. J. M. L. Martin and S. Parthiban, in *Quantum-Mechanical Prediction of Thermochemical Data*, J. Cioslowski (Ed.) Vol. 22 of, *Understanding Chemical Reactivity*, Kluwer Academic Publishers, 2001; chapter 2, pp. 31–66, W1 and W2 Theories, and Their Variants: Thermochemistry in the kJ/mol Accuracy Range.

251. K. Raghavachari and L. A. Curtiss, in *Quantum-Mechanical Prediction of Thermochemical Data*, J. Cioslowski (Ed.) Vol. 22 of, *Understanding Chemical Reactivity*, Kluwer Academic Publishers, 2001; chapter 3, pp. 67–98, Complete Basis Set Models for Chemical Reactivity: From the Helium Atom to Enzyme Kinetics.

252. L. A. Curtiss, K. Raghavachari, G. W. Trucks, and J. A. Pople, *J. Chem. Phys.*, **94**, 7221–7230 (1991). Gaussian-2 Theory for Molecular Energies of First- and Second-Row Compounds.

253. H. Haberland, C. Ludewigt, H. Schindler, and D. R. Worsnop, *Z. Phys. A*, **320**, 151–153 (1985). Field Detachment of the Negatively Charged Water Dimer.

254. H. Haberland, C. Ludewigt, H.-G. Schindler, and D. R. Worsnop, *Phys. Rev. A*, **36**, 967–970 (1987). Field Detachment of $(H_2 O)_2^-$ Clustered with Rare Gases.

255. J. H. Hendricks, H. L. de Clercq, S. A. Lyapustina, and K. H. Bowen, Jr., *J. Chem. Phys.*, **107**, 2962–2967 (1997). Negative Ion Photoelectron Spectroscopy of the Ground State, Dipole-Bound Dimeric Anion, $(HF)_2^-$.

256. P. Skurski, M. Gutowski, and J. Simons, *J. Chem. Phys.*, **114**, 7443–7449 (2001). *Ab Initio* Electronic Structure of HCN^- and HNC^- Dipole-Bound Anions and a Description of Electron Loss upon Tautomerization.

257. K. A. Peterson and M. Gutowski, *J. Chem. Phys.*, **116**, 3297–3299 (2002). Electron Binding Energies of Dipole-Bound Anions at the Coupled Cluster Level with Single, Double, and Triple Excitations: HCN^- and HNC^-.

258. T. Sommerfeld, A. DeFusco, and K. D. Jordan, *J. Phys. Chem. A*, **112**, 11021–11035 (2008). Model Potential Approaches for Describing the Interaction of Excess Electrons with Water Clusters: Incorporation of Long-Range Correlation Effects.

259. J. Xu and K. D. Jordan, *J. Phys. Chem. A*, **114**, 1364–1366 (2010). Application of the Diffusion Monte Carlo Method to the Binding of Excess Electrons to Water Clusters.

260. V. P. Vysotskiy, L. S. Cederbaum, T. Sommerfeld, V. K. Voora, and K. D. Jordan, *J. Chem. Theory Comput.*, **8**, 893–900 (2012). Benchmark Calculations of the Energies for Binding Excess Electrons to Water Clusters.

261. R. J. Bartlett and M. Musial, *Rev. Mod. Phys.*, **79**, 291–352 (2007). Coupled-Cluster Theory in Quantum Chemistry.

262. M. Nooijen and R. J. Bartlett, *J. Chem. Phys.*, **102**, 3629–3647 (1995). Equation of Motion Coupled Cluster Method for Electron Attachment.

263. J. F. Stanton and R. J. Bartlett, *J. Chem. Phys.*, **98**, 7029–7039 (1993). The Equation of Motion Coupled-Cluster Method. A Systematic Biorthogonal Approach to Molecular Excitation Energies, Transition Probabilities, and Excited State Properties.

264. V. G. Zakrezewski, O. Dolgounitcheva, A. V. Zakjevskii, and J. V. Ortiz, *Annu. Rep. Comput. Chem.*, **6**, 79–94 (2010). *Ab Initio* Electron Propagator Methods: Applications to Fullerenes and Nucleic Acid Fragments.

265. V. G. Zakrezewski, O. Dolgounitcheva, A. V. Zakjevskii, and J. V. Ortiz, *Adv. Quantum Chem.*, **62**, 105–136 (2011). *Ab Initio* Electron Propagator Calculations on Electron Detachment Energies of Fullerenes, Macrocyclic Molecules and Nucleotide Fragments.

266. O. Dolgounitcheva, V. G. Zakrzewski, and J. V. Ortiz, *Int. J. Quantum Chem.*, **112**, 184–194 (2012). *Ab Initio* Electron Propagator Calculations on Electron Detachment Energies of Nickel Phthalocyanine Tetrasulfonate Tetraanions.

267. G. Chałasiński and M. M. Szczęśniak, *Chem. Rev.*, **94**, 1723–1765 (1994). Origins of Structure and Energetics of van der Waals Clusters from ab Initio Calculations.

268. B. Jeziorski, R. Moszynski, and K. Szalewicz, *Chem. Rev.*, **94**, 1887–1930 (1994). Perturbation Theory Approach to Intermolecular Potential Energy Surfaces of van der Waals Complexes.

269. M. Gutowski, P. Skurski, K. D. Jordan, and J. Simons, *Int. J. Quantum Chem.*, **64**, 183–191 (1997). Energies of Dipole-Bound Anionic States.

270. J. M. Herbert and M. Head-Gordon, *Phys. Chem. Chem. Phys.*, **8**, 68–78 (2006). Accuracy and Limitations of Second-Order Many-Body Perturbation Theory for Predicting Vertical Detachment Energies of Solvated-Electron Clusters.

271. E. Fermi and E. Teller, *Phys. Rev.*, **72**, 399–408 (1947). The Capture of Negative Mesotrons in Matter.

272. A. D. Becke, *J. Chem. Phys.*, **98**, 1372–1377 (1993). A New Mixing of Hartree–Fock and Local Density-Functional Theories.

273. C. Lee, W. Yang, and R. G. Parr, *Phys. Rev. B*, **37**, 785–789 (1988). Development of the Colle-Salvetti Correlation-Energy Formula into a Functional of the Electron Density.

274. A. D. Becke, *Phys. Rev. A*, **38**, 3098–3100 (1988). Density-Functional Exchange-Energy Approximation with Correct Asymptotic Behavior.

275. J. Schirmer, *Phys. Rev. A*, **26**, 2395–2416 (1982). Beyond the Random-Phase Approximation: A New Approximation Scheme for the Polarization Propagator.

276. J. Schirmer, L. S. Cederbaum, and O. Walter, *Phys. Rev. A*, **28**, 1237–1259 (1983). New Approach to the One-Particle Green's Function for Finite Fermi Systems.

277. E. R. Davidson, S. A. Hagstrom, S. J. Chakravorty, V. M. Umar, and C. F. Fischer, *Phys. Rev. A*, **44**, 7071–7083 (1991). Ground-State Correlation Energies for Two- to Ten-Electron Atomic Ions.

278. G. K.-L. Chan and M. Head-Gordon, *J. Chem. Phys.*, **118**, 8551–8554 (2003). Exact Solution (within a Triple-Zeta, Double Polarization Basis Set) of the Electronic Schrödinger Equation for Water.

279. R. A. Kendall and H. A. Früchtl, *Theor. Chem. Acc.*, **97**, 158–163 (1997). The Impact of the Resolution of the Identity Approximate Integral Method on Modern ab Initio Algorithm Development.

280. H.-J. Werner, F. R. Manby, and P. J. Knowles, *J. Chem. Phys.*, **118**, 8149–8160 (2003). Fast Linear Scaling Second-Order Møller-Plesset Perturbation Theory (MP2) using Local and Density Fitting Approximations.

281. T. B. Pedersen, F. Aquilante, and R. Lindh, *Theor. Chem. Acc.*, **124**, 1–10 (2009). Density Fitting with Auxiliary Basis Sets from Cholesky Decomposition.

282. Y. Jung, A. Sodt, P. M. W. Gill, and M. Head-Gordon, *Proc. Natl. Acad. Sci. U.S.A.*, **102**, 6692–6697 (2005). Auxiliary Basis Expansions for Large-Scale Electronic Structure Calculations.

283. F. Weigend, M. Häser, J. Patzelt, and R. Ahlrichs, *Chem. Phys. Lett.*, **294**, 143–152 (1998). RI-MP2: Optimized Auxiliary Basis Sets and Demonstration of Efficiency.

284. R. Palangsuntikul, R. Polly, and B. Hartke, *Phys. Chem. Chem. Phys.*, **6**, 5456–5462 (2004). Global and Local Optimization of Auxiliary Basis Sets for RI-MP2 Calculations.

285. R. Polly, H.-J. Werner, F. R. Manby, and P. J. Knowles, *Mol. Phys.*, **102**, 2311–2321 (2004). Fast Hartree-Fock Theory using Local Density Fitting Approximations.

286. J. Bostrom, F. Aquilante, T. B. Pedersen, and R. Lindh, *J. Chem. Theory Comput.*, **5**, 1545–1553 (2009). Ab Initio Density Fitting: Accuracy Assessment of Auxiliary Basis Sets from Cholesky Decompositions.

287. R. P. Steele, R. A. DiStasio, Jr., Y. Shao, J. Kong, and M. Head-Gordon, *J. Chem. Phys.*, **125**(074108), 1–11 (2006). Dual-Basis Second-Order Møller-Plesset Perturbation Theory: A Reduced-Cost Reference for Correlation Calculations.

288. F. Weigend and R. Ahlrichs, *Phys. Chem. Chem. Phys.*, **1**, 4537–4550 (1999). Ab Initio Treatment of $(H_2O)_2^-$ and $(H_2O)_6^-$.

289. R. A. DiStasio, Jr. R. P. Steele, Y. M. Rhee, Y. Shao, and M. Head-Gordon, *J. Comput. Chem.*, **28**, 839–859 (2007). An Improved Algorithm for Analytical Gradient Evaluation in Resolution-of-the-Identity Second-Order Møller-Plesset Perturbation Theory: Application to Alanine Tetrapeptide Conformational Analysis.

290. J. Tomasi, B. Mennucci, and R. Cammi, *Chem. Rev.*, **105**, 2999–3093 (2005). Quantum Mechanical Continuum Solvation Models.

291. C. J. Cramer and D. G. Truhlar, *Acc. Chem. Res.*, **41**, 760–768 (2008). A Universal Approach to Solvation Modeling.

292. A. W. Lange and J. M. Herbert, *Chem. Phys. Lett.*, **509**, 77–87 (2011). Symmetric Versus Asymmetric Discretization of the Integral Equations in Polarizable Continuum Solvation Models.

293. D. M. Chipman, *Theor. Chem. Acc.*, **107**, 80–89 (2002). Comparison of Solvent Reaction Field Representations.

294. J. Tomasi, in *Continuum Solvation Models in Chemical Physics*, B. Mennucci and R. Cammi (Eds.), Wiley, Chichester, UK, 2007, pp. 1–28, Modern Theories of Continuum Models.

295. B. Mennucci, *WIREs Comput. Mol. Sci.*, **2**, 386–404 (2012). Polarizable Continuum Model.

296. J. D. Thompson, C. J. Cramer, and D. G. Truhlar, *J. Phys. Chem. A*, **108**, 6532–6542 (2004). New Universal Solvation Model and Comparison of the Accuracy of the SM5.42R, SM5.43R, C-PCM, D-PCM, and IEF-PCM Continuum Solvation Models for Aqueous and Organic Solvation Free Energies and for Vapor Pressures.

297. A. Klamt, B. Mennucci, J. Tomasi, V. Barone, C. Curutchet, M. Orozco, and F. J. Luque, *Acc. Chem. Res.*, **42**, 489–492 (2009). On the Performance of Continuum Solvation Methods. A Comment on "Universal Approaches to Solvation Modeling".

298. C. J. Cramer and D. G. Truhlar, *Acc. Chem. Res.*, **42**, 493–497 (2009). Reply to Comment on 'A Universal Approach to Solvation Modeling'.

299. A. V. Marenich, C. J. Cramer, and D. G. Truhlar, *J. Phys. Chem. B*, **113**, 4538–4543 (2009). Performance of SM6, SM8, and SMD on the SAMPL1 Test Set for the Prediction of Small-Molecule Solvation Free Energies.

300. A. Pomogaeva, D. W. Thompson, and D. M. Chipman, *Chem. Phys. Lett.*, **511**, 161–165 (2011). Modeling Short-Range Contributions to Hydration Energies with Minimal Parameterization.

301. R. A. Marcus, *J. Chem. Phys.*, **24**, 979–989 (1956). Electrostatic Free Energy and Other Properties of States Having Nonequilibrium Polarization. I.

302. M. M. Karelson and M. C. Zerner, *J. Phys. Chem.*, **96**, 6949–6957 (1992). Theoretical Treatment of Solvent Effects on Electronic Spectroscopy.

303. A. Klamt, *J. Phys. Chem.*, **100**, 3349–3353 (1996). Calculation of UV Vis Spectra in Solution.

304. H. Houjou, M. Sakurai, and Y. Inoue, *J. Chem. Phys.*, **107**, 5652–5660 (1997). Theoretical Evaluation of Medium Effects on Absorption Maxima of Molecular Solutes. I. Formulation of a New Method Based on the Self-Consistent Reaction Field Theory.

305. B. Mennucci, R. Cammi, and J. Tomasi, *J. Chem. Phys.*, **109**, 2798–2807 (1998). Excited States and Solvatochromatic Shifts within a Nonequilibrium Solvation Approach: A New

Formulation of the Integral Equation Formalism Method at the Self-Consistent Field, Configuration Interaction, and Multiconfiguration Self-Consistent Field Level.

306. M. Cossi and V. Barone, *J. Chem. Phys.*, **112**, 2427–2435 (2000). Solvent Effect on Vertical Electronic Transitions by the Polarizable Continuum Model.

307. M. Cossi and V. Barone, *J. Phys. Chem. A*, **104**, 10614–10622 (2000). Separation between Fast and Slow Polarizations in Continuum Solvation Models.

308. A. J. Cohen, P. Mori-Sanchez, and W. Yang, *Science*, **321**, 792–794 (2008). Insights into Current Limitations of Density Functional Theory.

309. P. Slavček, B. Winter, M. Faubel, S. E. Bradforth, and P. Jungwirth, *J. Am. Chem. Soc.*, **131**, 6460–6467 (2009). Ionization Energies of Aqueous Nucleic Acids: Photoelectron Spectroscopy of Pyrimidine Nucleosides and ab Initio Calculations.

310. J. P. Perdew and M. Ernzerhof, in *Electronic Density Functional Theory: Recent Progress and New Directions*, J. F. Dobson, G. Vignale, and M. P. Das (Eds.), Plenum, 1998, pp. 31–41, Density Functionals for Non-Relativistic Coulomb Systems.

311. R. Haunschild, T. M. Henderson, C. A. Jiménez-Hoyos, and G. E. Scuseria, *J. Chem. Phys.*, **133**, 134116:1–10 (2010). Many-Electron Self-Interaction and Spin Polarization Errors in Local Hybrid Density Functionals.

312. D. Lee and K. Burke, *Mol. Phys.*, **108**, 2687–2701 (2010). Finding Electron Affinities with Approximate Density Functionals.

313. M.-C. Kim, E.-J. Sim, and K. Burke, *J. Chem. Phys.*, **134**, 171103:1–4 (2011). Communication: Avoiding Unbound Anions in Density Functional Calculations.

314. J. P. Perdew and S. Kurth, in *Primer in Density Functional Theory*, C. Fiolhais, F. Nogueira, and M. A. L. Marques (Eds.), Vol. 620 of, *Lecture Notes in Physics*, Springer-Verlag, 2003, pp. 1–55, Density Functionals for Non-Relativistic Coulomb Systems in the New Century.

315. H. B. Shore, J. H. Rose, and E. Zaremba, *Phys. Rev. B*, **15**, 2858–2861 (1977). Failure of the Local Exchange Approximation in the Evaluation of the H^- Ground State.

316. K. Schwarz, *Chem. Phys. Lett.*, **57**, 605–607 (1978). Instability of Stable Negative Ions in the $X\alpha$ Method or Other Local Density Functional Schemes.

317. F. Jensen, *J. Chem. Theory Comput.*, **6**, 2726–2735 (2010). Describing Anions by Density Functional Theory: Fractional Electron Affinity.

318. E. J. Baerends, O. V. Gritsenko, and R. van Meer, *Phys. Chem. Chem. Phys.*, **15**, 16408–16425 (2013). The Kohn–Sham Gap, the Fundamental Gap and the Optical Gap: The Physical Meaning of Occupied and Virtual Kohn–Sham Orbital Energies.

319. Y. Zhao, N. E. Schultz, and D. G. Truhlar, *J. Chem. Theory Comput.*, **2**, 364–382 (2006). Design of Density Functionals by Combining the Method of Constraint Satisfaction with Parameterization for Thermochemistry, Thermochemical Kinetics, and Noncovalent Interactions.

320. Y. Zhao and D. G. Truhlar, *Theor. Chem. Acc.*, **120**, 215–241 (2008). The M06 Suite of Density Functionals for Main Group Thermochemistry, Thermochemical Kinetics, Noncovalent Interactions, Excited States, and Transition Elements: Two New Functionals and Systematic Testing of Four M06-Class Functionals and 12 Other Functionals.

321. R. Peverati, Y. Zhao, and D. G. Truhlar, *J. Phys. Chem. Lett.*, **2**, 1991–1997 (2011). Generalized Gradient Approximation That Recovers the Second-Order Density-Gradient Expansion with Optimized Across-the-Board Performance.

322. R. Peverati and D. G. Truhlar, *J. Phys. Chem. Lett.*, **3**, 117–124 (2012). M11-L: A Local Density Functional That Provides Improved Accuracy for Electronic Structure Calculations in Chemistry and Physics.

323. A. W. Lange and J. M. Herbert, *J. Am. Chem. Soc.*, **131**, 3913–3922 (2009). Both Intra- and Interstrand Charge-Transfer Excited States in B-DNA are Present at Energies Comparable to, or Just above, the $^1\pi\pi^*$ Excitonic Bright States.

324. M. A. Rohrdanz, K. M. Martins, and J. M. Herbert, *J. Chem. Phys.*, **130**, 054112:1–8 (2009). A Long-Range-Corrected Density Functional That Performs Well for Both Ground-State

Properties and Time-Dependent Density Functional Theory Excitation Energies, Including Charge-Transfer Excited States.

325. R. Peverati and D. G. Truhlar, *J. Phys. Chem. Lett.*, **2**, 2810–2817 (2011). Improving the Accuracy of Hybrid Meta-GGA Density Functionals by Range Separation.

326. Y. Zhao and D. G. Truhlar, *J. Chem. Theory Comput.*, **4**, 1849–1868 (2008). Exploring the Limit of Accuracy of the Global Hybrid Meta Density Functional for Main-Group Thermochemistry, Kinetics, and Noncovalent Interactions.

327. R. Peverati and D. G. Truhlar, *J. Chem. Phys.*, **135**, 191102:1–4 (2011). Communication: A Global Hybrid Generalized Gradient Approximation to the Exchange-Correlation Functional That Satisfies the Second-Order Density-Gradient Constraint and Has Broad Applicability in Chemistry.

328. J. P. Perdew and A. Zunger, *Phys. Rev. B*, **23**, 5048–5079 (1981). Self-Interaction Correction to Density-Functional Approximations for Many-Electron Systems.

329. O. A. Vydrov and G. E. Scuseria, *J. Chem. Phys.*, **121**, 8187–8193 (2004). Effect of the Perdew-Zunger Self-Interaction Correction on the Thermochemical Performance of Approximate Density Functionals.

330. T. Frigato, J. VandeVondele, B. Schmidt, C. Schütte, and P. Jungwirth, *J. Phys. Chem. A*, **112**, 6125–6133 (2008). Ab Initio Molecular Dynamics Simulation of a Medium-Sized Water Cluster Anion: From an Interior to a Surface-Located Excess Electron via a Delocalized State.

331. O. Marsalek, F. Uhlig, T. Frigato, B. Schmidt, and P. Jungwirth, *Phys. Rev. Lett.*, **105**, 043002:1–4 (2010). Dynamics of Electron Localization in Warm Versus Cold Water Clusters.

332. M. d'Avezac, M. Calandra, and F. Mauri, *Phys. Rev. B*, **71**, 205210:1–5 (2005). Density Functional Theory Description of Hole-Trapping in SiO_2: A Self-Interaction-Corrected Approach.

333. J. VandeVondele and M. Sprik, *Phys. Chem. Chem. Phys.*, **7**, 1363–1367 (2005). A Molecular Dynamics Study of the Hydroxyl Radical in Solution Applying Self-Interaction-Corrected Density Functional Methods.

334. I. Frank, J. Hutter, D. Marx, and M. Parrinello, *J. Chem. Phys.*, **108**, 4060–4069 (1998). Molecular Dynamics in Low-Spin Excited States.

335. I. Okazaki, F. Sato, T. Yoshihiro, T. Ueno, and H. Kashiwagi, *J. Mol. Struct. (THEOCHEM)*, **451**, 109–119 (1998). Development of a Restricted Open Shell Kohn–Sham Program and Its Application to a Model Heme Complex.

336. M. Filatov and S. Shaik, *Chem. Phys. Lett.*, **304**, 429–437 (1999). A Spin-Restricted Ensemble-References Kohn-Sham Method and Its Application to Diradicaloid Situations.

337. M. Sodupe, J. Bertran, L. Rodrguez-Santiago, and E. J. Baerends, *J. Phys. Chem. A*, **103**, 166–170 (1999). Ground State of the $(H_2O)_2^+$ Radical Cation: DFT Versus Post-Hartree–Fock Method.

338. J. Gräfenstein, E. Kraka, and D. Cremer, *Phys. Chem. Chem. Phys.*, **6**, 1096–1112 (2004). Effect of the Self-Interaction Error for Three-Electron Bonds: On the Development of New Exchange-Correlation Functionals.

339. Y. A. Mantz, F. L. Gervasio, T. Laino, and M. Parrinello, *J. Phys. Chem. A*, **111**, 105–112 (2007). Charge Localization in Stacked Radical Cation DNA Base Pairs and the Benzene Dimer Studied by Self-Interaction Corrected Density-Functional Theory.

340. O. Marsalek, T. Frigato, J. VandeVondele, S. E. Bradforth, B. Schmidt, C. Schütte, and P. Jungwirth, *J. Phys. Chem. B*, **114**, 915–920 (2010). Hydrogen Forms in Water by Proton Transfer to a Distorted Electron.

341. H. Iikura, T. Tsuneda, T. Yanai, and K. Hirao, *J. Chem. Phys.*, **115**, 3540–3544 (2001). A Long-Range Correction Scheme for Generalized-Gradient-Approximation Exchange Functionals.

342. Y. Tawada, T. Tsuneda, S. Yanagisawa, T. Yanai, and K. Hirao, *J. Chem. Phys.*, **120**, 8425–8433 (2004). A Long-Range Corrected Time-Dependent Density Functional Theory.

343. J.-W. Song, T. Hirosawa, T. Tsuneda, and K. Hirao, *J. Chem. Phys.*, **126**, 154105:1–7 (2007). Long-Range Corrected Density Functional Calculations of Chemical Reactions: Redetermination of Parameter.

344. R. Baer, E. Livshits, and U. Salzner, *Annu. Rev. Phys. Chem.*, **61**, 85–109 (2010). Tuned Range-Separated Hybrids in Density Functional Theory.

345. T. M. Henderson, B. G. Janesko, and G. E. Scuseria, *J. Phys. Chem. A*, **112**, 12530–12542 (2008). Range Separation and Local Hybridization in Density Functional Theory.

346. A. Savin and H.-J. Flad, *Int. J. Quantum Chem.*, **56**, 327–332 (1995). Density Functionals for the Yukawa Electron–Electron Interaction.

347. T. Leininger, H. Stoll, H.-J. Werner, and A. Savin, *Chem. Phys. Lett.*, **275**, 151–160 (1997). Combining Long-Range Configuration Interaction with Short-Range Density Functionals.

348. R. D. Adamson, J. P. Dombroski, and P. M. W. Gill, *J. Comput. Chem.*, **20**, 921–927 (1999). Efficient Calculation of Short-Range Coulomb Energies.

349. T. M. Henderson, B. G. Janesko, and G. E. Scuseria, *J. Chem. Phys.*, **128**, 194105:1–9 (2008). Generalized Gradient Approximation Model Exchange Holes for Range-Separated Hybrids.

350. M. Chiba, T. Tsuneda, and K. Hirao, *J. Chem. Phys.*, **124**, 144106:1–11 (2006). Excited State Geometry Optimizations by Analytical Energy Gradient of Long-Range Corrected Time-Dependent Density Functional Theory.

351. T. Sato, T. Tsuneda, and K. Hirao, *J. Chem. Phys.*, **126**, 234114:1–12 (2007). Long-Range Corrected Density Functional Study on Weakly Bound Systems: Balanced Descriptions of Various Types of Molecular Interactions.

352. O. A. Vydrov, J. Heyd, A. V. Krukau, and G. E. Scuseria, *J. Chem. Phys.*, **125**, 074106:1–9 (2006). Importance of Short-Range Versus Long-Range Hartree-Fock Exchange for the Performance of Hybrid Density Functionals.

353. O. A. Vydrov and G. E. Scuseria, *J. Chem. Phys.*, **125**, 234109:1–9 (2006). Assessment of a Long-Range Corrected Hybrid Functional.

354. R. M. Richard and J. M. Herbert, *J. Chem. Theory Comput.*, **7**, 1296–1306 (2011). Time-Dependent Density-Functional Description of the ^1L_a State in Polycyclic Aromatic Hydrocarbons: Charge-Transfer Character in Disguise?

355. M. A. Rohrdanz and J. M. Herbert, *J. Chem. Phys.*, **129**, 034107:1–9 (2008). Simultaneous Benchmarking of Ground- and Excited-State Properties with Long-Range-Corrected Density Functional Theory.

356. A. W. Lange, M. A. Rohrdanz, and J. M. Herbert, *J. Phys. Chem. B*, **112**, 6304–6308 (2008). Charge-Transfer Excited States in a π-Stacked Adenine Dimer, as Predicted Using Lon-Range-Corrected Time-Dependent Density Functional Theory.

357. P. Mori-Sánchez, A. J. Cohen, and W. Yang, *J. Chem. Phys.*, **125**, 201201:1–4 (2006). Many-Electron Self-Interaction Error in Approximate Density Functionals.

358. L. D. Jacobson, C. F. Williams, and J. M. Herbert, *J. Chem. Phys.*, **130**, 124115:1–18 (2009). The Static-Exchange Electron–Water Pseudopotential, in Conjunction with a Polarizable Water Model: A New Hamiltonian for Hydrated-Electron Simulations.

359. T. Tsuneda, T. Suzumura, and K. Hirao, *J. Chem. Phys.*, **110**, 10664–10678 (2008). A New One-Parameter Progressive Colle–Salvetti-Type Correlation Functional.

360. J. P. Perdew and M. Levy, *Phys. Rev. B*, **56**, 16021–16028 (1997). Comment on "Significance of the Highest Occupied Kohn-Sham Eigenvalue".

361. M. E. Casida, *Phys. Rev. B*, **59**, 4694–4698 (1999). Correlated Optimized Effective-Potential Treatment of the Derivative Discontinuity and of the Highest Occupied Kohn-Sham Eigenvalue: A Janak-Type Theorem for the Optimized Effective-Potential Model.

362. T. Tsudeda, J.-W. Song, S. Suzuki, and K. Hirao, *J. Chem. Phys.*, **133**, 174101:1–9 (2010). On Koopmans' Theorem in Density Functional Theory.

363. E. Livshits, R. S. Granot, and R. Baer, *J. Phys. Chem. A*, **115**, 5735–5744 (2010). A Density Functional Theory for Studying Ionization Processes in Water Clusters.

364. K. U. Lao and J. M. Herbert, *J. Phys. Chem. Lett.*, 3, 3241–3248 (2012). Accurate Intermolecular Interactions at Dramatically Reduced Cost: XPol+SAPT with Empirical Dispersion.

365. S. Jakobsen, K. Kristensen, and F. Jensen, *J. Chem. Theory Comput.*, 9, 3978–3985 (2013). Electrostatic Potential of Insulin: Exploring the Limitations of Density Functional Theory and Force Field Methods.

366. M. F. Falcetta and K. D. Jordan, *J. Phys. Chem.*, 94, 5666–5669 (1990). Assignments of the Temporary Anion States of the Chloromethanes.

367. M. F. Falcetta, Y. Choi, and K. D. Jordan, *J. Phys. Chem. A*, 104, 9605–9612 (2000). Ab Initio Investigation of the Temporary Anion States of Perfluoroethane.

368. H.-Y. Cheng and C.-W. Chen, *J. Phys. Chem. A*, 115, 10113–10121 (2011). Energy and Lifetime of Temporary Anion States of Uracil by Stabilization Method.

369. C.-Y. Juang and J. S.-Y. Chao, *J. Phys. Chem.*, 98, 13506–13512 (1994). Splitting Energies of π^* Anion States of 1,4-Cyclohexadiene via the Exponent Stabilization Method.

370. J. Berdys, I. Anusiewicz, P. Skurski, and J. Simons, *J. Am. Chem. Soc.*, 126, 6441–6447 (2004). Damage to Model DNA Fragments from Very Low-Energy (<1 eV) Electrons.

371. X. Li, M. D. Sevilla, and L. Sanche, *J. Am. Chem. Soc.*, 125, 13668–13669 (2003). Density Functional Theory Studies of Electron Interaction with DNA: Can Zero eV Electrons Induce Strand Breaks?

372. I. Anusiewicz, J. Berdys, M. Sobczyk, P. Skurski, and J. Simons, *J. Phys. Chem. A*, 108, 11381–11387 (2004). Effects of Base π-Stacking on Damage to DNA by Low-Energy Electrons.

373. T. Van Voorhis and M. Head-Gordon, *Mol. Phys.*, 100, 1713–1721 (2002). A Geometric Approach to Direct Minimization.

374. P. Pulay, *Chem. Phys. Lett.*, 73, 393–398 (1980). Convergence Acceleration of Iterative Sequences. The Case of SCF Iteration.

375. K. N. Kudin, G. E. Scuseria, and E. Cancès, *J. Chem. Phys.*, 116, 8255–8261 (2002). A Black-Box Self-Consistent Field Convergence Algorithm: One Step Closer.

376. A. T. B. Gilbert, N. A. Besley, and P. M. W. Gill, *J. Phys. Chem. A*, 112, 13164–13171 (2008). Self-Consistent Field Calculations of Excited States Using the Maximum Overlap Method (MOM).

377. N. A. Besley, A. T. B. Gilbert, and P. M. W. Gill, *J. Chem. Phys.*, 130, 124308:1–7 (2009). Self-Consistent-Field Calculations of Core Excited States.

378. C. Zener, *Proc. R. Soc. London A*, 137, 696–702 (1932). Non-Adiabatic Crossing of Energy Levels.

379. C. Wittig, *J. Phys. Chem. B*, 109, 8428–8430 (2005). The Landau–Zener Formula.

380. E. E. Nikitin, *Annu. Rev. Phys. Chem.*, 50, 1–21 (1999). Nonadiabatic Transitions: What We Learned from Old Masters and How Much We Owe Them.

381. W. Qian and S. Krimm, *J. Comput. Chem.*, 32, 1025–1033 (1992). Vibrational Studies of the Disulfide Group in Proteins. VI. General Correlations of SS and CS Stretch Frequencies with Disulfide Bridge Geometry.

382. W. P. Reinhardt, *Annu. Rev. Phys. Chem.*, 33, 223–255 (1982). Complex Coordinates in the Theory of Atomic and Molecular Structure and Dynamics.

383. E. Balslev and J. M. Combes, *Commun. Math. Phys.*, 22, 280–294 (1971). Spectral Properties of Many-Body Schrödinger Operators with Dilatation-Analytic Interactions.

384. N. Moiseyev, *Phys. Rep.*, 302, 211–293 (1998). Quantum Theory of Resonances: Calculating Energies, Widths and Cross-Sections by Complex Scaling.

385. N. Moiseyev, P. R. Certain, and F. Weinhold, *Mol. Phys.*, 36, 1613–1630 (1978). Resonance Properties of Complex-Rotated Hamiltonians.

386. P. Žd'ánská and N. Moiseyev, *J. Chem. Phys.*, 123, 194105:1–8 (2005). Hartree-Fock Orbitals for Complex-Scaled Configuration Interaction Calculation of Highly Excited Feshbach Resonances.

387. D. L. Yeager and M. K. Mishra, *Int. J. Quantum Chem.*, **104**, 871–879 (2005). Algebraic Modifications to Second Quantization for Non-Hermitian Complex Scaled Hamiltonians with Applications to a Quadratically Convergent Multiconfigurational Self-Consistent Field Method.

388. A. J. Krueger and N. T. Maitra, *Phys. Chem. Chem. Phys.*, **11**, 4655–4663 (2009). Autoionizing Resonances in Time-Dependent Density Functional Theory.

389. P. Elliott, S. Goldson, C. Canahui, and N. T. Maitra, *Chem. Phys.*, **391**, 110–119 (2011). Perspective on Double-Excitations in TDDFT.

390. C. W. McCurdy, Jr. and T. N. Rescigno, *Phys. Rev. Lett.*, **41**, 1364–1368 (1978). Extension of the Method of Complex Basis Functions to Molecular Resonances.

391. B. Simon, *Phys Lett. A.*, **71**, 211–214 (1979). The Definition of Molecular Resonance Curves by the Method of Exterior Complex Scaling.

392. N. Rom, E. Engdahl, and N. Moiseyev, *J. Chem. Phys.*, **93**, 3413–3419 (1990). Tunneling Rates in Bound Systems Using Smooth Exterior Complex Scaling within the Framework of the Finite Basis Set Approximation.

393. N. Moiseyev, *J. Phys. B*, **31**, 1431–1441 (1998). Derivations of Universal Exact Complex Absorption Potentials by the Generalized Complex Coordinate Method.

394. P. Balanarayan, Y. Sajeev, and N. Moiseyev, *Chem. Phys. Lett.*, **524**, 84–89 (2012). Ab-Initio Complex Molecular Potential Energy Surfaces by the Back-Rotation Transformation Method.

395. A. Ghosh, N. Vaval, and S. Pal, *J. Chem. Phys.*, **136**, 234110:1–6 (2012). Equation-of-Motion Coupled-Cluster Method for the Study of Shape Resonances.

396. E. Epifanovsky, I. Polyakov, B. Grigorenko, A. Nemukhin, and A. I. Krylov, *J. Chem. Theory Comput.*, **5**, 1895–1906 (2009). Quantum Chemical Benchmark Studies of the Electronic Properties of the Green Fluorescent Protein Chromophore. 1. Electronically Excited and Ionized States of the Anionic Chromophore in the Gas Phase.

397. A. Lange and J. M. Herbert, *J. Chem. Theory Comput.*, **3**, 1680–1690 (2007). Simple Methods to Reduce Charge-Transfer Contamination in Time-Dependent Density-Functional Calculations of Clusters and Liquids.

398. X. Chen and S. E. Bradforth, *Annu. Rev. Phys. Chem.*, **59**, 203–231 (2008). The Ultrafast Dynamics of Photodetachment.

399. A. Bernas, D. Grand, and E. Amouyal, *J. Phys. Chem.*, **84**, 1259–1262 (1980). Photoionization of Solutes and Conduction Band Edge of Solvents. Indole in Water and Alcohols.

400. A. U. Hazi and H. S. Taylor, *Phys. Rev. A*, **1**, 1109–1120 (1970). Stabilization Method of Calculating Resonance Energies: Model Problem.

401. C.-S. Chen, T.-H. Feng, and J. S.-Y. Chao, *J. Phys. Chem.*, **99**, 8629–8632 (1995). Stabilized Koopmans' Theorem Calculations on the π^* Temporary Anion States of Benzene and Substituted Benzenes.

402. M. Venuti and A. Modelli, *J. Chem. Phys.*, **113**, 2159–2167 (2000). Low-Energy Electron Attachment to Fused 1,4-Cyclohexadiene Rings by Means of Electron Transmission Spectroscopy and Exponent Stabilization Calculations.

403. K. D. Jordan, *Chem. Phys.*, **9**, 199–204 (1975). Construction of Potential Energy Curves in Avoided Crossing Situations.

404. A. D. Isaacson and D. G. Truhlar, *Chem. Phys. Lett.*, **110**, 130–134 (1984). Single-Root, Real-Basis-Function Method with Correct Branch-Point Structure for Complex Resonance Energies.

405. R. F. Frey and J. Simons, *J. Chem. Phys.*, **84**, 4462–4469 (1986). Resonance State Energies and Lifetimes via Analytic Continuation of Stabilization Graphs.

406. J. S.-Y. Chao, M. F. Falcetta, and K. D. Jordan, *J. Chem. Phys.*, **93**, 1125–1135 (1990). Application of the Stabilization Method to the $N_2^-(1\,^2\Pi_g)$ and $Mg^-(1\,^2P)$ Temporary Anion States.

407. K. Aflatooni, G. A. Gallup, and P. D. Burrow, *J. Phys. Chem. A*, **102**, 6205–6207 (1998). Electron Attachment Energies of the DNA Bases.

408. M. F. Falcetta and K. D. Jordan, *J. Phys. Chem.*, **113**, 2903–2999 (1991). Stabilization Calculations on the π^* Anion States of 1,4-Cyclohexadiene: Confirmation of the π^*_- and π^*_+ Orbital Ordering.

409. H.-Y. Cheng, C.-W. Chen, J.-T. Chang, and C.-C. Shih, *J. Phys. Chem. A*, **115**, 84–93 (2011). Application of the Stabilization Method to Temporary Anion States of CH_3CN, CH_3NC, CH_3SCN, and CH_3NCS in Density Functional Theory with Asymptotically Corrected Potentials.

410. J.-D. Chai and M. Head-Gordon, *Phys. Chem. Chem. Phys.*, **10**, 6615–6620 (2008). Long-Range Corrected Hybrid Density Functionals with Damped Atom–Atom Dispersion Corrections.

411. S. Grimme, *J. Chem. Phys.*, **124**, 034108:1–16 (2006). Semiempirical Hybrid Density Functional with Perturbative Second-Order Correction.

412. B. Nestmann and S. D. Peyerimhoff, *J. Phys. B*, **18**, 615–626 (1985). Calculation of the Discrete Component of Resonance States in Negative Ions by Variation of Nuclear Charges.

413. B. M. Nestmann and S. D. Peyerimhoff, *J. Phys. B*, **18**, 4309–4319 (1985). CI Method for Determining the Location and Width of Resonances in Electron–Molecule Collision Processes.

414. A. Whitehead, R. Barrios, and J. Simons, *J. Chem. Phys.*, **116**, 2848–2851 (2002). Stabilization Calculation of the Energy and Lifetime of Metastable SO_4^{2-}.

415. K. M. Ervin, I. Anusiewicz, P. Skurski, J. Simons, and W. C. Lineberger, *J. Phys. Chem. A*, **107**, 8521–8529 (2003). The Only Stable State of O_2^- is the $X\ ^2\Pi_g$ Ground State and It (Still!) Has an Adiabatic Electron Detachment Energy of 0.45 eV.

416. S. Feuerbacher, T. Sommerfeld, and L. S. Cederbaum, *J. Chem. Phys.*, **121**, 6628–6633 (2004). Extrapolating Bound State Data of Anions into the Metastable Domain.

417. A. Modelli, A. Foffani, M. Guerra, D. Jones, and G. Distefano, *Chem. Phys. Lett.*, **99**, 58–65 (1983). Electron Transmission Spectroscopy and MSXα Study of Closed-Shell and Open-Shell Metallocenes.

418. M. Guerra, *J. Phys. Chem.*, **94**, 8542–8547 (1990). Boxing Procedure for Estimating Shape Resonance Energies from Stabilization Graphs with the MSXα Method.

Index

Computer programs are denoted in boldface, databases and journals are in italics. Page numbers in *italics* refer to Figures; those in **bold** to Tables.

Reviews in Computational Chemistry, Volume 28, First Edition.
Edited by Abby L. Parrill and Kenny B. Lipkowitz.
© 2015 John Wiley & Sons, Inc. Published 2015 by John Wiley & Sons, Inc.